Handbook of Biomineralization

Edited by
Peter Behrens
and Edmund Bäuerlein

1807–2007 Knowledge for Generations

Each generation has its unique needs and aspirations. When Charles Wiley first opened his small printing shop in lower Manhattan in 1807, it was a generation of boundless potential searching for an identity. And we were there, helping to define a new American literary tradition. Over half a century later, in the midst of the Second Industrial Revolution, it was a generation focused on building the future. Once again, we were there, supplying the critical scientific, technical, and engineering knowledge that helped frame the world. Throughout the 20th Century, and into the new millennium, nations began to reach out beyond their own borders and a new international community was born. Wiley was there, expanding its operations around the world to enable a global exchange of ideas, opinions, and know-how.

For 200 years, Wiley has been an integral part of each generation's journey, enabling the flow of information and understanding necessary to meet their needs and fulfill their aspirations. Today, bold new technologies are changing the way we live and learn. Wiley will be there, providing you the must-have knowledge you need to imagine new worlds, new possibilities, and new opportunities.

Generations come and go, but you can always count on Wiley to provide you the knowledge you need, when and where you need it!

William J. Pesce
President and Chief Executive Officer

Peter Booth Wiley
Chairman of the Board

Handbook of Biomineralization

Biomimetic and Bioinspired Chemistry

Edited by
Peter Behrens and Edmund Bäuerlein

WILEY-VCH Verlag GmbH & Co. KGaA

The Editors

Prof. Dr. Peter Behrens
University of Hannover
Institute for Inorganic Chemistry
Callinstr. 9
30167 Hannover
Germany

Prof. Dr. Edmund Bäuerlein
Max-Planck-Institute for Biochemistry
Department of Membrane Biochemistry
Am Klopferspitz 18 A
82152 Planegg
Germany

Cover Illustration (designed by Felix Bäuerlein)
Top left: A Silicat-1 *Luffa* monolith, created in shape-preserving, *in situ* thermal reactions after complete coverage of the sponge *Luffa* as biotemplate with Silicalite-1.
(A. Zampieri et al., Chap. 14, Fig. 14.8 b)
Top right: This vaterite structure resulted from a rapid, kinetically controlled formation in presence of a macrocyclic polyacid with the highest charge density used.
(D. Volkmer, Chap. 4, Fig. 4.10, bottom right)
Bottom left: Formation of planar aragonite-type crystals of barium carbonate with silicate anions on a chitosan substrate
(H. Imai, Y. Oaki, Chap. 5, Fig. 5.9 b)
Bottom right: Calcium carbonate particle formed as a thin film over the membrane surface of polymer replicas of sea urchin skeletal plates
(F. Meldrum, Chap. 15, Fig. 15.4 c)

Handbook of Biomineralization
Biological Aspects and Structure Formation:
ISBN 978-3-527-31804-9
Biomimetic and Bioinspired Chemistry:
ISBN 978-3-527-31805-6
Medical and Clinical Aspects:
ISBN 978-3-527-31806-3
Set (3 volumes):
ISBN 978-3-527-31641-0

Library of Congress Card No.: applied for

British Library Cataloguing-in-Publication Data
A catalogue record for this book is available from the British Library.

Bibliographic information published by the Deutsche Nationalbibliothek
Die Deutsche Nationalbibliothek lists this publication in the Deutsche Nationalbibliografie; detailed bibliographic data are available in the Internet at ⟨http://dnb.d-nb.de⟩.

Printed in the Federal Republic of Germany
Printed on acid-free paper

Typesetting Asco Typesetters, North Point, Hong Kong
Printing betz-druck GmbH, Darmstadt
Binding Litges & Dopf GmbH, Heppenheim
Cover Design Felix Bäuerlein, Munich
Wiley Bicentennial Logo Richard J. Pacifico

ISBN 978-3-527-31805-6

Contents

Handbook of Biomineralization. Edited by P. Behrens and E. Bäuerlein

ISBN: 978-3-527-31805-6

Preface

On passing back through history and prehistory, one finds the Stone Age, the Bronze Age, and the Iron Age. Clearly, Man has in the past related the achievements of former generations to things solid and material – most likely because these materials have been rediscovered as artifacts, long after their day-to-day use. But today, we live in an era dominated by vast progress in the possibilities of information exchange, to a point where this period indeed may be referred to as the "Silicon Age". So, what might be the materials of the future?

Many well-established materials such as metals, ceramics, or plastics can no longer fulfill all of the needs of a technologically advanced society, and today a clear trend can be seen towards more complex functions that may be realized only with the use of materials of more complex composition and structure. At an early stage, engineers and scientists realized that, compared with their pure counterparts, mixtures of materials tend to show superior properties. Excellent examples of this are composite and hybrid structures, and especially those which contain an elastic, malleable organic component together with a hard, inorganic substance. For example, the addition of silica particles to the rubber used to make rubber tires enhances the tire's lifetime, while the properties of even a material as mundane as concrete can be "spiced up" with polymers,

Although, in terms of materials sciences, biominerals belong to this class of composites and hybrids, they also feature well-defined structures on several length scales, from the atomic scale to centimeter-sized functional arrangements of crystals in an organic matrix. In this way they demonstrate properties – or combinations of properties – which have not yet been achieved with synthetic hybrid materials. Thus, when using natural biominerals as a model, it is possible to prepare materials with both improved properties and applications.

Several different terms have been used to describe this approach, notably those appearing in the title of this Handbook, namely "biomimetic" and "bio-inspired". The process of mimicry involves constructing something that resembles the original as closely as possible, although this of course requires a detailed knowledge of the archetype. Inspiration, on the other hand, provides novel ideas which often appear as a "flash", even when observing the "original" only superficially. In fact, inspiration may even be hampered by too good a knowledge of the original, and by trying to follow it too closely. In this sense, we use the

Handbook of Biomineralization. Edited by P. Behrens and E. Bäuerlein

ISBN: 978-3-527-31805-6

term "biomimetic" when the scientific approach follows biomineralization so closely that it also allows conclusions to be drawn about the natural process. Bio-inspired materials are, correspondingly, much more loosely attached to archetype biominerals.

Part I of this volume begins with two chapters, by Behrens, Jahns, and Menzel, and by Brunner and Lutz, which focus on the biomimetic chemistry of silica deposition. On moving away from this amorphous material to crystalline substances, Cölfen's chapter on mesocrystals provides information on a novel model for crystallization processes and their products. Calcium carbonate is not only the most prominent natural biomineral, but also is highly suited to laboratory investigations, and this point is highlighted in chapters on the crystallization of calcium carbonate under monolayers (Volkmer), in layered organic systems resembling mollusk nacre (Imai and Oaki), or in eggshells (Arias, Arias, and Fernandez). de Masi and her co-workers then report on the biomimetic mineralization of protein aggregates. The next chapters, by Orme and Giocondi and by Faivre, describe the calcium phosphate and iron oxide systems, respectively.

Part II of the volume presents bio-inspired approaches to model certain structural features of biominerals, based either on a three-dimensional approach using ice crystals as a template (Deville, Saiz, and Tomsia), or by applying two-dimensional analogues (Helmecke, Behrens, and Menzel; and Tremel and co-workers). In Part III, we have collected "bio-supported" approaches, which do not aim to follow the self-aggregation and self-ordering processes of natural biomineralization, but rather use natural materials in shape-preserving reactions to imprint certain structural features on the resulting product. "Bioclastic" approaches using inorganic preforms are described by Sandhage et al. The use of organic plant tissues to generate inorganic macrostructures is summarized by Schwieger and Greil and their co-workers. In "bio-casting", the pores of biomineral skeletons are used as templates for the generation of macroporous structures (Meldrum).

Moving to smaller biological entities such as templates, Part IV contains details of studies conducted on the mineralization of protein cages (by Douglas and co-workers) and viruses (by Bittner). In Part V, the chapter of Boissière, Allouche, and Coradin on the medical applications of silica capsules formed with biopolymers, builds the first stages of a bridge to Volume 3 of this Handbook. The imaging methods described in the final part of this volume by Zolotoyabko and Stock, respectively, are of course important not only to the study of products of biomimetic, bio-inspired, bioclastic, biocast and synthetic products, but also to the investigation of biominerals themselves.

As editors, we hope that our collection of articles timely reflects the current importance of the field, and also highlights future trends – perhaps at the advent of the "Age of Complex Materials"? We thank all of the contributing authors for their commitment to this book, and especially Dr. Gudrun Walter from Wiley-VCH, without whose enthusiasm this Handbook of Biomineralization may not have materialized. We also thank Cornelia Meinertz-Bäuerlein and Birgit Förster for their ongoing secretarial support.

February 2007

Peter Behrens
Hannover
Germany

Edmund Bäuerlein
Munich/Martinsried
Germany

Foreword

Where does scientific creativity come from? If we knew the answer to this question then the world would be a very different place. Civilisation depends at its core on the creative inspiration found at the bedrock of science and art, yet we have very little understanding or control over the conditions and circumstances that set off the sparks of invention or reveal the sudden clarity of insight. Understandably, transforming existing scientific paradigms is an event that we expect to be very rare indeed, not only because of the fundamentally progressive, step-wise nature of the scientific endeavour but because the system is exemplified by its intellectual and empirical robustness. But most scientists do encounter moments of "jaw-dropping" originality that arise from time to time in their research fields through newly unveiled experiments and methodologies. (This is often recognised by a bemused state of half-envy/half-awe as one contemplates why one hadn't thought of it first!). Often, we tend to look to young scientists for this type of local scale revelation, which suggests that at least some aspects of the creative process fade with age.

If the essence of scientific creativity cannot be captured by prescriptive methods of teaching and practical experience, then perhaps we can get a hold of a little of its ephemeral quality by the process of inspiration. Nearly every scientist has encountered individuals that are worthy mentors, who express fun, excitement, and joy through their work, who take great delight in the unveiling of the unknown, and who are generous in sharing this knowledge and privilege. But do these fine attributes alone solve deep problems or uncover great secrets? Clearly, inspiration must also be stirred up at some fundamental practical level, where the sublime world of ideas and concepts slams hard against the prosaic experience of technical pragmatism demanded of scientific endeavour. Over many centuries, a tried and tested way of pump-priming this hard-wired type of inspiration is to look to Nature as a revelation not only with regard to the optimisation of solutions to specific functional problems, but as a treasure-trove of novel process, most of which are so unusual and strange that they fall outside the mind's imagination and have to be retained by fascination.

This book is another significant testament to the importance of inspiration as an engine of creativity in science. In this case, the primary archetype is the biological process of biomineralization, in which inorganic-based materials (calcium

Handbook of Biomineralization. Edited by P. Behrens and E. Bäuerlein

ISBN: 978-3-527-31805-6

carbonate, calcium phosphate, iron oxides, silica etc) are deposited in elaborate forms and structures for functional use. Each chapter illustrates striking advances in materials chemistry that have been inspired by a knowledge and understanding of biomineralization. This *biomimetic* (or *bioinspired*) approach involves a multitude of strategies that together illustrate the richness and fascination to be gained by studying a natural process within the context of a tangential field. As a natural phenomenon, biomineralization has been studied for many centuries, and whilst the importance of biomineral structure and hierarchy were recognised relatively early on as models for the design of materials with enhanced mechanical properties, the translation of molecular-based principles into synthetic strategies (chemistry) is very recent. In particular, biomineralization teaches that there are deep principles residing at the organic-inorganic interface – supramolecular preorganization, interfacial molecular recognition, vectorial regulation and multilevel processing for example – that exemplify new chemistry strategies based on confinement, template-directed nucleation, morphosynthesis, and crystal tectonics, respectively. Thus, we have chapters that highlight new model systems, novel concepts, innovative synthetic methods and applications, and in each case it is evident that there is a lot of fun and excitement going on in these research groups. Indeed, what is so clearly seen in this book is the remarkable inventiveness and creativity that scientists bubble with once a source of inspiration (biomineralization) has been identified.

I warmly congratulate the editors for bringing together such an outstanding group of scientists, and the authors themselves for their remarkable contributions to the advancement of this new and exciting field.

Bristol, March 2007 *Professor Stephen Mann FRS*

List of Contributors

Gul Ahmad
Georgia Institute of Technology
Materials Science and
Engineering
771 Ferst Drive
Atlanta, GA 30332
USA

Shawn M. Allan
Ceralink, Inc.
105 Jordan Road
Troy, NY 12180
USA

Mark A. Allen
Departments of Chemistry and
Biochemistry
Montana State University
108 Gaines Hall
Bozeman, MT 59717-3400
USA

Joachim Allouche
Chimie de la Matière Condensée
de Paris
CNRS, Université Pierre
et Marie Curie
4 place Jussieu
75252 Paris Cedex 05
France

José Ignacio Arias
Department of Animal Biology
Faculty of Veterinary and Animal
Sciences
University of Chile and CIMAT
Casilla 2 Correo 15
La Granja
Santiago
Chile

José Luis Arias
Department of Animal Biology
Faculty of Veterinary and Animal
Sciences
University of Chile and CIMAT
Casilla 2 Correo 15
La Granja
Santiago
Chile

Xiaolan Ba
The State University of New York at
Stony Brook
Materials Science Department
Old Engineering Building
Stony Brook, NY 11794-2275
USA

Handbook of Biomineralization. Edited by P. Behrens and E. Bäuerlein

ISBN: 978-3-527-31805-6

Mathias Balz
Institute for Inorganic Chemistry
and Analytical Chemistry
University of Mainz
Duesbergweg 10–14
55099 Mainz
Germany

Peter Behrens
Institute for Inorganic Chemistry
Leibniz-University of Hannover
Callinstraße 9
30167 Hannover
Germany

Alexander M. Bittner
Department of Nanoscale Science
Max-Planck-Institute for Solid
State Research
Heisenbergstr. 1
70569 Stuttgart
Germany

Michel Boissière
Chimie de la Matière Condensée
de Paris
CNRS, Université Pierre
et Marie Curie
4 place Jussieu
75252 Paris Cedex 05
France

Eike Brunner
Institute for Biophysics and
Physical Biochemistry
University of Regensburg
Universitätsstr. 31
93040 Regensburg
Germany

Ye Cai
Georgia Institute of Technology
Materials Science and Engineering
771 Ferst Drive
Atlanta, GA 30332
USA

Helmut Cölfen
Max-Planck-Institute of Colloids and
Interfaces
Colloid Chemistry
Research Campus Golm
14424 Potsdam
Germany

Thibaud Coradin
Chimie de la Matière Condensée de
Paris
CNRS, Université Pierre
et Marie Curie
4 place Jussieu
75252 Paris Cedex 05
France

Sylvain Deville
Laboratory of Synthesis and
Functionalization of Ceramics (LSFC)
FRE2770 CNRS/Saint-Gobain CREE
550, Avenue Alphonse Jauffret, BP 224
84306 Cavaillon Cedex
France

Matthew B. Dickerson
Georgia Institute of Technology
Materials Science and Engineering
771 Ferst Drive
Atlanta, GA 30332
USA

Elaine DiMasi
National Synchroton Light Source
Department
Brookhaven National Laboratory
Building 725 D
Upton, NY 11973-5000
USA

Trevor Douglas
Departments of Chemistry and Biochemistry
Montana State University
108 Gaines Hall
Bozeman, MT 59717-3400
USA

Eric M. Ernst
Georgia Institute of Technology
Materials Science and Engineering
771 Ferst Drive
Atlanta, GA 30332
USA

Damien Faivre
Max-Planck-Institute for Marine Microbiology
Department of Microbiology
Celsiusstraße 1
28359 Bremen
Germany

María Soledad Fernandez
Department of Animal Biology
Faculty of Veterinary and Animal Sciences
University of Chile and CIMAT
Casilla 2 Correo 15
La Granja
Santiago
Chile

Christopher S. Gaddis
Nanosphere, Inc.
4088 Commercial Avenue
Northbrook, IL 60062
USA

Jennifer L. Giocondi
Chemistry, Materials & Life Sciences Directorate
Lawrence Livermore National Laboratory
7000 East Avenue, L-396
Livermore, CA 94550
USA

Peter Greil
Department of Materials Science III
Glass and Ceramics
University of Erlangen-Nuremberg
Martensstr. 5
91058 Erlangen
Germany

Michael S. Haluska
Georgia Institute of Technology
Materials Science and Engineering
771 Ferst Drive
Atlanta, GA 30332
USA

Olaf Helmecke
Institute for Technical Chemistry
Technical University of Braunschweig
Hans-Sommer-Straße 10
38106 Braunschweig
Germany

Hiroaki Imai
Department of Applied Chemistry
Faculty of Science and Technology
Keio University
3-14-1 Hiyoshi
Kohoku-ku
233-8522 Yokohama
Japan

Michael Jahns
Institute for Inorganic Chemistry
Leibniz-University of Hannover
Callinstraße 9
30167 Hannover
Germany

Jörg Küther
Institute for Inorganic Chemistry
and Analytical Chemistry
University of Mainz
Duesbergweg 10–14
55099 Mainz
Germany

Seo-Young Kwak
The Forsyth Institute
Department of Biomineralization
140 Fenway
Boston, MA 02115
USA

Niklas Loges
Institute for Inorganic Chemistry
and Analytical Chemistry
University of Mainz
Duesbergweg 10–14
55099 Mainz
Germany

Katharina Lutz
Institute for Biophysics and
Physical Biochemistry
University of Regensburg
Universitätsstr. 31
93040 Regensburg
Germany

Fiona C. Meldrum
School of Chemistry
University of Bristol
Cantock's Close
Bristol BS8 1TS
UK

Yizhi Meng
The State University of New York at
Stony Brook
Department of Bioengineering
343D Psychology Building A
Stony Brook, NY 11794-2580
USA

Henning Menzel
Institute for Chemical Engineering
Technical University of Braunschweig
Hans-Sommer-Straße 10
38106 Braunschweig
Germany

Yuya Oaki
Department of Applied Chemistry
Faculty of Science and Technology
Keio University
3-14-1 Hiyoshi
Kohoku-ku
233-8522 Yokohama
Japan

Christine A. Orme
Chemistry, Materials & Life Sciences
Directorate
Lawrence Livermore National
Laboratory
7000 East Avenue, L-350
Livermore, CA 94550
USA

Nadine Pernodet
The State University of New York at
Stony Brook
Materials Science Department
Old Engineering Building
Stony Brook, NY 11794-2275
USA

M. Matthew Prissel
Departments of Chemistry and Biochemistry
Montana State University
108 Gaines Hall
Bozeman, MT 59717-3400
USA

Miriam Rafailovich
The State University of New York at Stony Brook
Materials Science Department
Old Engineering Building
Stony Brook, NY 11794-2275
USA

Eduardo Saiz
Materials Sciences Department
Lawrence Berkeley National Laboratory
Building 62, Room 351
1 Cyclotron Road
Berkeley, CA 94720
USA

Kenneth H. Sandhage
Georgia Institute of Technology
Materials Science and Engineering
771 Ferst Drive
Atlanta, GA 30332
USA

Wilhelm Schwieger
Institute of Chemical Reaction Engineering
University of Erlangen-Nuremberg
Egerlandstr. 3
91058 Erlangen
Germany

Samuel Shian
Georgia Institute of Technology
Materials Science and Engineering
771 Ferst Drive
Atlanta, GA 30332
USA

Robert L. Snyder
Georgia Institute of Technology
Materials Science and Engineering
771 Ferst Drive
Atlanta, GA 30332
USA

Stuart R. Stock
Department of Molecular Pharmacology and Biological Chemistry
Northwest University
Feinberg School of Medicine
303 E. Chicago Ave.
Chicago, IL 60611-3008
USA

Karthikeyan Subburaman
The State University of New York at Stony Brook
Materials Science Department
Old Engineering Building
Stony Brook, NY 11794-2275
USA

Antoni P. Tomsia
Materials Sciences Department
Lawrence Berkeley National Laboratory
Building 62, Room 351
1 Cyclotron Road
Berkeley, CA 94720
USA

Wolfgang Tremel
Institute for Inorganic Chemistry
and Analytical Chemistry
University of Mainz
Duesbergweg 10–14
55099 Mainz
Germany

Raymond R. Unocic
The Ohio State University
Department of Materials Science
and Engineering
2041 College Road
Columbus, OH 43210
USA

Dirk Volkmer
Inorganic Chemistry II
University of Ulm
Albert-Einstein-Allee 11
89081 Ulm
Germany

Michael R. Weatherspoon
Georgia Institute of Technology
Materials Science and
Engineering
771 Ferst Drive
Atlanta, GA 30332
USA

Stephan E. Wolf
Institute for Inorganic Chemistry
and Analytical Chemistry
University of Mainz
Duesbergweg 10–14
55099 Mainz
Germany

Mark J. Young
Department of Plant Sciences
and Plant Pathology
Montana State University
PO Box 173150
Bozeman, MT 59717-3400
USA

Frank M. Zalar
The Ohio State University
Department of Materials Science and
Engineering
2041 College Road
Columbus, OH 43210
USA

Alessandro Zampieri
Institute of Chemical Reaction
Engineering
University of Erlangen-Nuremberg
Egerlandstr. 3
91058 Erlangen
Germany

Vladimir Zeitsev
The State University of New York at
Stony Brook
Materials Science Department
Old Engineering Building
Stony Brook, NY 11794-2275
USA

Cordt Zollfrank
Department of Materials Science III
Glass and Ceramics
Friedrich-Alexander-University of
Erlangen-Nuremberg
Martensstr. 5
91058 Erlangen
Germany

Emil Zolotoyabko
Department of Materials
Engineering
Technion-Israel Institute of
Technology
Technion-City, Promenade,
Blg. 590
Haifa 32000
Israel

Part I
Biomimetic Model Systems in Biomineralization

Handbook of Biomineralization. Edited by P. Behrens and E. Bäuerlein

ISBN: 978-3-527-31805-6

1
The Polyamine Silica System: A Biomimetic Model for the Biomineralization of Silica

Peter Behrens, Michael Jahns, and Henning Menzel

Abstract

From laboratory studies and industrial applications, it has long been known that polyamines influence the condensation process of silica. Interestingly, it has been found that polyamines are also involved in the formation of the silica exoskeletons of diatoms. This biomineralization process yields intricately patterned silica shells. Thus, a large variety of polyamines have been studied with regard to their influence on the kinetics of silica condensation, and with regard to possible mechanisms producing the patterns, among them polymers as polylysine and polyallylamine. In this chapter these studies are reviewed, with emphasis placed on the behavior of poly(ethyleneimine) and poly(propyleneimine) in silica solutions undergoing a condensation reaction. These amines have an architecture which closely resembles that of the polyamines occurring in diatoms. Therefore, model studies involving these synthetic polyamines can provide valuable additional information on the biosilicification process in diatoms.

Key words: silica, biosilicification, polyamines, diatoms, condensation reactions, dynamic light scattering, kinetic investigations.

1.1
Introduction

A large number of organisms form silica as a biomineral, including animals (e.g., sponges) as well as lower and higher plants (e.g., rice, horsetail, cereals) [1–5]. There appears to be evidence that silica is required in vertebrates for the proper development of cartilage and bone [6–9] but, judged by the mass of silica produced, the most important biosilicifying organisms are lower aquatic life forms, especially diatoms and radiolaria. Diatoms are unicellular algae which form silica exoskeletons; these shells are patterned by pores, and their arrangement often leads to beautiful ornate patterns. The scales of the patterns reach from 10 to

Handbook of Biomineralization. Edited by P. Behrens and E. Bäuerlein

ISBN: 978-3-527-31805-6

1000 nm, thus qualifying the shells as truly nanostructured materials. As the patterns are faithfully reproduced in a species-specific manner in sequential generations, their formation is clearly under genetic control.

With regard to the formation mechanisms of biominerals, biosilicification is special among the biomineralization processes, as the silica formed is always amorphous [2]. Thus, basic arguments valid for the formation of biominerals with crystalline inorganic phases – for example, possible epitaxial nucleation on pre-formed organic matrices, preferred formation of certain crystal faces or the building principle of mesocrystals [2, 10, 11] – do not apply. The formation of solid amorphous silica rather relies on condensation reactions, occurring largely in a random fashion, between mono-, oligo-, and polysilicic acids. Whereas in some higher plants, larger silica particles (sol particles) may be incorporated directly from soil fluids, for many other biosilicifying organisms, monosilicic acid $Si(OH)_4$ is the assumed starting point for the formation of silica biominerals.

1.2
Mechanisms of Biomineralization in Diatoms

In recent years much progress has been made in the study of the biomineralization processes in diatoms [12–31]. The formation of the biosilica structures takes place during cell division in specialized organelles, the *silica deposition vesicles* (SDVs), and it has been shown that the pH value in these vesicles is around 5.5 [14]. In spite of many attempts, it has until now not been possible to isolate these SDVs and thus to analyze their chemical contents. Most of our knowledge of the biochemistry of biomineralization therefore relies on studying the organic molecules which are enshrouded within the silica shells. For this purpose, the silica cell walls are first isolated and then dissolved by applying HF or NH_4F, so that their contents in bioorganic molecules can be analyzed. It is reasonable to assume that these silica-encased molecules take part in the control of silica deposition and nanopatterning.

In this way, Kröger, Sumper and co-workers were able to isolate two types of ingredients from diatom shells, namely silaffins and polyamines. *Silaffins* are a family of polypeptides (silaffin-1A, silaffin-1B, silaffin-2) that carry unusual modifications on their amino acids [13, 17, 20, 24, 26]. The chemical structure of nat-Sil-$1A_1$ is depicted in Figure 1.1(a) [20]. The serine residues are phosphorylated; all lysine residues are modified, appearing either as ε-*N*,*N*-dimethyllysine, phosphorylated ε-*N*,*N*,*N*-trimethyl-δ-hydroxylysine, or carrying polyamine modifications. The polyamine modifications are linear chains of between six and 11 propyleneimine units. While the phosphorylations introduce negative charges to the molecule, the basic polyamines bear positive charges, thus rendering the silaffin zwitterionic. In contrast, nat-Sil-2, whilst also carrying the lysine modifications present in nat-Sil-$1A_1$, has a polyanionic character due to numerous phosphate, sulfate and glucuronate modifications [22]. Apart from their role as modifications of the silaffins, free polyamines also occur in the extracts from diatom shells, and

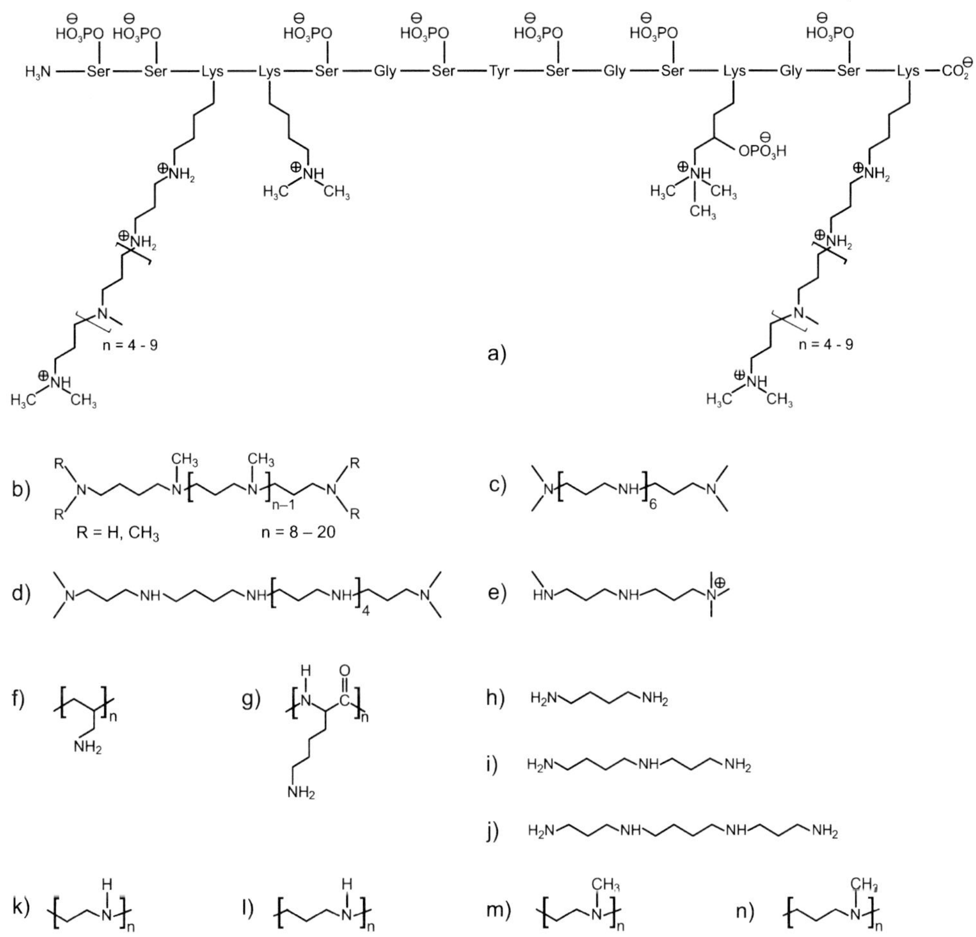

Fig. 1.1 Chemical formulae of molecules involved in the biomineralization of silica in diatoms (a–e) and of model compounds (f–n). (a) native silaffin 1A (nat-Sil-1A_1). (b) Polyamines from diatoms of the genus *Coscinodiscus*. (c–e) Amine compounds of the diatom *Thalassiosira pseudonana*. Polymers used as model compounds: (f) poly(allylamine); (g) polylysine. Low molecular-mass compounds used as model compounds: (h) putrescine; (i) spermidine; (j) spermine. Compounds depicted in (f–j) are available commercially. Synthetic linear polyamines used as model compounds: (k) poly(ethyleneimine) (PEI); (l) poly(propyleneimine) (PPI); (m) poly(*N*-methylethyleneimine) (PMEI); (n) poly(*N*-methylpropyleneimine) (PMPI).

appear to be at least as abundant as the silaffins. In diatoms of the genus *Coscinodiscus*, the silaffins may also be the only organic component associated with the biosilica [15, 19, 28]. It may therefore be assumed that, at least in some cases, polyamines are the decisive chemical component acting in the structuring process of diatom shells. The polyamines from diatoms have been characterized as polypropyleneimine chains linked to a putrescine residue. The amine groups may be methylated, so that the general structure depicted in Figure 1.1(b) results. The chain lengths of these polyamines range from $n = 8$ to 20. The fact that different diatom species which exhibit different cell wall patterns display different, species-specific compositions of the polyamine mixtures [15] furthermore underlines the presumed importance of polyamines in the formation of the nanopatterns. For the diatom *Thalassiosira pseudonana*, recent investigations have shown that other types of polyamines can also occur (Fig. 1.1(c–e)) [28].

At a pH of approximately 5.5 within the silica deposition vesicle [14], the amine groups of the polyamines are partly protonated, giving them some amphiphilic character. Based on this fact, Sumper has developed a phase-separation model which involves consecutive phase separations in polyamine-silica solutions within the SDV [19]. It is noteworthy that this model reduces the formation of the ornamental patterns on diatom shells to a purely physico-chemical process of aggregation and self-structuring [19, 30, 31], occurring on hierarchical lengths scales. Later, the phase-separation process was found to require an anionic component, which bears either multiple negative charges or can also undergo hydrogen bonding interactions [23, 26, 29]. In model experiments, mono-/dihydrogen phosphate ions can play this role; *in vivo*, silaffin-2 could be that partner, whereas in *Coscinodiscus* species other as-yet unidentified components might be involved [26].

1.3
Polyamine-Silica Systems

Even before the essential role of polyamines in the formation of silica nanostructures was noted, interactions between polyamines and silica solutions were studied intensively [26, 32]. The importance of organic amines and ammonium ions in the synthesis of crystalline aluminosilicate zeolites and pure-silica zeosils has been known since the 1950s [33]. The flocculation of silica particles with polyethyleneimine was investigated in 1976 [34], and the formation of mesoporous materials using alkyltrimethylammonium ions was discovered in 1992 [35]. In 1998, it was found that the addition of oligoamines containing three to five amine functionalities to silica solutions undergoing a sol–gel transition increased the gelation times [36].

With the finding that polyamines play a decisive role in the formation of the silica shells of diatoms, interest in the interactions between polyamines and silica solutions increased considerably [36]. Two of the polymeric amines that have been studied in recent years are depicted in Figure 1.1(f and g); both of these are commercially available.

The effect of poly(allylamine) (Fig. 1.1(f)) on silica formation has been investigated in several studies [37–39]. The system can yield spherical particles, as do natural polyamines, but can under certain circumstances also develop special morphologies which are vaguely reminiscent of a honeycomb-like patterning, as observed in diatom shells [40]. Among the poly-homopeptides, poly-L-lysine

(Fig. 1.1(g)) has been the most studied [35, 40–45]. Silica spheres of various sizes and also other morphologies can be obtained. Special morphologies can also be observed when linear poly(ethyleneimine) (PEI) is used [46, 47]. Significantly, fibrous silica is produced which consists of silica particles coating central PEI filaments.

A large number of commercially available non-macromolecular amines were also tested [40, 42–45, 48–53], many of which lead to an increase in the amount of precipitated silica. In some cases, special morphologies can be observed [40, 43, 44]. Of special interest are molecules such as putrescine, spermine, and spermidine (Fig. 1.1(h–j)), which in chemical terms are quite close to the polyamines isolated from diatom shells, but still have a much lower degree of polymerization. These molecules have been thoroughly investigated with regard to their possibilities of tailoring the properties of silicas [50, 51].

It is to be noted that although these polyamines share one or more of the characteristics with the polyamines occurring in diatoms, they cannot serve as true models for these. Differences exist with regard to:

- *Chemical make-up*: often, the amine functionalities are not part of the main polymer chain (as in polyamines isolated from diatoms), but are located in side chains; also, differences exist in the type of amine functionalities (primary, secondary, ternary, quaternary ammonium ions).
- *Architecture*: the polyamines isolated from diatoms are always linear; however, commercially poly(ethyleneimine), for example, has a branched architecture.
- *Degree of polymerization* (P_n): the diatoms' polyamines are special in that they contain between eight and 20 propyleneimine units; commercially available polyamines often have much larger P_n values, whereas available oligoamines possess fewer nitrogen atoms per molecule.

Whereas these differences are often not harmful when biomineralization processes are used as models for the preparation of materials in the sense of bio-inspired materials chemistry, it can be misleading to draw conclusions about the natural processes using imperfect model compounds. In a truly biomimetic approach, we have therefore chosen to use polyamines as model compounds, which are very similar to their natural counterparts. These are linear PEIs and poly(propyleneimines) (PPIs) of appropriate P_n, as depicted in Figure 1.1(k–n). As such PEIs and PPIs are not commercially available, we have chosen to synthesize these molecules, including the methylated variants poly(*N*-methylethyleneimine) (PMEI) and poly(*N*-methylpropyleneimine) (PMPI) (see Section 1.4) [54, 55]. These investigations have provided interesting insights into the biosilicification in diatoms. The results gain from the possibility of easily varying the characteristics of the polyamines (PEIs versus PPIs, non-methylated versus methylated amine groups, adjustment of P_n) – something which is not possible when natural extracts isolated from diatom cell walls are used.

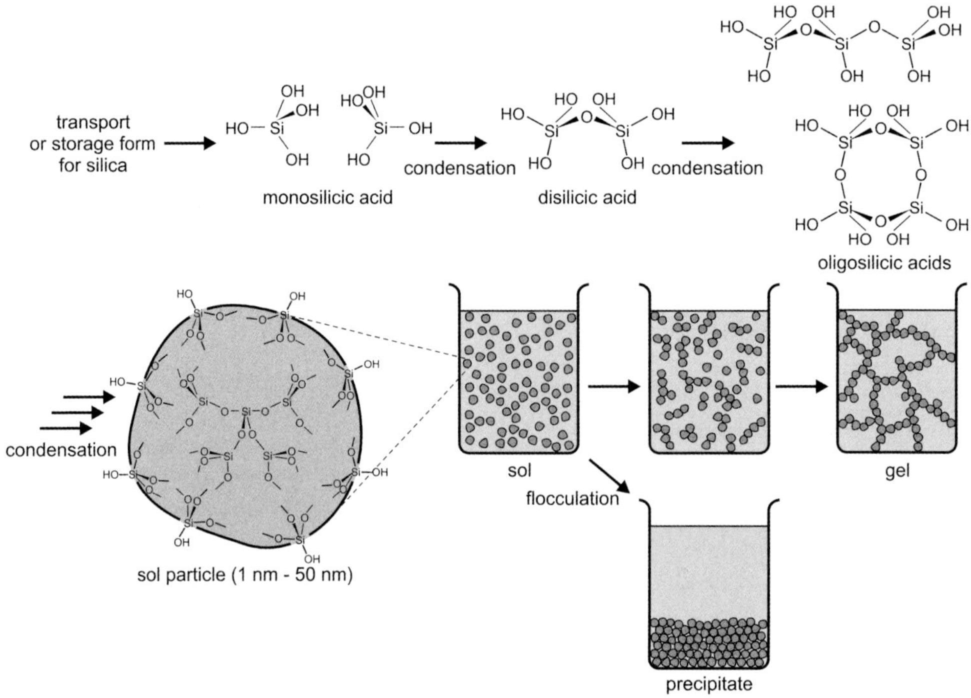

Fig. 1.2 A schematic representation of the silica sol–gel process.

It is clear from the previous studies on polyamine-silica solutions that polyamines strongly influence the solidification of silica from aqueous solutions. However, this solidification process in itself is very complicated as it includes steps of condensation, aggregation, agglomeration, gelation or flocculation, involving different chemical species [56, 57]. This is depicted in Figure 1.2. Starting from an aqueous solution of monosilicic acid, the first condensation step leads to disilicic acid; subsequently, oligomeric silicic acids form, until the state of polysilicic acid particles is reached. At this point, a colloidal solution – the silica sol – is formed. The rates of the consecutive condensation reactions depend on many factors such as temperature and ionic strength of the solution, but most notably upon pH. Also, in dependence of these factors, a sol may be stable for an infinite time, but it can also solidify to form a gel, corresponding to a network of sol particles. The repulsive interactions between the sol particles can be overcome by chemical agents that cause coagulation and flocculation of the sol particles, leading to a precipitate.

With regard to the complexity of the solidification of silica from solution, the notion that polyamines "catalyze" the condensation of silica – as often cited in the literature – must therefore be evaluated carefully. The question to be asked is whether it is the condensation reaction which is catalyzed, the aggregation of the silica sol particles, or the flocculation of larger particles? Moreover, when the polyamines are consumed during the reaction, are they true catalysts?

As the condensation, particle formation and flocculation of silica comprises several length scales from that of molecules up to the micrometer range, we have investigated the reaction of the synthetic linear PEIs and PPIs with silica using different methods. The molybdate reaction allows us to follow the kinetics of the early stages of the condensation reaction. Dynamic light scattering (DLS) is a suitable method to follow the formation of larger aggregates.

1.4 Synthesis of Linear Polyamines

In order to model the polyamines found in diatoms, which are all linear and have a relatively low P_n of eight to 20 monomer units, a polymerization method offering reliable control about the product characteristics is needed. We have therefore chosen the ring-opening polymerization of oxazolines or 1,3-oxazines for the synthesis of PEI and PPIs, respectively. The strategy is exemplified here for the synthesis of PPI and PMPI (Fig. 1.3); the procedure for the synthesis of PEI and PMEI has been described in Ref. [54].

For the synthesis of linear poly(propylene imines), the methyloxazine monomer is synthesized by reaction of 3-amino-1-propanol with acetonitrile [58]. Cationic ring-opening polymerization gives polyacetylpropyleneimine (PAPI) [59],

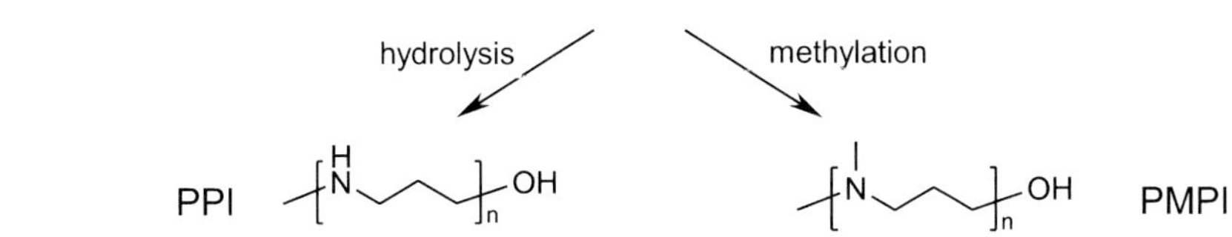

Fig. 1.3 Synthesis of methyloxazine, its cationic ring-opening polymerization to polyacetylpropylene imine (PAPI), and the subsequent hydrolysis or combined hydrolysis/methylation yielding linear poly(propyleneimine) (PPI) or poly(*N*-methylpropyleneimine) (PMPI), respectively.

and subsequent hydrolysis yields PPI. Alternatively, the polymer can be simultaneously hydrolyzed and methylated employing a Leuckart–Wallach reaction, resulting in PMPI [54]. This polymerization has been adapted here to yield polymers with average P_n of 8 to 40 and with a narrow distribution of chain lengths which is comparable to that found for the polyamines isolated from diatoms [15]. By using this synthetic approach, we are able to prepare linear polyamines with either ethylene imine or propylene imine repeating units, as methylated or non-methylated variants and with well-defined chain lengths.

1.5
Kinetic Investigations on Polyamine-Silica Systems

Formation of the yellow molybdosilicate complex ion between a molybdate source and monosilicic acid can be used to photometrically determine the concentration of monosilicic acid [5, 60]. This analytical method involves a well-defined reaction time (of 10 min in our procedure) during which additional monosilicic acid molecules can be formed from larger silicic acid entities, namely from disilicic and, probably, from lower oligosilicic acids. When a sol–gel process is followed using this method, the concentration of monosilicic acid will of course decrease with time. The application of this method to study the kinetics of silica formation processes was elucidated by Perry and co-workers [5].

In our initial investigations [54, 55], we observed a strong dependence of the rate of decrease in monosilicic acid concentration on the type of amine added (Fig. 1.4). The strongest acceleration was found for the diamines diaminoethane $(PEI)_2$ and diaminopropane $(PPI)_2$, whereas the effects caused by synthetic linear polyamines $(PEI)_{8-9}$ and diaminopropane $(PPI)_{12-13}$, as well as those due to their methylated counterparts $(PMEI)_{8-9}$ and $(PMPI)_{12-13}$, were less pronounced [54, 55]. However, it was realized that the different reaction rates were direct consequences of the pH values at which the reactions were running (see Fig. 1.4). These in turn were influenced by the basicities of the amines, as no buffer was added to the reaction solutions intentionally. The amines (bases of intermediate strength) then regulate the pH together with the silicic acid (a weak acid). The pH-dependence of the condensation reaction of silica is well known, as the rates increase strongly with increasing pH in a region from pH 3 to pH 9. The stronger the basicity of a base, the higher the pH value of the reaction and, correspondingly, the faster the condensation reaction. The behavior of all polyamines, whether bearing ethylene linkers or propylene linkers, could be explained in this way, without further reference to their chemical make-up.

In all reactions, an induction period can be observed which lasts for approximately 15 to 20 min at a pH of 5.5. During this time, the concentration of monosilicic acid appears to be constant, as determined by the present method. This induction period is caused by the fact that the molybdate method detects not only monosilicic acid molecules present in the solution, but in addition also

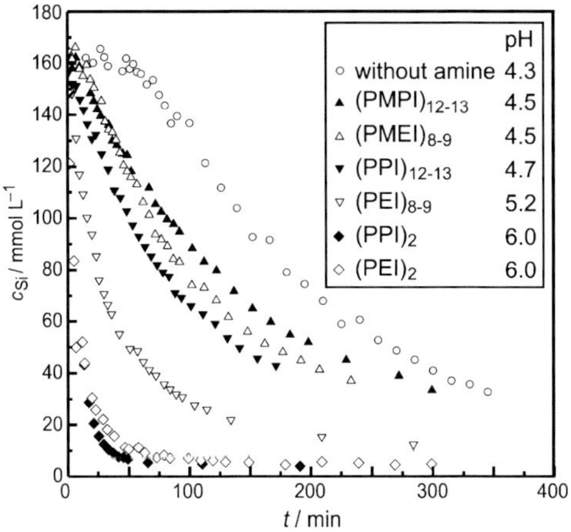

Fig. 1.4 Kinetic curves on the silica condensation process determined using the molybdate method. Unbuffered solutions containing different polyamines. The mean pH value of the solution is also given; the error in pH values is ca. 0.2. With increasing pH, the reactions become faster.

monosilicic acid molecules that are formed by the hydrolysis of oligosilicic acid molecules during the analysis time [5]. Thus, during the induction period, the silica is present as mono-, di-, or oligosilicic acids; only thereafter are larger oligosilicic acids and polysilicic acids formed.

Further investigations were therefore carried out in buffered solutions [61]. As a buffer for the pH region of ca. 5.5, we chose the malonate/dimalonate system. The results obtained then were striking: Independent of the chemical make-up of the polyamine – chain length, PEI versus PPI, methylated versus non-methylated polyamines – the kinetic curves were practically identical (Fig. 1.5(a)). Any minor deviations were due to small variations that occur despite the presence of a buffer (a low buffer concentration was chosen in order not to disturb the system; consequently, buffer capacities were slightly overstretched). The same behavior is observed when the solutions are buffered at pH values of 3.8 (glycolic acid/glycolate buffer) or 4.7 (acetic acid/acetate buffer) [61]. Also, when the concentration of linear polypropyleneimine $(PPI)_{20}$ was increased 14-fold, the kinetics of the reaction were not accelerated (Fig. 1.5(b)). Bearing in mind that, by using the molybdate reaction, we investigate the primary step(s) of the condensation reaction, it is clear that polyamines do not have any direct catalytic effect on the first steps of the condensation reaction. It then appears very unlikely that further steps within the condensation sequence are being directly catalyzed by PPIs, for example according to a mechanism which had once been proposed in the first edition of this book [16, 21] (but no longer in the second edition [24]).

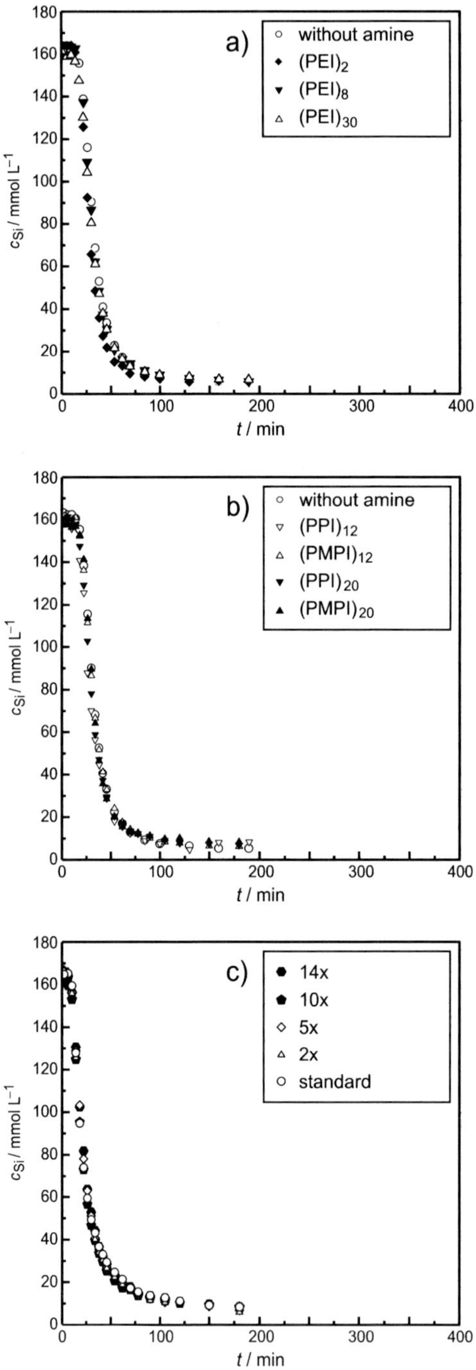
a)
without amine
(PEI)2
(PEI)8
(PEI)30
c_Si / mmol L^-1
t / min
b)
without amine
(PPI)12
(PMPI)12
(PPI)20
(PMPI)20
c_Si / mmol L^-1
t / min
c)
14x
10x
5x
2x
standard
c_Si / mmol L^-1
t / min

1.6 Investigations of the Aggregation Behavior in Polyamine-Silica Systems

Apart from the fact that polyamines do not influence the condensation reaction of monosilicic acid, it is nevertheless clear that the polyamines strongly interact with the silica. We propose that the polyamine molecules, which will carry several positive charges due to protonation at a pH of 5.5, interact with silicate species, either by ionic interactions with deprotonated oligosilicate ions or by hydrogen bonding. It should be noted that the strength of such interactions will increase with the degree of condensation of the silicic acid molecules, due to the fact that the acidity of these species increases with increasing mass. A polyamine molecule will then gather several silicate species around it (preferably those with an already higher degree of condensation), and will thus facilitate further condensation reactions between these species. A similar process was proposed to be involved in the formation of mesostructured silica materials [62]. In this way, the earlier observations that amines shorten the gelation time of silica sol–gel systems [36] can also be reconciled with our results that the primary condensation is not accelerated by polyamines – that is, that no direct catalysis occurs.

Sumper and co-workers have shown that the additional presence of ions as phosphate in polyamine-silica sol–gel systems is necessary to form special morphologies, especially spherical particles of varying size with well-defined particle size distributions [23, 26, 29]. As the size of these particles is between 50 and 1000 nm, the possible mechanism by which their formation occurs must be based on the combined aggregation of polyamine molecules, negatively charged ions, and silica primary particles. Brunner et al. have found that the negative ions responsible for aggregation of the polyamines should possess the capability to bear multiple negative charges and/or to form hydrogen bonds [29]. This is also true for the malonate (mononegative, hydrogen bond donor and acceptor) and the dimalonate ion (two negative charges, hydrogen bond acceptor) which we use in our pH 5.5 buffer system. Therefore, when polyamine-silica reactions are carried out in this buffer system, a similar chemistry as in phosphate-containing polyamine-silica systems [23, 26, 29] can be observed.

Visual inspection of the aggregation behavior allows one to determine whether such a system tends to rapidly precipitate silica (typically starting 20 min after the start of the reaction at pH 5.5) or whether a gel is formed (typically after 8 h at pH 5.5). Additional information can be obtained by using dynamic light scattering (DLS); this technique determines the hydrodynamic radius of the particles and thus provides a measure of particle size. Results from our measurements using

Fig. 1.5 Kinetic curves on the silica condensation process determined using the molybdate method. Solutions buffered at pH 5.5 using the malonate/dimalonate system. (a,b) The resulting curves are practically identical for different polyamines, independent of their chemical make-up: (a) Polyethyleneimines; (b) polypropyleneimines. (c) The resulting curves are also practically identical when different concentrations of the linear polypropyleneimine $(PPI)_{20}$ are used (standard concentration: 132 mg L^{-1}).

synthetic long-chain polyamines are depicted in Figure 1.6, where the development of the diameters of relevant particle populations are shown over time. When no polyamine is applied, a slow increase in particle size is observed over 6 h which is typical of the formation of a silica sol. Results obtained after this time and through the sol–gel transition (which takes place after 8 h) are meaningless, as the method relies on the mobility of the particles which becomes increasingly restricted during gelation. A similar curve is obtained in the presence of $(PEI)_8$. Clearly, this rather short-chain polyamine does not influence the aggregation behavior on the length scales investigated here. No precipitate is observed in the $(PEI)_8$-silica system; rather, this system gels after 8 h, as does the amine-free system.

Polyamines with slightly longer chains (e.g., $(PPI)_{20}$) exhibit a different behavior, however. Shortly after the start of the reaction (ca. 20 min), very large particles can be observed (ca. 1300 nm diameter); then, after about a further 1 to 2 min, the precipitation can be observed visually. This time period coincides with the end of the induction period observed in the kinetic investigations; that is, large aggregates form only after the condensation reaction produces larger silica entities (and not only disilicic or lower oligosilicic acids, as occurs typically during the induction period). The size of the observed particles does not increase further – a circumstance which is ascribed to the precipitation of larger particles that are subsequently removed from the region where they are probed by the laser beam. In fact, the mean particle size seems to decrease after a reaction time of ca. 100 min. This decrease may be due to the fact that, during the entire reaction, larger particles have a greater tendency to precipitate, such that the mean diameter of those particles which remain in solution continuously decreases. It should be noted that, in addition to the rapid formation of large particles and their removal from solution by precipitation, a slow sol-gel process also occurs which clearly involves the silica species not affected by aggregation with the polyamines. In nature, these two parallel processes may be responsible for the dense and glass-like appearance of diatom silica. First, the larger polyamine-silica particles are formed (these may undergo patterning processing according to the model of Sumper [19]), with the silica gel solidifying at a later stage and filling out the interstices between them.

The behavior of the methylated variant $(PMPI)_{20}$ is similar, although larger particles can be observed when using DLS [1800 nm for $(PMPI)_{20}$ versus 1300 nm for $(PPI)_{20}$]. This trend is also observed for the $(PPI)_{12}/(PMPI)_{12}$ pair. Whereas $(PMPI)_{12}$ can stabilize in solution particles with diameters of 1500 nm for extended times, the particles remain much smaller (<700 nm) when $(PPI)_{12}$ is used as an additive. The peak intensities obtained with the DLS measurements are also much smaller in the latter case. In line with these results, precipitation can clearly be observed in the case of the $(PMPI)_{12}$-silica system, whereas no precipitate can be discerned visually for the $(PPI)_{12}$-silica system. Finally, it should be noted that this different behavior does not influence the concentration of monosilicic acid as determined by the molybdate method, as long as it can be assumed that the precipitated silica remains in equilibrium with monosilicic acid.

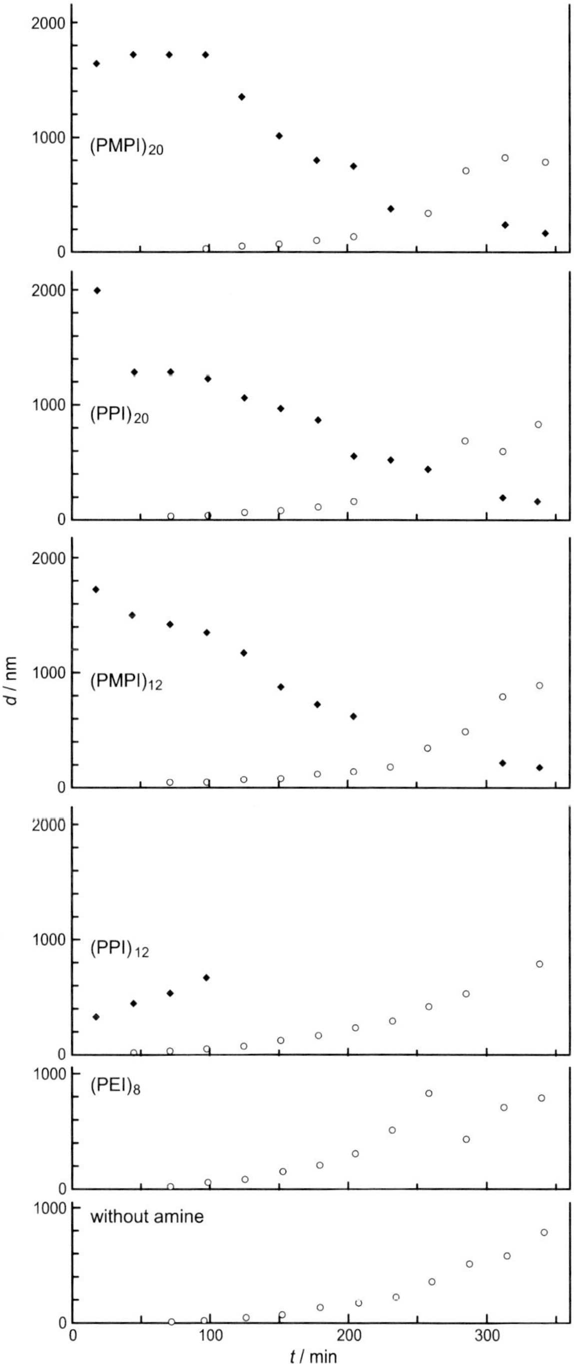

Fig. 1.6 Dynamic light scattering measurements on reacting polyamine-silica solutions (particle diameter *d* versus time). Open symbols indicate sol formation; closed symbols indicate the formation of large aggregates.

In Nature, most amine groups of the extracted polyamines are methylated and, interestingly, in the present investigations this has been the variant that allows the formation of larger polyamine-silica particles. On the basis of these results, it is also possible to state a minimum size for polyamines above which they influence the aggregation of silica to form large particles. This must be fixed at approximately eight to 12 nitrogen atoms in the linear chain, which corresponds nicely with the smallest polyamines isolated from diatom shells (eight nitrogen atoms) [15, 26].

1.7 Conclusions

When the principles of biomineralization are used to prepare materials in bio-inspired syntheses, it is not necessary strictly to adhere to natural processes. If, however, model studies are aimed at investigating such natural processes in order to generate relevant results that complement the findings of biological and biochemical studies on biomineralization mechanisms, then it is advantageous to mimic the chemical and physico-chemical features of the natural systems as closely as possible. This is shown by the present results that are based on synthetic PEIs and PPIs, which closely resemble the naturally occurring linear polyamines, the reactions being conducted at a "natural" pH of 5.5. Under these circumstances such biomimetic approaches can provide valuable additional insight into the biomineralization mechanisms, mainly because – in contrast to the true agents acting in Nature – synthetic model compounds can be obtained in larger quantities than from bioextracts, and the effect of slight chemical variations on the formation of solids can be tested. Based on the results of the present case study, the following points can be transferred to the natural biomineralization processes:

- There is no direct catalysis of the condensation reactions of silica.
- Polyamines must contain at least eight to 12 imine units in order to form large aggregates.
- In addition to a rapid precipitation of silica together with polyamine, residual silica present in the solution can undergo a sol–gel process – which might explain why diatom biosilica is dense and has a smooth surface.

References

1 H.A. Lowenstam, *Science* **1981**, *211*, 1126.
2 S. Mann, *Biomineralization*. Oxford University Press, **2001**.
3 E. Baeuerlein, *Biomineralization*. Wiley-VCH, Weinheim, **2004**.
4 M. Epple, *Biomaterialien und Biomineralisation*. Teubner, Stuttgart, **2003**.

5 C.C. Perry, T. Keeling-Tucker, *J. Biol. Inorg. Chem.* **2000**, *5*, 537–550.

6 E.M. Carlisle, *Science* **1970**, *167*, 279–280.

7 E.M. Carlisle, *Science* **1972**, *178*, 619–621.

8 E.M. Carlisle, *J. Nutr.* **1980**, *110*, 352–359.

9 E.M. Carlisle, *J. Nutr.* **1980**, *110*, 1046–1055.

10 H. Cölfen, M. Antonietti, *Angew. Chem. Int. Ed.* **2005**, *44*, 5576–5591.

11 H. Cölfen – see Chapter 3, this volume.

12 N. Kröger, C. Bergsdorf, M. Sumper, *Eur. J. Biochem.* **1996**, *239*, 259–264.

13 N. Kröger, R. Deutzmann, M. Sumper, *Science* **1999**, *286*, 1129–1132.

14 E.G. Vrieling, W.W.C. Gieskes, T.P.M. Beelen, *J. Phycol.* **1999**, *35*, 548–559.

15 N. Kröger, R. Deutzmann, C. Bergsdorf, M. Sumper, *Proc. Natl. Acad. Sci. USA* **2000**, *97*, 14133–14138.

16 N. Kröger, M. Sumper, in: E. Baeuerlein (Ed.), *Biomineralization*, 1st edn. Wiley-VCH, Weinheim, **2000**, pp. 151–170.

17 N. Kröger, R. Deutzmann, M. Sumper, *J. Biol. Chem.* **2001**, *276*, 26066–26070.

18 S. Scala, C. Bowler, *Cell. Mol. Life Sci.* **2001**, *58*, 1666–1673.

19 M. Sumper, *Science* **2002**, *295*, 2430–2433.

20 N. Kröger, S. Lorenz, E. Brunner, M. Sumper, *Science* **2002**, *298*, 584–586.

21 G. Pohnert, *Angew. Chem. Int. Ed.* **2002**, *41*, 3167–3169.

22 N. Poulsen, M. Sumper, N. Kröger, *Proc. Natl. Acad. Sci. USA* **2003**, *100*, 12075–12080.

23 M. Sumper, S. Lorenz, E. Brunner, *Angew. Chem. Int. Ed.* **2003**, *42*, 5192–5195.

24 N. Kröger, M. Sumper, in: E. Baeuerlein (Ed.), *Biomineralization*, 2nd edn. Wiley-VCH, Weinheim, **2004**, pp. 137–158.

25 N. Poulsen, N. Kröger, *J. Biol. Chem.* **2004**, *279*, 42993–42999.

26 M. Sumper, N. Kröger, *J. Mater. Chem.* **2004**, *14*, 2059–2065.

27 N. Poulsen, N. Kröger, *FEBS J.* **2005**, *272*, 3413–3423.

28 M. Sumper, E. Brunner, G. Lehmann, *FEBS Lett.* **2005**, *579*, 3765–3769.

29 K. Lutz, C. Gröger, M. Sumper, E. Brunner, *Phys. Chem. Chem. Phys.* **2005**, *7*, 2812–2815.

30 R. Gordon, R.W. Drum, *Int. Rev. Cytol.* **1994**, *150*, 243.

31 T. Coradin, P.J. Lopez, *ChemBioChem* **2003**, *3*, 1–9.

32 S.V. Patwardhan, S.J. Clarson, C.C. Perry, *Chem. Commun.* **2005**, 1113–1121.

33 R.M. Barrer, *The Hydrothermal Chemistry of Zeolites*. Academic Press, London, **1982**.

34 G.M. Lindquist, R.A. Stratton, *J. Colloid Interf. Sci.* **1976**, *57*, 132.

35 C. Kresge, M. Leonowicz, W. Roth, C. Vartuli, J. Beck, *Nature* **1992**, *359*, 710.

36 T. Mitzutani, H. Nagase, N. Fujiwara, H. Ogoshi, *Bull. Chem. Soc. Jpn.* **1998**, *71*, 2017.

37 S.V. Patwardhan, N. Mukherjee, S.J. Clarson, *J. Inorg. Organomet. Polym.* **2001**, *11*, 117.

38 S.V. Patwardhan, N. Mukherjee, S.J. Clarson, *Silicon Chem.* **2002**, *1*, 47.

39 M. Sumper, *Angew. Chem. Int. Ed.* **2004**, *43*, 2251.

40 S.V. Patwardhan, N. Mukherjee, S.J. Clarson, *J. Inorg. Organomet. Polym.* **2001**, *11*, 193.

41 S.V. Patwardhan, S.J. Clarson, *Silicon Chem.* **2002**, *1*, 207.

42 T. Coradin, O. Durupthy, J. Livage, *Langmuir* **2002**, *18*, 2331.

43 S.V. Patwardhan, N. Mukherjee, M. Steinitz-Kannan, S.J. Clarson, *Chem. Commun.* **2003**, 1122.

44 S.V. Patwardhan, C. Raab, N. Hüsing, S.J. Clarson, *Silicon Chem.* **2003**, *2*, 279.

45 D. Belton, G. Paine, S.V. Patwardhan, C.C. Perry, *J. Mater. Chem.* **2004**, *14*, 2231.

46 R.-H. Jin, J.-J. Yuan, *Chem. Commun.* **2005**, 1399.

47 R.-H. Jin, J.-J. Yuan, *Adv. Mater.* **2005**, *17*, 885.

48 F. Noll, M. Sumper, N. Hampp, *Nano Lett.* **2002**, *2*, 91.

49 K.M. Roth, Y. Zhou, W. Yang, D.E. Morse, *J. Am. Chem. Soc.* **2005**, *127*, 325.
50 D. Belton, S.V. Patwardhan, C.C. Perry, *Chem. Commun.* **2005**, 3475.
51 D. Belton, S.V. Patwardhan, C.C. Perry, *J. Mater. Chem.* **2005**, *15*, 4629–4638.
52 T. Coradin, J. Livage, *Colloids Surf. B* **2001**, *21*, 329.
53 L. Sudheendra, A.R. Raju, *Mater. Res. Bull.* **2002**, *37*, 151.
54 H. Menzel, S. Horstmann, P. Behrens, P. Bärnreuther, I. Krueger, M. Jahns, *Chem. Commun.* **2003**, 2994–2995.
55 P. Bärnreuther, M. Jahns, I. Krueger, P. Behrens, S. Horstmann, H. Menzel, in: N. Auner, J. Weis (Eds.), *Organosilicon VI*. Wiley-VCH, Weinheim **2005**, pp. 949–954.
56 R.K. Iler, *The Chemistry of Silica*. Wiley, New York, **1979**.
57 C.J. Brinker, G.W. Scherer, *Sol-Gel Science: The Physics and Chemistry of Sol-Gel Processing*. Academic Press, Boston, **1990**.
58 J. Lee, K. Lee, H. Kim, *Bull. Korean Chem. Soc.* **1996**, *17*, 115–116.
59 T. Saegusa, S. Kobayashi, Y. Nagura, *Macromolecules* **1974**, *7*, 713.
60 B.G. Alexander, *J. Am. Chem. Soc.* **1953**, *75*, 5655.
61 P. Behrens, M. Jahns, F. Cornelius, P. Cordes, H. Menzel, unpublished results.
62 G.D. Stucky, A. Monnier, F. Schüth, Q. Huo, D. Margolese, D. Kumar, M. Krishnamurty, P. Petroff, A. Firouzi, M. Janicke, B.F. Chmelka, *Mol. Cryst. Liq. Cryst.* **1994**, *240*, 187.

2
Solid-State NMR in Biomimetic Silica Formation and Silica Biomineralization

Eike Brunner and Katharina Lutz

Abstract

During the past decades, modern solid-state NMR spectroscopic techniques have found an overwhelming variety of applications in materials science as well as biology. The unicellular, photosynthetic diatoms are well known for their nanostructured silica-based cell walls. These cell walls exhibit intricate species-specific patterns of high regularity and beauty. Diatom cell walls are made up of a composite material containing silica as well as certain biomolecules. It has been shown that solid-state NMR spectroscopy is useful for characterizing both, the silica material as well as the organic molecules embedded within the silica. Isotope enrichment of rare nuclei such as ^{13}C or ^{29}Si allows for the direct detection of the corresponding spectra. Based on the understanding of the physico-chemical principles used in natural biosilica formation, silica materials can be synthesized under improved conditions. Such materials can also be studied using solid-state NMR spectroscopic methods.

Key words: NMR spectroscopy, solid-state, liquid-state, silica biomineralization, biomimetic silica synthesis, silaffins, polyamines, phase separation.

2.1
Introduction

During the past decades, solid-state NMR spectroscopy has found an overwhelming variety of applications in materials science as well as biology. Silica-based materials such as the amorphous silica gels or amorphous mesoporous materials (MCM materials) as well as crystalline compounds such as zeolites, are frequently studied with solid-state NMR spectroscopy (e.g., [1–7]). Furthermore, solid-state NMR spectroscopy becomes increasingly important for the investigation of biomolecules such as proteins and peptides (e.g., [8–12]).

Handbook of Biomineralization. Edited by P. Behrens and E. Bäuerlein

ISBN: 978-3-527-31805-6

The unicellular, photosynthetic diatoms [13] are well known for their nanostructured, silica-based cell walls [14]. These cell walls exhibit intricate species-specific patterns of high regularity and beauty. Diatom cell walls are made up of a composite material containing silica as well as certain biomolecules. Thus, solid-state NMR spectroscopy (see Section 2.2) should be useful for characterizing both, the silica material as well as the organic molecules embedded within the silica (see Section 2.3).

Diatom cell walls currently attract increasing interest for two reasons:

- The physico-chemical processes underlying the formation of the nanostructured amorphous silica are studied with respect to improved, so-called biomimetic materials synthesis. Numerous attempts have been made recently to mimic biomineralization processes and to utilize methods of bio-inspired materials synthesis (see Section 2.4).
- In addition to their nanopatterning, diatom cell walls exhibit several other very interesting materials properties. Their mechanical stability is extraordinary [15]; furthermore, diatom cell walls are likely to act as so-called photonic crystals [16].

Diatom cell walls are coated with organic molecules [14]. For example, special proteins known as *frustulins* were found to cover the frustule of several diatom species. This organic coating can be removed by treatment with ethylene diamine tetra-acetic acid (EDTA) and sodium dodecyl sulfate (SDS) [17]. In contrast to the coat proteins, however, certain biomolecules are tightly attached to, or even embedded within, the silica, thus creating an extremely interesting composite material. Some of these biomolecules which cannot be removed by EDTA and SDS were shown to be involved in the process of silica precipitation and pattern formation. Such biomolecules are the so-called *silaffins* [18–20], a family of polypeptides with characteristic post-translational modifications. Often, long-chain polyamines are covalently bound to the side chains of lysine residues. Furthermore, silaffins are usually phosphorylated [19]. Peptide-free long-chain polyamines [21, 22] could also be found in diatom cell walls. Following acid hydrolysis of the cell wall silica, these biomolecules can be extracted and later purified; the purified material is then accessible to the established methods of biomolecular liquid-state NMR spectroscopy (see Sections 2.3 and 2.4).

2.2 General Remarks on Solid-State NMR Spectroscopy

The NMR spectra of static solids typically exhibit relatively broad signals [23, 24] without spectral resolution. Very limited structural information is available from such spectra. The width of solid-state NMR signals arises from so-called internal magnetic interactions. For spin-1/2 nuclei such as ^{1}H, ^{13}C, ^{29}Si, or ^{31}P, the line width is mainly determined by the simultaneous influence of the chemical shift anisotropy (CSA) and homonuclear as well as heteronuclear magnetic dipole–dipole interactions with neighboring spins (through-space interactions). Indirect spin–spin couplings (through-bond couplings, *J*-couplings) are usually several orders of magnitude smaller than the broadening associated with the aforementioned anisotropic interactions.

The resolution and sensitivity of the NMR spectra of solid samples may be improved significantly by magic angle spinning (MAS). This involves a rapid rotation of the sample around an axis tilted by the "magic angle" $\Theta_m = 54.7°$ with respect to the external magnetic field, B_0 [25]. This MAS of the sample causes the orientation-dependent anisotropic spin interactions such as chemical shift anisotropy and magnetic dipole–dipole interaction to become time-dependent [26]. The secular parts of these interactions, which govern the spectral line shape, are averaged out completely if the sample spinning rate, ν_r, is large compared to the static line width (fast spinning limit) – the result is a narrow, "liquid-like" spectrum. It is, however, often impossible to fulfill the fast-spinning condition. As the MAS NMR signal is periodic, with $t_r = 1/\nu_r$, Fourier transformation yields a spectrum with intensities only at the resonance frequency, ν_0, plus integer multiples of the sample spinning rate, $n \times \nu_r$ ($n = 0, \pm1, \pm2, \ldots$). This is demonstrated in Figure 2.1, which shows the ^{31}P {^{1}H} CP MAS NMR spectrum of diatom cell walls extracted from the species *Stephanopyxis turris*. During signal acquisition, ^{1}H decoupling sequences such as two pulse phase-modulated (TPPM) decoupling [27] must be applied in order to obtain the maximum spectral resolu-

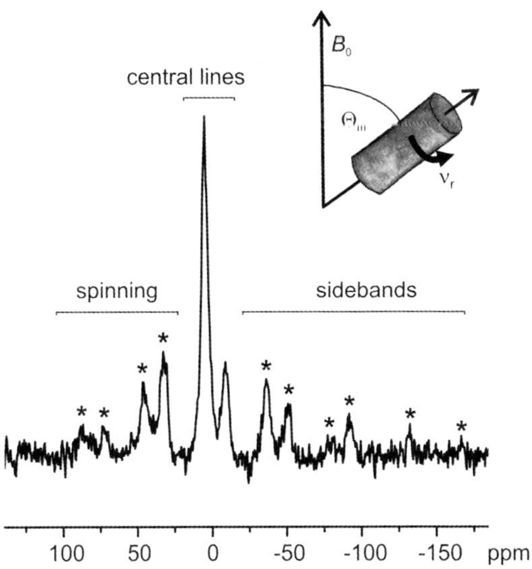

Fig. 2.1 ^{31}P {^{1}H} CP MAS NMR spectrum of cell walls isolated from *S. turris* measured at 5 kHz sample spinning rate. * denotes spinning sidebands.

tion; that is, the minimum residual line width (defined as the full width at half maximum) for the central lines of the considered signals. The ^{31}P NMR signals observed in Figure 2.1 are likely to be due to the presence of phosphorylated proteins (silaffins; see Section 2.3.2) in the cell walls, as well as inorganic phosphate which eventually interacts with the polyamines embedded in the cell walls (see Section 2.4). The so-called cross-polarization (CP) [28–30] technique is applied in order to enhance the otherwise small signal intensity as the concentration of ^{31}P in the samples is low. CP relies on the transfer of spin polarization from sensitive nuclei (^{1}H in most cases) to dipolarly coupled insensitive nuclei (here ^{31}P), which results in an enhanced spin polarization for the latter species. In addition, the repetition time of the CP experiment is dictated by the longitudinal relaxation time, T_1, of ^{1}H which is usually considerably shorter than that of the insensitive nuclei such as ^{13}C, ^{29}Si, or ^{31}P. Thus, a much higher number of scans can be accumulated within a given measurement time if CP is applied, and this results in a further sensitivity gain. In addition, CP is often used as the fundamental heteronuclear magnetization transfer step in so-called multidimensional HETeronuclear CORrelation (HETCOR) experiments. Examples of ^{29}Si $\{^{1}H\}$ HETCOR experiments on diatom cell wall silica are provided in Section 2.3.

Biominerals such as diatom cell walls, sponge skeletons, shells, bones, and others are often amorphous composite materials, or composites made from microcrystalline regions embedded in an amorphous matrix. As a result of the distribution of local parameters (such as bond angles and bond distances), amorphous compounds usually exhibit relatively large residual line widths for the MAS NMR signals, in contrast to those of crystalline compounds. This is demonstrated in Figure 2.2,

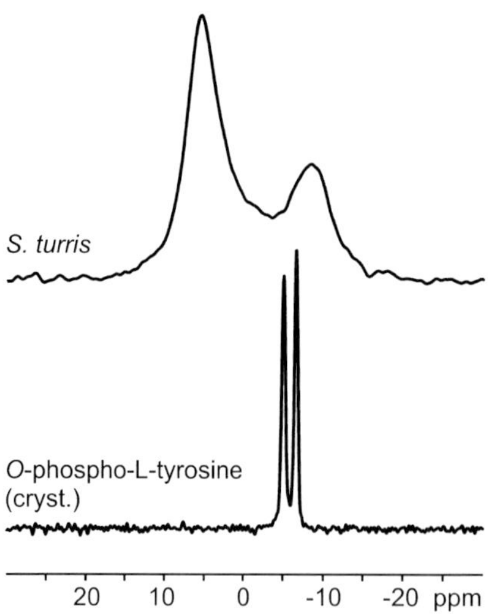

Fig. 2.2 ^{31}P $\{^{1}H\}$ CP MAS NMR spectra (central lines only) of cell walls isolated from *S. turris* (top) and of crystalline *O*-phospho-L-tyrosine (bottom).

which compares the central lines of the $^{31}P\ \{^1H\}$ CP MAS NMR spectra of *S. turris* cell walls and polycrystalline *O*-phospho-L-tyrosine. The differences in residual line width are striking. The relatively large residual line widths typically observed for amorphous systems sometimes limit the application of advanced multi-dimensional techniques. However, solid-state NMR spectroscopy – when used in combination with liquid-state NMR spectroscopy and other biochemical and physico-chemical methods – significantly contributes to our understanding of biominerals and biomimetically synthesized materials, as will be shown in the following section.

2.3
Multinuclear NMR Studies of Diatom Cell Walls

2.3.1
Studies with Solid-State ^{29}Si NMR Spectroscopy

Diatom cell wall silica can be studied using solid-state ^{29}Si NMR spectroscopy [31–34]. As the natural abundance of the isotope ^{29}Si $(I = \frac{1}{2})$ amounts to only 4.7 % and T_1 tends to be rather long, ^{29}Si isotope labeling is necessarily required especially for multidimensional experiments. A typical one-dimensional ^{29}Si MAS NMR spectrum of diatom cell walls from *Thalassiosira pseudonana* is shown in Figure 2.3. The spectrum exhibits the well-known signals at $\delta = -92$ ppm (Q^2 groups), $\delta = -102$ ppm (Q^3 groups), and $\delta = -111$ ppm (Q^4 groups) characteristic of amorphous silica [1, 2]. It is suggested that the organic material embedded

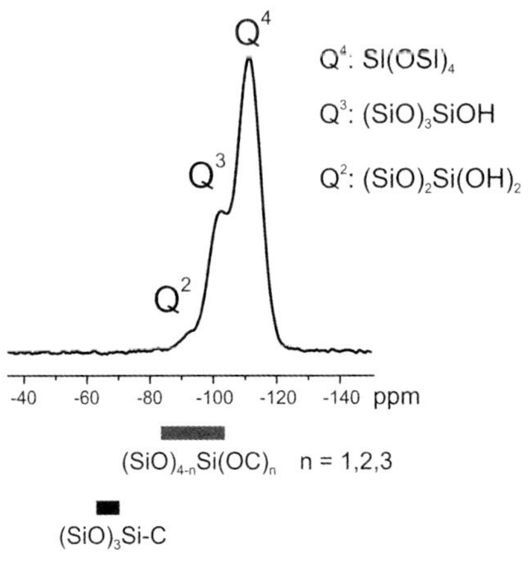

Fig. 2.3 ^{29}Si MAS NMR spectrum of ^{29}Si isotope-labeled cell walls isolated from *T. pseudonana*. Chemical shift ranges characteristic for ^{29}Si involved in covalent bonds with organic material are given at the bottom for comparison (chemical shift data are taken from Ref. [33]).

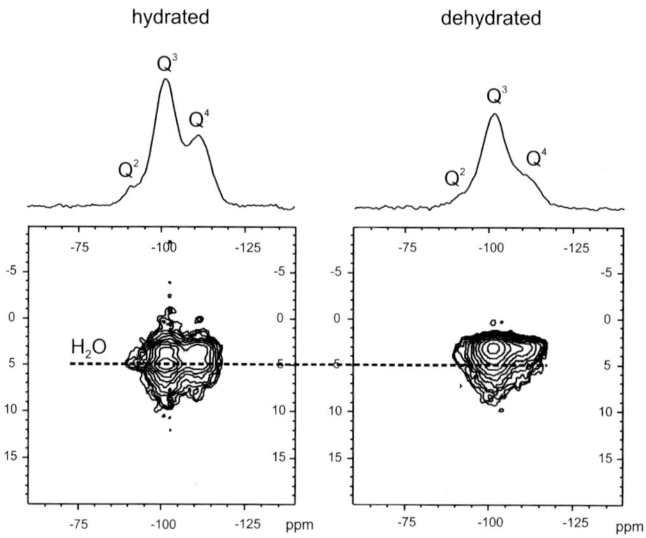

Fig. 2.4 $^{29}Si\ \{^1H\}$ HETCOR spectra of hydrated and dehydrated ^{29}Si isotope-labeled diatom cell walls from *T. pseudonana* measured under phase-modulated Lee–Goldburg decoupling [35] during ^1H evolution and with a CP contact time of 7 ms. The ^{29}Si projections which correspond to the $^{29}Si\ \{^1H\}$ CP MAS NMR spectra are given on top of the 2D spectra.

in the cell walls may form covalent bonds with the surrounding silica. Therefore, characteristic ranges for the ^{29}Si chemical shifts of such species are shown at the bottom of Figure 2.3. However, the existence of organic species covalently bound to the cell wall silica of diatoms has not yet been detected [33]. The $^{29}Si\ \{^1H\}$ CP MAS NMR spectrum of *T. pseudonana* cell walls exhibits the same signals as the directly excited ^{29}Si MAS NMR spectrum. However, the relative signal intensities are different (see Fig. 2.4).

As the Q^2 and Q^3 groups exhibit ^1H nuclei in their spatial neighborhood, these signals are preferentially enhanced in the CP experiment [2, 31, 33]. A further interesting effect is found [34] when comparing the 2D $^{29}Si\ \{^1H\}$ HETCOR spectra of hydrated and dehydrated diatom cell walls from ^{29}Si isotope-labeled *T. pseudonana* (see Fig. 2.4). For the hydrated sample, ^1H magnetization is transferred to ^{29}Si from two resolved ^1H moieties at ca. 3 and 5 ppm, as well as a relatively broad "shoulder" ranging beyond 5 ppm. For the dehydrated cell walls, however, transferred ^1H magnetization originates from the ^1H NMR signal at 3 ppm and the broad "shoulder" only, as described previously [33]. The intense signal at 5 ppm is only present in the hydrated sample, and can be assigned to physisorbed water. The existence of this CP polarization transfer from water to ^{29}Si shows that the physisorbed water molecules do not move isotropically. This signal completely disappears from the spectrum after dehydration. The remaining signals in ^1H dimension are due to SiOH groups. ^1H nuclei located at the embedded organic material may also contribute. It is surprising that a significant ^{29}Si $\{^1H\}$ cross-polarization from the adsorbed water is also observed in Q^4 groups. These units are expected to be located in the bulk SiO_2 which should be inaccessible for water. This observation can be explained by two possible assumptions: (i) water penetrates into the bulk SiO_2 phase of diatom cell walls; or (ii) ^{29}Si-^{29}Si spin diffusion takes place from the Q^2 and Q^3 groups located at the outer surface to the Q^4 groups in the bulk during the CP contact time as the samples are ^{29}Si isotope-labeled. In any case, a pronounced influence of sample hydration upon the ^{29}Si $\{^1H\}$ CP MAS NMR spectra must be stated.

2.3.2
Studies of the Embedded Organic Material by NMR Spectroscopy

Another interesting application of solid-state NMR spectroscopy is the characterization of organic material in diatom cell walls [33, 34]. Figure 2.5 shows the ^{13}C MAS NMR spectra of ^{13}C isotope-labeled cell walls extracted from *T. pseudonana* in the hydrated (upper line) and dehydrated (lower line) state. Although the cell walls were treated with EDTA and SDS, a certain amount of organic material could not be removed by this treatment. These biomolecules are obviously bound to, or even embedded within, the siliceous cell wall material. Dehydration causes line broadening which can be interpreted by the assumption that sample hydration results in an increased mobility, at least for the water-accessible part of the

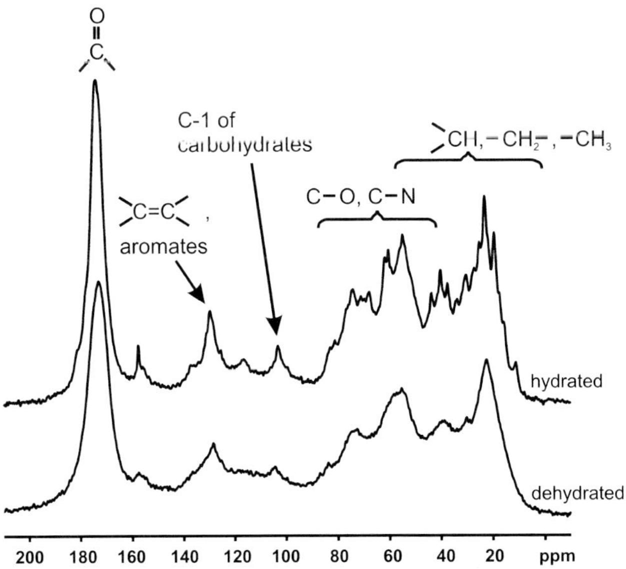

Fig. 2.5 ^{13}C MAS NMR spectra of isolated cell walls from *T. pseudonana*. The spectra of the hydrated (top) and dehydrated (bottom) cell wall samples are shown [34].

organic material. In general, it must be stated that the spectrum consists of the superimposition of various signal components as the organic material exhibits a variety of different molecules. Due to this heterogeneity of the organic material, solid-state NMR spectroscopic analyses of diatom cell walls yield an overall picture, but no specific information concerning a special type of biomolecules. It is, however, possible to extract the organic material from the cell walls by dissolution of the silica using hydrogen fluoride [18, 21] or ammonium fluoride [19]. The biomolecules can then be separated into the various fractions and become accessible to the well-established methods of biomolecular liquid-state NMR spectroscopy.

Figure 2.6 displays a typical ^1H NMR spectrum of long-chain polyamines isolated from *T. pseudonana*. Together with mass-spectroscopic data, this spectrum allowed the determination of the molecular structure of the corresponding polyamine [22]. The polyamines isolated from *T. pseudonana* are chains of various lengths consisting of [–CH_2–CH_2–CH_2–NX–] repetitive units (X: H or CH_3) which are attached to a small basis (1,3-diaminopropane or spermidine). Additionally, they may contain terminal quaternary ammonium functionalities, as

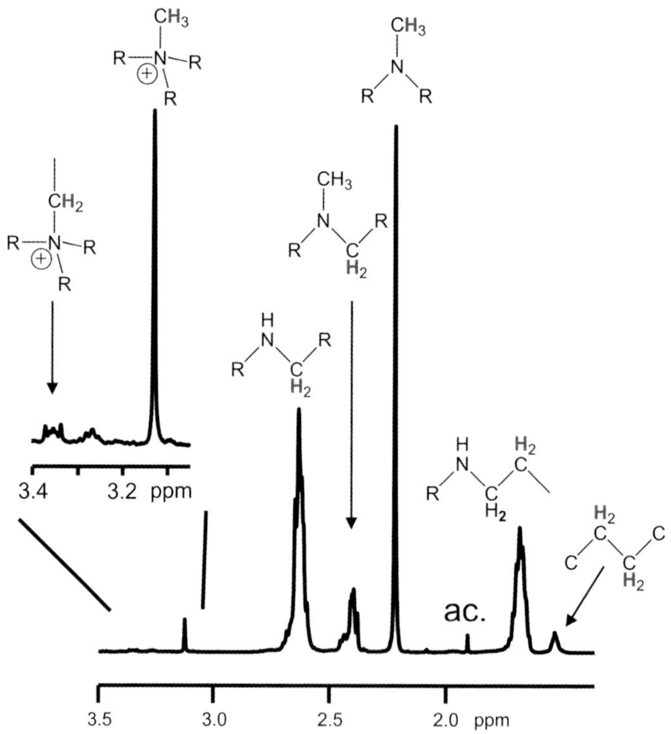

Fig. 2.6 ^1H NMR spectrum of the native long-chain polyamine population isolated from *T. pseudonana* measured at 298 K, pH 11 [22] (reproduced by permission of Federation of the European Biochemical Societies). The signal denoted by "ac." is due to spurious amounts of acetate ions introduced during sample purification.

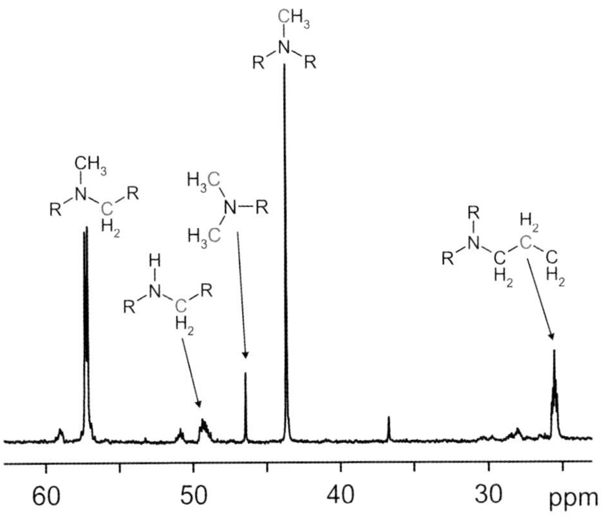

Fig. 2.7 ^{13}C NMR spectrum of a polyamine fraction isolated from the diatom *S. turris* recorded at 298 K, pH 11.

indicated in Figure 2.6. ^{13}C NMR spectroscopy is another helpful method which can be used to resolve structural features of polyamines, as ^{13}C isotope labeling has been accomplished successfully. As an example, Figure 2.7 shows the directly detected ^{13}C NMR spectrum of a polyamine fraction isolated from the diatom *S. turris* recorded at $B_0 = 18.79$ T. The excellent spectral resolution and signal/noise ratio obtained for ^{13}C isotope-labeled polyamines at this field strength make liquid-state ^{13}C NMR spectroscopy a valuable tool for the detailed investigation of diatom cell wall biomolecules.

Solid-state ^{31}P NMR spectroscopy allows the detection of phosphate moieties in diatom silica (see Figs 2.1 and 2.2). Both, phosphate covalently bound to silaffins as well as free phosphate ions which are not covalently attached to the polyamine matrix, could be detected [32, 34]. It should be noted within this context that marine sediments are known to contain large amounts of diatom silica. Interestingly, such sediments were found to store relatively high amounts of phosphate, preferentially as orthophosphate esters and diesters [36]. The chemical shifts of their ^{31}P MAS NMR signals are in the same range as the signals found for diatom cell wall samples. Therefore, phosphate bound to the organic material enclosed within sedimented diatom cell walls may be a major contributor to the phosphate pool stored in marine sediments.

Liquid-state ^{31}P NMR experiments confirmed the phosphorylation of the native form of silaffins. One particularly interesting molecule is native silaffin-1A (natSil-1A) extracted from the diatom *Cylindrotheca fusiformis* [19], as its amino acid sequence could be determined. The ^{31}P NMR signals (Fig. 2.8) of this highly phosphorylated peptide could be assigned to the phosphorylated serine and trimethylhydroxylysine residues [19].

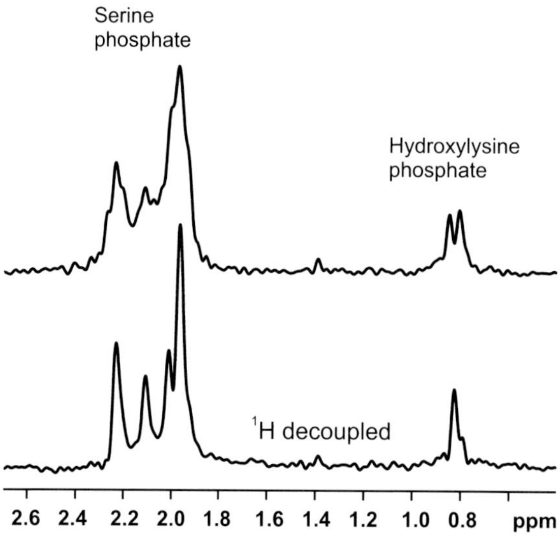

Fig. 2.8 ^{31}P NMR spectrum of the silaffin natSil-1A from *C. fusiformis* measured in 200 mM sodium citrate at pH 5.0 without (top) and with (bottom) ^{1}H decoupling [19].

2.4 Silica Precipitation and Self-Assembly of Silaffins and Polyamines

2.4.1 Silica Precipitation Activity of Natural Polyamines and Silaffins

Silica can be dissolved as ortho-silicic acid ($Si(OH)_4$), the autocondensation of this monomer finally resulting in a polymer with a maximum number of siloxane bonds (Si–O–Si) and a minimum number of non-condensed SiOH-groups [37, 38]. Depending on the pH and ionic strength of the solution, formation of different intermediates is favored whose subunits are spherical and which – at neutral pH – exhibit negative charges on their surface [37–39].

As described previously by Iler [37], cationic and hydrogen bond-forming polymers are capable of inducing SiO_2 precipitation from silicic/oligosilicic acid solutions *in vitro*. In particular, polyamines are able to induce silica polymerization [40]. This is also true for silaffins and polyamines isolated from diatoms. *In-vitro* experiments have shown that polyamines from diatom cell walls are able to induce silica precipitation from monosilicic acid solutions within a few minutes, and the polyamines co-precipitate at the same time. Clearly, the positively charged polyamine molecules come into contact with the negatively charged polysilica molecules via electrostatic interactions. The resulting silica networks are composed of spherical aggregates with diameters between 1 and 1000 nm, depending on the polyamine composition and the reaction conditions [21].

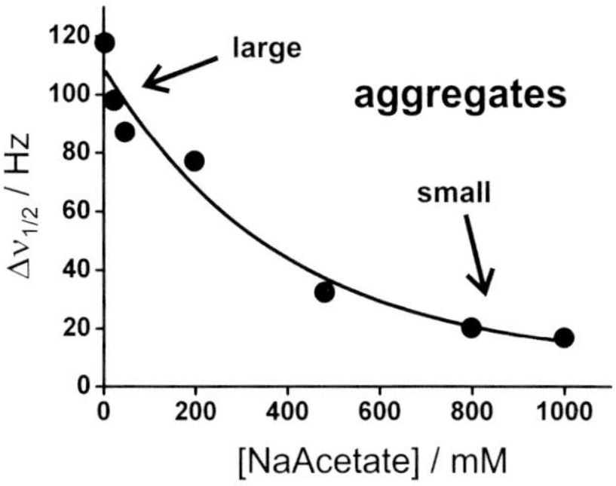

Fig. 2.9 Line width (full width at half-maximum) of the ^{31}P NMR signal of trimethylhydroxylysine phosphate in natSil-1A (cf. Fig. 2.8) measured with ^{1}H decoupling as a function of the sodium acetate concentration at pH 5.5 [19].

In order to achieve a better understanding of the role of polyamines and silaffins in the biosilicification process of diatoms, the properties of these molecules were analyzed extensively. The ^{31}P NMR experiments on natSil-1A described in Section 2.3.2 revealed a very interesting behavior [19], namely that the line width of the ^{31}P NMR signals was found to depend heavily on the salt concentration (Fig. 2.9). The signals are very broad at low salt concentrations, indicating the formation of large aggregates; however, increasing the salt concentration leads to a decrease in line width, indicating the formation of smaller aggregates. This self-assembling process of the zwitterionic molecules is driven by electrostatic interactions between the negative charges of the phosphate residues and the positive charges of the protonated amino groups of the polyamine chains. Interestingly, self-assembly of silaffins was shown to be a necessary precondition for their ability to induce the formation of silica nanospheres [19]. This result is supported by silica precipitation experiments performed with silaffin-1A, a molecule where the covalently bound phosphate groups are removed by the HF extraction procedure [18, 41]. In contrast to natSil-1A, the polycationic silaffin-1A molecules show no silica precipitation activity. Following the addition of phosphate ions, however, silica precipitation occurs in analogy to natSil-1A. The phosphate ions clearly provide the negative charges necessary for the self-aggregation of silaffin-1A [19]. As mentioned above, polyamines are also capable of inducing the silicic acid polycondensation, although *in-vitro* experiments revealed that the silica precipitation activity of polyamines extracted from the diatom *S. turris* only occurs in the presence of phosphate. An increase in phosphate concentration leads to an increase in SiO_2 particle diameter [42], and the replacement of orthophosphate by pyrophosphate enormously enhances this effect. Hence, it was suggested that the polyamines might be able to self-assemble in the presence of phosphate via electrostatic interactions [42], in analogy to the silaffins.

2.4.2
Self-Assembly of Polyamines: Poly(allylamine) as a Model Compound

The self-assembly (aggregation) of silaffins and long-chain polyamines from diatom cell walls is necessary for silica precipitation. In order to understand this interesting behavior better, detailed model studies on synthetic long-chain polyamines – namely poly(allylamine hydrochloride) (PAA) – were performed. Among the numerous long-chain polyamines used in biomimetic silica synthesis [43–49], PAA is a favorite model compound. Furthermore, PAA can be purchased in sufficiently large quantities, in contrast to the polyamines extracted from diatoms. Amine-terminated dendrimers [50] could also be used for biomimetic silica synthesis experiments. Other investigators have used poly(amino acids) such as poly(L-histidine), poly(L-lysine), and poly(L-arginine) [51–54] or other polypeptides [54–56]. It should be noted within this context that poly(amino acids) such as poly(L-lysine) are also known to self-assemble in the presence of phosphate ions [57].

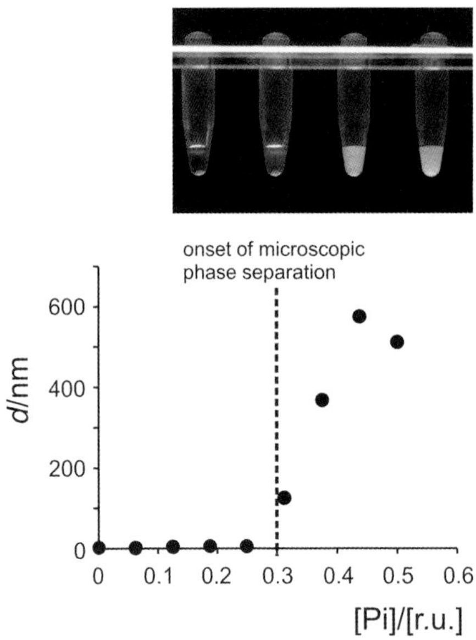

Fig. 2.10 Phase-separation behavior of PAA/phosphate solutions at pH 5.8 [58]. Top: Photographs of a 0.2 mM PAA solution at various phosphate contents. Phosphate contents from left to right: 0.15, 0.31, 0.47, 0.62 [Pi]/[r.u.] (r.u. = repetitive unit of PAA). At ca. 0.3 [Pi]/[r.u.], the samples become cloudy, indicating microscopic phase separation. The small bright spot at the bottom of the vial is due to a reflection and is not related to the sample solution. Bottom: Diameter, *d*, of the PAA aggregates determined by dynamic light scattering as a function of the phosphate concentration. PAA concentration: 1 mM.

2.4.2.1 The Dependence of PAA Aggregation on the Phosphate Concentration

Phosphate ions clearly play a central role in the aggregation process of natural polyamines (see Section 2.4.1). The same is true for PAA, as can be seen from Figure 2.10: the PAA aggregate diameter in aqueous solution was measured using dynamic light scattering (DLS) as a function of phosphate concentration at pH 5.8 [58]. The diameter was small at low phosphate concentrations, but beyond a threshold value of about 0.3 phosphate ions per repetitive unit of the PAA molecule ([Pi]/[r.u.]), the aggregate diameter increased continuously with increasing phosphate concentration up to approximately 600 nm at 0.44 [Pi]/[r.u.]. In addition, the initially clear samples become cloudy, indicating that microscopic phase separation had occurred. If the phosphate concentration is further increased, the microscopic phase separation proceeds, finally leading to macroscopic phase separation beyond approximately 0.6 [Pi]/[r.u.]. As a consequence, the increasingly large PAA droplets sink to the bottom of the vials where they form a visibly separated phase. Hence, the detection of PAA aggregates in the top part of the sample tubes is no longer possible by DLS. In complete agreement with the observations made for silaffins and polyamines from diatom cell walls, it was shown that the precipitation of silica nanospheres from silicic acid solutions only occurred if the samples exhibited microscopic phase separation (Fig. 2.11). The size of the silica

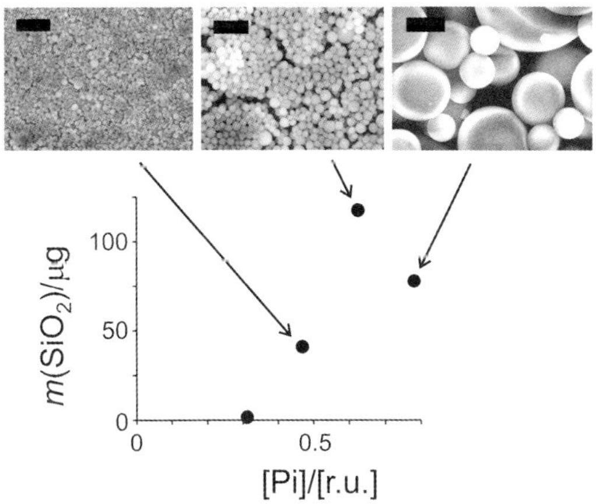

Fig. 2.11 Silica precipitated from aqueous PAA/silicic acid solutions as a function of the phosphate concentration [58]. Top: Scanning electron microscopy images of the precipitates. The black bar shown in the images defines a length of 2 μm. For the samples shown here, the following average particle diameters could be determined: Left: 170 nm (70 nm); Middle: 290 nm (150 nm); Right: 2400 nm (1100 nm). The standard deviations of the particle diameters are given in parentheses. Note, that the particle diameters exhibit relatively broad distributions for all samples. The average particle diameter of the precipitated silica nanospheres is correlated with the phosphate concentration. Bottom: Amount of precipitated silica as a function of phosphate concentration. Note that silica precipitation starts at 0.3 [Pi]/[r.u.], the threshold phosphate concentration for microscopic phase separation (see Fig. 2.10).

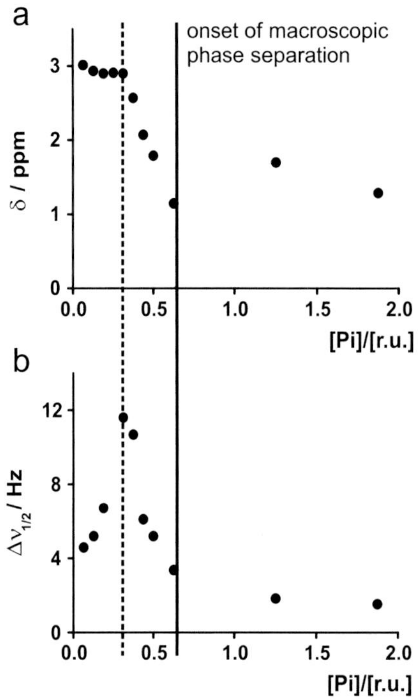

Fig. 2.12 ^{31}P NMR chemical shift, δ, (a) and full width at half-maximum, $\Delta\nu_{1/2}$, (b) of the phosphate signal as a function of the phosphate concentration [58]. PAA concentration: 1 mM, pH 5.8. The solid vertical line shows the onset of macroscopic phase separation. The threshold concentration for microscopic phase separation (i.e., for silica precipitation) is indicated by a dashed vertical line.

nanospheres is strictly determined by the phosphate concentration, in analogy to the behavior described in Section 2.4.1 for polyamines from diatom cell walls. It should be noted within this context that the size of the silica nanospheres precipitated from solutions of amine-terminated dendrimers was also found to be controlled by the phosphate buffer concentration [59].

The formation of PAA aggregates was also proven by chemical shift and line width analysis of the ^{31}P NMR spectra of the PAA/phosphate solutions (Fig. 2.12) [58]. At low phosphate concentrations, all phosphate ions are bound to the PAA molecules which is reflected by a constant chemical shift of about 3 ppm (Fig. 2.12a). Beyond a concentration of 0.3 [Pi]/[r.u.], corresponding to the onset of microscopic phase separation, the chemical shift decreases and finally reaches a constant minimum of about 1.2 ppm for 0.6 [Pi]/[r.u.], where the macroscopic phase separation starts. This can be explained as follows. Increasing the phosphate concentration results in an increasing amount of free phosphate ions which rapidly exchange with phosphate ions interacting with the PAA molecules. The observed chemical shift then corresponds to the weighted average of the chemical shifts of phosphate bound to the polyamines and free phosphate. At high phosphate concentrations, all PAA/phosphate aggregates can be found in the macroscopically separated phase, whereas the aqueous solution at the top only contains free phosphate ions. This interpretation is further corroborated by observing the line width, $\Delta\nu_{1/2}$, of the ^{31}P NMR signals of these solutions (Fig. 2.12b). The line width is correlated to the size of the particles. Increasing the aggregate diameters results in higher correlation times for reorientation of the aggregates and, therefore, in increasing line widths. The line width increase observed up to 0.3 [Pi]/[r.u.], therefore, reflects the growth of the polyamine aggregates. The detected line width is again a weighted average of the line widths which would be observed for bound and free phosphate ions. For further increasing phosphate concentrations – which means for an increasingly large amount of free phosphate ions – the line width begins to fall. When macroscopic phase separation has occurred, the small line width which is characteristic for rapidly reorienting free phosphate is observed in the aqueous part of the solution.

2.4.2.2 The Dependence of PAA Aggregation on the pH Value

Another important parameter determining the self-assembly in PAA/phosphate solutions is pH. Figure 2.13 shows graphically the aggregate diameter measured on PAA/phosphate solutions by DLS as a function of pH [32]. At low pH values (up to pH 4), aggregation does not occur; it is likely that repulsion of the positively charged PAA molecules then exceeds the attractive forces caused by the interaction of PAA with the negatively charged phosphate ions. Increasing the pH leads to a decreasing positive charge on the PAA molecules and an increasing

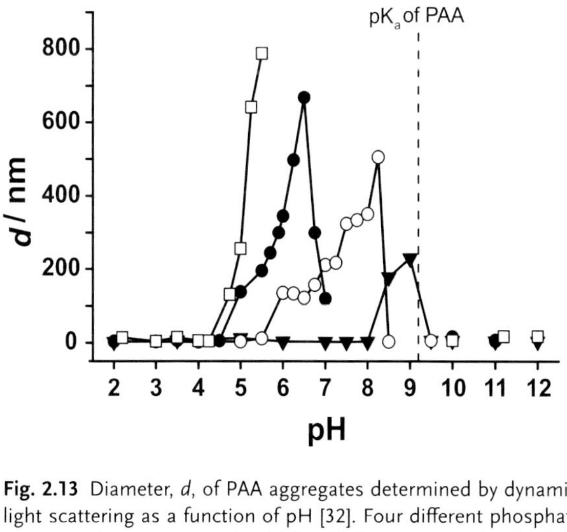

Fig. 2.13 Diameter, *d*, of PAA aggregates determined by dynamic light scattering as a function of pH [32]. Four different phosphate concentrations were examined: 70 mM (open squares), 55 mM (filled circles), 40 mM (open circles), and 25 mM (filled triangles). PAA concentration: 1 mM.

negative charge on the phosphate ions. The attractive forces between PAA and phosphate ions exceed the repulsive forces at a certain threshold pH value and aggregate formation is then initiated. This threshold pH must, of course, depend on the phosphate concentration as it is indeed observed (see above). As the aggregates keep growing with increasing pH, sedimentation of the aggregates and – finally – macroscopic phase separation is observed. At about pH 9 – which corresponds to the pK_a of PAA – the molecules lose their positive charge and become neutral. Therefore, no aggregates are observed beyond this pH.

Other anions were also examined with respect to their ability to induce PAA self-assembly [32, 58]. Similar to phosphate, the addition of sulfate to the PAA solutions causes microscopic phase separation beyond 0.3 [sulfate]/[r.u.] and macroscopic phase separation at higher sulfate concentrations. In contrast, the addition of monovalent chloride anions only results in the formation of very small aggregates (ca. 9 nm), even at high concentrations (20 [Cl]/[r.u.]). No microscopic or macroscopic phase separation could be observed.

Taken together, these experiments show that the formation of PAA aggregates must be related to the electrostatic interactions between the positively charged polyamine molecules and the negatively charged phosphate ions. However, PAA self-assembly cannot be considered to be a purely electrostatic effect [32]. PAA aggregation does not occur in the presence of methylphosphonate (MeP) or phosphite, two multivalent anions with similar pK_a values to phosphate. Consequently, these anions should behave similarly to phosphate from an electrostatic point of view. Their failure to induce PAA aggregation leads to the conclusion that PAA self-assembly is not determined exclusively by electrostatic interactions but also depends on the geometry of the anions and – possibly – also on their ability to form hydrogen bonds. Obviously, the assembly of the PAA/phosphate network is interrupted if one of the four oxygen atoms of phosphate is substituted. In summary, it is concluded that the phosphate-induced self-assembly of polyamines – which is a necessary prerequisite for silica precipitation from silicic-acid containing solutions – is caused by the formation of a defined polyamine/phosphate network properly balanced by electrostatic interactions.

2.4.3
Microscopic Phase Separation Mediates Cell Wall Biogenesis

Microscopic phase separation of the polyamine/phosphate solutions is necessary for their silica precipitation activity *in vitro*. However, polyamines not only induce the silica precipitation, they may also be critically involved in the pattern formation. Sumper [60] has suggested a model which explains the structure formation of the diatom cell walls of the genus *Coscinodiscus*. This model relies on the assumption that the hierarchically arranged silica patterns originate from consecutive phase separation events during cell wall biosynthesis [60, 61]. In order to verify this model, solid-state ^{31}P NMR studies were performed [32] on cell walls isolated from *C. granii*. Silaffins do not occur in the cell walls of this species, but polyamines are found. If phosphate-induced polyamine aggregation plays a role, one would expect to find a signal due to phosphate ions attached to the polyamines entrapped in the silica material. A corresponding signal could indeed be detected at a chemical shift of 2.5 ppm, which is close to the chemical shift of 2.0 ppm observed for phosphate entrapped in a synthetic silica precipitate prepared from a silicic-acid containing PAA/phosphate solution [32]. This observation supports the idea that the biomineralization process in diatoms is indeed controlled by phase separation processes of the silica-precipitating biomolecules during cell wall formation.

The biomimetically synthesized silica precipitates discussed so far have been exclusively silica nanospheres obtained from solutions containing monosilicic acid as silica source. Sumper [46] has shown, however, that elongated silica structures can be produced if a polyamine-stabilized silica sol is used as the silicon source instead of monosilicic acid. The silica patterns then resemble the structure of a diatom cell wall, although they are far from the perfection found in natural diatom silica [46]. We have also measured ^{29}Si MAS NMR spectra of the resulting two types of synthetic silica precipitates obtained from PAA/phosphate solutions using monosilicic acid or a polyamine-stabilized silica sol as silicon source (Fig. 2.14) [32, 61]. The signals appearing in the ^{29}Si MAS NMR spectra can be assigned to the different Q^n units. Comparison with Figure 2.3 shows that the degree of silica condensation for both synthetic precipitates is lower than for diatom biosilica. The fraction of Q^2 and Q^3 units found in the precipitates is higher than in diatom cell walls. However, the degree of silica condensation in the elongated silica structures obtained from the polyamine-stabilized silica sol is closer to that of the natural cell walls than in the silica nanospheres prepared from monosilicic acid solutions.

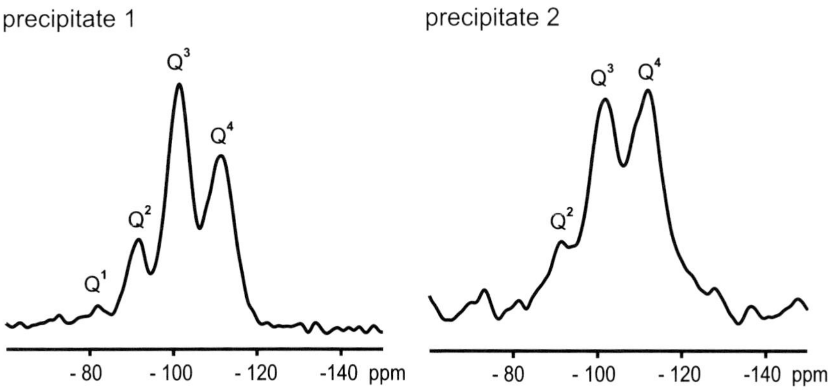

Fig. 2.14 ^{29}Si MAS NMR spectra of two different synthetic silica precipitates at a sample spinning rate of 10 kHz. For precipitate 1, an aqueous solution of monosilicic acid served as silicon source. Precipitate 2 was prepared from a polyamine-stabilized silica sol [61].

2.5 Summary

In summary, we can state that solid-state NMR spectroscopy in combination with liquid-state NMR and other physico-chemical techniques such as DLS allows a detailed characterization not only of biosilica but also of biomimetically synthesized materials. Enrichment with isotopes such as ^{13}C, ^{15}N, and ^{29}Si has become feasible, and this greatly enhances the sensitivity of NMR spectroscopy. In particular, multidimensional NMR techniques can now successfully be applied. NMR spectroscopy yields information at the molecular level and allows concentration measurements to be conducted. *In-vitro* experiments have shown that microscopic phase separation processes of the relevant biomolecules are likely to direct the silica precipitation and nanopattern formation during diatom cell wall biosynthesis.

Acknowledgments

The authors gratefully acknowledge the fruitful collaboration with Prof. Manfred Sumper (Regensburg), as well as financial support from the Deutsche Forschungsgemeinschaft (Br 1278/12-1). They also thank Ms. Ingrid Cuno for carefully proofreading the manuscript.

References

1 G. Engelhardt, D. Michel, *High-Resolution Solid-State NMR of Silicates and Zeolites*. Wiley VCH, Chichester, **1987**.
2 D.W. Sindorf, G.E. Maciel, *J. Am. Chem. Soc.* **1983**, *105*, 1487–1493.
3 J.S. Beck, J.C. Vartuli, W.J. Roth, M.E. Leonowicz, C.T. Kresge, K.D. Schmitt, C.T.-W. Chu, D.H. Olson, E.W. Sheppard, S.B. McCullen, J.B. Higgins, J.L. Schlenker, *J. Am. Chem. Soc.* **1992**, *114*, 10834–10843.
4 H. Pfeifer, H. Ernst, *Annu. Rep. NMR Spectrosc.* **1994**, *28*, 91–187.
5 C.A. Fyfe, G.T. Kokotailo, in: G.E. Maciel (Ed.), *Nuclear Magnetic Resonance in Modern Technology*. NATO ASI Series C, Vol. 447, Kluwer Academic Publishers, Dordrecht, Boston, London, **1994**, pp. 277–337.
6 M. Hunger, E. Brunner, *Mol. Sieves* **2004**, *4*, 201–293.
7 J.D. Epping, B.F. Chmelka, *Curr. Opin. Coll. Interf. Sci.* **2006**, *11*, 81–117.
8 R. Tycko, *Curr. Opin. Chem. Biol.* **2000**, *5*, 500–506.
9 F. Castellani, B. van Rossum, A. Diehl, M. Schubert, K. Rehbein, H. Oschkinat, *Nature* **2002**, *420*, 98–102.
10 A.E. McDermott, *Curr. Opin. Struct. Biol.* **2004**, *14*, 554–561.
11 C.E. Hughes, M. Baldus, *Annu. Rep. NMR Spectrosc.* **2005**, *55*, 121–158.
12 D. Reichert, *Annu. Rep. NMR Spectrosc.* **2005**, *55*, 159–203.
13 F. Round, R. Crawford, D. Mann, *The Diatoms*. Cambridge University Press, Cambridge, **1990**.
14 B.E. Volcani, in: T.L. Simpson, B.E. Volcani (Eds.), *Silicon and siliceous structures in biological systems*. Springer, New York, **1981**, pp. 157–200.

15 C.E. Hamm, R. Merkel, O. Springer, P. Jurkojc, C. Maier, K. Prechtel, V. Smetacek, *Nature* **2003**, *421*, 841–843.

16 T. Fuhrmann, S. Landwehr, M. El Rharbi-Kucki, M. Sumper, *Appl. Phys. B* **2004**, *78*, 257–260.

17 N. Kröger, C. Bergsdorf, M. Sumper, *Eur. J. Biochem.* **1996**, *239*, 259–264.

18 N. Kröger, R. Deutzmann, M. Sumper, *Science* **1999**, *286*, 1129–1132.

19 N. Kröger, S. Lorenz, E. Brunner, M. Sumper, *Science* **2002**, *298*, 584–586.

20 N. Poulsen, N. Kröger, *J. Biol. Chem.* **2004**, *279*, 42993–42999.

21 N. Kröger, R. Deutzmann, C. Bergsdorf, M. Sumper, *Proc. Natl. Acad. Sci. USA* **2000**, *97*, 14133–14138.

22 M. Sumper, E. Brunner, G. Lehmann, *FEBS Lett.* **2005**, *579*, 3765–3769.

23 A. Abragam, *Principles of Nuclear Magnetism*. Oxford University Press, **1961**.

24 M. Mehring, *Principles of High-Resolution NMR in Solids*. Springer, Berlin, **1983**.

25 E.R. Andrew, A. Bradbury, R.G. Eades, *Nature* **1958**, *182*, 1659.

26 M.M. Maricq, J.S. Waugh, *J. Chem. Phys.* **1979**, *70*, 3300–3316.

27 A.E. Bennett, C.M. Rienstra, M. Auger, K.V. Lakshmi, R.G. Griffin, *J. Chem. Phys.* **1995**, *103*, 6951–6958.

28 A. Pines, J.S. Waugh, M.G. Gibby, *J. Chem. Phys.* **1972**, *56*, 1776–1777.

29 J. Schaefer, E.O. Stejskal, *J. Am. Chem. Soc.* **1976**, *98*, 1031–1032.

30 G. Metz, X. Wu, S.O. Smith, *J. Magn. Reson. A* **1994**, *110*, 219–227.

31 R. Bertermann, N. Kröger, R. Tacke, *Anal. Bioanal. Chem.* **2003**, *375*, 630–634.

32 K. Lutz, C. Gröger, M. Sumper, E. Brunner, *Phys. Chem. Chem. Phys.* **2005**, *7*, 2812–2815.

33 S.C. Christiansen, N. Hedin, J.D. Epping, M.T. Janicke, Y. del Amo, M. Demarest, M. Brzezinski, B.F. Chmelka, *Solid State Nucl. Magn. Reson.* **2006**, *29*, 170–182.

34 C. Gröger, K. Lutz, M. Sumper, E. Brunner, unpublished results.

35 E. Vinogradov, P.K. Madhu, S. Vega, *Chem. Phys. Lett.* **1999**, *314*, 443–450.

36 E.D. Ingall, P.A. Schroeder, R.A. Berner, *Geochim. Cosmochim. Acta* **1990**, *54*, 2617–2620.

37 R.K. Iler, *The Chemistry of Silica*. Wiley, New York, **1979**.

38 C.C. Perry, T. Keeling-Tucker, *J. Biol. Inorg. Chem.* **2000**, *5*, 537–550.

39 M. Hildebrand, R. Wetherbee, in: W.E.G. Müller (Ed.), *Progress in Molecular and Subcellular Biology*, Vol. 33. Springer, Berlin, Heidelberg, **2003**.

40 T. Mizutani, H. Nagase, N. Fujiwara, H. Ogoshi, *Bull. Chem. Soc. Jpn.* **1998**, *71*, 2017–2022.

41 N. Kröger, R. Deutzmann, M. Sumper, *J. Biol. Chem.* **2001**, *276*, 26066–26070.

42 M. Sumper, S. Lorenz, E. Brunner, *Angew. Chem. Int. Ed.* **2003**, *42*, 5192–5195.

43 S.V. Patwardhan, S.J. Clarson, *Silicon Chem.* **2002**, *1*, 207–214.

44 S.V. Patwardhan, N. Mukherjee, S.J. Clarson, *J. Inorg. Organomet. Polym.* **2001**, *11*, 117–121.

45 S.V. Patwardhan, N. Mukherjee, S.J. Clarson, *Silicon Chem.* **2002**, *1*, 47–55.

46 M. Sumper, *Angew. Chem. Int. Ed.* **2004**, *43*, 2251–2254.

47 J.-J. Yuan, R.-H. Jin, *Adv. Mater.* **2005**, *17*, 885–888.

48 D.J. Kim, K.-B. Lee, Y.S. Chi, W.-J. Kim, H.-J. Paik, I.S. Choi, *Langmuir* **2004**, *20*, 7904–7906.

49 H. Menzel, S. Horstmann, P. Behrens, P. Bärnreuther, I. Krueger, M. Jahns, *Chem. Commun.* **2003**, 2994–2995.

50 M.R. Knecht, D.W. Wright, *Langmuir* **2004**, *20*, 4728–4732.

51 T. Coradin, O. Durupthy, J. Livage, *Langmuir* **2002**, *18*, 2331–2336.

52 S.V. Patwardhan, S.J. Clarson, *J. Inorg. Organomet. Polym.* **2002**, *12*, 109–116.

53 S.V. Patwardhan, S.J. Clarson, *J. Inorg. Organomet. Polym.* **2003**, *13*, 49–53.

54 S.V. Patwardhan, S.J. Clarson, C.C. Perry, *Chem. Commun.* **2005**, 1113–1121.

55 L.L. Brott, R.R. Naik, D.J. Pikas, S.M. Kirkpatrick, D.W. Tomlin, P.W. Whitlock, S.J. Clarson, M.O. Stone, *Nature* **2001**, *413*, 291–293.
56 B.J. McKenna, H. Birkedal, M.H. Bartl, T.J. Deming, G.D. Stucky, *Angew. Chem. Int. Ed.* **2004**, *43*, 5652–5655.
57 F.J. Padden, Jr., H.D. Keith, G. Giannoni, *Biopolymers* **1969**, *7*, 793–804.
58 E. Brunner, K. Lutz, M. Sumper, *Phys. Chem. Chem. Phys.* **2004**, *6*, 854–857.
59 M.R. Knecht, S.L. Sewell, D.W. Wright, *Langmuir* **2005**, *21*, 2058–2061.
60 M. Sumper, *Science* **2002**, *295*, 2430–2433.
61 M. Sumper, E. Brunner, *Adv. Funct. Mater.* **2006**, *16*, 17–26.

3
Mesocrystals: Examples of Non-Classical Crystallization

Helmut Cölfen

Abstract

This chapter highlights mesocrystals as an interesting example of particle-mediated, non-classical crystallization processes. Mesocrystals – the shortened name for mesoscopically structured crystals – are superstructures composed of nanoparticles, being arranged three-dimensionally in crystallographic register. Mesocrystals are often only intermediate structures in a non-classical crystallization pathway leading to a final single crystal by nanoparticle fusion. Therefore, they are difficult to detect. Although mesocrystals were initially described for synthetic systems, recent investigations have revealed an increasing number of biomineral systems which appear to be mesocrystals, but which so far have been considered to be single crystalline, including nacre and sea urchin spines. This chapter briefly defines non-classical crystallization processes, provides some examples of synthetic mesocrystals and mesocrystals in biomineralization, and attempts to provide some insight into their formation mechanisms, despite their being as yet largely unexplored.

Key words: mesocrystal, mesoscale transformation, particle-mediated crystallization, non-classical crystallization, self-organization, polymer, adsorption, colloids, biomineralization.

3.1
Introduction

Biominerals provide us not only with fascinating examples for organic–inorganic hybrid materials with complex morphology, hierarchical structuration and optimized materials properties, but can also teach us important lessons in how effectively to control crystallization events. Based on such inspiration, a rapidly developing research field has evolved, which could be summarized as *bio-inspired* or *biomimetic materials chemistry* [1, 2]. Bio-inspired morphosynthesis, dedicated to

Handbook of Biomineralization. Edited by P. Behrens and E. Bäuerlein

ISBN: 978-3-527-31805-6

the creation of superstructures resembling naturally existing biominerals [3–6] with their unusual shapes and complexity, is an important branch in the broad area of biomimetics [7–11]. It provides an important and environmentally friendly route to generate materials with controlled morphology by using self-assembled organic superstructures, inorganic or organic additives, and/or templates with complex functionalization patterns [12]. During the past decade, the exploration – as well as the application of these bio-inspired synthesis strategies – has resulted in the generation of complex materials with specific size, shape, orientation, composition, and hierarchical organization [13–21].

In addition to the generation of inorganic nanoparticle superstructures by directed self-organization, increasing experimental evidence has been found, based on synthetic systems, that crystallization does not always proceed along the pathway of atom/ion/molecule addition to a critical crystal nucleus, as is discussed as the mechanism of growth for single crystals in classical textbooks of crystallization. Instead, nanoparticle-based crystallization pathways have been revealed, which generally involve mesoscopic transformation, as recently reviewed in Refs. [16, 22, 23]. These particle-mediated crystallization pathways involve the self-organization of metastable or amorphous precursor particles into nanoparticulate superstructures, and are generally considered as "non-classical" crystallization routes. In the case of mesoscale transformation, not only are single crystals with complex morphologies produced, but also superstructures consisting of nanoparticles interspaced by organic additives or, more generally, a second phase such as amorphous matter. In general, these mesocrystals consist of highly oriented building units that are considered to be formed solely through a self-organization approach. Mesocrystal formation has been identified in biominerals such as sea urchin spicules [24] or aragonite tablets in nacre [25, 26], but has also been observed for various other synthetic systems, either as kinetically metastable species or as intermediates in a crystallization reaction, leading to single crystals with typical defects and inclusions [22].

In this chapter, selected examples of mesocrystals in a non-classical crystallization process will be presented, thereby highlighting the importance and the versatility of this non-classical crystallization process not only for biomineralization but also for materials science. Clearly, this represents an intriguing development of new strategies for crystal morphogenesis.

3.2
Classical and Non-Classical Crystallization

Mesocrystal formation, in contrast to atom/ion/molecule-based classical crystallization pathways, is a particle-based, non-classical crystallization mechanism. A schematic representation of classical versus non-classical crystallization is provided in Figure 3.1. The classical crystallization model (Fig. 3.1a) starts from primary building blocks such as atoms, ions or molecules, and forms clusters which may either grow or disintegrate again, depending on the counterplay of the sur-

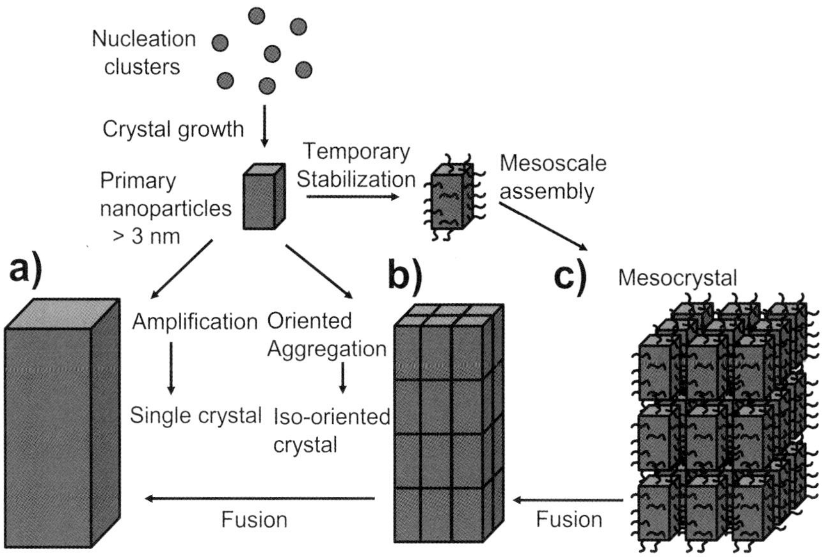

Fig. 3.1 Schematic representation of classical and non-classical crystallization. (a) Classical crystallization pathway. (b) Oriented attachment of primary nanoparticles forming an iso-oriented crystal followed by nanoparticle fusion. (c) Mesocrystal formation via self-assembly of primary nanoparticles covered with organics. For details, see text. (Reprinted with permission of Wiley-VCH from Ref. [87].)

face and crystal lattice energies. Eventually, some clusters reach the size of a so-called *critical crystal nucleus*. At this stage, the free enthalpy of the system becomes negative upon further particle growth, because the gain in lattice energy overcompensates the loss in surface energy. These primary nanoparticles grow further via ion-by-ion attachment and unit cell replication.

In contrast to the atom/ion/molecule-mediated classical crystallization pathway, non-classical crystallization events are always particle-mediated and involve a mesoscopic transformation process [16]. To date, the main pathways of non-classical crystallization known are based on crystalline nanoparticle precursors, as summarized in Figure 3.1, and are referred to as *oriented attachment* and *mesocrystal formation*. In Figure 3.1, pathway (b) involves the arrangement of primary nanoparticles into an iso-oriented crystal via oriented attachment, which can form a single crystal upon face-selective crystallographic fusion of the nanoparticles gaining energy by elimination of two high-energy surfaces by their fusion. If the three-dimensional (3-D) crystallographically aligned nanoparticles are coated by some organic components, they can form a mesocrystal via mesoscale assembly (pathway c), possibly followed by fusion of the already aligned subunits to an iso-oriented crystal, and finally to a single crystal [22]. Thus, mesocrystal formation and oriented attachment are related mechanisms, and initial mesocrystal formation can lead to a single crystal via oriented attachment and fusion of the nanoparticle building units to a single crystal (as shown in Fig. 3.1). Typical of such crystallization pathways are polymer inclusions in a single crystal, which are found in many biominerals that are considered as single crystals, such as nacre or sea urchin spines. Although related to mesocrystal formation, oriented attachment will not be discussed here, and the reader is referred to reviews which at least in part discuss the process [23, 27] as presented by Penn [28] and by Leite et al. [29, 30]. More specialized literature is also available relating to oriented attachment for systems as diverse as TiO_2 [31–36], $BaSO_4/BaCrO_4$ [37–43], FeOOH [44–46], Co_3O_4 [47], CuO [48, 49], ZnO [50], MnO [51], CoOOH [52], SnO_2 [53], $LaMn_2O_{5+\delta}$, CeO_2 [54], Ag [55], ZnS [56, 57], $CaCO_3$ [58, 59], PbSe [60] and Pt [61].

Non-classical crystallization offers certain peculiar advantages with respect to crystal morphogenesis; typical benefits include the near-independence of solubility products and indifference to pH and osmotic pressure, a lack of any need to remove large amounts of water associated with the precipitation of sparingly soluble minerals from slightly supersaturated solutions, and the availability of large amounts of building materials in the form of precursor particles. These features are particularly relevant in biological systems and, accordingly, non-classical crystallization processes are relevant in the situation of biomineralization. These potential advantages for biomineral-forming organisms have long been recognized, and in recent years interest has focused on investigations into the role of amorphous precursor particles [16, 62–68] rather than on investigations into oriented attachment or mesocrystal formation [24–26]. Nevertheless, very recent reports have revealed that these new crystallization routes also play a role in biomineralization processes. This is especially valid for biominerals such as nacre, sea urchin skeletons or marine sponges, which until now have been considered as single crystals, albeit with complex shapes.

3.3 Mesocrystals

Mesocrystals are colloidal crystals composed of individual nanocrystals that are aligned in a common crystallographic register with scattering properties similar to a single crystal [69]. The nanoparticle building units can be stabilized by a polymer layer, as shown in Figure 3.1c, although the phase surrounding the nanocrystals can in fact be any solid phase. Important examples for biomineralization can be crystalline nanoparticles embedded in an amorphous phase, or in a phase consisting of small and unordered crystalline nanoparticles. The existence of an amorphous layer between the aligned nanocrystals of a mesocrystal has not yet been demonstrated experimentally, although the existence of such thin amorphous layers has recently been seen in a high-resolution transmission electron microscopy (HRTEM) study on synthetic aragonite [69] and on nacre of the gastropod *Haliotis laevigata* [70]. In fact, it was proposed that the amorphous layer was formed by an accumulation of impurities, which prevented further crystallization, rather analogous to the zone melting process in metallurgy.

In contrast to common colloidal crystals, which are built from spherical subunits, mesocrystals offer additional degrees of freedom and can display anisotropic properties. The formation of superstructures using nanoparticulate building units with anisotropic shape offers completely new possibilities for crystal morphology control [22]. There are many reasons to believe that mesocrystals are much more common than has been assumed to date, and the number of reported examples is steadily increasing, despite the clear difficulties encountered in their detection. On the one hand, mesocrystals may be misinterpreted as single crystals due to their single crystal scattering properties, whilst on the other hand they are often characterized by a well-facetted appearance. Furthermore, a mesocrystal can easily transform to a single crystal by oriented attachment and fusion (Fig. 3.1).

This chapter cannot provide an exhaustive overview on mesocrystals, especially in view of the rapidly increasing number of examples, and consequently the reader requiring additional information in this respect is referred to relevant reviews [22, 23, 71]. Rather, I will concentrate on examples where the mesocrystal concept appears relevant in biominerals.

It is interesting to note that mesocrystals can be obtained not only for inorganic but also for organic crystals; these have the advantage of being molecular crystals, where both a dipole moment and anisotropic polarizability can be encoded into the molecule. Thus, mesocrystals are especially well suited to demonstrate the mesocrystal concept as they can be obtained without any further additives. An example here is that of DL-alanine mesocrystals. Alanine itself has a molecular dipole moment, whilst in addition the DL-alanine default needle morphology exposes a positively and a negatively charged tip along the needle c-axis [72, 73]. It is possible, therefore, to create DL-alanine nanoparticles with counter-charged (001) faces.

Conditions of high supersaturation favor the nucleation of nanoparticles and, in relation, to particle-mediated reaction pathways, whereas low supersaturation leads to the classical molecule-based growth mechanism. At high supersaturation, amorphous nanoparticles form, crystallize and build up the mesocrystals by nanoparticle alignment, presumably through dipolar forces. However, if the crystallization driving force becomes too high, the nanoparticle units of the mesocrystal fuse together to form a single crystal. In fact, the nanoparticulate units in the mesocrystal can become aligned with such perfection that even a single crystal analysis (which can be used to investigate the crystal in all directions) yields a single crystal diffraction pattern (Fig. 3.2) [74]. An evaluation of these data yields the unit cell constants of DL-alanine, in analogy with the common single crystal grown by classical crystallization. Figure 3.2 illustrates the typical features of a mesocrystal, including (a) the facetted nanoparticle superstructure; (b) the single crystal scattering behavior; and (c) nanoparticle subunits that are either directly visible on the external faces or on a fracture surface, as shown in Figure 3.2. This example shows a mesocrystal with well-facetted external faces.

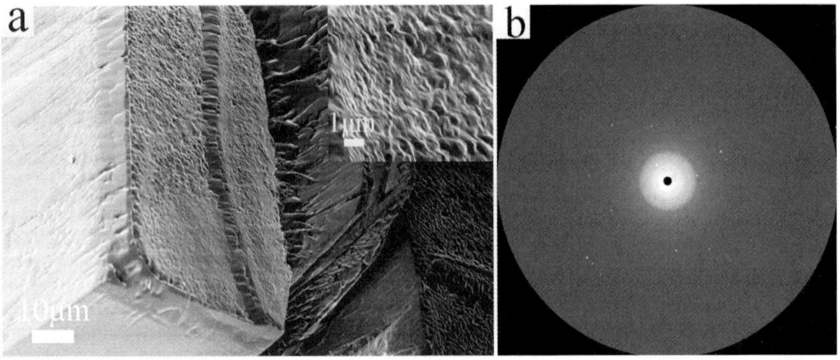

Fig. 3.2 DL-Alanine mesocrystals obtained in supersaturated solution. (a) Scanning electron microscopy image of a cleavage plane showing the inner nanoparticulate mesostructure. (b) Typical image of single crystal X-ray diffraction analysis. (Figure reproduced from Ref. [74] with permission of the American Chemical Society.)

A good example of a mesocrystal with clearly polycrystalline morphology is that of hematite particles; these have a disc-like shape, and are about 1 µm in diameter and 250 nm thick (Fig. 3.3a) [75]. Clearly, each particle consists of many plate-like crystallites, sometimes with interpenetration, though this does not influence the size and shape of the particle superstructures (Fig. 3.3b). This observation was confirmed by TEM measurements on a single particle (Fig. 3.3c). The HRTEM image provides clear evidence of the internal composite nature of the particles (Fig. 3.3d), whereby the small crystallites are perfectly aligned and a hexagonal pattern of dark spots can be seen in the whole region. Selected area electron diffraction studies performed on the particle reveals a spot pattern which is characteristic of the hexagonal lattice of a single crystalline hematite particle (Fig. 3.3e), although these particles are clearly nanoparticle superstructures [75].

In addition to the above-described cases of mesocrystal generation by direct nanoparticle synthesis in solution at high supersaturation, gels – with their limited diffusion of reaction partners for a crystallization event – were also found to serve as a suitable medium to induce mesocrystal formation. Crystal growth in gels takes place under high supersaturation [76], which leads to an increased nucleation of the small clusters which are the building units for the mesocrystals, similar to mesocrystal formation in free solution. Simultaneously, convection or turbulence throughout crystallization can be suppressed, thus allowing the mutual interaction potentials to dominate the mutual alignment of particles. This is why many of the most defined mesocrystals are indeed observed in gels. Here, one conclusion which can be drawn from these results is that a high supersaturation and a high particle nucleation rate preferentially induce mesocrystal formation rather than an ion-mediated classical crystallization process.

Mesocrystals with high definition were reported for $CaCO_3$ made in silica gels [78], and were constructed from a set of cleaved calcite rhombohedra arranged

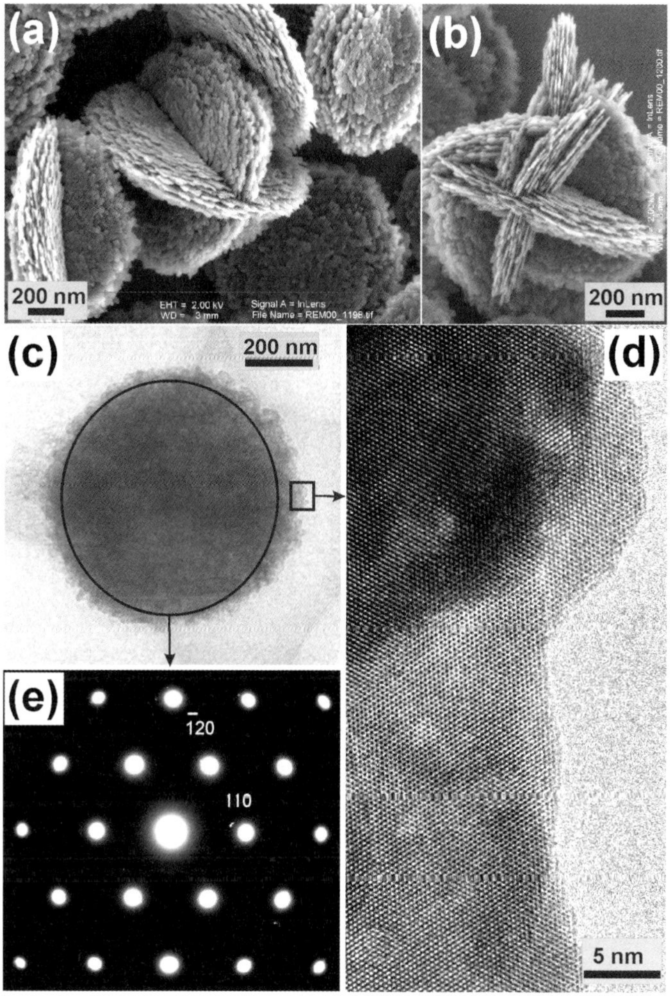

Fig. 3.3 (a) Scanning electron microscopy image of hematite colloids. (b) Side view of interpenetrated particles. (c) Transmission electron microscopy (TEM) top view image of one particle. (d) High-resolution TEM image of the edge of one particle. (e) Selected area electron diffraction of one particle along the c-axis. (Images reproduced from Ref. [75] with permission of the American Chemical Society.)

along their c-axis. Although each fiber was an aggregate of crystalline subunits, the behavior observed using polarization microscopy was that of a single crystal, indicating the high orientational alignment of the subunits in this one-dimensional mesocrystal.

One of the most investigated synthetic mesocrystals to date is the hexagonal prismatic seed crystal of fluoroapatite; this is formed in a gelatin gel and grows

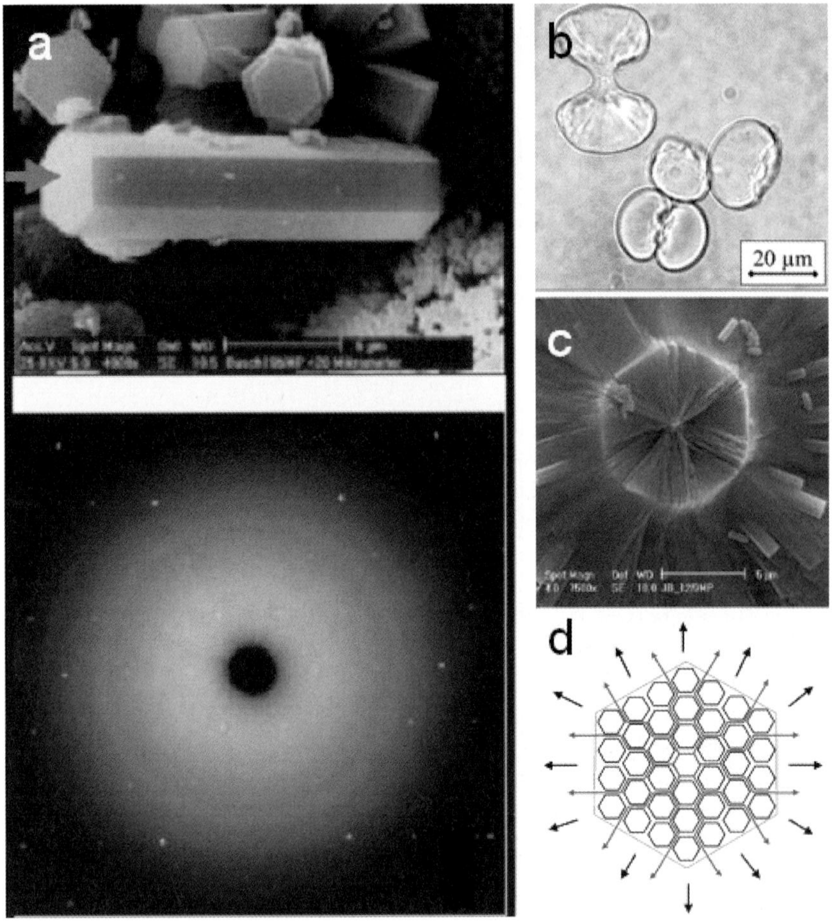

Fig. 3.4 Fluoroapatite grown in a gelatin gel. (a) Scanning electron microscopy (SEM) image of the hexagonal prismatic seed, together with the corresponding diffraction pattern. The arrow indicates the direction of the incident X-ray beam. (b) Gelatinous residue of the composite. (c) SEM image of the fracture area of a central seed. (d) Arrangement of hexagonal nanoparticles forming a superstructure. The lines and arrows indicate the preferred cleaving directions. (Figs. a, b and d reproduced from Ref. [80] with permission of Wiley-VCH; Fig. c reprinted from Ref. [82] with permission of the Royal Society of Chemistry.)

further to spherical particles via dumbbell intermediates (Fig. 3.4a and b) [78, 79]. The hexagonal seed crystal shows all typical features of a mesocrystal, and is therefore a good example for demonstrating the basic properties of a mesocrystal – as well the problems encountered in its identification (Fig. 3.4a). The hexagonal seed crystal was not directly recognizable as a mesocrystal, as it showed a well-facetted, single crystal-like morphology (Fig. 3.4a). Even X-ray diffraction showed features of a fluoroapatite single crystal oriented along the c-axis [80], due to the very high vectorial order of its nanoparticulate building units (Fig. 3.4a). This example illustrates clearly the difficulties encountered in recognizing mesocrystals, and the high probability of erroneously considering such crystals to be single crystals. However, Kniep et al. elucidated the radial inner structure by a hexagonal cross-section perpendicular to the seed axis, thereby disproving the existence of a classical single crystal (Fig. 3.4c) [79]. It was concluded that the hexagonal seed crystal with single crystalline appearance and scattering behavior is a hierarchically ordered inorganic–organic composite superstructure with periodic orientation of hexagonal primary apatite nanocrystals [81] or, in other words, a mesocrystal.

After gelatin crosslinking and apatite removal, polymer distribution inside the crystal could be visualized as replicating the structure of the former dumbbell-shaped hybrid particle (Fig. 3.4b). The internal structure of the hexagonal seed crystals (Fig. 3.4c) revealed a radial pattern and a superstructure periodicity of 10 nm, which was in good agreement with a primary nanoparticle size of about 10 nm [82]. In addition, structural defects attributed to a collagen triple-helix strand, as well as to self-similar nano-subunits nucleated by gelatin, were detected [83]. On basis of this evidence, the growth model for the observed radial outgrowth in the hexagonal seed was developed (Fig. 3.4d). This agreed with the mesocrystal scheme presented in Figure 3.1c, though with hexagonal building units. The stiffness of the gelatin molecules, which was tunable by interaction with Ca^{2+} or $PO_4{}^{3-}$, was found to have a pronounced influence on the scenario of morphogenesis following formation of the hexagonal seed mesocrystal [84].

Although this system has been extensively investigated, the formation process of the mesocrystal has only very recently been partly revealed (see Section 3.4) [85]. Indeed, for most other reported mesocrystal systems the formation mechanism is as yet unknown. Although precursor nanoparticles have been identified experimentally for mesocrystals [86], and dipole fields have been suggested as being responsible for their almost perfect alignment [79, 87], the full growth mechanism involving hierarchical structuration – as well as its underlying mechanism – remain to be revealed.

Whereas gelatin gels can be considered to interact with inorganic crystals at least via their charged groups, polyacrylamide gels are thought to be essentially inert. The growth of $CaCO_3$ in the latter gels led to remarkable pseudo-octahedral calcite mesocrystal morphologies built up from rhombohedral primary nanocrystallites [88, 89]. The external faces of the superstructure could even be indexed, and a growth model based on hierarchical aggregation of rhombohedral subunits was proposed [89].

As identified previously for the fluoroapatite hexagonal seeds, the crystallographic orientation of the subcrystallites is almost perfect, and the organic matrix appears to be intercalated between the individual crystallites [90]. However, whereas the fluoroapatite diffraction pattern was that of a single crystal (see Fig. 3.4b) [80], the calcite mesocrystals indicated a slight orientational distortion of the diffraction spots, corresponding to an average mosaic spread of 3.9 ± 1.1 degrees. This observation nevertheless confirms a high orientational order of the subunits in the mesocrystal [89]. This difference is attributed to the fact that the calcite system contains many vacancies and is potentially twinned (to allow the construction of an octahedron from rhombohedra), whereas the particles reported by Kniep et al. possess a higher symmetry and are clearly rather tightly packed.

Variation of the polyacrylamide hydrogels by co-polymerization with charged acrylamidopropanesulfonate (AMPS) allowed the morphology of the calcite mesocrystals to be tuned from a pseudo-octahedral [89] towards a cubo-octahedral morphology, with an increased amount of the charged AMPS in the copolymer gel [90]. The substructure of the mesocrystals with aligned smaller crystals could be clearly visualized. One important conclusion of this study was that alteration of the functional groups in the hydrogel did not change the mesocrystal formation as such, but rather the morphogenesis process. This shed some light onto the question of why mesocrystals exhibit defined outer faces, and how they could be influenced [90].

Apart from gels, mesocrystals can also be found in a precipitation reaction in solution. In many cases, polymer additives can stabilize the mesocrystals so that they can be recognized. A mesocrystal formation process in solution was reported for copper oxalate [91], in which case the nanoparticles were found to arrange almost perfectly to a mesocrystal, and that this could be influenced morphologically by a hydroxymethylpropylcellulose (HPMC) additive. The polymer was found to influence nucleation, nanocrystal growth and aggregation by its selective interaction with the more hydrophobic lateral (110)/(110) ε-faces of a nanocrystal elongated along [001] as compared to the hydrophilic (001) α-face. Upon aggregation of the nanocrystals, a mesocrystal is formed as intermediate, but is apparently not stable due to the low repulsive electrostatic and steric forces. Depletion flocculation of the weakly adsorbed polymer layers was suggested as a reason for displacement of the polymer from the inner mesocrystal surfaces to the outer mesocrystal faces, resulting in an attraction of the nanoparticles with subsequent nanoparticle fusion towards an iso-oriented crystal. Typical of mesocrystals, electron diffraction indicated a minor (but detectable) orientational disorder, which further supported the nanoparticle self-assembly-based mechanism for mesocrystal formation. Later, the "brick-by-brick" self-assembly mechanism was revealed experimentally in a time-dependent study [92]. Oxalate mesocrystals can also be formed without any additive, as has been found for the related cobalt oxalate dihydrate [93].

In a series of related studies, Oaki and Imai reported on various mesocrystal systems. These authors discussed the existence of mineral bridges between the individual nanoparticle building units [94]; these bridges would serve as a plausible explanation for a perfect crystallographic alignment of the building units, but their existence is extremely difficult to prove experimentally, and therefore it remains unclear as to what determines the high (Fig. 3.4b) or low crystallographic order in a mesocrystal. Face-selective adsorption of poly(acrylic acid) (PAA) onto orthorhombic K_2SO_4 crystals led to the formation of tilted unit crystals. These were assembled in a diffusion-limited process, and resulted in various complex morphologies such as helices or zigzag assembly of twinned crystals [95]. A conclusive explanation of the various possibilities of particle growth in an anisotropic diffusion field was provided in these studies. It is remarkable that the K_2SO_4-PAA system was found to have six hierarchical levels, from the nanometer scale to that of several hundreds of micrometers, which is a typical feature of biominerals and to date has been reported only rarely for a synthetic system. The superstructure design at each level could be controlled by changing the polymer concentration, and the observed hierarchy was attributed to the interaction between crystals and polymers and the diffusion-controlled conditions [96]. A similar hierarchical system was recently identified for potassium hydrogen phthalate and PAA [94].

The same research group reported on the emergence of acute calcite morphologies on a surface consisting of nanobricks, which were suggested to be interconnected by mineral bridges [97]. These mesocrystals also showed the typical brick and mortar motif as displayed in Figure 3.1c. In this case, it was argued that the calcite nanobricks were held together by PAA mortar and small mineral bridges. The instability of a surface in a diffusion field was given as a reason for the generation of acute morphologies by growth in a concentration gradient.

Nanoparticles were also suggested as building units of biominerals in combination with a brick-by-brick formation mechanism [26, 98], and these findings are consistent with the mesocrystal illustrated in Figure 3.1. Although indications for the formation of mesocrystals in biomineralization have long existed, detailed and targeted research on this topic has begun only recently [24, 25], possibly as a result of previously missing corresponding concepts. In this context, the investigation of the sea urchin spicule is a good example (Fig. 3.5a). The spicule consists of calcite, and shows single crystalline behavior in polarization microscopy and scattering experiments. However, a fracture surface of the brittle spicule does not show the typical cleavage plane expected for a single crystal, but rather a chonchoidal fracture surface, which is typical of an amorphous body (Fig. 3.5b). A closer microscopic examination of the fracture surface reveals the clear presence of nanoparticles, which is supportive of a mesocrystal rather than a single crystal (Fig. 3.5c,d).

These apparently contradicting results remained the subject of debate for several decades, as to whether the spine is in fact a single crystal [99, 100], or not [101, 102]. A solution to this problem was recently suggested by Weiner et al., who considered the chonchoidal fracture surface to be the result of occluded proteins in an otherwise single crystalline calcite spicule [103]. Indeed, nanoindentation experiments showed that the calcite became less brittle and fractured when proteins were occluded in the crystals. However, a mesocrystal composed of calcite nanocrystals (Fig. 3.1c) may also explain why the spicule scatters like a single crystal (in mutually and crystallographically aligned calcite nanocrystals) and fractures like an amorphous body. Subsequently, Sethmann et al. found that a sea urchin skeleton is indeed composed of nanocrystals [104]. Further support for a sea urchin spicule mesocrystal was recently provided by Oaki and Imai [24], who produced higher magnification scanning electron microscopy (SEM) images of the skeleton that illustrated the radial alignment of the building blocks and their nanoparticulate morphology (in analogy to Fig. 3.5). TEM images also confirmed the nanoparticulate character of the skeleton, and provided clear evidence for the

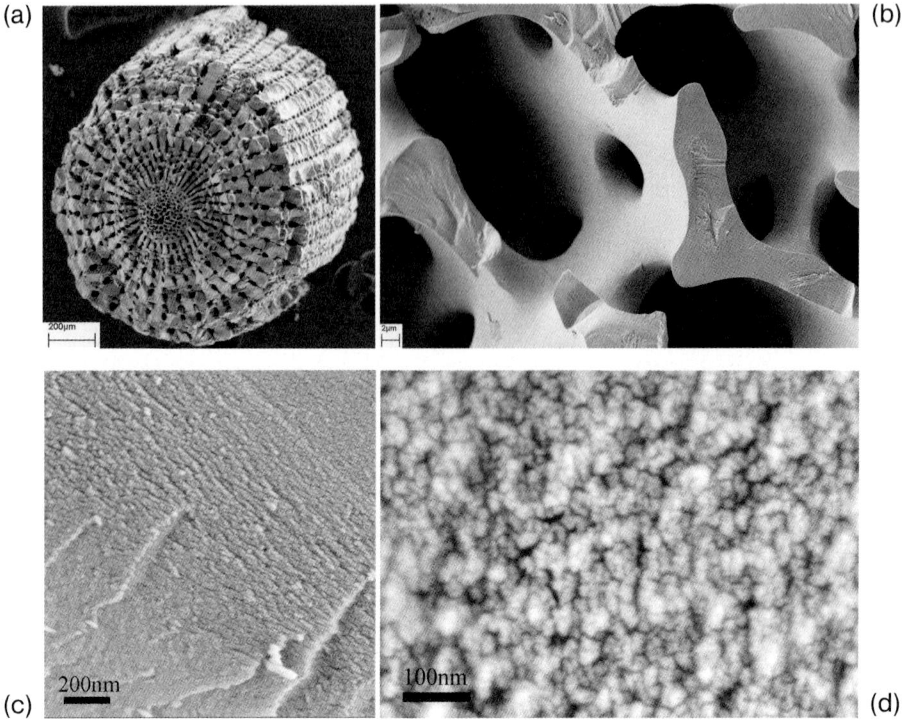

Fig. 3.5 Scanning electron microscopy images of (a) sea urchin spicule from *Anthocidaris crassispina*. (b–d) Conchoidal fracture surface, at various magnifications. (Figs. a and b reproduced from Ref. [27] with permission of the Royal Society of Chemistry.)

mutual alignment of the nanoparticles resulting in a SAED pattern similar to a single crystal. In this case, the authors speculated on mineral bridges between the individual nanoparticle units, which would make alignment of the nanoparticles an oriented attachment or epitaxial growth phenomenon (see Section 3.4).

Aligned nanoparticle building units were also found in sponge spicules consisting of calcite (Fig. 3.6) [105]. Similar to the sea urchin spine, the scattering behavior was that of a single crystal (Fig. 3.6d), although various microscopy studies [atomic force microscopy (AFM), HRTEM; see Fig. 3.6b and c] clearly revealed the nanoparticulate building units of the sponges, which is consistent with the definition of a mesocrystal. Combined high-resolution and energy-filtering TEM studies revealed carbon enrichments located in between the crystal domain boundaries, which pointed strongly to the presence of an intercalated, network-like, proteinaceous organic matrix. The organic species is involved in the nanoclustered calcite precipitation via a transient phase that may enable brick-by-brick formation of composite spicules with elaborate morphologies and single crystalline scattering behavior. On the basis of these results, the authors proposed the argument that nanoclustered crystal growth, induced by organic matrices, is a

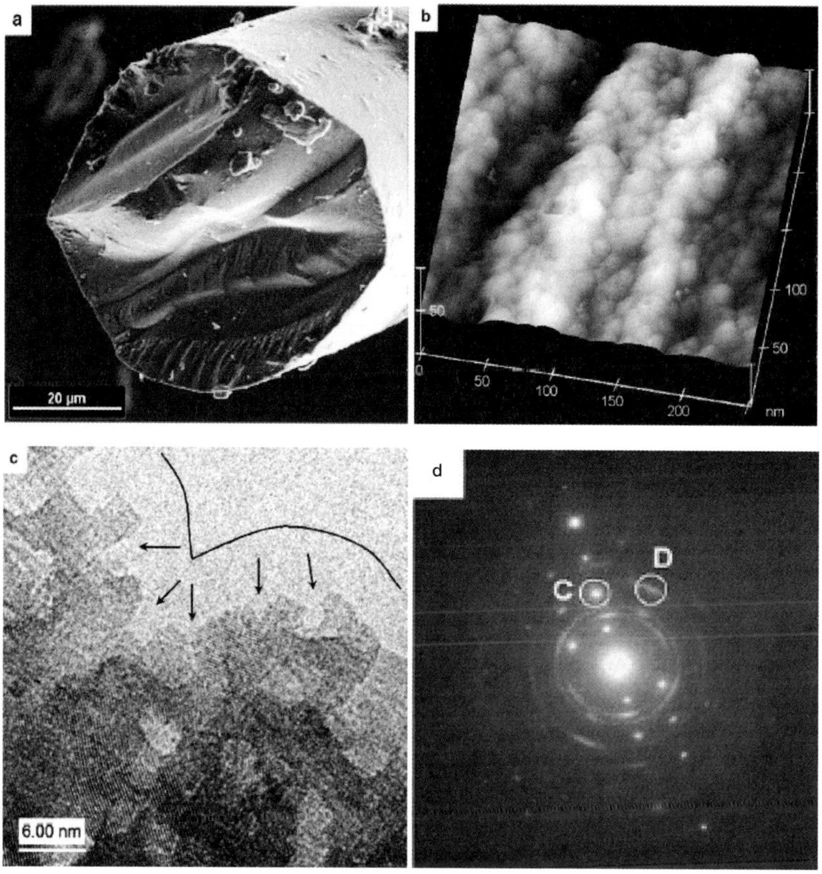

Fig. 3.6 (a) Fractured spicule actine of the calcareous sponge *Pericharax heteroraphis* with conchoidal fracture pattern (FESEM). (b) Atomic force microscopy height image of a conchoidal fracture surface, revealing the nanocluster structure of the spicule material. (c) HRTEM image of the fracture pattern in relation to the crystal texture. The trace of fracture is clearly influenced by the crystal domain boundaries (arrows). The black line represents a rounded fracture trace, which may clarify the structural relation between crystal domains and cluster units as displayed in (b). (d) A selected area electron diffraction pattern of a *Pericharax* spicule (polished, ion-milled, and carbon-coated). (Figure reproduced from Ref. [105] with permission of Elsevier.)

basic characteristic of biomineralization that enables the production of composite materials with elaborate morphologies [105].

Indeed, mutually aligned nanoparticulate building units were also found in the crystalline tablets of nacre. Until now, the aragonite tablets of nacre were believed to be aligned single crystals, with their c-axis orientation perpendicular to the interspacing organic sheets. However, Dauphin [106], Chang et al. [107], Oaki and Imai [26] and Rousseau et al. [25, 108] found that these tablets consisted of

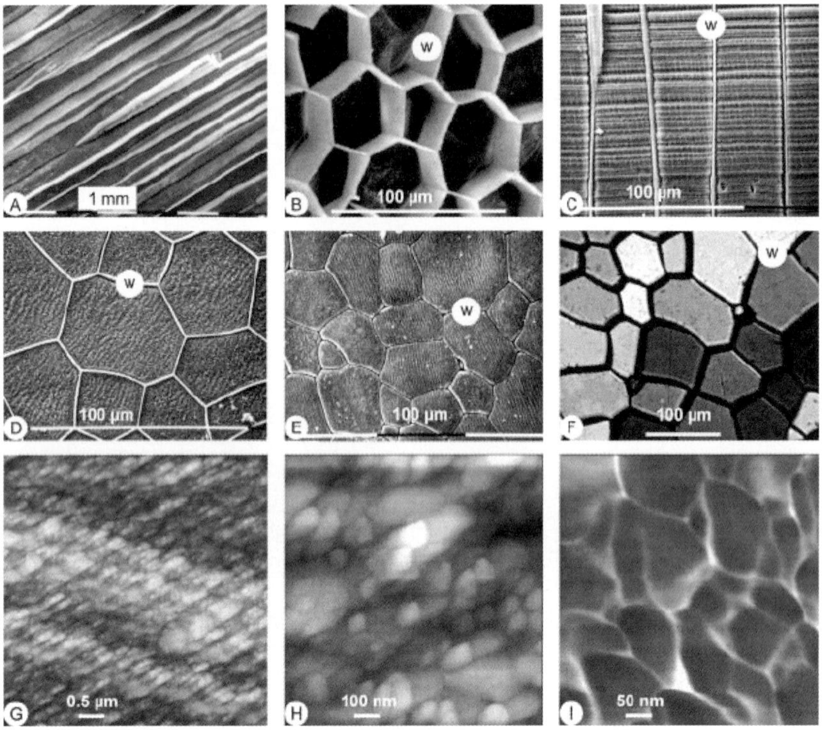

Fig. 3.7 Microstructures and nanostructures of the calcitic prisms of *Pinna nobilis*. (A) Vertical fracture in the outer layer showing the long parallel regular prismatic units. No pattern is visible on the outer surface of the prisms. (B) Transverse polished, fixed, and etched section showing the thick organic walls (w) (fixation conditions: 6% glutaraldehyde, 1 h; rapid decalcifier, 30 min). (C) Vertical polished and etched section showing the thick interprismatic walls (w) and the regular growth lines. (D) Transverse polished, fixed, and etched section showing the interprismatic wall (w) and a pattern of parallel crests (glutaraldehyde + formic acid + Alcian blue solution). (E) Inner surface showing the same pattern of parallel crests and the walls (w) (pronase in Tris buffer, pH 8, at 30 °C for 3 h). (F) Thin section observed in transmitted light (cross-nicols) showing the walls (w) and the monocrystalline extinction of each prismatic unit. (G) Vertical polished and etched section showing the aligned acicular crystallites (AFM tapping mode; 2% glutaraldehyde, 0.1% formic acid, 10 s). (H) Detail of (G). (I) Transverse polished and etched section of the prisms showing the small crystallites surrounded by an organic thin layer (AFM image phase; 2% glutaraldehyde, 0.1% formic acid, 10 s). (Figure reproduced from Ref. [109] with permission of the American Society for Biochemistry and Molecular Biology, Inc.)

aligned nanoparticles. The calcitic prism of *Pinna nobilis* was also found to be a mesocrystal (see Fig. 3.7) [109]. It could be shown that the prisms are not compact structures but rather are composed of oblique and elongated crystallites (Fig. 3.7G–H), the width of which varied from 150 to 180 nm. These crystallites could be subdivided into smaller rounded units with distinct boundaries in AFM height and phase images, which suggested that they were surrounded by organic envelopes (Fig. 3.7I). Nevertheless, the prisms behave like a single crystal under crossed polarizers (Fig. 3.7F), from which evidence it was concluded that the single crystal prism concept was not applicable in this case.

Another example of mesocrystals in biomineralization are the fibrous subunits in corals; these show single crystalline domains in the polarization microscope, but are composed of nanoparticles interspaced by organic matrix, which separates the nanograins [110]. It was suggested that, the crystal lattice of fibers may act as a template during the first step of the self-assembly process, which would result in an orientation of the matrix organic framework in conformity with the crystallographic orientation of each of the underlying fibers. After having been oriented by the underlying mineral surface, the organic framework itself may ensure the crystallographic coherence of the observed crystalline nanograins, an essential requirement for maintaining the overall crystallinity of fibers. From this viewpoint, it is worth noting that close SEM examination of fibers shows that the monocrystalline status of fibers is not as perfect as might be expected if they were to be continuously growing, pure mineral units [110].

These examples imply that mesocrystal formation plays a role in biomineralization. Currently, many more examples of mesocrystals are available from the synthetic world (for a review, see [22], but research investigations into biomineral mesocrystals have only recently started.) Thus, it is likely that many more examples for this intriguing concept for the synthesis of polycrystalline nanoparticle superstructures with single crystalline properties of the assembly will follow in the future.

3.4 Mesocrystal Formation Mechanisms

Although, today, numerous examples of mesocrystals have been observed, their mechanisms of formation still have not been elucidated, other than direct matrix-mediated nanoparticle growth on a fibrous organic matrix, as has been suggested earlier for the growth of corals [110]. To date, three principal possibilities have been suggested which might account for the striking mutual 3-D alignment of nanocrystals in crystallographic register, as illustrated in Figure 3.8.

The first possibility requires the existence of ordering directional physical fields such as electric, magnetic or dipole fields, as well as polarization forces (Fig. 3.8a). In this situation the nanoparticles must be anisotropic with respect to the physical fields, so that an ordering can take place. Such anisotropy could, for example, include the oppositely charged counterfaces of a crystal, a magnetic or dipole moment along one nanocrystal axis, or differences in the polarizability of different faces. Moreover, the anisotropy might already be present in the nanocrystal itself, or be induced by face-selective polymer adsorption.

Nanoparticles with dipole or magnetic moments will create local dipole/magnetic fields, and can attract each other mutually in crystallographic register.

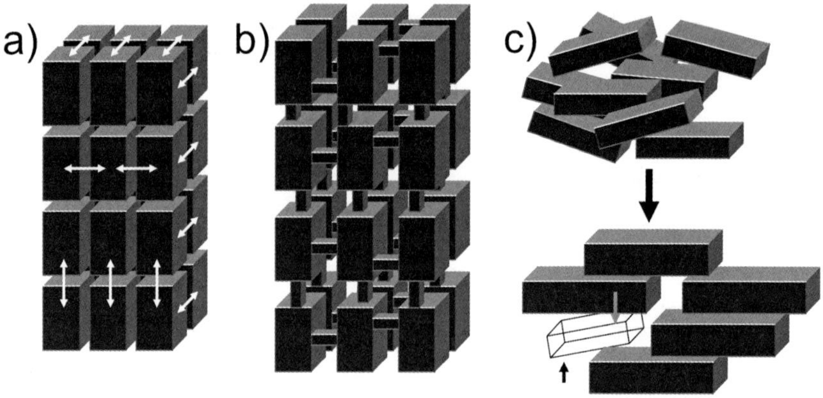

Fig. 3.8 The three principal possibilities for a 3-D mutual alignment of nanoparticles to a mesocrystal. (a) Nanoparticle alignment by physical fields. The arrows indicate the mutual alignment by physical fields. (b) Epitaxial growth of a nanoparticle on a mineral bridge connecting the two nanoparticles. (c) Nanoparticle alignment by spatial constraints. Upon growth in a constrained environment, the particles will align upon growth according to the space restrictions as indicated by the open drawn particle in the arrangement of already grown and aligned particles in the lower image of (c).

The same is true for anisotropic particle polarization, where particle surfaces with equal polarizability attract each other by directed van der Waals forces [111]. This concept requires the nucleation of a large number of nanoparticles of about the same size, with the requirement for an anisotropy along at least one crystallographic axis. This anisotropy can be inherent to the crystal system, as is observed for the case of amino acid crystals [75, 112], or it might be induced by selective polyelectrolyte adsorption.

A second possibility for the 3-D nanoparticle alignment in crystallographic register is the crystallographic connection of nanoparticles by mineral bridges (Fig. 3.8b). This concept was first used to explain the mutual c-axis orientation of aragonite tablets in nacre, and some experimental evidence for the mineral bridges has been provided [113]. An assembly of mineral bridge-connected crystals would in fact be a single crystal, however, so that this concept is the most straightforward attempt to explain the mutual crystallographic orientation of the nanocrystals.

Although the mineral bridge concept can easily explain the crystallographic orientation of the nanocrystals in a mesocrystal, there are two major problems associated with this concept. The first problem is that these mineral bridges have not yet been detected in a mesocrystal, and must be assumed to be only a few nanometers in diameter. The second problem is the reason why a mineral bridge should grow anisotropically on an existing nanoparticle, rather than an isotropic growth of the entire face. It is also difficult to explain why the second nanoparticle should grow in the perpendicular plane to the direction of the mineral bridge, which can act as a nucleation center. In addition, the size of the nanoparticles should be difficult to control in this situation. Therefore, it appears more plausible that mesocrystals are formed by a nucleation burst of a large number of similarly sized nanoparticles, which then self-assemble by physical or geometric forces, although these mesocrystal formation processes have not yet been identified.

A third possibility for mesocrystal formation would be a simple geometric argument, as outlined in Figure 3.8c. When crystalline nanoparticles grow – most likely via the solid-state transition of an amorphous precursor phase – and these nanoparticles are not spherical in shape, then simple geometric arguments can explain their alignment in a constrained reaction environment upon particle growth, as shown in Figure 3.8c for the open drawn particle. Initially, the particles would crystallize into a relatively unordered and loose structure of crystalline nanoparticles, but upon further growth in a confined reaction space they would have to align mutually (as shown for the open drawn particle in Fig. 3.8c). This argument for mesocrystal formation requires only the presence of a constrained reaction environment. This does not necessarily require a pre-formed crystallization environment, but mutual alignment of the first crystallizing particles will lead to confined spaces for the alignment of further growing nanoparticles. This means that an assembly of anisotropic nanoparticles – presumably in the form of rods or plates – may be sufficient for a mutual alignment in the crystallographic register, as the exposed faces would naturally be attracted more by van der Waals forces than the minor faces, as with the tips of rod-like particles. Moreover, this mechanism would be universal to all crystalline systems, as long as the primary nanoparticle building units are anisotropic.

Here, some experimental evidence is provided for the above-mentioned nanoparticle orientation to a mesocrystal by dipole fields, supporting the mechanism illustrated in Figure 3.8a. However, some of these results can also be explained by the purely geometric ordering suggested in Figure 3.8c, which simply requires nanoparticle building units with anisotropic shape.

One example of the ordering of nanoparticles by dipole fields is the study on $CaCO_3$ mesocrystals, which are formed by selective adsorption of poly(styrene sulfonate) (PSS) onto the highly polar calcite (001) nanoparticle faces (Fig. 3.9) [86, 114]. The important difference to previously reported results is that, in the initial particle formation stages, the PSS adsorption only occurred on one (001) face and not the counterface (00-1). The reason for this observation is shown in Figure 3.9.

In the initial step, Ca^{2+} forms a complex with PSS through electrostatic attraction. When CO_3^{2-} is added, the CO_3^{2-} groups compete with PSS for the Ca^{2+}, and $CaCO_3$ is nucleated in the aggregates as amorphous precursor particles. When it crystallizes, the polyanion is attracted by the charged (001) face and blocks this face from further growth, leaving the overall charge of this face positive. If the nanocrystal is considered as a dielectric medium, adsorption of PSS onto one (001) face results in repulsion of cations from the opposite (00-1) face, so that this face remains negative and a charge separation of the opposite crystal faces is generated. The important point here is that the non-screened coulomb

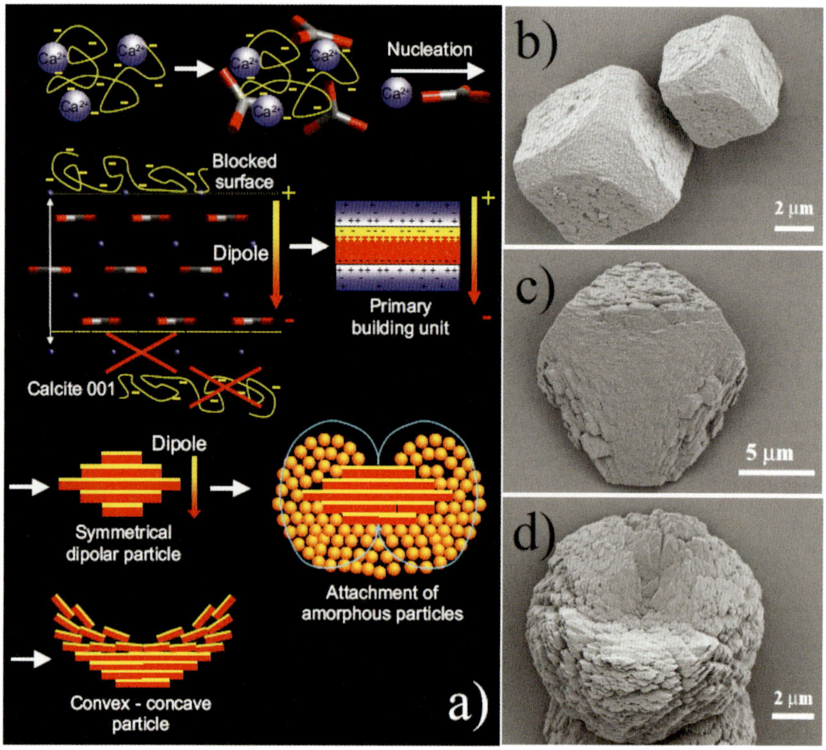

Fig. 3.9 (a) Schematic representation of the formation of $CaCO_3$ mesocrystals. Mechanism of final morphology change in calcite crystals due to selective adsorption of poly(styrene sulfonate) (PSS) to one (001) face and the resulting generation of an inner dipole moment within the crystal in the c-direction. The primary crystals are asymmetric as they bind the polymer only onto one side. These primary blocks assemble to flat, pseudo-symmetric mesocrystal structures. By exceeding a certain size, not only primary platelets but also amorphous intermediates are attracted. By recrystallization of those species, bent crystalline structures without translational order can develop. For further explanation, see text. (b–d) Typical SEM images of calcite mesocrystals obtained on a glass slide by the gas diffusion reaction after 1 day in 1 mL of solution with different concentrations of Ca^{2+} and PSS: (b) $[Ca^{2+}] = 1.25$ mmol L^{-1}, [PSS] $= 0.5$ g L^{-1}; (c) $[Ca^{2+}] = 2.5$ mmol L^{-1}, [PSS] $= 1.0$ g L^{-1}; (d) $[Ca^{2+}] = 5$ mmol L^{-1}, [PSS] $= 0.5$ g L^{-1}. (Figure reproduced from Ref. [114] with permission of Wiley-VCH.)

law can act throughout the crystal, whereas the fields in solution are screened by ions. The primary platelet (Fig. 3.9, red) carries the polyelectrolyte on one side (Fig. 3.9, yellow), but is negatively charged on the counterface. Such a structure is colloidally stable with respect to direct charge interactions, and carries the typical electrostatic double layer (Fig. 3.9, blue). In this way, a dipole field is built up throughout the crystal. This dipole field for the ca. 40 nm-thick platelets cannot be compensated by the ca. 2 nm-thick electrical double layer, so that individual nanoplates attract each other and begin to stack. This process can lead to a controlled arrangement of the nanoparticle subunits to mesocrystals, and the dipole moment is further increased as it is proportional to the distance of charge separation. Above a critical size, due to the linear dependence of dipole moment on thickness, a new effect comes into play. The inner dipole moment is becoming so large that not only the dipolar nanoplatelets (as for the first set of morphologies) but also polymer-stabilized amorphous intermediates are attracted. As these small particles have a homogeneous surface charge of one type, the potential of attracting amorphous intermediates defines the moment, when the mesostructure develops curvature and dissymmetry: the one, similarly charged pole in the middle of one mesocrystal (001) face repels all intermediates from further crystal growth (and a hole is formed), whereas the other oppositely charged pole attracts the particles. As there is a continuous change from attractive to repulsive interaction from pole to pole, the continuous variation of aggregation probability leads to the finally observed, nicely curved concave/convex particle morphology (Fig. 3.9) [114]. The opposite charge of the upper and lower crystal surfaces could be demonstrated by staining experiments with charged dyes for the convex–concave-shaped particles [115]. These mesocrystals are highly porous, but show their common mutual high orientation by an almost perfect tensorial birefringence under crossed polarizers, as calcite is highly birefringent except when observed from the [001] direction.

The effects of the dipole fields on the final mesocrystal morphology can be systematically tuned by the $CaCO_3$/PSS ratio (Fig. 3.9b–d) [114]. Higher Ca^{2+}/PSS ratios lead to increasing exposition of the highly polar (001) face, finally resulting in a multi-curved convex–concave structure with broken symmetry along the (001) direction at high Ca^{2+} and PSS concentrations (Fig. 3.9). This observation manifests the importance of dipole fields as ordering forces for mesocrystal formation [86, 114].

A striking direct experimental proof for the importance of dipole fields on the formation of fluoroapatite mesocrystals in a gelatin gel (see Fig. 3.10 and corresponding text) was recently reported by Kniep et al. [85]. By using electron holography, it was possible to detect the phase of the image in vacuum, which contains information about the electrical potential distribution and thus, the electric field around the hexagonal seed particle (Fig. 3.10a,b). The results could be compared with computer calculations for a dipole consisting of aligned gelatin triple helical dipolar nano units with given length and aligned along the c-axis of the seed (Fig. 3.10c). Although the absolute agreement between measurement and calculation was not good due to various uncertainties in the calculation (pH-dependent gelatin charge, length and spacing of the helices, etc.), the qualitative agreement between prediction and experiment is good (Fig. 3.10).

The model calculations show that the tendency for branching at the ends of the elongated seed is caused by an energetically favored orientation of triple helices, significantly of the hexagonal c-axis direction. Thus, the role of the intrinsic dipole fields in structuring inorganic crystals was revealed experimentally for the first time, even if the situation in a gel may have differed from the investigated vacuum environment [85].

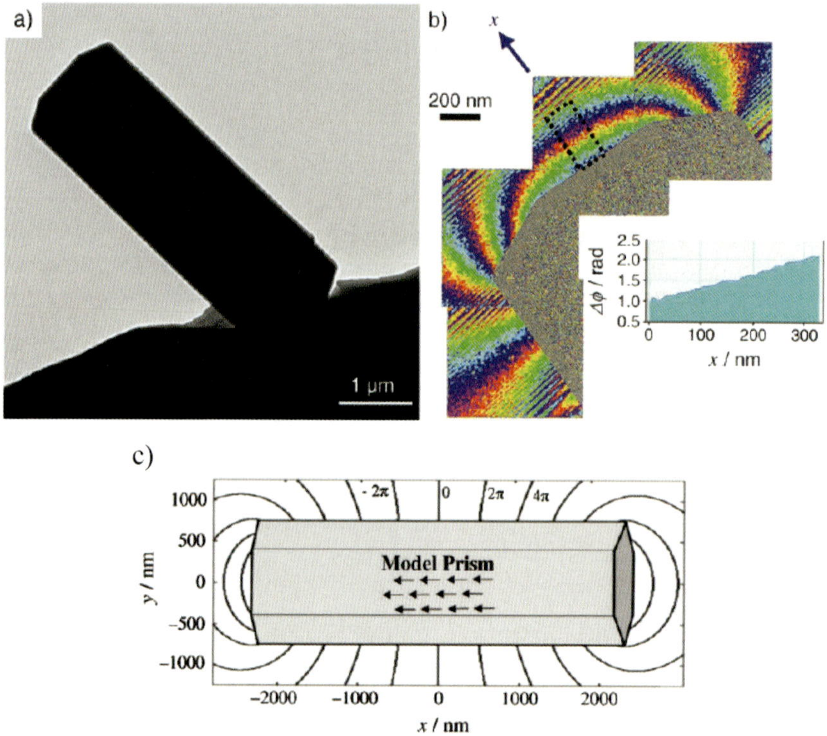

Fig. 3.10 Hexagonal-prismatic fluoroapatite–gelatin nanocomposite seed. (a) Conventional TEM micrograph. (b) Retrieved phase image of an electron hologram (amplified eight-fold, composed of four single images), showing the electric potential distribution around a seed. The color code denotes a phase shift of 2π from green to green. Fresnel fringes of the interferograms appear as striation patterns at the corners of the phase images. The observed projected potential corresponds to a mesoscopic dipole. Inset: The phase profile reveals a phase increase of about 1 rad per 300 nm. (c) Simulation of the phase image around a nanocomposite seed based on an ideal arrangement of nanodipoles. The model is constructed from triple helices in parallel orientation along [001] within the seed. Triple helices represent dipoles and are depicted as arrows. For further details, see text. The contour plot of the phase shift shows a good qualitative agreement with the electron holographic experimental data in (b). (Reprinted from Ref. [85] with permission of Wiley-VCH.)

A model example of a mesocrystal formed by anisotropic van der Waals attraction was reported by Taden et al. for dyes with an anisotropic polarizability [111]. Amorphous particle precursors were prepared in a size-controlled fashion by cooling the liquid nanodroplets of a miniemulsion, whereupon spontaneous rearrangement of many nanodroplets towards well-defined linear mesocrystal aggregates was observed. These rod-like particles could be ripened to larger 3-D mesocrystals. Almost perfect molecular orientation was retained in the nanoparticle aggregate, which was obvious when turning the ripened mesocrystals under crossed polarizers.

The fact that the dipole moment of such dyes is oriented along the molecular axis and coupled to the maximum color allows differentiation to be made between polarizability and dipole moment as the driving force of organization. As the dye molecules are clearly perfectly aligned perpendicular to the main axis of the mesocrystal, these experiments revealed tensorial differences in the polarizability of various nanocrystal faces as the driving force for mesocrystal formation with the observed high orientational order. Whether or not the nanocrystalline building units in the mesocrystals fused together by oriented attachment to form a single crystal was not revealed, however.

To date, the reported examples of possible mechanisms of mesocrystal formation support the alignment of nanoparticles to a mesocrystal by mutual physical fields. Nevertheless, alignment by pure geometric constraints (see Fig. 3.8c) might also account for the observed alignment of some mesocrystals.

3.5 Conclusions

In this chapter, some examples have been provided of non-classical crystallization mechanisms, with focus placed on the currently existing examples of biominerals. The universal applicability of the classical crystallization mechanism was first questioned a few years ago [115], partly as a result of the careful observation of biominerals. Hence, the development of alternative crystallization mechanisms is an important and challenging task. To date, although oriented attachment and mesocrystal formation have been identified as active non-classical crystallization mechanisms, this list seems incomplete and, indeed, may be far from complete. For example, the liquid $CaCO_3$ and other mineral precursors reported by Gower et al. show an additional pathway, of how complex mineral morphologies including curved surfaces can be obtained from a liquid precursor [116–118]. Overall, much evidence exists for a number of alternative crystallization pathways to the classical ion/atom/molecule-mediated crystallization mechanism, which is considered by current textbooks of crystallization to be the favored pathway.

The relevant parameters for classical crystallization are molecular solubility (defining the speed of crystallization) and crystal face-specific surface energies defining growth rates, relative exposure and, accordingly, the morphology of the crystal. All of these characteristics are certainly not equally valid for oriented attachment and mesocrystal formation, which are particle-mediated crystallization pathways involving mesoscopic transformation [16].

The use of nanoparticles as building blocks for the bottom-up fabrication of superstructures represents a powerful means of obtaining novel materials with collective properties. On the one hand, nanoparticles covering a broad range of chemical and physical properties are readily available, whilst on the other hand, combining these nanoparticles (and mixtures thereof) provides access to an almost indefinite number of possibilities to tailor the properties of a material [119]. Mesocrystals, as one of the most interesting examples of non-classical crystallization, are usually intermediates in the formation process of a single crystal, and are observed especially in cases of very low molecular solubilities or high supersaturations. The nanoparticle building units can be molded into any shape, and the mutual crystallographic alignment of the nanoparticle building units leads to the observed single crystalline scattering behavior. Mesocrystals are currently more important from a fundamental scientific viewpoint, but potentially exciting applications – perhaps in construction materials (as exerted by Nature with biominerals) or as functional ceramics (e.g., color or magnetism) – can be imagined.

Today, the numbers of mesocrystals being reported are rapidly increasing, indicating that they have always been an integral part of crystalline systems. The morphology of a mesocrystal superstructure is encoded in the shape of the nanoparticles, colloidal stabilization and vectorial long-range interaction potentials, as well as the corresponding constraints upon nanoparticle growth. Nanoparticle surface interactions appear to be important for the mesocrystal formation, and may also be responsible for the formation of external faces, which can be indexed like those of a single crystal. The reason for the remarkable, almost perfect order of the nanocrystalline subunits resulting in diffraction patterns similar to single crystals is still unknown to a large extent, and this is similarly true for the exact ordering principles, although in this respect tensorial polarization forces and dipole fields have been discussed [79, 85, 86]. Indeed, theoretical studies consider that a non-spherical charged object in an electrolyte creates a screened electrostatic potential that is anisotropic at any distance [120], so that mutual ordering of the nanoparticle building units can be induced. In order to create such anisotropic electrostatic potentials, the adsorption of a polyelectrolyte to a single crystal face can be applied. This process leads to opposite charges of the counter face and polyelectrolyte adsorption face, and results in charge separation throughout the crystal and the emergence of a dipole field. The detection of such ordering fields and their effects is difficult, but it can be achieved in some cases [85].

To summarize, mesocrystal formation is an especially promising example of the controlled self-organization of nanoparticles to 3-D superstructures which, potentially, can demonstrate new physical properties. Mesocrystal formation seems to be an active crystallization pathway in several biomineral systems, and is indeed a potential "treasure trove" in appreciating the formation of biominerals such as sea urchin spicules. In addition, an understanding of the alignment forces for mesocrystals represents a key step towards our comprehension of non-classical crystallization.

Acknowledgments

The author thanks the Max Planck Society and the DFG (SPP 1117 *Principles of Biomineralization*) for financial support of his studies related to non-classical crystallization. Thanks are also extended to Dr. Yannicke Dauphin, Université Paris, for providing the illustrations of *Pinna nobilis* mesocrystals, and for useful discussions. The sea urchin spicule images in Figure 3.5 were kindly provided by Dr. Yurong Ma.

References

1 S. Mann, *Biomimetic Materials Chemistry*. VCH Publishers, New York, **1996**.
2 H. Cölfen, S.H. Yu, *MRS Bull.* **2005**, *30*, 727.
3 H.A. Lowenstam, S. Weiner, *On Biomineralization*. Oxford University Press, New York, **1989**.
4 S. Mann, J. Webb, R.J.P. Williams, *Biomineralization: Chemical and Biochemical Perspectives*. VCH, Weinheim, **1989**.
5 S. Mann, *Biomineralization. Principles and Concepts in Bioinorganic Materials Chemistry*. Oxford University Press, **2002**.
6 E. Bäuerlein, *Biomineralization*. Wiley-VCH, Weinheim, **2000**.
7 S.H. Yu, H. Cölfen, *J. Mater. Chem.* **2004**, *14*, 2124.
8 S. Mann, *Nature* **1993**, *365*, 499.
9 S.A. Davis, M. Breulmann, K.H. Rhodes, B. Zhang, S. Mann, *Chem. Mater.* **2001**, *13*, 3218.
10 H. Cölfen, *Curr. Opin. Colloid Interface Sci.* **2003**, *8*, 23.
11 F.C. Meldrum, *Int. Mater. Rev.* **2003**, *48*, 187.
12 S. Mann, G.A. Ozin, *Nature* **1996**, *382*, 313.
13 W. Jones, C.N.R. Rao, *Supramolecular Organization and Materials Design*. Cambridge University Press, Cambridge, **2002**.
14 M. Antonietti, C. Göltner, *Angew. Chem. Int. Ed.* **1997**, *36*, 910.
15 D.M. Dabbs, I.A. Aksay, *Annu. Rev. Phys. Chem.* **2000**, *51*, 601.
16 H. Cölfen, S. Mann, *Angew. Chem. Int. Ed.* **2003**, *42*, 2350.
17 J. Aizenberg, *Adv. Mater.* **2004**, *15*, 1295.
18 D.D. Archibald, S. Mann, *Nature* **1993**, *364*, 430.
19 H. Yang, N. Coombs, G.A. Ozin, *Nature* **1997**, *386*, 692.
20 M. Li, H. Schnablegger, S. Mann, *Nature* **1999**, *402*, 393.
21 L.A. Estroff, A.D. Hamilton, *Chem. Mater.* **2001**, *13*, 3227.
22 H. Cölfen, M. Antonietti, *Angew. Chem. Int. Ed.* **2005**, *44*, 5576.
23 H. Cölfen, in: K. Naka (Ed.), *Topics in Current Chemistry*. Springer, Berlin, Heidelberg, **2007**, Volume 271, pp. 1–77.
24 Y. Oaki, H. Imai, *Small* **2006**, *2*, 66.
25 M. Rousseau, E. Lopez, P. Stempfle, M. Brendle, L. Franke, A. Guette, R. Naslain, X. Bourrat, *Biomaterials* **2005**, *26*, 6254.
26 Y. Oaki, H. Imai, *Angew. Chem. Int. Ed.* **2005**, *44*, 6571.
27 M. Niederberger, H. Cölfen, *Phys. Chem. Chem. Phys.* **2006**, *8*, 3271.
28 R.L. Penn, *J. Phys. Chem. B* **2004**, *108*, 12707.
29 C. Ribeiro, E.J.H. Lee, E. Longo, E.R. Leite, *Chem. Phys. Chem.* **2006**, *328*, 229–235.
30 Z. Tang, N.A. Kotov, *Adv. Mater.* **2005**, *17*, 951.
31 R.L. Penn, J.F. Banfield, *Am. Miner.* **1998**, *83*, 1077.
32 R.L. Penn, J.F. Banfield, *Geochim. Cosmochim. Acta* **1999**, *63*, 1549.
33 R.L. Penn, J.F. Banfield, *Science* **1998**, *281*, 969.
34 J.F. Banfield, S.A. Welch, H.Z. Zhang, T.T. Ebert, R.L. Penn, *Science* **2000**, *289*, 751.
35 J. Polleux, N. Pinna, M. Antonietti, M. Niederberger, *Adv. Mater.* **2004**, *16*, 436.
36 J. Polleux, N. Pinna, M. Antonietti, C. Hess, U. Wild, R. Schlögl, M. Niederberger, *Chem. Eur. J.* **2005**, *11*, 3541.
37 L.M. Qi, H. Cölfen, M. Antonietti, *Angew. Chem. Int. Ed.* **2000**, *39*, 604.
38 L.M. Qi, H. Cölfen, M. Antonietti, *Chem. Mater.* **2000**, *12*, 2392.

39 L.M. Qi, H. Cölfen, M. Antonietti, M. Li, J.D. Hopwood, A.J. Ashley, S. Mann, *Chem. Eur. J.* **2001**, *7*, 3526–3532.

40 S.H. Yu, H. Cölfen, M. Antonietti, *Chem. Eur. J.* **2002**, *8*, 2937.

41 S.H. Yu, M. Antonietti, H. Cölfen, J. Hartmann, *Nano Lett.* **2003**, *3*, 379.

42 M. Li, S. Mann, H. Cölfen, *J. Mater. Chem.* **2004**, *14*, 2269.

43 S.H. Yu, H. Cölfen, M. Antonietti, *Adv. Mater.* **2003**, *15*, 133.

44 R.L. Penn, G. Oskam, T.J. Strathmann, P.C. Searson, A.T. Stone, D.R. Veblen, *J. Phys. Chem. B* **2001**, *105*, 2177.

45 Y. Guyodo, A. Mostrom, R.L. Penn, S.K. Banderjee, *Geophys. Res. Lett.* **2003**, *30*, 19.

46 D.J. Burleson, R.L. Penn, *Langmuir* **2006**, *22*, 402.

47 T. He, D. Chen, X. Jiao, *Chem. Mater.* **2004**, *16*, 737.

48 B. Liu, H.C. Zeng, *J. Am. Chem. Soc.* **2004**, *126*, 8124.

49 W.-T. Yao, S.-H. Yu, Y. Zhou, J. Jiang, Q.-S. Wu, L. Zhang, J. Jiang, *J. Phys. Chem. B* **2005**, *109*, 14011.

50 C. Pacholski, A. Kornowski, H. Weller, *Angew. Chem. Int. Ed.* **2002**, *41*, 1188.

51 D. Zitoun, N. Pinna, N. Frolet, C. Belin, *J. Am. Chem. Soc.* **2005**, *127*, 15034.

52 R.L. Penn, A.T. Stone, D.R. Veblen, *J. Phys. Chem. B* **2001**, *105*, 4690.

53 E.J.H. Lee, C. Ribeiro, E. Longo, E.R. Leite, *J. Phys. Chem. B* **2005**, *109*, 20842.

54 R. Si, Y.-W. Zhang, L.-P. You, C.-H. Yan, *J. Phys. Chem. B* **2006**, 110, 5994–6000.

55 M. Giersig, I. Pastoriza-Santos, L.M. Liz-Marzan, *J. Mater. Chem.* **2004**, *14*, 607.

56 F. Huang, H. Zhang, J.F. Banfield, *Nano Lett.* **2003**, *3*, 373.

57 J.H. Yu, J. Joo, H.M. Park, S.-I. Baik, Y.W. Kim, S.C. Kim, T. Hyeon, *J. Am. Chem. Soc.* **2005**, *127*, 5662.

58 N. Gehrke, H. Cölfen, N. Pinna, M. Antonietti, N. Nassif, *Cryst. Growth Des.* **2005**, *5*, 1317.

59 A.W. Xu, M. Antonietti, H. Cölfen, Y.P. Fang, *Adv. Funct. Mater.* **2006**, 16, 903–908.

60 K.-S. Cho, D.V. Talapin, W. Gaschler, C.B. Murray, *J. Am. Chem. Soc.* **2005**, *127*, 7140.

61 Y. Yamauchi, T. Momma, M. Fuziwara, S.S. Nair, T. Ohsuna, O. Terasaki, T. Osaka, K. Kuroda, *Chem. Mater.* **2005**, *17*, 6342.

62 E. Beniash, J. Aizenberg, L. Addadi, S. Weiner, *J. Roy. Soc. Lond. B* **1997**, *264*, 461.

63 E. Beniash, L. Addadi, S. Weiner, *J. Struct. Biol.* **1999**, *125*, 50.

64 L. Addadi, S. Raz, S. Weiner, *Adv. Mater.* **2003**, *15*, 959.

65 S. Raz, P.C. Hamilton, F.H. Wilt, S. Weiner, L. Addadi, *Adv. Funct. Mater.* **2003**, *13*, 480.

66 J. Aizenberg, G. Lambert, S. Weiner, L. Addadi, *J. Am. Chem. Soc.* **2002**, *124*, 32.

67 Y. Politi, T. Arad, E. Klein, S. Weiner, L. Addadi, *Science* **2004**, *306*, 1161.

68 S. Weiner, I. Sagi, L. Addadi, *Science* **2005**, *309*, 1027.

69 N. Nassif, N. Gehrke, N. Pinna, N. Shirshova, K. Tauer, M. Antonietti, H. Cölfen, *Angew. Chem. Int. Ed.* **2005**, *44*, 6004.

70 N. Nassif, N. Pinna, N. Gehrke, M. Antonietti, C. Jäger, H. Cölfen, *Proc. Natl. Acad. Sci. USA* **2005**, *102*, 12653.

71 M. Niederberger, H. Cölfen, *Phys. Chem. Chemical Phys.* **2006**, *8*, 3271.

72 L.J.W. Shimon, F.C. Wireko, J. Wolf, I. Weissbuch, L. Addadi, Z. Berkovitchyellin, M. Lahav, L. Leiserowitz, *Mol. Cryst. Liq. Cryst.* **1986**, *137*, 67.

73 L.J.W. Shimon, M. Vaida, L. Addadi, M. Lahav, L. Leiserowitz, *J. Am. Chem. Soc.* **1990**, *112*, 6215.

74 Y.R. Ma, H. Cölfen, M. Antonietti, *J. Phys. Chem. B* **2006**, *110*, 10822.

75 M. Niederberger, F. Krumeich, K. Hegetschweiler, R. Nesper, *Chem. Mater.* **2002**, *14*, 78.

76 A. Putnis, M. Prieto, L. Fernandez-Diaz, *Geol. Mag.* **1995**, *132*, 1.

77 S.D. Bella, J.M. Garcia-Ruiz, *J. Crystal Growth* **1986**, *79*, 236.
78 R. Kniep, S. Busch, *Angew. Chem. Int. Ed.* **1996**, *35*, 2624.
79 S. Busch, H. Dolhaine, A. DuChesne, S. Heinz, O. Hochrein, F. Laeri, O. Podebrad, U. Vietze, T. Weiland, R. Kniep, *Eur. J. Inorg. Chem.* **1999**, *10*, 1643.
80 S. Busch, U. Schwarz, R. Kniep, *Adv. Funct. Mater.* **2003**, *13*, 189.
81 S. Busch, U. Schwarz, R. Kniep, *Chem. Mater.* **2001**, *13*, 3260.
82 P. Simon, W. Carillo-Cabrera, P. Formanek, C. Göbel, D. Geiger, R. Ramlau, H. Tlatlik, J. Buder, R. Kniep, *J. Mater. Chem.* **2004**, *14*, 2218.
83 P. Simon, U. Schwarz, R. Kniep, *J. Mater. Chem.* **2005**, *15*, 4992.
84 H. Tlatlik, P. Simon, A. Kawska, D. Zahn, R. Kniep, *Angew. Chem. Int. Ed.* **2006**, *45*, 1905.
85 P. Simon, D. Zahn, H. Lichte, R. Kniep, *Angew. Chem. Int. Ed.* **2006**, *45*, 1911.
86 T.X. Wang, H. Cölfen, M. Antonietti, *J. Am. Chem. Soc.* **2005**, *127*, 3246.
87 S. Wohlrab, N. Pinna, M. Antonietti, H. Cölfen, *Chem. Eur. J.* **2005**, *11*, 2903.
88 O. Grassmann, G. Müller, P. Löbmann, *Chem. Mater.* **2002**, *14*, 4530.
89 O. Grassmann, R.B. Neder, A. Putnis, P. Löbmann, *Am. Miner.* **2003**, *88*, 647.
90 O. Grassmann, P. Löbmann, *Chem. Eur. J.* **2003**, *9*, 1310.
91 N. Jongen, P. Bowen, J. Lemaitre, J.C. Valmalette, H. Hofmann, *J. Colloid Interface Sci.* **2000**, *226*, 189.
92 L.C. Soare, PhD thesis. Ecole Polytechnique Federale de Lausanne, Lausanne, **2004**.
93 O. Pujol, P. Bowen, P.A. Stadelmann, H. Hofmann, *J. Phys. Chem. B* **2004**, *108*, 13128.
94 Y. Oaki, H. Imai, *Chem. Commun.* **2005**, 6011.
95 Y. Oaki, H. Imai, *Langmuir* **2005**, *21*, 863.
96 Y. Oaki, H. Imai, *Adv. Funct. Mater.* **2005**, *15*, 1407.
97 T. Miura, A. Kotachi, Y. Oaki, H. Imai, *Cryst. Growth Des.* **2006**, *6*, 612.
98 K. Takahashi, H. Yamamoto, A. Onoda, M. Doi, T. Inaba, M. Chiba, A. Kobayashi, T. Taguchi, T. Okamura, N. Ueyama, *Chem. Commun.* **2004**, 996.
99 C.D. West, *J. Paleontol.* **1937**, *11*, 458.
100 G. Donnay, *Carnegie Inst. Yearbook* **1956**, *55*, 205.
101 J. Garrido, J. Blanco, *Compt. Rend.* **1947**, *224*, 485.
102 H. Nissen, *Jahrb. Minerl. Abhand.* **1963**, *117*, 230.
103 S. Weiner, L. Addadi, H.D. Wagner, *Mater. Sci. Eng. C* **2000**, *11*, 1.
104 I. Sethmann, A. Putnis, O. Grassmann, P. Löbmann, *Am. Mineral.* **2005**, *90*, 1213.
105 I. Sethmann, R. Hinrichs, G. Wörheide, A. Putnis, *J. Inorg. Biochem.* **2006**, *100*, 88.
106 Y. Dauphin, *Paläontol. Z.* **2001**, *75*, 113.
107 W.C. Chang, X. Li, J. Chao, R. Wang, M. Chang, *Nano Lett.* **2004**, *4*, 613.
108 M. Rousseau, E. Lopez, A. Coute, G. Mascarel, D.C. Smith, R. Naslain, X. Bourrat, *J. Struct. Biol.* **2005**, *149*, 149.
109 Y. Dauphin, *J. Biol. Chem.* **2003**, *278*, 15168.
110 J.P. Cuif, Y. Dauphin, *Biogeosciences* **2005**, *2*, 61.
111 A. Taden, K. Landfester, M. Antonietti, *Langmuir* **2004**, *20*, 957.
112 Y.R. Ma, H.G. Börner, J. Hartmann, H. Cölfen, *Chem. Eur. J.* **2006**, *12*, 7882–7888.
113 T.E. Schaffer, C. Ionescu-Zanetti, R. Proksch, M. Fritz, D.A. Walters, N. Almqvist, C.M. Zaremba, A.M. Belcher, B.L. Smith, G.D. Stucky, D.E. Morse, P.K. Hansma, *Chem. Mater.* **1997**, *9*, 1731.
114 T.X. Wang, M. Antonietti, H. Cölfen, *Chem. Eur. J.* **2006**, *12*, 5722.
115 A.P. Alivisatos, *Science* **2000**, *289*, 736.
116 X.G. Cheng, L.B. Gower, *Biotechnol. Prog.* **2006**, *22*, 141.

117 L.A. Gower, D.A. Tirrell, *J. Crystal Growth* **1998**, *191*, 153.
118 M.J. Olszta, D.J. Odom, E.P. Douglas, L.B. Gower, *Connect. Tissue Res.* **2003**, *44*, 326.
119 E.V. Shevchenko, D.V. Talapin, N.A. Kotov, S. O'Brien, C.B. Murray, *Nature* **2006**, *439*, 55.
120 R. Agra, F. von Wijland, E. Trizac, *Phys. Rev. Lett.* **2004**, *93*, 018304.

4
Biologically Inspired Crystallization of Calcium Carbonate beneath Monolayers: A Critical Overview

Dirk Volkmer

Abstract

In the literature, the growth of inorganic materials below negatively charged monolayers is frequently considered to be a suitable model system for biomineralization processes. The fact that some monolayers give rise to oriented overgrowth of calcium carbonate crystals has been interpreted in terms of a geometrical and stereochemical complementarity between the arrangement of headgroups in the monolayer and the position of Ca ions in the crystal plane that attaches to the monolayer. Recent investigations into the mechanisms of nacre formation in mollusk shells, as well as comparative studies on suitably adapted model systems, suggest that this commonly held view of a structure-directing organic template matrix represents an oversimplified concept of the complex crystallization process.

Key words: calcium carbonate, calcite, aragonite, biomineralization, mollusk shell, nacre, template mechanism, epitaxy, monolayers, calixarenes, resorcarenes, charge density.

4.1
Introduction

Many organisms have developed sophisticated strategies for directing the growth of the inorganic constituents of their mineralized tissues. Active control mechanisms are effective at almost all levels of structural hierarchy, ranging from the nanoscopic regime – the nucleation of a crystallite at a specific site – up to the macroscopic regime, where the biophysical properties of the mineralized tissue are matched to certain functions [1]. Among the many open questions, one of the most challenging scientific problems is to gain greater insight into the molecular interactions occurring at the interface between the inorganic mineral and the macromolecular organic matrix. Biogenic crystals often express exceptional habits

Handbook of Biomineralization. Edited by P. Behrens and E. Bäuerlein

ISBN: 978-3-527-31805-6

that are seemingly unrelated to the morphology of the same type of crystals grown under equilibrium conditions. The morphology of the calcified tissue is ultimately encoded in the genome governing the biosynthesis of required materials at the cellular level of structural hierarchy. Therefore, biomineralization as a highly complex phenomenon involving living organisms, cannot be reduced to a single mechanistic aspect. The following representation of $CaCO_3$ mineralization in mollusks, and its mimesis by simple model systems, is admittedly a crude simplification that concentrates mainly on structural aspects, while at the same time ignoring the dynamic character of the biological process. Special emphasis here is placed on $CaCO_3$ crystal nucleation – that is, the early stages of crystal growth where the system properties might be described by (supra-)molecular recognition events occurring at the mineral–matrix interface. At this level, a common feature seems to exist for many mineralizing organisms: the interaction of highly specialized acidic macromolecules with different surfaces of the growing single crystal [2]. For the most widespread calcified tissues it is frequently assumed that a structurally rigid composite matrix consisting of fibrous proteins and acidic macromolecules adsorbed thereon acts as a "supramolecular blueprint" that templates nucleation of the inorganic phase.

In this chapter, special emphasis is placed upon the crystallization of calcium carbonate beneath insoluble monolayers – a model system that is often regarded as a straightforward experimental approach toward biomimetic crystallization. As will be shown, the numerous enthusiastic early reports of epitaxial growth of inorganic crystals beneath membrane-like monolayers might require some profound revision of the suggested mechanisms. Experimental evidence from recent investigations suggests that the commonly held view of a structure-directing organic template matrix represents an oversimplified concept of the complex crystallization process.

4.2
Nacre Formation

Mollusks, which are among the most thoroughly investigated organisms in biomineralization studies, build concrete shells from $CaCO_3$ [3]. The mollusk shell may be regarded as a microlaminate composite consisting of layers of highly oriented $CaCO_3$ crystals interspersed with thin sheets of an organic matrix. Crystals within separate shell layers usually consist of either pure aragonite or pure calcite. Vaterite, when present, is usually associated with shell repair. Shell formation occurs in two principal phases. The first phase involves the cellular processes of ion transport and organic matrix synthesis, which occur in different compartments of the molluskan mineralizing system. The second phase consists of a series of crystal nucleation and growth processes taking place in a specialized mineralization compartment, the so-called *extrapallial space* (Fig. 4.1, left) [4].

In the past, special attention has been paid to the microstructure of nacre – the iridescent inner layer of mollusk shells – which exhibits an exceptionally regular

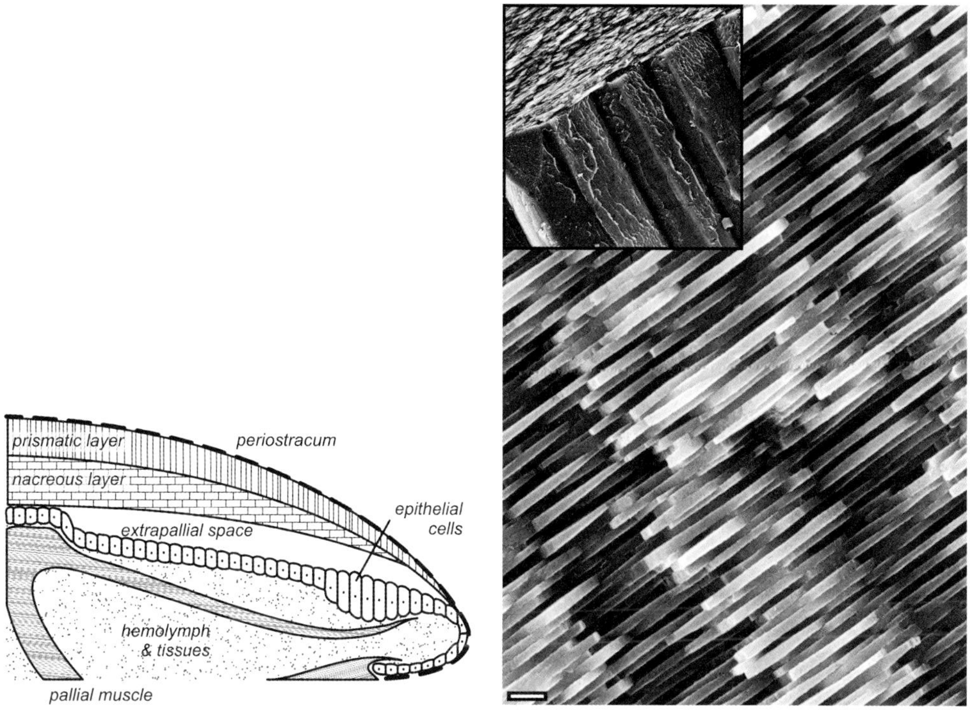

Fig. 4.1 Left: Transverse section of the mantle edge of a bivalve showing the system of compartments. Right: Fractured surface of the nacreous layer of the bivalve mollusk *Atrina rigida*. The inset shows the inner nacreous layer of tabular aragonite crystals (top) and the outer prismatic layer of columnar calcite crystals (bottom). SEM micrographs; scale bar = 1 μm. (Reprinted from Ref. [4], with permission.)

arrangement of tabular aragonite crystals (Fig. 4.1, right). From a materials scientist's point of view, nacre can be regarded as a *hierarchical* biological nanocomposite [5]. Although nacre is mainly composed of a brittle inorganic material, its highly organized design (Fig. 4.2) leads to extraordinary mechanical performance owing to an excellent combination of stiffness, strength, impact resistance, and toughness [6]. The structure of nacre has been reviewed repeatedly [7], and numerous reports indicate that it is composed of pseudo-hexagonal, polygonal, or rounded aragonite tablets having lateral dimensions of ~5 to 20 μm and a thickness of ~0.3 to 1.5 μm (Fig. 4.2) [8]. For gastropod nacre, scanning electron microscopy (SEM) images suggest that each aragonite tablet is subdivided by radial vertical organic membranes into a varying number of sectors (Fig. 4.2c), which have been interpreted to be polysynthetically twinned crystalline lamellae [9]; this interpretation, however, has been at least partly refuted by subsequent investigations [10].

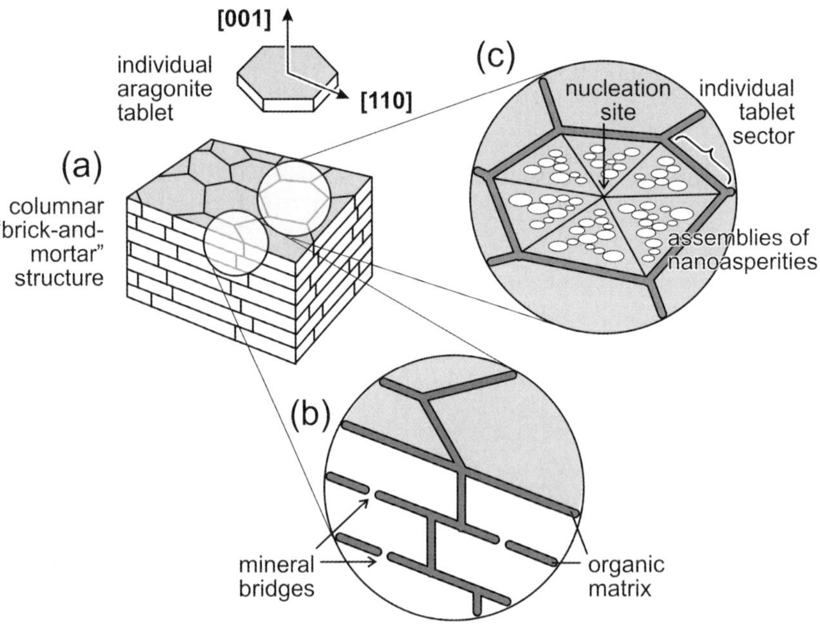

Fig. 4.2 Schematic illustration of the multiscale hierarchical structure of nacre. (a) 10-μm length scale; (b,c) 2-μm length scale, showing individual tablet features at a 200-nm length scale consisting of assemblies of nanoasperities. (Scheme redrawn after Ref. [8].) In "columnar" nacre, the tablets are relatively uniform in size and stacked vertically along the *c*-axis direction (with a slightly staggered arrangement laterally), thereby forming microlaminate sheets and tessellated bands; a domain structure has been found parallel to the *c*-axis direction consisting of three to ten tablets with *a*-axes parallel to each other, while in the plane of the sheet, the *a*-axis orientation is variable. (Reprinted from Ref. [34], with permission.)

The surface of nacre tablets (Fig. 4.2c) from California red abalone (*Haliotis rufescens*) possesses nanoasperities (~30–100 nm diameter and ~10 nm in height) [6a, 11], and mineral "bridges" (Fig. 4.2b) (~25–34 nm in size, ~91–116 μm^{-2} surface density) between sheets that pass through the organic intertablet layers [12, 13]. The organic matrix constitutes the remaining ~5 wt% of the material. The intertabular polymer layer has a thickness varying between ~30 and 300 nm [14], with pores that allow the mineral bridges to pass through [12], and intracrystalline proteins present within the tablets themselves [15].

A natural adhesive protein (*Lustrin A*) has been isolated from the organic matrix of California red abalone nacre, and was found to have a "modular" structure – that is, a multidomain architecture composed of folded, nanometer-sized modules, covalently linked together in series along a single macromolecular chain [16]. While the primary structure of Lustrin A has approximately ten alternating Cys- and Pro-rich domains, it contains only a very short polyelectrolyte ("acidic") segment that could be responsible for attachment to the mineral surface [17]. A biophysical model of its elastic properties has been proposed that relates its modular structure to the mechanical toughness and fracture resistance of nacre [18].

Considered as a whole, the physiological processes that ultimately lead to the formation of a complete mineralized mollusk shell are largely unknown. However, the general consensus is that the crystal nucleation and growth events are strictly regulated by a number of highly specialized organic macromolecules. Unfortunately, a deeper understanding of the biomineralization processes at the molecular level of structural hierarchy is hampered by our lack of knowledge of the three-dimensional (3-D) structures of macromolecules that are directly associated with the mineral layer.

Traditionally, macromolecules isolated from mollusk tissues have been distinguished into two different classes based on solubility properties [19]. Chemical analysis showed that the water-insoluble fraction consists mainly of fibrous proteins (collagen, chitin) and/or polysaccharides. These macromolecules together build a rigid framework upon which specific macromolecules from the soluble fraction may become adsorbed. The macromolecules present in the soluble fraction share common sequence motifs consisting of repeating oligomeric units of acidic residues. The primary function of the insoluble organic matrix is to subdivide the mineralization compartment into an organized network of microcompartments, and thus to delimit the available space for crystal growth and/or to constrain the crystal packing arrangement to a certain extent. The surface of this macromolecular assembly may serve as a supramolecular template for oriented nucleation of single crystals, and in fact crystallization experiments employing reconstructed matrices of purified mollusk shell macromolecules have shown that it is possible to switch between different $CaCO_3$ polymorphs, and to rebuild *in vitro* the gross structural features of the nacreous layer [20].

For the induction of calcite and aragonite nucleation, systematic investigations on biological and suitably assembled artificial systems have shed some light on the structural requirements of a putative nucleation site, especially in mollusk shells [21]. The model proposes the existence of structurally pre-organized domains of acidic residues, which could serve as a supramolecular template for oriented crystal nucleation. Such highly ordered domains could result from acidic macromolecules being adsorbed onto a rigid scaffold of insoluble matrix proteins (Fig. 4.3).

In fact, investigations of demineralized mollusk shells have shown that the interlamellar organic sheets of nacre consist of thin layers of β-chitin [22] sandwiched between thicker layers of silk fibroin-like proteins [23]. Silk fibroin itself possesses microcrystalline domains of repeating $[\text{Gly-Ala-Gly-Ala-Gly-Ser}]_n$ units that adopt an antiparallel β-pleated sheet conformation. These domains have a highly regular and hydrophobic surface upon which acidic macromolecules are adsorbed from solution. In the course of adsorption, the acidic macromolecule must fold into the appropriate conformation in order to maximize its hydrophobic interactions with the silk fibroin surface. Possible candidates for acidic macromolecules that interact with silk fibroin in the described way are oligopeptides containing sequence motifs of $[\text{Asp-X}]_n$, ($X = \text{Gly, Ser}$), which have a strong ten-

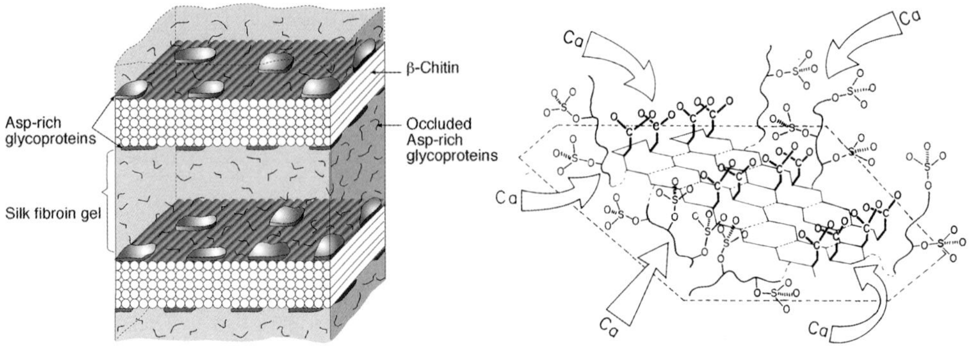

Fig. 4.3 Left: Schematic representation of the organic matrix in the nacreous layer of *Atrina* according to Weiner and Addadi. (Reprinted from Ref. [23], with permission.) The β-chitin lamellae are interspersed in a highly hydrated silk fibroin gel. The gel contains soluble Asp-rich glycoproteins, which can bind to the β-chitin surface by means of hydrophobic or electrostatic interactions. Right: Structure model of a putative nucleation site in molluskan tissues. The sulfate groups, linked to flexible oligosaccharide side chains, concentrate Ca^{2+} ions on an Asp-rich oligopeptide domain that is assumed to adapt a highly regular β-sheet conformation. A first layer of Ca^{2+} ions may thus be fixed and oriented in space upon which further mineral growth ensues.

dency to fold into a β-sheet conformation in the presence of Ca^{2+} ions [24]. As a consequence, the aspartic acid residues of $[\text{Asp-X}]_n$ sequences would be positioned at only one side of the β-pleated sheet, resulting in an organized two-dimensional (2-D) array of carboxylate ligands.

It is tempting to assume that carboxylate residues coordinate a first layer of Ca^{2+} ions, which would in turn become the first layer of an epitaxially growing $CaCO_3$ crystal. However, studies have so far failed to provide evidence for an epitaxial growth mechanism or a close stereochemical complementarity between the nucleating macromolecules and the incipient $CaCO_3$ crystal surface. Much is expected to be learned from the first 3-D structure of a "crystal nucleating" macromolecule, although its "active" conformation may depend on the accompanying insoluble organic matrix in the biological tissues. It is noteworthy to mention that there exists a (genealogically unrelated) group of crystallization-inhibiting proteins, the so-called antifreeze proteins (AFPs) [25], for some of which precise X-ray structural data are now available. Some of these proteins assume rigid molecular conformations and prevent ice crystal nuclei from growing into larger crystals by specific adsorption to specific crystal faces. However, as Evans pointed out in a critical review [26], the structural aspects of AFPs and proteins involved in biomineralization processes are quite different: in contrast to ice-interaction proteins, which typically adopt stable secondary structures (α-helix, β-sheet, β-helix, etc.), biomineral-interaction proteins typically adopt unfolded, open conformations, and, where mineral binding motifs have been identified, these sequences exhibit structural trends toward extended, random coil, or other unstable secondary structures.

This short summary of mollusk shell formation and the morphological properties of nacre indicates that the biological growth processes involved are much more complex than the seemingly primitive, brickwork-like structure might, at first glance, suggest. Nevertheless, in the past different types of bottom-up approaches have been sought to mimic different aspects of the molluskan mineralization system [27]. Especially, the quest for a possible epitaxial relation between the organic matrix and the overgrowing inorganic crystals has been subject of many experimental studies.

4.3 Biomimetic Crystallization of $CaCO_3$ beneath Monolayers: Experimental Set-Up

Several studies have been devoted to the putative mechanisms underlying the formation of highly organized mineral structures in mollusks and similar organisms. Due to the various experimental difficulties of observing crystal growth in living mollusks, suitable model systems have been sought for biomineralization processes reflecting those mineral/matrix interactions that are active at the atomic or molecular level of structural complexity. Seminal contributions were made by Mann and Heywood, who studied the influence of charged monolayers on crystal nucleation [28]. This group made use of a Langmuir film balance that allows for the spreading of surfactant molecules as a water-insoluble monolayer at the air–water interface. The basic principle of the experimental set-up is shown in Figure 4.4, in which the surfactant molecules are symbolized by a wire-model of hydrocarbon chains and the charged head groups (symbolized by triangles) point toward the aqueous subphase.

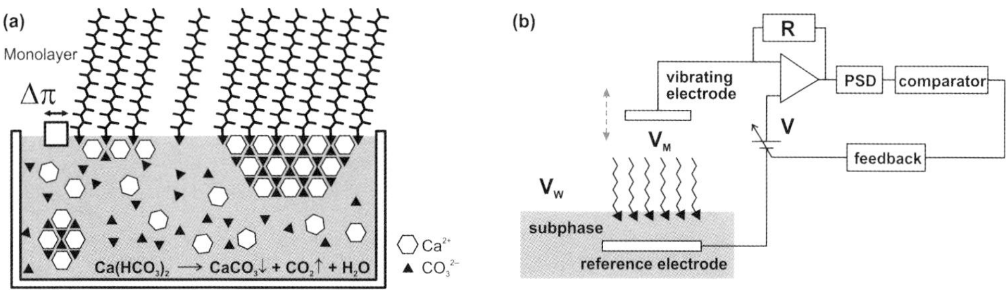

Fig. 4.4 (a) A Langmuir film balance used for biomimetic crystallization of $CaCO_3$ and other inorganic phases beneath insoluble monolayers of amphiphilic molecules.
(b) Working principle of a Kelvin probe apparatus for measuring the electrostatic surface potential of a monolayer. From the Langmuir and surface potential isotherms the pressure-dependent phase behavior of the monolayer and average orientation of the molecules in the monolayer can often be deduced.
PSD = position-sensitive detector; surface potential of pure water subphase (V_W) and with monolayer (V_M).

In studies relating to the biomimetic crystallization of $CaCO_3$, a freshly prepared ~9–10 mM aqueous solution of $Ca(HCO_3)_2$ is often used, which becomes unstable with respect to the dissociation of bicarbonate anions into CO_2 and carbonate anions. Due to slow decomposition of the intermediate carbonic acid – yielding CO_2 and a water molecule – $CaCO_3$ precipitation is relatively slow (in the optical microscope, the first macroscopic crystals become visible within some minutes to hours, depending on the experimental conditions). The slowness of crystal formation allows for adjusting the compression state of the monolayer – that is, to limit the average surface area available for a single molecule in the monolayer and/or to induce a monolayer phase transformation [29].

From a technical standpoint, the monolayer crystallization method has both advantages and disadvantages. As a general advantage, the use of a Langmuir film balance allows the creation of a smooth and clean interface and for the deposition of a measured quantity of amphiphilic molecules in a controlled manner. The average area per molecule is adjusted by a movable barrier, and the surface pressure or tension is monitored by a pressure sensor. The simple technical set-up can be augmented by various analytical characterization methods, for example, by Brewster-angle or fluorescence microscopy for imaging domain structures of the monolayer, or by surface potential measurements [29]. In many cases it is straightforward to obtain a first-approximation model of the arrangement and packing of molecules in the monolayer. This technique has been used over many years, mainly by the group of Leiserowitz and Lahav, to investigate epitaxy and chiral discrimination of molecular (non-ionic) crystals growing beneath specially designed monolayers [30].

One technical *disadvantage* of the Langmuir set-up is the mobility and the thinness of the monolayer, which renders structural investigations at high resolution (i.e., at the atomic or molecular level) a difficult task. In principle, it is possible to deposit a monolayer as a Langmuir–Blodgett film on a substrate surface and to employ surface analytical techniques, such as scanning force or tunneling microscopy, in order to obtain high-resolution images of the interface structure. However, this strategy has rarely been utilized in studies related to biomimetic crystallization [31].

One particular *advantage* of crystallizing calcium carbonate beneath monolayers is that the three main crystal polymorphs (i.e., calcite, aragonite, and vaterite) can be distinguished easily by their characteristic crystal habits. While aragonite and vaterite crystals normally adopt complex acicular or floret-type morphologies (see examples shown in Fig. 4.10), calcite crystals often appear as sharply defined single crystals displaying the shape of a (truncated) {10.4} cleavage rhombohedron.

The orientations of calcite crystals with respect to the monolayer can be deduced from optical or scanning electron micrographs. A projection of rhombohedral faces onto the image plane yields characteristic interfacial angles, based upon which crystal orientations can be determined by comparing the measured angles and the general outline evident in micrograph images to computer-generated crystal models cleaved along various crystallographic planes. Owing to the fact that only a very limited number of crystal specimens can be analyzed in this manner within a reasonable time frame, the method suffers from the inherent disadvantage that crystals tend to be neglected when their orientations are difficult to determine (e.g., because the crystals are small), or when the crystals grow as polycrystalline aggregates. Unfortunately, there are no simple-to-use analytical techniques for simultaneously determining the phase composition, average crystallite size and orientation distribution function (ODF) of the loose assembly of macroscopic crystals that is typically encountered in monolayer investigations. Some dedicated methods are available for these tasks, such as X-ray pole-figure measurements or automated analysis of Kikuchi line patterns using electron backscatter diffraction (EBSD) systems [32]. However, these techniques are not readily available for *in-situ* measurements of crystal growth beneath monolayers, and they normally impose strict requirements on sample preparation, such as a uniform sample thickness or a flat (= polished) sample surface. Given the technical difficulties described above, any statements in the literature referring to the epitaxial growth of inorganic crystals beneath monolayers should be interpreted with caution, as the presentation and evaluation of crystal orientation data are somewhat arbitrary. Moreover, many studies describing the heteroepitaxy of *calcite* crystals focus explicitly on the particular crystal texture of nacre, which consists of aragonite and not calcite. The reason for the latter contradiction might lie in the fact that aragonite single crystals are difficult to prepare at ambient conditions [33], and they rarely manifest the pseudohexagonal shape of the tabular aragonite crystals that constitute the brick-and-mortar structure of nacre (see Fig. 4.2).

4.4 $CaCO_3$ Crystallization beneath Monolayers of Macrocylic Amphiphiles

Since the first reports appeared of $CaCO_3$ crystallization beneath monolayers, many groups have made use of the Langmuir trough set-up for investigating the different effects of structurally dissimilar amphiphilic molecules on $CaCO_3$ crystal growth (for a review, see [34]). The partially contradictory experimental findings led us to design a small library of macrocyclic amphiphiles, which was based on calix[*n*]arene ($n = 4, 6, 8$) and resorc[4]arene moieties, respectively. The use of these particular compounds might be rationalized in terms of their modular structure, which allowed us to synthesize amphiphilic compounds of different sizes and shapes, varying charge distributions and stereochemical arrangements of the coordinating head groups A (Scheme 4.1). In order to distinguish as much as possible between the various physical parameters that are known to influence the epitaxial growth of inorganic crystalline materials, several working hypotheses were defined and subsequently tested in comparative crystal growth studies.

In a first set of experiments, the growth of calcite crystals beneath monolayers of the tetracarboxy-calix[4]arenes – namely, 5,11,17,23-tetrakis-(1,1,3,3-tetramethylbutyl)-25,26,27,28-tetra(carboxymethoxy)calix[4]arene (**1**) and 5,11,17,23-tetra-

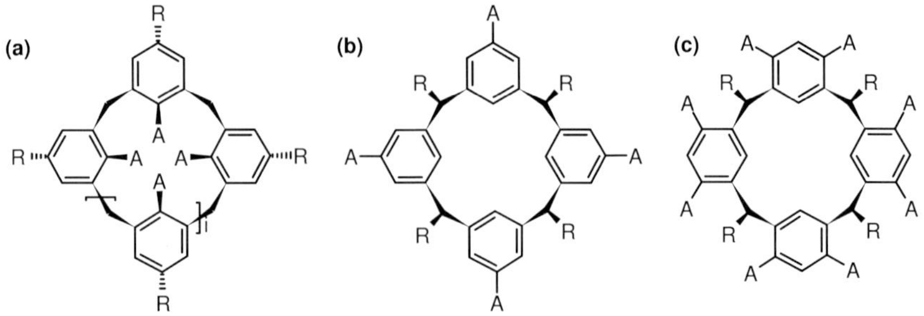

Scheme 4.1 Structural diagrams of calixarene (a, $i = 0, 2$, or 4), and resorcarene macrocycles (b,c) used in monolayer investigations, where R represents a strongly hydrophobic residue, and A represents an acidic (= metal ion-coordinating) group.

t-butyl-25,26,27,28-tetrakis(carboxymethoxy)calix[4]arene (**2**) – were investigated. These compounds differ from each other by *the bulkiness of hydrophobic substituents* at the upper-rim of the macrocyclic backbone (Scheme 4.2).

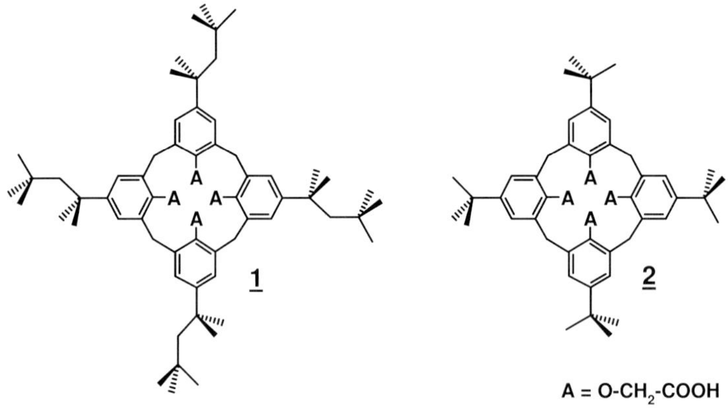

Scheme 4.2 Structural formulae of tetracarboxy-calix[4]arenes (**1**, **2**) bearing different hydrophobic substituents. The average area/molecule of **1** in a compressed monolayer is more than 15% larger than the corresponding value of **2** (at equal surface pressure).

Crystal growth experiments have shown that the same uniformly (01.2) oriented calcite single crystals grow beneath monolayers of **1** as well as **2** (Fig. 4.5) [35, 36]. The fact that identically oriented calcite crystals are obtained under monolayers of differently sized calix[4]arenes gives a strong indication *against* the assumption that an epitaxial match between the monolayer and the juxtaposed crystal surface is required for inducing a certain crystal growth behavior.

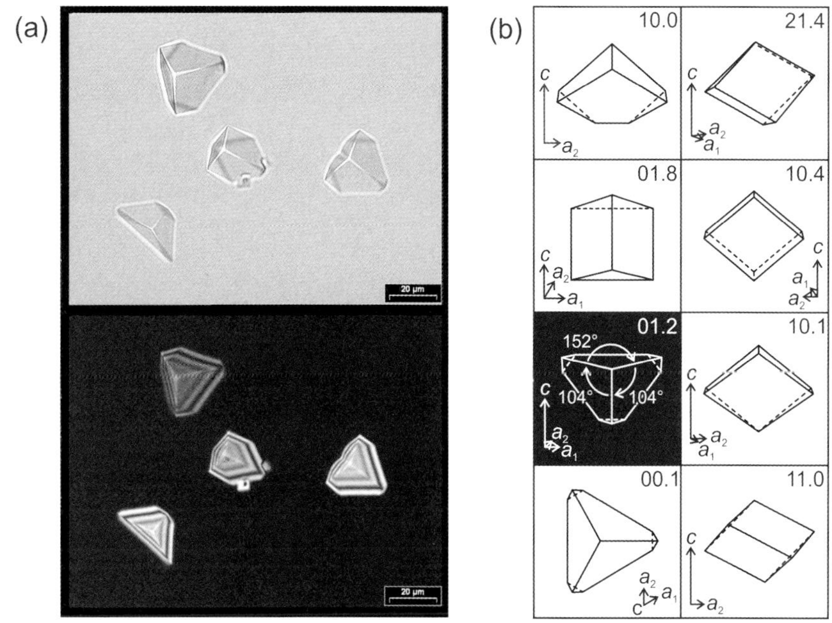

Fig. 4.5 (a) Upper: Optical micrograph (bright field) of (01.2)-oriented calcite single crystals grown under a monolayer of **1** and **2**, respectively, after 3 h ($\pi = 0.1$ mN m^{-1}, $[Ca(HCO_3)_2]_{t=0} = 9$ mM). Lower: the same crystal specimen observed in plane-polarized light. The viewing direction is parallel to the monolayer surface normal, and crystals are observed from below the aqueous subphase. (b) Outlines of calcite single crystals cleaved along various crystallographic planes (specified by their Miller–Bravais indices). Models of crystals are oriented such that their cleavage planes coincide with the image plane. The highlighted model corresponds to a truncated calcite {10.4} rhombohedron cut along its (01.2) plane. The displayed numbers represent the theoretical interfacial angles of crystal edges projected onto the image plane, which can be determined experimentally from optical micrographs of the crystals, such as those shown on the left side.

However, according to these results it was still possible that the calix[4]arene ligands **1** and **2** might afford a common, highly specific coordination motif by which they could bind to the {01.2} crystal face of the overgrown calcite crystals.

In order to answer this critical question, a second type of macrocyclic ligand, the tetracarboxy-resorc[4]arene *rccc*-5,11,17,23-tetracarboxy-4,6,10,12,16,18,22,24-octa-*O*-methyl-2,8,14,20-tetra(*n*-undecyl)resorc[4]arene (**3**; Scheme 4.3), was designed and used in crystallization assays [37].

The single crystal X-ray structure analysis of a number of Ca^{2+} complexes of **1**, **2**, and **3** showed that the typical Ca^{2+} coordination motifs of these ligands are largely different (Fig. 4.6) [35–37]. However, in spite of these fundamental differences in molecular structure, crystallization assays employing monolayers of **3**

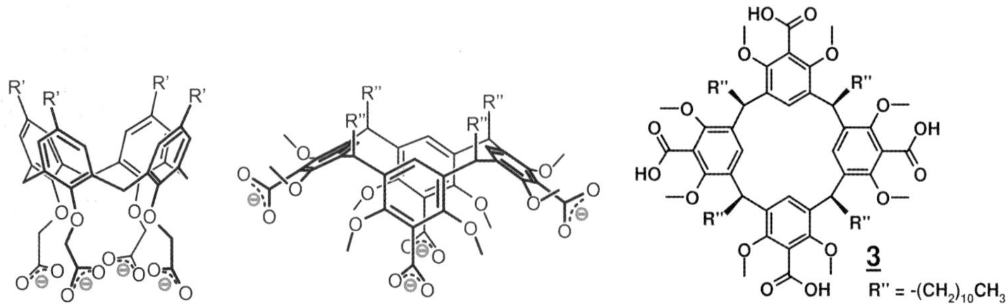

Scheme 4.3 Left: typical arrangements of carboxylate moieties in calix[4]arene-type (**1**, **2**) and resorc[4]arene-type (**3**) ligands (R′: *tert*-butyl or *tert*-octyl, R″: *n*-undecyl). Right: structural formula of tetracarboxy-resorc[4]arene **3** which, according to X-ray analytical investigations and NMR studies, assumes a C_{2v}-symmetric *boat* conformation [37].

gave (01.2)-oriented calcite crystals that were virtually indistinguishable from those that grew beneath monolayers of **1** or **2**.

This unexpected experimental observation led us to examine the phase behavior of **1**, **2**, and **3** monolayers in more detail, for which we simultaneously recorded Langmuir and surface potential isotherms (Fig. 4.7). Moreover, Brewster-angle microscopy (BAM) images were taken for monolayers at different compression states.

In summary, the results of these investigations demonstrated that neither calixarene nor resorcarene derivatives have a tendency to form highly ordered liquid-crystalline phases during compression. Moreover, the growth of (01.2)-oriented calcite crystals exclusively occurred at low surface pressure ($\pi = 0.1$–1 mN m^{-1}), where the monolayers show a transition from the liquid-expanded to the liquid-condensed state of matter. The morphology of the monolayer in this pressure regime is characterized by a high mobility of the molecules, as can be concluded from the typical foamy texture observed in BAM micrographs.

In order to gain additional insight into the interactions between monolayers and hydrated calcium and carbonate ions, the *non-charged* amphiphilic alcohol 5,11,17,23-tetrakis-(1,1,3,3-tetramethylbutyl)-25,26,27,28-tetra(2-hydroxyethoxy)calix[4]arene (**4**) was employed. Monolayers of **4** were found strongly to inhibit the heterogeneous nucleation of $CaCO_3$ crystals [38]. The surface potential of a monolayer arises from the dipole moment of the constituent molecules, from the change of orientation of water molecules, and also from interactions between the headgroups and electrolytes dissolved in the subphase [39]. In this study, electrostatic interactions between monolayers of structurally related amphiphilic calix[4]arene derivatives **1** and **4** and subphase ions was examined. In the low-pressure region ($\pi = 0.0$–0.5 mN m^{-1}) of the monolayer phase diagram, the experimental surface potential values for monolayers of **1** and **4** on H_2O are almost identical. On a Ca-containing subphase (10 mM $CaCl_2$), the surface potential

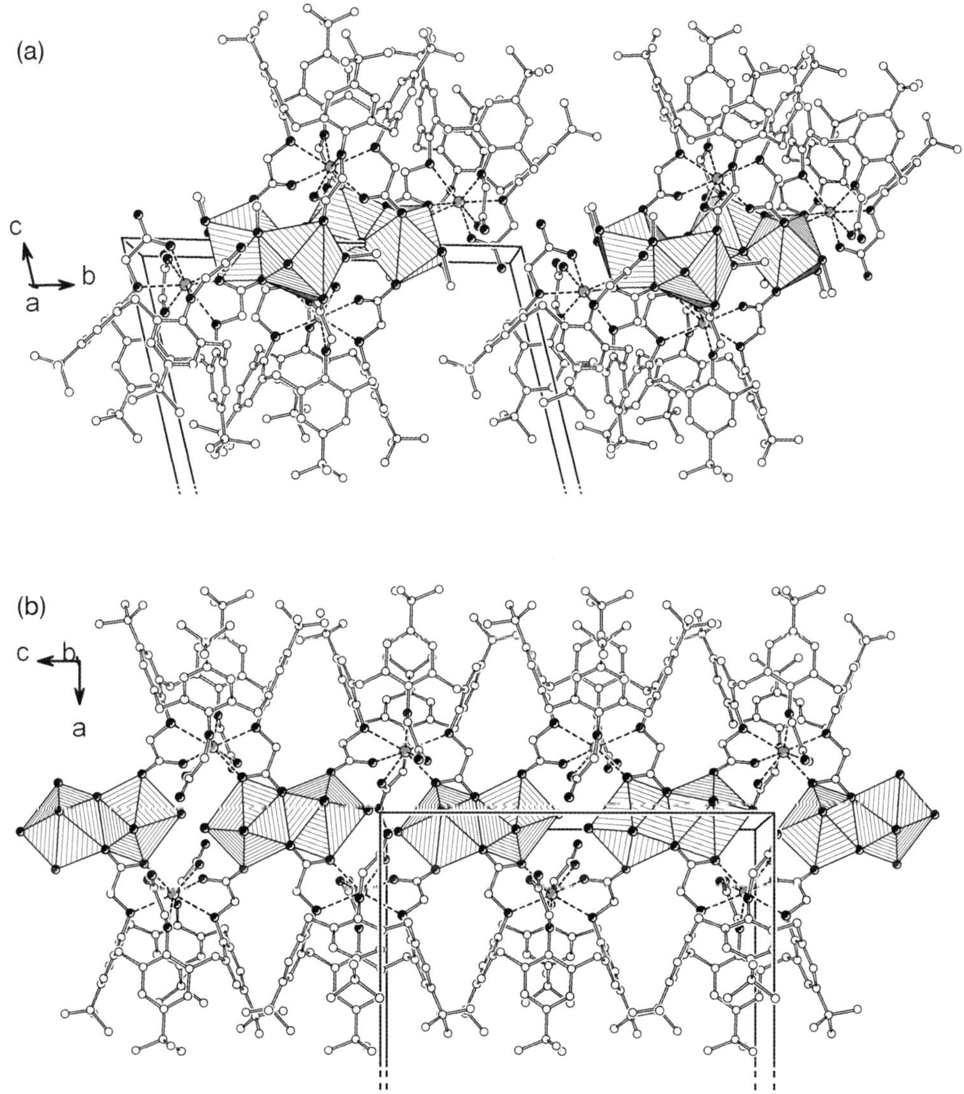

Fig. 4.6 (a,b) Models of different calcium coordination compounds of **2**, showing the packing arrangements of one-dimensional polymeric strands in the crystal lattices [36]. Solvent molecules occluded in the crystal lattice and hydrogen atoms are omitted for clarity. Coordination polyhedrons are displayed for interconnecting Ca ions only.

shows a significant increase only for monolayers of **1**, which clearly demonstrates that the calix[4]arene derivative **4** is unable to bind Ca ions by virtue of electrostatic and/or coordinative interactions. Furthermore, the experimental results indicated that a *low* surface pressure ($\pi = 0.1$–0.5 mN m^{-1}) is a necessary condition

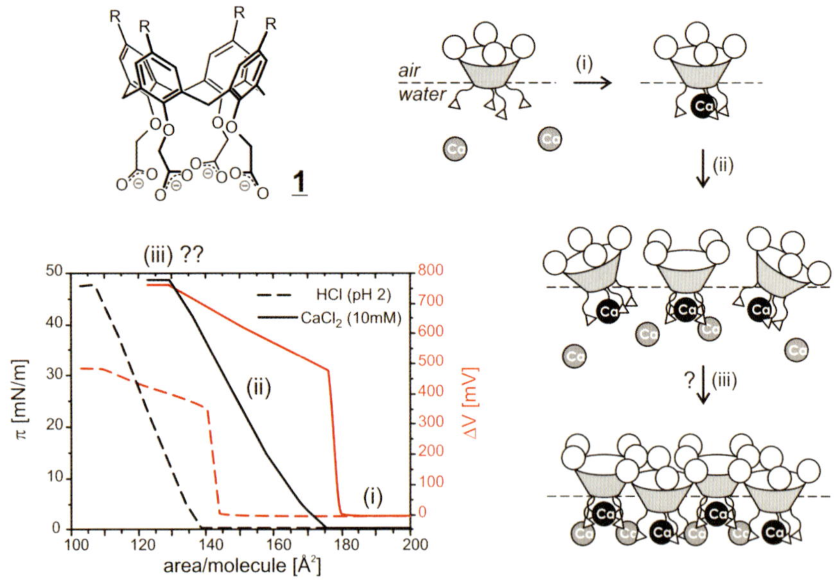
1
air
water
(i)
(ii)
?
(iii)
Ca
(iii) ??
HCl (pH 2)
CaCl2 (10mM)
π [mN/m]
ΔV [mV]
area/molecule [Å²]
(i)
(ii)

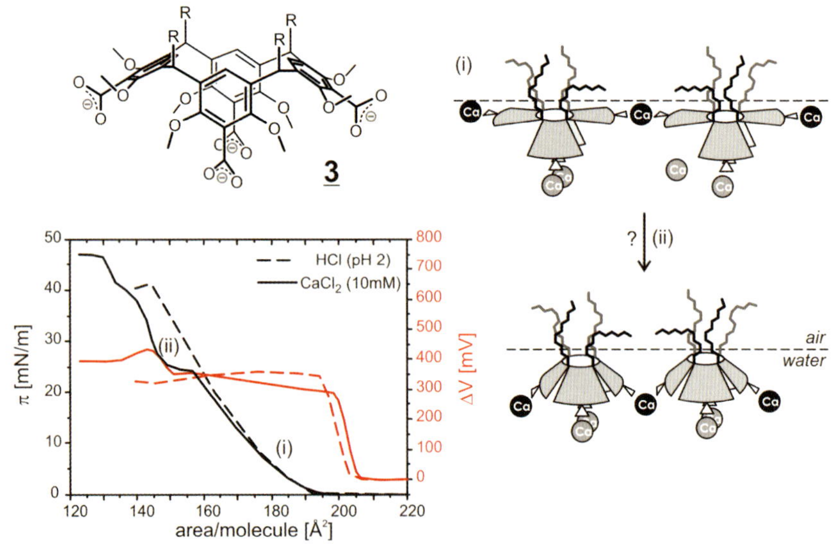
3
(i)
?
(ii)
air
water
Ca
HCl (pH 2)
CaCl2 (10mM)
π [mN/m]
ΔV [mV]
area/molecule [Å²]
(i)
(ii)

for the growth of uniformly (01.2)-oriented calcite crystals. The phase diagram of **1** and previous investigations of the monolayer structure show no indication for long-range order in the monolayer – that is, the x and y-components of the molecular dipole moments are laterally uncorrelated.

The growth of calcite single crystals showing a preferential (01.2) orientation has been reported for other systems, including polymeric Langmuir–Schaefer films of 10,12-pentacosadiynoic acid [40], hydrogen-bonded molecular ribbons [41], as well as short-chain "acidic" model peptides (e.g., H–(Phe–Asp)$_4$–OH) [42]. Much effort has been spent on correlating the periodicity of the calcite {01.2} cleavage plane with the proposed monolayer structures. However, it is difficult to believe that the coincident growth of (01.2)-oriented calcite crystals beneath structurally highly dissimilar monolayers is a consequence of a strict geometrical or even a stereochemical matching. In fact, this coincidence (that we laxly designate the "012-syndrome" in our group), indicated to us that non-directional interactions between the monolayer and the juxtaposed crystal face might be the most important factor.

In order to further support this hypothesis, the crystallization of $CaCO_3$ below monolayers of two structurally different macrocyclic octacarboxylic acids, *rccc*-4,6,10,12,16,18,22,24-octakis-*O*-(carboxymethyl)-2,8,14,20-tetra(*n*-undecyl)resorc[4]arene (**5**) and 5,11,17,23,29,35,41,47-octakis-(1,1,3,3-tetramethylbutyl)-49,50,51,52,53,54,55,56-octa(carboxymethoxy)-calix[8]arene (**6**), were investi-

Fig. 4.7 Langmuir (black lines) and surface potential isotherms (red lines) of monolayers of **1** and **3** spread at 24 °C on different aqueous subphases. Isotherm data indicate different dynamic monolayer phase behaviors, as shown on the right-hand side. Top: For a monolayer of **1** spread on 10 mM $CaCl_2$ solution, isotherm data are in agreement with a rapid sequestration of a Ca ion by the calix[4]arene ligand which occurs at low surface pressure (i). The Ca complex thus formed represents a rigid entity which behaves like a stiff molecular dipole at the air–water interface (ii). The long-range order in this monolayer of dipoles improves, if an external pressure is imposed on the monolayer (by closing the barriers of the trough). At point (iii) of the isotherms, a maximum order is reached and most dipoles are co-aligned with respect to each other. Further compression probably leads to a collapse of the monolayer. Bottom: The phase behavior of the monolayers of **3** is more complicated. Crystallographic investigations on the solid-state structures of the Ca complex of **3** show that the resorc[4]arene macrocycle prefers a C_{2v}-symmetric *boat* rather than the C_{4v}-symmetric *crown* conformation. If the more extended boat conformation prevails in the monolayer, two out of four benzene rings would be parallel, whereas the other ones would be perpendicular to the air–water interface (i). Thus, linear one-dimensional strands of Ca coordination polymers could form even at low surface pressure (i), and the long-range order in the monolayer would not significantly improve upon further compression. A phase transition is observed at point (ii) of the isotherms, which might be tentatively ascribed to a geometrical change of **3** to adopt the more compact crown conformation at high surface pressure. Despite these fundamental differences in Ca complex formation and monolayer phase behavior, $CaCO_3$ crystals grow with the same orientation beneath both monolayers.

(5)

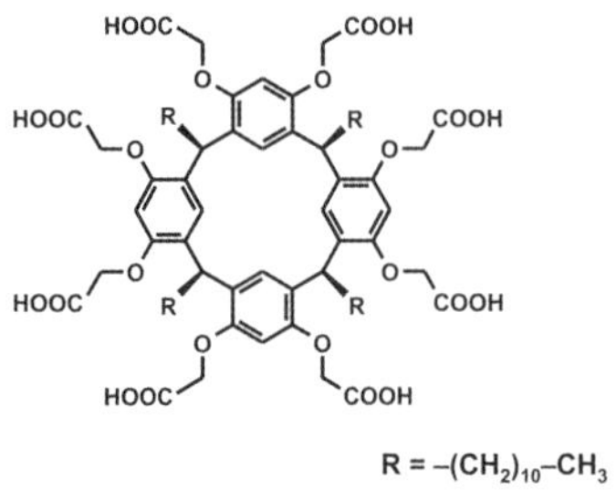

(6)

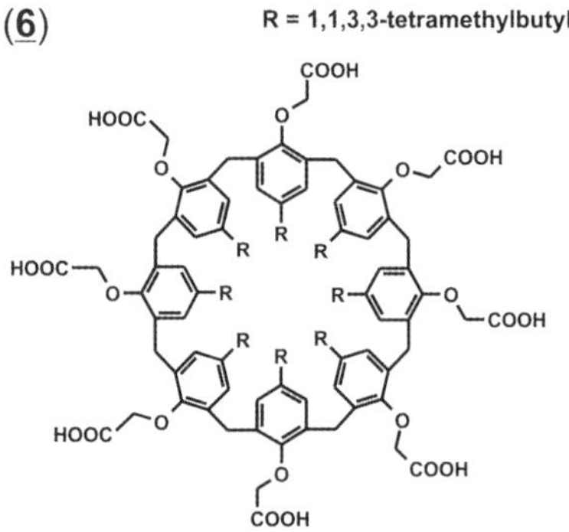

Fig. 4.8 Left: Structural formulae of octacarboxy-resorc[4]arene (**5**) and octacarboxy-calix[8]arene (**6**) having the same number (= 8) of carboxylate residues but differing largely in molecular size. Right: Scanning electron micrographs (crystals collected after 6 h) of acicular aragonite crystals grown under a monolayer of **5** ($\pi \sim 20$ mN m^{-2}, $CaCl_2/NaHCO_3$ (ca. ~9/18 mM). Approximate viewing directions are: (a) from below; (b) from above; and (c) from the side of the monolayer.

gated (Fig. 4.8) [43]. Compounds **5** and **6** were chosen for these investigations because they differ by a factor of 2 in molecular surface area. It is thus possible to adjust the surface charge density underneath monolayers by a proper molecular design.

Interestingly, $CaCO_3$ crystallization beneath monolayers of **6** led to the formation of (01.2)-oriented calcite crystals, despite the fact that the monolayer of **5** manifests a completely different growth characteristic: namely, the formation of acicular aragonite crystal aggregates at high surface pressure and at an ionic composition of the aqueous subphase that would normally yield the thermodynamically more stable calcite (Fig. 4.8).

In summary, these investigations provide compelling experimental evidence that the crystallization of $CaCO_3$ below monolayers of amphiphilic polyacids is largely controlled by the *surface charge density* that accumulates underneath the monolayer. Unfortunately, this functional correlation is difficult to confirm, as there exists (to the best of our knowledge) no analytical technique for measuring the absolute value of the surface charge density beneath the monolayer. The measurable *surface potential* of a monolayer is dependent on many factors, including the molecular structure of the surfactant, the packing arrangement, and the composition of the aqueous subphase [39]. Moreover, it is not clear whether control over crystal nucleation by an increased surface charge density is gained through polarity matching between the surface monolayer and the incipient crystal surface, or if crystal nucleation and growth processes are kinetically controlled, or if both factors are equally important. In order to test whether monolayers generating a sufficiently high surface charge density are able to induce and stabilize metastable crystal phases at the air–water interface, the highly charged amphiphilic dendron-calix[4]arene, 5,17-di[(2-carboxyacetyl)amino]-11,23-di(*tert*-butyl)-25,26,27,28-tetradodecyloxycalix[4]arene [2]:(2-aza-3-oxopentylidyn):propanacid (**7**) was employed [44]. Indeed, crystallization of $CaCO_3$ beneath monolayers of **7** selectively led to growth of the metastable polymorph vaterite at low surface pressure. Monolayers of **7** induce the formation of *vaterite* at a surface charge density corresponding to 6.7 to 7.2 COO^- nm^{-2}. On the other hand, surface charge densities in the range of 4.65 to 5.00 COO^- nm^{-2} led to selective crystallization of *aragonite*, as was shown for monolayers of **5**. The formation of uniformly oriented *calcite* crystals with the highly polar {01.2} face oriented toward the monolayer was observed on many structurally different monolayers, all sharing similar charge densities of 2.0 to 2.4 COO^- nm^{-2}. (For a complete summary of the various carboxylic amphiphiles used by our group as crystallization templates and the $CaCO_3$ growth morphologies that they induce, see Fig. 4.10.)

4.5 Formation of Tabular Aragonite Crystals via a Non-Epitaxial Growth Mechanism

As the number of investigations targeting the molecular blueprinting beneath monolayers steadily increased, a notable change of the current template model of biomineralization took place, such that the vital role of *amorphous precursors* in biologically induced mineralization processes was emphasized [45]. This change in paradigm was stimulated by the discovery of thin layers of amorphous calcium carbonate preceding the formation of crystalline materials in different organisms [46]. These reports – together with the fact that, to date, no monolayer model system has been able to produce the specific crystal morphology of pseudohexagonal tabular aragonite crystals found in nacre – has led us to modify the experimental conditions in monolayer investigations.

Previous studies performed using arachidic acid [47] or amphiphilic tricarboxyphenylporphyrin iron (III) μ-oxo dimers [48] had already shown that a thin film of amorphous calcium carbonate might form below the monolayer if the subphase contains a soluble polyelectrolyte that acts as (non-specific) crystal growth inhibitor. In presence of polyacrylic acid, the fatty acid per calcium ratio is about

14 to 25, according to grazing-incidence X-ray diffraction measurements, and tablet morphologies of $CaCO_3$ most often resulted in the formation of calcite films.

Recently, the formation of single-crystalline aragonite (as well as calcite) tablets and films was accomplished by precipitating an amorphous precursor under conditions exploiting the synergistic effects of a monolayer of tetracarboxy-resorc[4]arene (**3**) and water-soluble additives, such as polyacrylic acid and/or magnesium ions [49]. Using appropriate concentrations of the latter, an initially amorphous film can be deposited beneath the monolayer that later crystallizes into polycrystalline films with either single-crystalline mosaic or spherulitic structure (Fig. 4.9a). Of particular importance is the synthesis of single-crystalline mosaic films of aragonite, which resemble the tablets of aragonite crystals in the nacreous layer of mollusk shells (Fig. 4.9b).

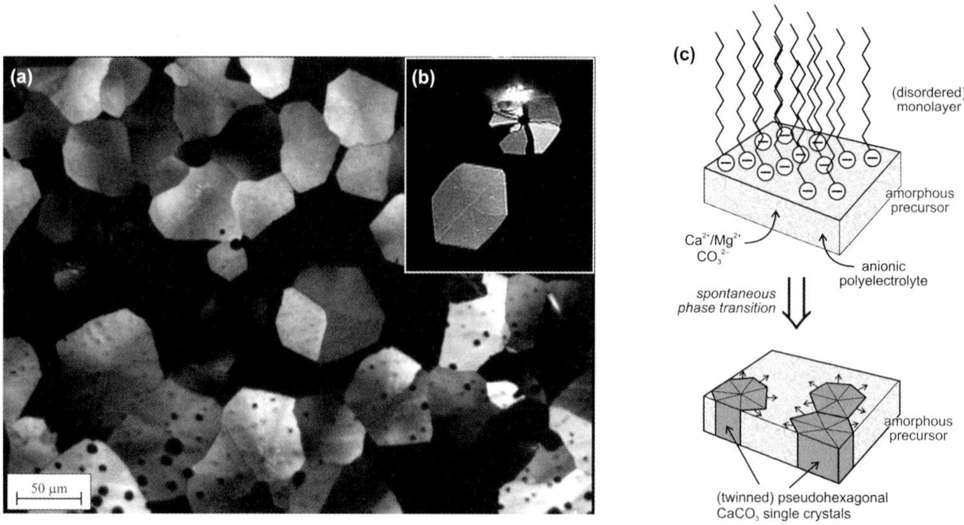

Fig. 4.9 (a) Polarization optical micrograph of polycrystalline calcium carbonate crystals grown from an amorphous thin film that forms beneath a monolayer of **3** when the aqueous subphase contains a growth inhibitor (polyacrylate). Holes inside the crystals are presumably due to the evolution of CO_2 bubbles from the aqueous subphase. (b) Close-up of two crystal specimens. The intact hexagonal platelet manifests sectors that might result from interpenetration twinning. The upper crystal has disintegrated into triangular sectors, the breakage presumably originating in the twinned contact planes of the crystal. Note that the pseudohexagonal shape and the twinning pattern are quite similar to those of the aragonitic platelets in biogenic nacre [49]. (c) Formation of polycrystalline thin films comprised of tabular pseudohexagonal calcium carbonate crystals via transformation of an amorphous precursor thin film. Recent investigations indicate that a similar phase transition occurs during the biological formation of nacre.

4.6 Conclusions

In the literature, the growth of inorganic materials below negatively charged monolayers is frequently considered to be a suitable model system for biomineralization processes. The fact that some monolayers give rise to oriented overgrowth of calcium carbonate crystals has been interpreted in terms of a geometrical and stereochemical complementarity between the arrangement of headgroups in the monolayer and the position of Ca ions in the crystal plane that attaches to the monolayer. The concept of the "molecular blueprinting" of ionic solids has partly arisen from the numerous astonishingly perfect crystal architectures found in biomineralizing organisms, the most prominent example of which is certainly provided by the "brick-and-mortar" structure of nacre. Until recently, the most widely accepted model for the emergence of nacre has suggested an epitaxy mechanism, according to which a highly ordered composite of crystal-nucleating acidic proteins and structural framework macromolecules (mostly chitin) acts as a supramolecular template matrix governing the site-selective and orientation-selective nucleation of the inorganic crystals. However, recent investigations of the structure of nacre provide strong evidence against this simple template hypothesis [50]. It transpires that the biosynthesis of an amorphous calcium carbonate precursor precedes the formation of aragonite crystals. The mechanisms by which the amorphous phase is switched into a specific calcium carbonate polymorph and into crystals of particular shape are still largely unknown. However, by adapting model systems to this new paradigm, specific aspects of the complex biological growth processes might be suitably addressed.

Comparative investigations of the growth of calcium carbonate beneath monolayers of macrocyclic polyacids, for instance, have demonstrated that non-directional electrostatic parameters, such as the average charge density or the mean dipole moment of the monolayer, determine the orientation and the polymorph of the overgrowing crystals. A summary of the monolayer systems we have employed so far and the crystal morphologies they induce is provided in Figure 4.10.

The present results show that it is possible to control the surface charge densities in monolayers by the appropriate design of amphiphilic molecules. A polymorph switch occurs above a critical monolayer charge density at which aragonite or vaterite nucleation ensues presumably due to a kinetically controlled precipitation process (Fig. 4.10). Below this critical charge density value (for which, based on numerous monolayer investigations, we may tentatively assign a value range of 3.5–4.5 $COO^- \ nm^{-2}$), calcite single crystals form. The sharply defined rhombohedral shape of these crystals – and the fact that they show an island-like distribution beneath the monolayer – provide a strong indication against epitaxial lattice correlations between the monolayer matrix and the incipient crystal face. Moreover, monolayers consisting of structurally very dissimilar amphiphiles often lead to identical crystal orientations, for which nucleation from the {01.2} crystal face of calcite is most often observed. We suggest that crystal nucleation occurs

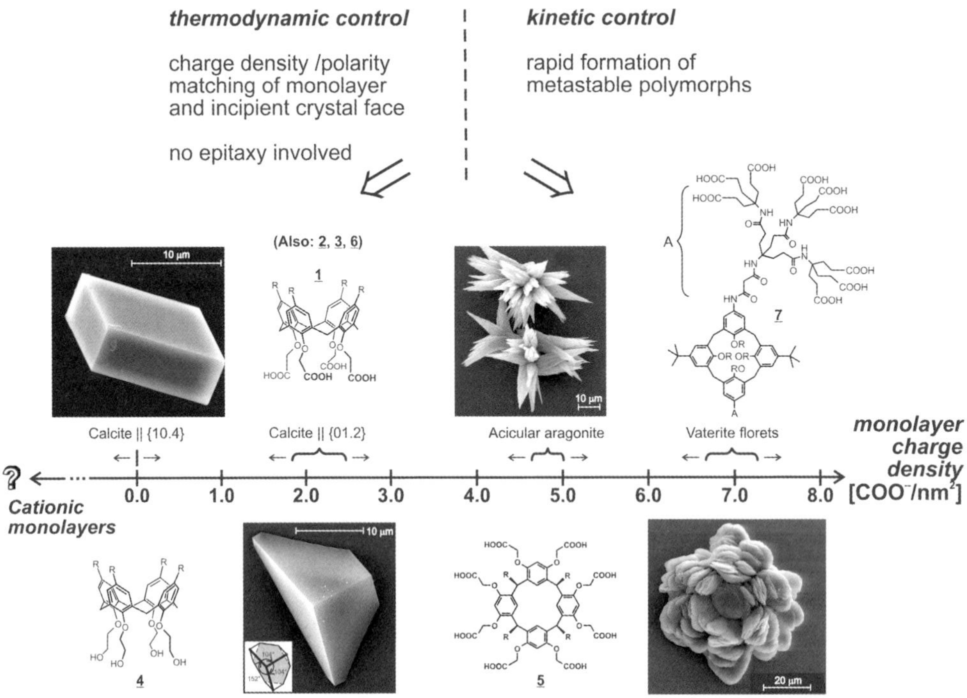

Fig. 4.10 Upper: Overview of macrocyclic polyacids employed in previous investigations of the growth of calcium carbonate beneath monolayers. Polyacids are arranged according to increasing (negative) charge density, which is expressed here as the number of carboxylate residues per unit area. Experimental ranges of charge density leading to a characteristic calcium carbonate crystal habit or orientation are indicated with braces. (Charge density values are directly gained from Langmuir isotherms.) Lower: Scheme of charge density controlling the inorganic crystal polymorph that forms beneath the monolayer. For calcium carbonate, a switch from thermodynamically stable calcite to less-stable aragonite or vaterite occurs at a charge density ranging from 3.5 to 4.5 COO^- nm^{-2}.

preferentially from this particular face because of charge density/polarity matching at the interface, considering that the {01.2} face is the most strongly polar (low-index) crystal face of the calcite crystal lattice. This hypothesis is further supported by recent modeling studies, which question the idea of stereochemical or geometrical complementarity between the monolayer and the nucleated crystal face [51].

There are notable exceptions, such as monolayers of simple monofunctional surfactants, which lead to exceptional crystal growth orientations (e.g., alkyl sulfates leading to nucleation from the {00.1} crystal face or alkyl carboxylates nucleating calcite from a {11.0} face). However, more in-depth studies employing these simple surfactants once again have challenged the simple picture of interfacial complementarity [52]. To give an example, the crystallization of $CaCO_3$ beneath fatty acid monolayers was studied by *in-situ* grazing incidence X-ray diffraction [53]. It was found that in monolayer systems of stearic acid or arachidic acid, spread on 4.5 or 9 mM calcium carbonate solutions, there are only some four to eight surfactant molecules per calcium ion. A strict epitaxial correlation between the monolayer and the nucleated crystal would require a 1:1 or 2:1 ratio.

These results demonstrate that the structure of the monolayer/crystal interface is very likely not as simple as has been suggested in preceding reports. Thus, it is safe to predict that many more investigations will be conducted in the near future targeting the fundamental mechanisms of crystal nucleation and growth beneath structurally mobile monolayers (without having to emphasize any model character that such systems may have for biomineralization processes!).

On the other hand, one of the most annoying and puzzling aspects of monolayer studies claimed as biomimetic model systems has been the fact that, until very recently, none of the systems has produced crystal morphologies reminiscent of aragonitic nacre, although this was implicit in the motivation of numerous studies. As stressed in this chapter, the main reason for this discrepancy is most likely an imperfect working hypothesis regarding the biological process of nacre formation itself. In fact, in recent model systems explicitly taking into account the multi-stepped formation process of biogenic nacre – including a phase transition from an amorphous precursor to a solid – many characteristic features of nacre could be reproduced in a quite simple model system [54].

Acknowledgments

Financial support for these studies was provided by the Deutsche Forschungsgemeinschaft (DFG Priority Program 1117, *Principles of Biomineralization*, DFG grant Vo829/3), and is gratefully acknowledged.

References

1 S. Weiner, L. Addadi, *J. Mater. Chem.* **1997**, *7*, 689.

2 (a) B.A. Gotliv, L. Addadi, S. Weiner, *ChemBioChem* **2003**, *4*, 522; (b) J.L. Arias, A. Neira-Carrillo, J.I. Arias, C. Escobar, M. Bodero, M. David, M.S. Fernandez, *J. Mater. Chem.* **2004**, *14*, 2154.

3 K. Simkiss, K.M. Wilbur, in: *Biomineralization. Cell Biology and Mineral Deposition*. Academic Press, San Diego, **1989**, p. 230.

4 D. Volkmer, in: M. Cooke, C.F. Poole (Eds.), *Encyclopedia of Separation Science, Vol. 2, Crystallization*. Academic Press, **2000**, p. 940.

5 A.P. Jackson, J.F.V. Vincent, R.M. Turner, *Proc. R. Soc. London B, Biol. Sci.* **1988**, *234*, 415.

6 (a) R.Z. Wang, Z. Suo, A.G. Evans, N. Yao, I.A. Aksay, *J. Mater. Res.* **2001**, *16*, 2485; (b) A.P. Jackson, J.F.V. Vincent, R.M. Turner, *J. Mater. Sci.* **1990**, *25*, 3173; (c) J.D. Currey, *Proc. R. Soc. London B, Biol. Sci.* **1977**, *196*, 443.

7 (a) H.K. Erben, *Biomineralisation* **1974**, *7*, 14; (b) L. Addadi, D. Joester, F. Nudelman, S. Weiner, *Chem. Eur. J.* **2006**, *12*, 980; (b) A. Lin, M.A. Meyers, *Mater. Sci. Eng. A* **2005**, *390*, 27.

8 B.J.F. Bruet, H.J. Qi, M.C. Boyce, R. Panas, K. Tai, L. Frick, C. Ortiz, *J. Mater. Res.* **2005**, *20*, 2400.

9 H. Mutvei, *Zool. Scr.* **1978**, *7*, 287.

10 C. Hedegaard, H.-R. Wenk, *J. Moll. Stud.* **1998**, *64*, 133.

11 X. Li, W.-C. Chang, Y.J. Chao, R. Wang, M. Chang, *Nano Lett.* **2004**, *4*, 613.

12 T.E. Schäffer, C. Ionescu-Zanetti, R. Proksch, M. Fritz, D.A. Walters, N. Almqvist, C.M. Zaremba, A.M. Belcher, B.L. Smith, G.D. Stucky, D.E. Morse, P.K. Hansma, *Chem. Mater.* **1997**, *9*, 1731.

13 S. Manne, C.M. Zaremba, R. Giles, L. Huggins, D.A. Walters, A.M. Belcher, D.E. Morse, G.D. Stucky, J.M. Didymus, S. Mann, P.K. Hansma, *Proc. R. Soc. London B* **1994**, *256*, 17.

14 N. Watabe, *J. Ultrastruct. Res.* **1965**, *12*, 351.

15 (a) A.M. Belcher, X.H. Wu, R.J. Christensen, P.K. Hansma, G.D. Stucky, D.E. Morse, *Nature* **1996**, *381*, 56; (b) G. Falini, S. Albeck, S. Weiner, L. Addadi, *Science* **1996**, *271*, 67.

16 X. Schen, A.M. Belcher, P.K. Hansma, G.D. Stucky, D.E. Morse, *J. Biol. Chem.* **1997**, *272*, 32472.

17 B.A. Wustman, D.E. Morse, J.S. Evans, *Langmuir* **2002**, *18*, 9901.

18 B.L. Smith, T.E. Schäffer, M. Viani, J.B. Thompson, N.A. Frederick, J. Kindt, A.M. Belcher, G.D. Stucky, D.E. Morse, P.K. Hansma, *Nature* **1999**, *399*, 761.

19 F. Marin, G. Luquet, *C. R. Palevol.* **2004**, *3*, 469.

20 T. Samata, N. Hayashi, M. Kono, K. Hasegawa, C. Horita, S. Akera, *FEBS Lett.* **1999**, *462*, 225.

21 (a) L. Addadi, S. Weiner, in: S. Mann, J. Webb, R.J.P. Williams (Eds.), *Biomineralization*. VCH, Weinheim, **1989**, p. 133; (b) L. Addadi, J. Moradian, E. Shay, N.G. Maroudas, S. Weiner, *Proc. Natl. Acad. Sci. USA* **1987**, *84*, 2732; (c) Y. Levi, S. Albeck, A. Brack, S. Weiner, L. Addadi, *Chem. Eur. J.* **1998**, *4*, 389.

22 *β*-chitin is a water-insoluble $(1 \rightarrow 4)$-linked 2-acetamido-2-deoxy *β*-D-glucan: I.M. Weiss, C. Renner, M.G. Strigl, M. Fritz, *Chem. Mater.* **2002**, *14*, 3252.

23 Y. Levi-Kalisman, G. Falini, L. Addadi, S. Weiner, *J. Struct. Biol.* **2001**, *135*, 8.

24 M. Bertrand, A. Brack, *Origin Life Evol. B* **1997**, *27*, 589.

25 (a) Y. Yeh, R.E. Feeney, *Chem. Rev.* **1996**, *96*, 601; (b) P.L. Davies, B.D. Sykes, *Curr. Opin. Struct. Biol.* **1997**, *7*, 828.

26 J.S. Evans, *Curr. Opin. Coll. Int. Sci.* **2003**, *8*, 48.

27 F.C. Meldrum, *Int. Mater. Rev.* **2003**, *48*, 187.

28 S. Mann, B.R. Heywood, S. Rajam, J.D. Birchall, *Nature* **1998**, *334*, 692.

29 (a) V.M. Kaganer, H. Möhwald, P. Dutta, *Rev. Mod. Phys.* **1999**, *71*, 779; (b) P. Dynarowicz-Latka, A. Dhanabalan, O.N. Oliveira, Jr., *Adv. Colloid. Interface Sci.* **2001**, *91*, 221.

30 (a) E.M. Landau, M. Levanon, L. Leiserowitz, M. Lahav, J. Sagiv, *Nature* **1985**, *318*, 353; (b) I. Weissbuch, M. Lahav, L. Leiserowitz, *Cryst. Growth Des.* **2003**, *3*, 125.

31 Y. Zhang, R. Jin, L. Zhang, M. Liu, *New J. Chem.* **2004**, *28*, 614.

32 H.-R. Wenk, P. Van Houtte, *Rep. Prog. Phys.* **2004**, *67*, 1367.

33 (a) Y. Kitano, D.W. Hood, K. Park, *J. Geophys. Res.* **1962**, *67*, 4873; (b) O. Sohnel, J.W. Mullin, *J. Cryst. Growth* **1982**, *60*, 239; (c) M. Kitamura, H. Konno, A. Yasui, H. Masuoka, *J. Cryst. Growth* **2002**, *236*, 323.

34 M. Ficke, D. Volkmer, *Top. Curr. Chem.* **2007**, *270*, 1.

35 D. Volkmer, M. Fricke, D. Vollhardt, S. Siegel, *J. Chem. Soc. Dalton Trans.* **2002**, 4547.

36 D. Volkmer, M. Fricke, *Z. Anorg. Allg. Chem.* **2003**, *629*, 2381.

37 D. Volkmer, M. Fricke, C. Agena, J. Mattay, *Cryst. Eng. Commun.* **2002**, *4*, 288.

38 D. Volkmer, M. Fricke, M. Gleiche, L. Chi, *Mater. Sci. Eng. C* **2005**, *2*, 161.

39 (a) M.J. Lochhead, S.R. Letellier, V. Vogel, *J. Phys. Chem. B* **1997**, *101*, 10821; (b) D.M. Taylor, *Adv. Colloid. Interface Sci.* **2000**, *87*, 183; (c) H.

Brockman, *Chem. Phys. Lipids* **1994**, *73*, 57.

40 A. Berman, D.J. Ahn, A. Lio, M. Salmeron, A. Reichert, D. Charych, *Science* **1995**, *269*, 515.

41 S. Champ, J.A. Dickinson, P.S. Fallon, B.R. Heywood, M. Mascal, *Angew. Chem. Int. Ed.* **2000**, *39*, 2716.

42 D. Volkmer, M. Fricke, T. Huber, N. Sewald, *Chem. Commun.* **2004**, 1872.

43 D. Volkmer, M. Fricke, C. Agena, J. Mattay, *J. Mater. Chem.* **2004**, *14*, 2249.

44 M. Fricke, D. Volkmer, C.E. Krill, III, M. Kellermann, A. Hirsch, *Cryst. Growth Des.* **2006**, *6*, 1120.

45 (a) L. Addadi, S. Raz, S. Weiner, *Adv. Mater.* **2003**, *15*, 959; (b) M.J. Olszta, D.J. Odom, E.P. Douglas, L.B. Gower, *Connect. Tiss. Res.* **2003**, *44* (Suppl. 1), 326.

46 (a) I.M. Weiss, N. Tuross, L. Addadi, S. Weiner, *J. Exp. Zool.* **2002**, *293*, 478; (b) Y. Politi, T. Arad, E. Klein, S. Weiner, L. Addadi, *Science* **2004**, *306*, 1161.

47 E. DiMasi, V.M. Patel, M. Sivakumar, M.J. Olszta, Y.P. Yang, L.B. Gower, *Langmuir* **2002**, *18*, 8902.

48 G.F. Xu, N. Yao, I.A. Aksay, T.J. Groves, *J. Am. Chem. Soc.* **1998**, *120*, 11977.

49 F.F. Amos, D.M. Sharbaugh, D.R. Talham, L.B. Gower, M. Fricke, D. Volkmer, *Langmuir* **2007**, *23*, 1988.

50 N. Nassif, N. Pinna, N. Gehrke, M. Antonietti, C. Jäger, H. Cölfen, *Proc. Natl. Acad. Sci. USA* **2005**, *102*, 12653.

51 (a) D.M. Duffy, J.H. Harding, *Langmuir* **2004**, *20*, 7630; (b) D.M. Duffy, J.H. Harding, *Langmuir* **2004**, *20*, 7637.

52 D. Volkmer, N. Mayr, M. Fricke, *J. Chem. Soc. Dalton Trans.* **2006**, 4889–4895.

53 E. DiMasi, M.J. Oltsza, V.M. Patel, L. Gower, *Cryst. Eng. Commun.* **2003**, *5*, 346.

54 D. Volkmer, M. Harms, L. Gower, A. Ziegler, *Angew. Chem. Int. Ed.* **2005**, *44*, 639.

5
The Hierarchical Architecture of Nacre and its Mimetic Materials

Hiroaki Imai and Yuya Oaki

Abstract

The nacreous layer has been found to be composed of a three-level hierarchical architecture consisting of aragonite nano-building blocks. The nanometric crystals in submicrometric oriented platy units were bridged and covered with organic molecules. Nanostorage, an additional nanoscopic function leading to the incorporation of versatile organic dye molecules, resided in the aragonite/biopolymer composites. Nacre-mimetic architectures in terms of the oriented assembly of nanocrystals with nanostorage were successfully produced through the crystal growth of alkali-earth carbonates in aqueous systems containing soluble polymeric species, such as poly(acrylic acid) (PAA) and silicate anions. A hierarchy similar to the nacreous layer could be formed by an appropriate combination of potassium sulfate and PAA. The nanometric and submicrometric stepwise growth of inorganic crystals accompanied with specific adsorption of soluble polymeric molecules is associated with the formation of a hierarchical architecture consisting of bridged nanocrystals.

Key words: hierarchy, nanocrystal, polymer, composite, mineral bridge, carbonate, poly(acrylic acid), silicate.

5.1
Introduction

Biominerals fascinate many people because of their seemingly well-designed morphologies and hierarchical structures [1–4]. The nacreous layer (mother-of-pearl) has attracted research interest over a broad range of chemistry disciplines [5–24], especially in terms of its detailed structure [9, 10], its structure relevant to its mechanical properties [11–13], the incorporated macromolecules [14, 15], mineral bridges [16, 17], and the formation process via an amorphous phase [18, 19]. Recently, many investigations have been focused on nanoscopic structures

Handbook of Biomineralization. Edited by P. Behrens and E. Bäuerlein

ISBN: 978-3-527-31805-6

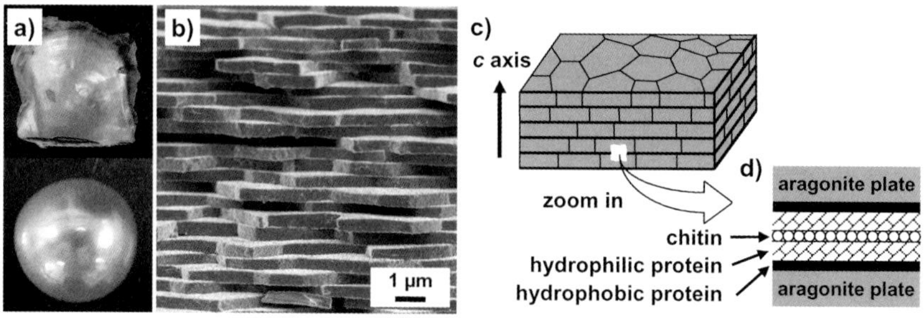

Fig. 5.1 An overview of the nacreous layer, showing (a) the pearly oyster and artificially induced pearl and (b) the nacreous layer. (c,d) A schematic illustration of the layered structure consisting of aragonite plates and organic matrices

[20–25]. As shown in Figure 5.1, the macroscopic assembly and orientation of aragonite plates of 200 to 600 nm thickness in the nacreous layer have been well investigated. The nacre of shells forms a layered composite structure consisting of flat crystals of calcium carbonate ($CaCO_3$) and biomacromolecules, such as hydrophobic proteins and chitin, which can be compared to bricks and mortar, respectively. The high mechanical strength and pearl luster of these materials originate from their layered composite structures.

It has been reported that various carbonate-based biominerals consist of a large single crystalline material [26–29]. However, as in the case of other biominerals, this raises the simple question of why a wide variety of curved and complex morphologies emerge from the same combination of carbonate crystals and biopolymers. The answer might be hidden in the nanoscopic structures in the nacreous layer and the sea urchin spine [20–25, 30, 31]. Unfortunately, in earlier studies the nanoscopic structures of biominerals were not fully understood, particularly in terms of the morphology and orientation of the nanometric building blocks. The general strategy for morphogenesis based on crystal growth may lie in the nature of the various biominerals, and an elucidation of the biological strategy for versatile morphologies is requisite in order to obtain an improved understanding of biomineralization and to further develop the topic of materials science. In the past, biominerals have also served as a sophisticated model for versatile organic-inorganic composite materials produced through self-organization under ambient conditions. An exquisite association of organics and inorganic substances is generally required for the construction of biogenetic superstructures, and many research groups have demonstrated biomimetic crystal design with the assistance of organic polymers. Although various unique structures and morphologies have been reported previously, the hierarchy associated with nanoscale structures has not been studied in detail.

In this chapter, the hierarchical architectures – including the nanometric structure of the nacreous layer and other biominerals – are described on the basis of their in-depth characterization. An understanding of the nanoscale structures of biominerals represents a significant challenge in materials chemistry for the preparation and organization of biomimetic architectures. The reproduction of a hierarchical architecture consisting of nanoscale building blocks is then illustrated by the crystal growth of an inorganic substance in artificial aqueous systems containing soluble polymeric species.

5.2 The Hierarchical Structures of the Nacreous Layers

The Japanese pearl oyster (*Pinctada fucata*) and its artificially induced pearl were characterized using a field-emission scanning electron microscope (FESEM), a field-emission transmission electron microscope (FETEM) with selected area electron diffraction (SAED), and powder X-ray diffraction (XRD). Fractured samples were used for characterization of the real nanoscopic structure and morphology in order to avoid damage and contamination during sample preparation by using a focused ion beam, ion milling, and microtome techniques.

As shown in Figure 5.2, mother-of-pearl had a hierarchical structure, with tiers 1 to 3 mediating twice the oriented assembly. As reported previously, the layered structure (tier 1, Fig. 5.2a,d) consisted of aragonite plates (tier 2, Fig. 5.2b,e) each approximately 1 to 5 μm wide and 200 to 700 nm thick. The magnified FESEM image (Fig. 5.2b) clearly indicates the presence of smaller components in each aragonite plate. An FETEM image of a nanocrystal exhibits the pseudo-hexagonal habit of aragonite (tier 3, Fig. 5.2c,f). The high-resolution image on the fringe of the nanocrystal shows a lattice spacing of 0.423 nm, corresponding to the (110) plane of aragonite (Fig. 5.2g). The appearance of nanocrystals was attributable to neither the sample preparation process nor to the radiation damage on the FETEM observation, because the preceding FESEM observation clearly showed the presence of nanocrystals in a plate (Fig. 5.2b). Through several FETEM and FESEM observations, the size of a nanocrystal was found to range between 20 and 180 nm.

The hierarchical structure was associated with the oriented architectures and mineral bridges in two different scales (Fig. 2j,k) [25]. As reported previously, XRD analysis indicated that the layers were perpendicular to the *c*-axis. Thus, the layered structure of mother-of-pearl (tier 1) was considered to be an oriented architecture of the aragonite plates (tier 2) in the *c*-axis (Fig. 5.2d), although the orientation of the *a*- and *b*-axes in a layer remains unclear. Schaffer et al. reported the presence of a mineral bridge between each aragonite plate (Fig. 5.2j), implying that the *a*- and *b*-axes should be perfectly oriented in all layers in tier 1 [16, 17]. However, Sarikaya et al. recently showed by using dark-field TEM images, that the *a*- and *b*-axes were not perfectly aligned in all of the layers [10].

An aragonite plate (tier 2) was also an oriented architecture of the nanocrystals (tier 3). A spot pattern of SAED for a submicrometric fragment (Fig. 5.2h) indicates that the inside of a platy unit was a highly aligned assembly of the nanocrystals. Aggregations of nanocrystals exhibiting a hexagonal habit were commonly

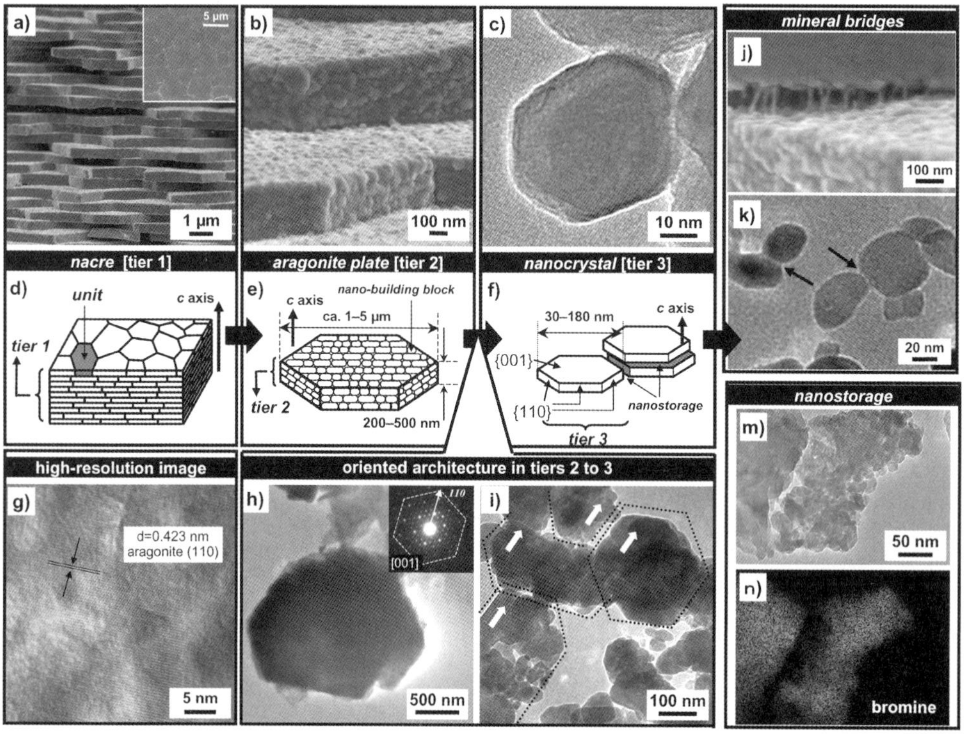

Fig. 5.2 A hierarchical architecture of the nacreous layer. (a) A layered structure (tier 1); (b) aragonite plates (tier 2); (c) a nanocrystal (tier 3); (d–f) schematic illustrations of a layered structure of platy units consisting of nanocrystals; (g) high-resolution TEM image of the nanocrystal; (h,i) the oriented architecture of the nanocrystals; (j,k) mineral bridges in submicrometric and nanometric scales; (m,n) unfiltered normal FETEM image and corresponding EF-TEM image with bromine mapping for the incorporation of dye molecules [25].

observed; these were roughly arranged in the same direction (Fig. 5.2i). The peak broadening due to the crystallite size effect was not recognized in the XRD profiles of the powdered sample. As nanoscopic mineral bridges between two adjacent nanocrystals were observed in the FETEM images (Fig. 5.2k), the nanocrystals could be connected crystallographically through the bridges. These findings support the suggestion that a platy unit (tier 2) is composed of nanocrystals (tier 3) with an oriented architecture [25]. This concept has been implied in previous reports [20, 21], but the presence of nanocrystals and their oriented architecture have been quite unclear.

Organic dye molecules can be introduced in the nanoscopic structures of the nacreous layer, whereby powdered samples of the nacreous layer are immersed in an ethanol solution of organic dyes for one day. Anionic dyes of eosin Y (EY) and rhodamine B (RB), and a hydrophobic dye of pyrene (PY), were introduced into the aragonite/biopolymer composite. The powders were then washed using ethanol and dried at room temperature, whereupon emission from the included dye molecules was observed using ultraviolet-light excitation. The aragonite/biopolymer composite incorporated not only hydrophilic molecules but also hydrophobic dyes, such as PY. Whereas the luminescent color of an aragonite/biopolymer composite was blue [32], the EY, RB, and PY including an aragonite/biopolymer produced strong orange, yellow, and blue luminescences, respectively, to the naked eye with ultraviolet excitation. The incorporation of dye molecules suggests that an additional nanoscopic function resided in the aragonite/biopolymer composite. As a model, we performed observations using FETEM with energy-filtered mapping (EF-TEM) and an electron energy loss spectrum (EELS) to investigate the dispersion and inclusion behaviors of the EY molecules. The presence of a bromine substituent in the EY molecules was clearly detected and recognized in the EF-TEM images (Fig. 5.2m,n). These images suggest that the EY molecules were dispersed homogeneously and incorporated into the aragonite/biopolymer composite on a nanoscopic scale. In the case of RB and PY, similar homogeneous incorporation would lead to a strong emission. This dye-incorporating behavior is attributed to the composite structure of nanocrystals and biopolymers. Moreover, these results imply that an emergent function – a host for organic molecules – resided in the aragonite/biopolymer composite in the nanoscopic scale.

5.3
Hierarchical Structures of Other Biominerals

Various biominerals consisting of $CaCO_3$ were analyzed, as is the case with the nacreous layers: the specimens were sponge skeletons of echinoderms, a coral, the skeletal structure of a starfish, a foraminifer, a commercial eggshell, and an eggshell of an emu. The hierarchical architectures of these biominerals are summarized in Figure 5.3 [33]; Figure 5.3a and insets illustrate the characteristic morphologies of each biomineral on a macroscopic scale. Magnified FESEM images of a fractured surface revealed the presence of nanoscopic structures (Fig. 5.3b); the nanocrystals were also recognizable on the FETEM images, indicating that the inside consisted of nanocrystal bricks (Fig. 5.3c). The magnified FETEM images showed the respective units corresponding to aragonite, calcite, and magnesium-bearing calcite (Fig. 5.3d). The sizes of the aragonite- and calcite-type nanocrystals were within a range of approximately 20 to 180 nm and 10 to 80 nm, respectively. Each nanobrick was a single crystal because the lattice fringes corresponding to each crystal phase were commonly observed in the high-resolution images (Fig. 5.3e). Thus, nanocrystals were resident in various carbonate-based biominerals, even though their morphologies and crystal structures were different.

According to a spot diffraction pattern of SAED for FETEM images (Fig. 5.4a,c), the nanocrystals were arranged in the same crystallographic direction. At a level

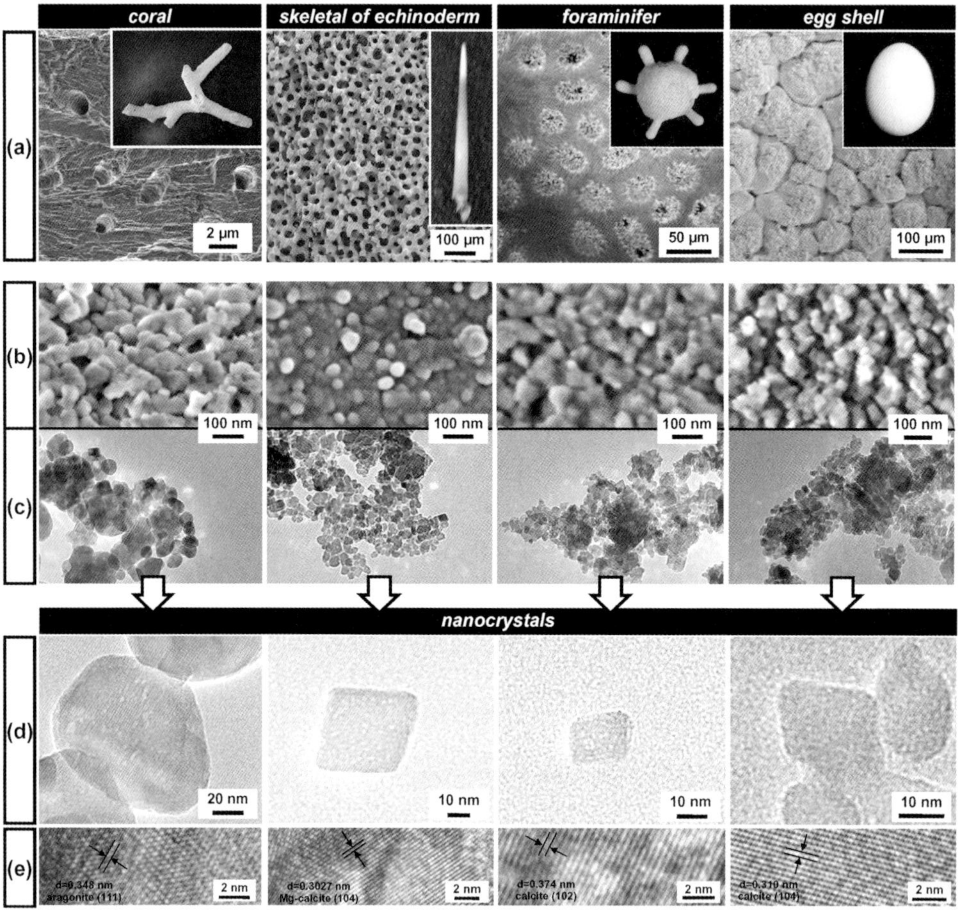

Fig. 5.3 Summarized FESEM (a,b) and FETEM (c–e) images of various biominerals. (a) The macroscopic appearance (inset) and FESEM images of the characteristic morphologies. (b) Magnified FESEM images on the fractured surface. (c) Corresponding FETEM images on the same scale as (b). (d) FETEM images of each nanocrystal exhibiting a specified facet. (e) High-resolution FETEM images of the nanocrystals [33].

of several hundred nanometers, an assembly reflecting the crystal habit of a nanocrystal was observed in many samples (Fig. 5.4b,d). Peak broadening in the XRD profile was not recognized in the powdered samples, which indicated that the oriented architecture of the nanocrystals was formed, at least, within a certain area that is several hundred nanometers in scale, regardless of the species and crystal phases. Moreover, nanoscopic mineral bridges between two adjacent nanocrystals

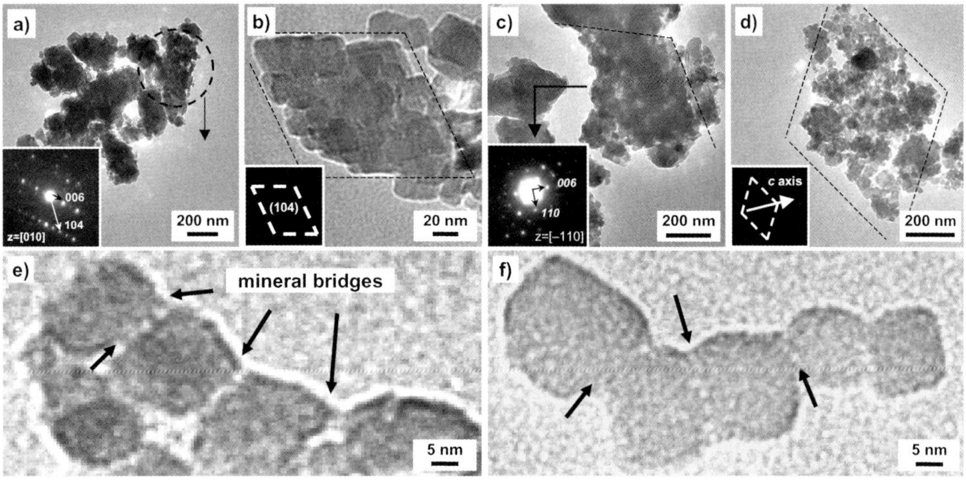

Fig. 5.4 FETEM analysis of the oriented architecture of the nanocrystals. (a,c) FETEM images of the oriented architecture showing the spotted SAED pattern (insets) in sea urchin spine and eggshell, respectively. (b,d) Corresponding FETEM images of the assembly with a similar morphology to each nanocrystal. (e,f) FETEM images of the mineral bridge in sea urchin spine and eggshell, respectively [33].

were acquired (Fig. 5.4e,f), and therefore the nanocrystals were connected with one or more mineral bridges. The inclusion of dye molecules was also demonstrated in a specimen of sea urchin spine, as was the case with the nacreous layer.

These results indicate that typical carbonate-based biominerals have a hierarchical architecture that ranges from the nanoscopic to the macroscopic scale [33]. A schematic representation of an oriented architecture made by bridged nanocrystals and biopolymers is illustrated in Figure 5.5. As described above, a macroscopic mineral bridge was proposed in the interlayer of mother-of-pearl [17]. As the daughter aragonite plate is directed towards the mother plate through the mineral bridge, an orientation of the *c*-axis perpendicular to the plates is inherited at a late stage of growth. This idea can be applied to the nanoscopic structures of various biominerals. An oriented architecture consisting of bridged nanocrystals with the incorporation of biopolymers is represented in Figure 5.5a, where nanoscopic mineral bridges are seen to direct the orientation of nanocrystals.

Nanostorage, an additional nanoscopic space for organic molecules, can be formed in the architecture made by the bridged nanocrystals and polymers. Various organic molecules were incorporated into the nanostorage, regardless of the hydrophilic or hydrophobic dyes (Fig. 5.5b). A similar strategy for crystal growth lies in the nanoscopic architectures of various biominerals, even though the specific macroscopic shapes are genetically controlled.

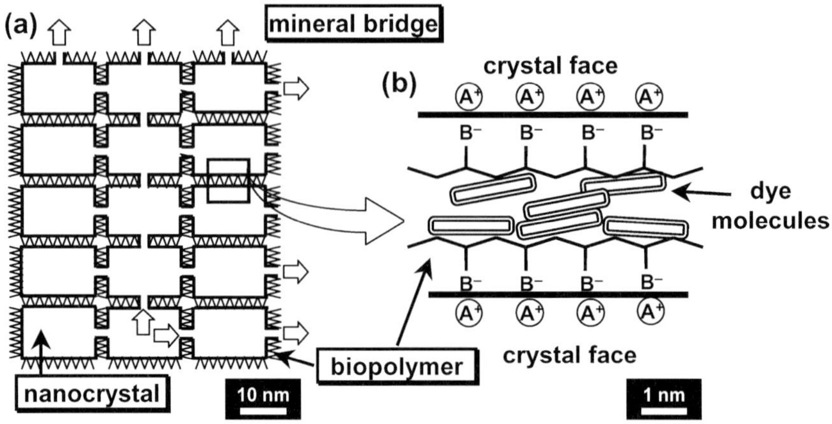

Fig. 5.5 Schematic illustrations of the oriented architecture made by bridged nanocrystals (a) and the nanoscopic space for dye molecules in the architecture consisting of the nanocrystals and organic polymers (nanostorage) (b).

5.4
Nacre-Mimetic $CaCO_3$ with Organic Polymers

5.4.1
Strategy for the Synthesis of $CaCO_3$ Planar Films with Soluble Agents and Insoluble Matrices

In the mineralization processes that take place in living organisms, the cooperation of soluble proteins, insoluble hydrophobic proteins, and polysaccharides is important [1–3, 5, 6]. The formation of the nacreous layer occurs through the selective crystallization of aragonite plates inside pre-organized insoluble matrices consisting of chitin/silk-fibroin-like proteins with soluble proteins (see Fig. 5.1). With a view to mimicking biomineralization, many approaches for the preparation of thin films of $CaCO_3$ have been reported using soluble agents and insoluble matrices (Fig. 5.6a). Inorganic crystals are formed on an insoluble matrix as a platform for crystallization under the control of soluble agents. Nucleation can be controlled on the surface of solid matrices, such as Langmuir monolayers [34], insoluble polymers, and self-assembled monolayers (SAMs), which have regular arrays of functional groups. In particular, well-controlled thin films of $CaCO_3$ were successfully obtained by the combination of soluble acidic polymers and insoluble macromolecules (Fig. 5.6b) [35]. Chitin, chitosan, and cellulose were chosen as solid insoluble matrices in the presence of soluble polymeric additives, such as poly(acrylic acid) (PAA), poly(glutamic acid), and poly(aspartic acid). A planar morphology could be achieved by macroscopic two-dimensional (2-D) growth, which was promoted by the capping effect of the soluble agents and the template effect of the insoluble substrate (Fig. 5.6a). However, hierarchical archi-

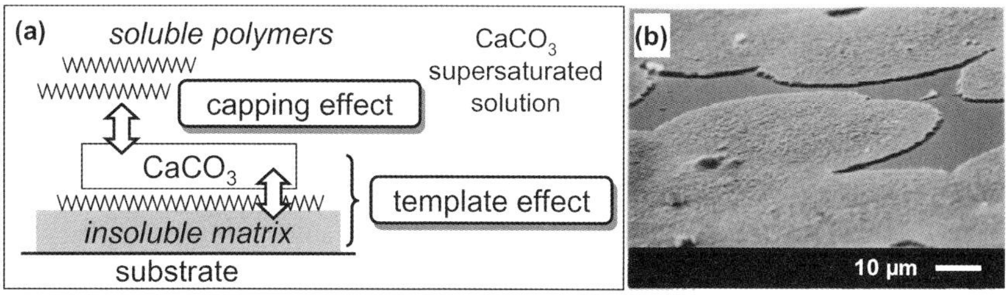

Fig. 5.6 Preparation of nacre-mimetic $CaCO_3$ films. (a) A general concept with combination of insoluble and soluble polymers. (b) Pioneer study of nacre-mimetic thin films. Figure reproduced from Ref. [35a], with permission from the Chemical Society of Japan.)

tectures including nanoscopic structures similar to the natural nacreous layer have not been recognized in these systems. The following section focuses on the nanoscopic structures in biomimetically grown alkali-earth carbonate crystals.

5.4.2
Reproduction of Bridged Nanocrystals with Biogenic Agents

It has been reported that organic molecules extracted from biominerals induced the macroscopic morphological variation of inorganic crystals [28, 36–38]. Recently, retrosynthesis of the nacreous layer was also successfully achieved with a combination of biological and synthetic polymers [38]. Although the macroscopic morphologies were well observed in these reports, the structures in nanoscale were not fully investigated. The reproduction of nanoscopic objects in sea urchin spines is described in the following section [33].

Biopolymers that are rich in water-soluble proteins consisting mainly of aspartic acid and glutamic acid interact strongly with nanocrystals in the oriented architecture of various biominerals [29]. A clear solution containing calcium ions and biopolymers was obtained by dissolution of a powdered sample of a sea urchin spine (*Echinometra mathaei* (*Blainville*)) in hydrochloric acid. A white precipitate of $CaCO_3$ was observed in the solution following the introduction of carbon dioxide generated by the decomposition of fresh ammonium carbonate at 25 °C. A hopper-shaped calcite with a rhombohedral habit (Fig. 5.7a, inset) and disk-like vaterite were contained in the precipitation. The magnified FESEM images showed that the skeletal morphology consisted of nanocrystals (Fig. 5.7a,b), while a spotted SAED pattern indicated that the nanocrystals formed the oriented architecture, as observed in real biominerals (Fig. 5.7c). Each nanocrystal with a size of 20 to 50 nm was deduced to be a single calcite crystal because lattice spacing corresponding to calcite crystal was observed (Fig. 5.7d,e). These results indicated that the nanoscopic structure reproduced was the oriented architecture composed of bridged nanocrystals, as is the case with real biominerals. However, the characteristic shapes of the spine on the macroscopic scale, such as sponge

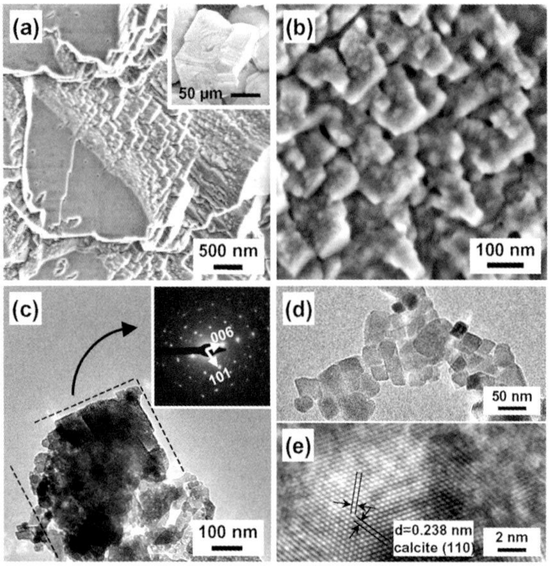

Fig. 5.7 Reproduction of the sea urchin spine. (a,b) FESEM images of the macroscopic and nanoscopic structure; (c) FETEM and the corresponding SAED pattern (inset); (d) FETEM image of the bridged nanocrystals (ca. 20 nm in size); (e) HRTEM image showing the nanocrystal to be a single calcite crystal [33].

and acute morphology, were not reproduced because genetic factors and/or genetically induced insoluble organic components were not included in this experimental system.

5.4.3
Synthesis of Planar Films Consisting of Bridged Nanocrystals with Synthetic Polymeric Agents

As mentioned above, $CaCO_3$ thin films were crystallized from a supersaturated aqueous solution containing a soluble acidic polymer, such as PAA, on a specific solid matrix, such as chitosan. However, the presence of a high-molecular-weight PAA (MW 90 000 or 250 000) resulted in the formation of $CaCO_3$ films even on a bare glass substrate without a coating of insoluble organic species (Fig. 5.8a) [33, 39]. As the resultant films consisted of nanocrystals of 5 to 10 nm in size (Fig. 5.8b–e) and contained 3–4 wt% of organic compounds, nanometric organic/inorganic composites were constructed on a platform. According to XRD analysis and HRTEM images (Fig. 5.8e), the products were assigned to calcite, the *c*-axis of which was perpendicular to the substrate. The formation of an oriented architecture of bridged nanocrystals similar to biominerals was suggested by the FETEM images and spotted SAED patterns (Fig. 5.8c,d). The specific adsorption

Fig. 5.8 Synthetic $CaCO_3$/PAA thin films. (a) FESEM images of a thin film; (b) magnified FESEM image; (c) FETEM and the corresponding SAED pattern (inset); (d) FESEM image of the bridged nanocrystals; (e) HRTEM image of the lattice fringes corresponded to calcite; (f,g) lozenge shaped films with combination of low- and high-molecular-weight PAA and its the cross-sectional view [33].

of PAA chains suppressed the regular growth of calcite crystals, and the morphology then changed from polyhedral to a bridged architecture of miniaturized units (Fig. 5.8b,d). Two-dimensional growth of the nanocrystals could be induced on a glass substrate covered with a high-molecular-weight PAA.

A combination of low-molecular-weight (MW 2000) and high-molecular-weight (MW 250 000) PAAs produced lozenge-shaped calcite films with definite edges on a glass substrate (Fig. 5.8f,g) [39]. Most of the parallelograms showed angles of 104° and 120°, corresponding to the shape of the {10–14} and {0001} faces, respectively. On the other hand, nanometric crystal grains were observed on the surface of the faceted films. These findings indicate that the nanocrystals were arranged with a specific crystallographic orientation in the lozenge-shaped films. The hierarchical morphology of the faceted plates is similar to that of a platy unit in the nacreous layer, even though the polymorph of the artificial films was different from that of the natural crystals. The macroscopic habit was clearly observed with an increase in the size of the nanograins with the addition of a low-molecular-weight PAA. Thus, the suppression of the crystal growth was weakened with a decrease in the average molecular weight of PAA.

5.5
Nacre-Mimetic Aragonite-Type Carbonate Crystals with Organic and Inorganic Polymeric Agents

Calcite and vaterite are commonly produced in artificial aqueous systems, although the nacreous layer is composed of aragonite nanocrystals. Recently, the polymorph control of calcium carbonate has been investigated using an organic template [40–42]. The nacre-mimetic structures of aragonite-type barium and strontium carbonates are described in the following section [43]. Planar films of strontium carbonate were prepared on an insoluble chitosan matrix in a supersaturated solution containing PAA (MW 90 000) (Fig. 5.9a). As the nanoscale building blocks were observed in the planar films, a hierarchical architecture was achieved with the aragonite-type crystals. The presence of silicate anions also induced the formation of planar crystals of aragonite-type carbonates on the insoluble polyalcohol substrate [43]. Here, the silicate-mediated carbonate films were mainly shown because the nanostructure of the composite films was clearly characterized.

Barium and strontium carbonates were deposited by the introduction of carbon dioxide into barium and strontium chloride aqueous solutions, respectively. Silicate anions were provided from silica gel placed in the chloride solutions at pH 10.5. In these solutions, the formation of planar crystals of the carbonates was observed on a chitosan substrate (Fig. 5.9b). A laminated architecture was constructed by stacking the units in which the *c*-axis was arranged perpendicularly to the surface (Fig. 5.9c). At the growing edge of the planar crystals, small grains of less than 100 nm diameter were seen to spread over the surface of the substrate, and the presence of a silica "skin" was confirmed after dissolution of the carbonate crystals in an EDTA solution. These *c*-axis-arranged aragonite layers covered with a skin are similar to the nacre in shells. The specific adsorption of silicate anions, and their subsequent polymerization on the inorganic crystals,

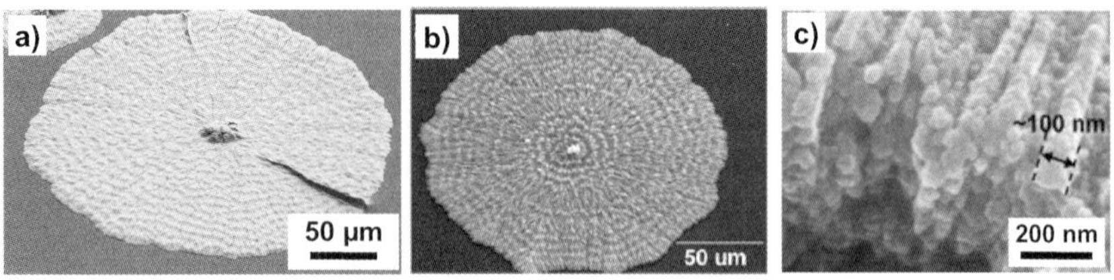

Fig. 5.9 Formation of planar aragonite crystals. (a) Strontium carbonate with PAA on a chitosan substrate; (b) barium carbonate with silicate anions on a chitosan substrate; (c) magnified image of the barium carbonate film [43].

produced the nanostructure of the composite material. The planar morphology could be formed through 2-D growth with oriented hexagonal units covered with a silica skin (Fig. 5.9c). The effects of the inorganic soluble species were found to be almost the same as that of organic soluble polymers, such as PAA.

5.6 Nacre-Mimetic Hierarchical Structure of Potassium Sulfate and PAA

Whereas nacre-mimetic platy morphologies consisting of oriented nanocrystals were successfully reproduced by the combination of alkaline-earth carbonates and various soluble agents, macroscopic layered structures could not be realized. In the following section, the details are provided of a hierarchically organized architecture (Fig. 5.10) similar to the natural nacreous layer emerging from potassium sulfate (K_2SO_4) in the presence of PAA [25, 44]. Although the formation of mother-of-pearl requires several macromolecules, the two roles of PAA comprise its mimetic structure.

Formation of the K_2SO_4/PAA composite proceeded as water evaporated from the precursor solution, thereby dissolving a certain amount of K_2SO_4 and PAA (MW 250 000) over several days. The layered morphology in tier 1 was composed of platy units that were approximately 0.5 to 1.0 μm thick (tier 2) (Fig. 5.10a,d). An oriented assembly of the units in the *b*-axis led to the layered structures, while nanocrystals of 20 nm diameter comprised the platy unit (tiers 2 and 3) (Fig. 5.10b,e). The oriented assembly of the nanocrystals was suggested by a spot pattern of the corresponding SAED (Fig. 5.10c,f).

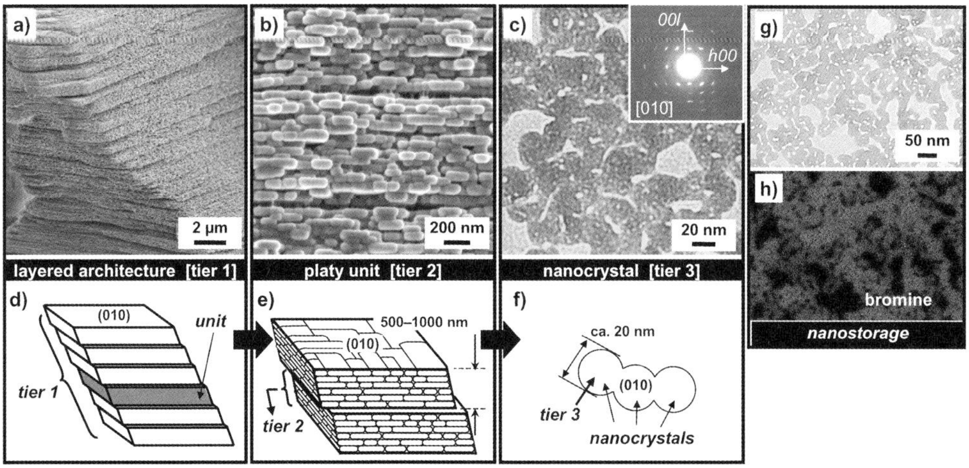

Fig. 5.10 A hierarchical structure of K_2SO_4/PAA composite. (a) Layered morphology (tier 1); (b) a platy unit (tier 2); (c) nanocrystals with an SAED pattern (tier 3); (d–f) schematic model; (g,h) FESEM and FETEM images, respectively, for the incorporation of dye molecules [44a].

Dye molecules were initially introduced into the K_2SO_4/PAA composite by the dissolution of EY and RB in the precursor solution. The incorporation of dye molecules was also achieved by subsequent immersion of the resultant K_2SO_4/PAA composite in an ethanol solution of EY, RB, or PY. Homogeneous incorporation of EY molecules on a nanoscopic scale was confirmed by the FETEM and corresponding EF-TEM images (Fig. 5.10g,h). Thus, nanostorage for organic molecules was seen to reside in the K_2SO_4/PAA composite, as is the case with the nacreous layer. Similar hierarchical structures were obtained in the PAA-mediated mineralization of potassium hydrogen phthalate instead of K_2SO_4 [45].

5.7
Self-Organization of Nacre-Mimetic Crystal Growth

5.7.1
Bridged Nanocrystals Leading to an Oriented Architecture

Oriented architectures consisting of bridged nanocrystals were found both in natural biominerals and their mimetic materials with the association of polymeric species. The hierarchical architectures of these composite products can be distinguished from a perfect rigid single crystal. Macroscopic curved and complex shapes could be achieved in the morphogenesis of the architecture with oriented nanocrystals. The morphological design can be compared to a "nanoscale Lego" construction, in that it has no restrictions [31, 33].

Recent reports have suggested the existence of various types of oriented architecture, such as oriented attachment and a mesocrystal (Fig. 5.11) [46–52]. These ideas are based on the subsequent assembly of as-grown crystalline particles. Oriented attachment was observed under hydrothermal conditions, with the perfect crystallographic connection of metal oxide nanoparticles forming quasi-single

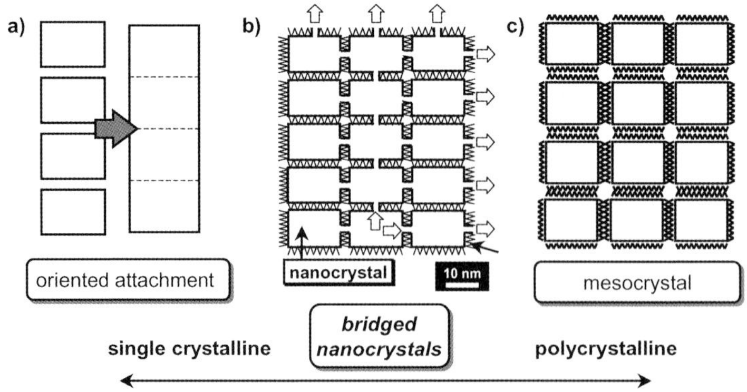

Fig. 5.11 Schematic representation of various types of oriented architecture [31, 33, 48, 51].

crystal structures [46–48]. Antonietti, Cölfen, and Niederberger have recently proposed a non-classical crystallization theory whereby the concept of a mesocrystal could be interpreted as the coupled synthesis and assembly of modular crystals stabilized with organic molecules [48, 49, 51]. Fusion results in the formation of iso-oriented crystals, and a single crystal is eventually formed via oriented attachment. The size of each nanocrystal can be determined by nucleation and subsequent growth processes. However, the orientation of as-grown nanoparticles remains an unclear issue in this model because the nanocrystals are completely isolated with organic molecules.

Although the nacre-mimetic structures described here may be similar to a mesocrystal, bridged nanocrystals are characterized as an intermediate structure between oriented attachment and a mesocrystal (Fig. 5.11). It is inferred that the oriented architectures were generated from the growth of bridged nanocrystals with the incorporation of polymers [33]. The formation process can be ascribed to crystal growth distinguished from the aggregation of nanoparticles, and the bridges would play an important role in the sequential growth of the oriented assembly. A nanocrystal of a specific size would be formed during the initial stage (Fig. 5.12a), after which the specific adsorption of polymeric species having an

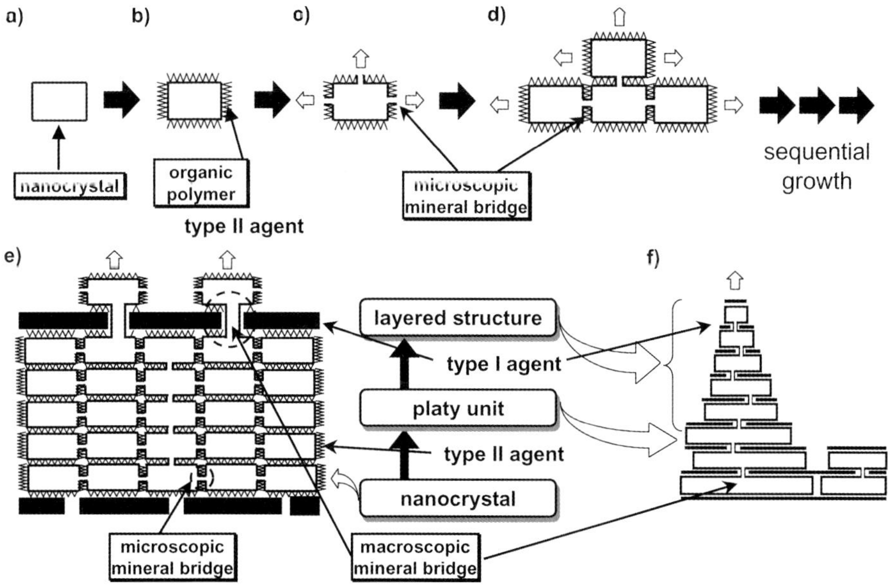

Fig. 5.12 Schematic representation for the growth of bridged nanocrystals, leading to the hierarchical oriented architecture. (a) Formation of a nanocrystal; (b) growth inhibition with adsorption of a type II agent; (c) growth resumption with microscopic mineral bridges; (d) growth of adjacent nanocrystals; (e) formation of a platy unit with adsorption of a type I agent; (f) formation of a layered structure with microscopic mineral bridges [31, 33].

affinity to the crystal surface would result in an inhibition of crystal growth (Fig. 5.12b). Although the crystals are miniaturized by the capping effect, growth restarts via the mineral bridge from defective sites of the capping layer under a supersaturated condition (Fig. 5.12c). In this way, the repetition of inhibition and resumption leads to stepwise crystal growth, and finally produces an oriented architecture of nanocrystals with the incorporation of polymers (Fig. 5.12d). In the artificial system, the grain size composing the bridged structure decreased with an increase in the molecular weight of the PAA used. Thus, the size of nanocrystals is dependent on the balance between the driving force of the crystallization and the adsorbability of the polymeric agent. Nanostorage as a host for organic molecules is generated by electrostatic interaction between the nanocrystals and the organic capping agent, with the carbon main chains providing a versatile nanoscopic environment for the dye molecules (see Fig. 5.5b).

Recently, the presence of a transient amorphous phase of calcium carbonate (ACC) has been suggested [52–55]. This stepwise growth model can be applied to crystallization through the transient ACC phase, and in this case an oriented architecture of bridged nanocrystals is produced by sequential growth with dissolution of the metastable phase.

5.7.2
Formation of Hierarchical Architectures

In biomineralization, several types of organic molecules are involved in the construction of inorganic superstructures under ambient conditions. Insoluble organic matrices play an important role in the determination of the mineral phase and the direction of macroscopic morphogenesis. Thus, the combination of soluble and insoluble biopolymers induces hierarchical architectures including oriented nanocrystals and nanostorage. The generation of the nacreous layer requires at least two types of agents (I and II) in the mineralization process (Fig. 5.12e). As chitin and hydrophobic proteins (type I) form the layered sheets, an aragonite plate eventually forms in the compartment. The next layer is then induced through the formation of a macroscopic mineral bridge from the small pore on the sheets (Fig. 5.12e,f) [16, 17]. Soluble acid proteins (type II) that consist of amino acids having carboxy and hydroxy groups interact strongly with the carbonate crystal and then produce nanocrystals with a controlled polymorphism and orientation.

Nacre-mimetic composite structures including $CaCO_3$/PAA, $BaCO_3$/silicate, and K_2SO_4/PAA were achieved in a simple system containing a single soluble species. Strongly interacting polymeric molecules (type II) formed the nanocrystals and directed the oriented architectures through stepwise growth. Furthermore, an excess amount of polymeric molecules concurrently limited the diffusion of inorganic ions and inhibited the consecutive growth; in this way, a platy unit was eventually formed. A planar morphology of $CaCO_3$/PAA and $BaCO_3$/silicate composites was formed through the 2-D growth of bridged nanocrystals guided by a solid platform as a template. In this case, the excess soluble species suppressed 3-D growth as a type I agent. Although the presence of polyalcoholic molecules [e.g., chitosan and poly(vinyl alcohol)] was effective in the heterogeneous nucleation on the platform, a high-molecular-weight PAA promoted 2-D growth even on a bare glass substrate. An oriented architecture consisting of nanocrystals could be constructed with a weak guidance on the solid matrix.

A platy shape of K_2SO_4/PAA was produced by the specific adsorption of an excess amount of the polymeric agent on the (010) face without a platform. As growth perpendicular to the (010) face could not be perfectly inhibited by the type I agent of the attached polymer, the next layer was mediated on a basal platy unit through the formation of a mineral bridge (Fig. 5.12e). In this way, a macroscopic periodic architecture was ascribed to repetition of the growth inhibition and the subsequent restart, as is the case with bridged nanocrystals (Fig. 5.12f). A diffusion field could be formed around the growing crystal because the viscosity of the precursor solution containing polymeric species increased with the evaporation of water. The presence of the diffusion field is essential for the formation of a layered architecture through macroscopically periodic construction with the repetition of inhibition with the polymer and restarting of the crystal growth.

5.8 Conclusions

The nacreous layer was found to be a biogenic hierarchical structure, from the nanoscopic to the macroscopic level. Nanocrystals constitute the oriented architecture with biopolymers through bridged growth on a mesoscopic scale, while macroscopic mineral bridges mediate the layered morphology. An exquisite association of biopolymers realizes the hierarchically organized structures under ambient conditions. Nacre-mimetic composite materials were also prepared through a biomimetic pathway in terms of the cooperation of polymers. Bioinorganic superstructure and suprabiomineral materials were generated by the multiple roles of various polymeric agents in the mineralization process. An improved understanding of real and mimetic biominerals holds promise for the further development of chemical, biological, and materials sciences.

References

1 (a) S. Mann, *Biomineralization*. Oxford University Press, Oxford, **2001**; (b) S. Mann, *Angew. Chem. Int. Ed.* **2000**, *39*, 3392–3406.

2 (a) S. Weiner, L. Addadi, *J. Mater. Chem.* **1997**, *7*, 689–702; (b) L. Addadi, S. Weiner, *Angew. Chem. Int. Ed. Engl.* **1992**, *31*, 153–169.

3 E. Bäuerlein, *Angew. Chem. Int. Ed.* **2003**, *42*, 614–641.

4 (a) J. Aizenberg, A. Tkachenko, S. Weiner, L. Addadi, G. Hendler, *Nature* **2001**, *412*, 819–822; (b) J. Aizenberg, J.C. Weaver, M.S. Thanawala, V.C. Sundar, D.E. Morse, P. Fratzl, *Science* **2005**, *309*, 275–278.

5 N. Watabe, *J. Ultrastruct. Res.* **1965**, *12*, 351.

6 T. Kato, A. Sugawara, N. Hosoda, *Adv. Mater.* **2002**, *14*, 869–877.

7 T. Kato, *Adv. Mater.* **2000**, *12*, 1543–1546.

8 P. Calvert, S. Mann, *J. Mater. Sci.* **1988**, *23*, 3801–3815.

9 M. Sarikaya, *Microsc. Res. Tech.* **1994**, *27*, 360–375.

10 E. DiMasi, M. Sarikaya, *J. Mater. Res.* **2004**, *19*, 1471–1476.

11 R.Z. Wang, Z. Suo, A.G. Evans, N. Yao, I.A. Aksay, *J. Mater. Res.* **2001**, *16*, 2485–2493.

12 A.P. Jackson, J.F.V. Vincent, R.M. Turner, *Proc. R. Soc. Lond. B* **1988**, *234*, 415–440.

13 Q.L. Feng, F.Z. Cui, G. Pu, R.Z. Wang, H.D. Li, *Mater. Sci. Eng. C* **2000**, *11*, 19–25.

14 B.A. Gotliv, L. Addadi, S. Weiner, *ChemBioChem* **2003**, *4*, 522–529.

15 Y.L. Kalisman, G. Falini, L. Addadi, S. Weiner, *J. Struct. Biol.* **2001**, *135*, 8–17.

16 L. Addadi, S. Weiner, *Nature* **1997**, *389*, 912–913.

17 T.E. Schäffer, C. Ionescu-Zanetti, R. Proksch, M. Fritz, D.A. Walters, N. Almqvist, C.M. Zaremba, A.M. Belcher, B.L. Smith, G.D. Stucky, D.E. Morse, P.K. Hansma, *Chem. Mater.* **1997**, *9*, 1731–1740.

18 L. Addadi, D. Joester, F. Nudelman, S. Weiner, *Chem. Eur. J.* **2006**, *12*, 980–987.

19 N. Nassif, N. Pinna, N. Gehrke, M. Antonietti, C. Jager, H. Cölfen, *Proc. Natl. Acad. Sci. USA* **2005**, *102*, 12653–12655.

20 X. Li, W.C. Chang, Y.J. Chao, R. Wang, M. Chang, *Nano Lett.* **2004**, *4*, 613–617.

21 K. Takahashi, H. Yamamoto, A. Onoda, M. Doi, T. Inaba, M. Chiba, A. Kobayashi, T. Taguchi, T. Okamura, N. Ueyama, *Chem. Commun.* **2004**, 996–997.

22 M. Rousseau, E. Lopez, P. Stempfle, M. Brendle, L. Franke, A. Guette, R. Naslain, X. Bourrat, *Biomaterials* **2005**, *26*, 6254–6262.

23 Y. Dauphin, *Zoology* **2006**, *109*, 85–95.

24 M. Rousseau, E. Lopez, A. Coute, G. Mascarel, D.C. Smith, R. Naslain, X. Bourrat, *J. Struct. Biol.* **2005**, *149*, 149–157.

25 Y. Oaki, H. Imai, *Angew. Chem. Int. Ed.* **2005**, *44*, 6571–6575.

26 (a) K.M. Towe, *Science* **1967**, *157*, 1048–1050; (b) J.D. Currey, D. Nichols, *Nature* **1967**, *214*, 81–83; (c) G. Donnay, D.L. Pawson, *Science* **1969**, *166*, 1147–1150; (d) H.U. Nissen, *Science* **1969**, *166*, 1150–1152.

27 A. Berman, L. Addadi, S. Weiner, *Nature* **1988**, *331*, 546–548.

28 (a) A. Berman, J. Hanson, L. Leiserowitz, T.F. Koetzle, S. Weiner, L. Addadi, *Science* **1993**, *259*, 776–779; (b) J. Aizenberg, J. Hanson, T.F. Koetzle, L. Leiserowitz, S. Weiner, L. Addadi, *Chem. Eur. J.* **1995**, *1*, 414–422.

29 (a) J. Aizenberg, J. Hanson, T.F. Koetzle, S. Weiner, L. Addadi, *J. Am. Chem. Soc.* **1997**, *119*, 881–886; (b) S. Albeck, J. Aizenberg, L. Addadi, S. Weiner, *J. Am. Chem. Soc.* **1993**, *115*, 11691–11697.

30 (a) I. Sethmann, A. Putnis, O. Grassmann, P. Löbmann, *Am. Mineral.* **2005**, *90*, 1213–1217; (b) I. Sethmann, R. Hinrichs, G. Wörheide, A. Putnis, *J. Inorg. Biochem.* **2006**, *100*, 88–96.

31 Y. Oaki, H. Imai, *Small* **2006**, *2*, 66–70.

32 (a) T. Miyoshi, Y. Matsuda, S. Akamatsu, *Jpn. J. Appl. Phys.* **1988**, *27*, 151–152; (b) T. Miyoshi, Y. Matsuda, H. Komatsu, *Jpn. J. Appl. Phys.* **1987**, *26*, 578–581; (c) T. Miyoshi, Y. Matsuda, H. Komatsu, *Jpn. J. Appl. Phys.* **1986**, *25*, 1606–1607.

33 Y. Oaki, A. Kotachi, T. Miura, H. Imai, *Adv. Funct. Mater.* **2006**, *16*, 1633–1639.

34 G. Xu, N. Yao, I.A. Aksay, J.T. Groves, *J. Am. Chem. Soc.* **1998**, *120*, 11977–11985.

35 (a) T. Kato, T. Suzuki, T. Irie, *Chem. Lett.* **2000**, 186–187; (b) T. Kato, T. Suzuki, T. Amamiya, T. Irie, M. Komiyama, H. Yui, *Supramol. Sci.* **1998**, *5*, 411–415; (c) T. Kato, T. Amamiya, *Chem. Lett.* **1999**, 199–200; (d) N. Hosoda, T. Kato, *Chem. Mater.* **2001**, *13*, 688–693.

36 G. Falini, S. Fermani, S. Vanzo, M. Miletic, G. Zaffino, *Eur. J. Inorg. Chem.* **2005**, 162–167.

37 A. Sugawara, T. Nishimura, Y. Yamamoto, H. Inoue, H. Nagasawa, T. Kato, *Angew. Chem. Int. Ed.* **2006**, *45*, 2876–2879.

38 N. Gehrke, N. Nassif, N. Pinna, M. Antonietti, H.S. Gupta, H. Cölfen, *Chem. Mater.* **2005**, *17*, 6514–6516.

39 A. Kotachi, T. Miura, H. Imai, *Chem. Mater.* **2004**, *16*, 3191–3196.

40 N. Hosoda, A. Sugawara, T. Kato, *Macromolecules* **2003**, *36*, 6449–6452.

41 P.K. Ajikumar, R. Lakshminarayanan, S. Valiyaveettil, *Cryst. Growth Des.* **2004**, *4*, 331–335.

42 A. Kotachi, H. Imai, *Cryst. Growth Des.* **2006**, *6*, 1636–1641.

43 A. Kotachi, T. Miura, H. Imai, *Cryst. Growth Des.* **2004**, *4*, 725–729.

44 (a) Y. Oaki, H. Imai, *Adv. Funct. Mater.* **2005**, *15*, 1407–1414; (b) Y. Oaki, H. Imai, *Langmuir* **2005**, *21*, 863–869.

45 Y. Oaki, H. Imai, *Chem. Commun.* **2005**, 6011–6013.

46 (a) R.L. Penn, J.F. Banfield, *Science* **1998**, *281*, 969–971; (b) R.L. Penn, J.F. Banfield, *Geochim. Cosmochim. Acta* **1999**, *63*, 1549–1557.

47 (a) C. Pacholski, A. Kornowski, H. Weller, *Angew. Chem. Int. Ed.* **2002**, *41*, 1188–1191; (b) J. Polleux, N. Pinna, M. Antonietti, C. Hess, U. Wild, R. Schlögl, M. Niederberger, *Chem. Eur. J.* **2005**, *11*, 3541–3551.

48 M. Niederberger, H. Cölfen, *Phys. Chem. Chem. Phys.* **2006**, *8*, 3271–3287.

49 H. Cölfen, M. Antonietti, *Angew. Chem. Int. Ed.* **2005**, *44*, 5576–5591.

50 H. Cölfen, S. Mann, *Angew. Chem. Int. Ed.* **2003**, *42*, 2350–2365.

51 S. Wohlrab, N. Pinna, M. Antonietti, H. Cölfen, *Chem. Eur. J.* **2005**, *11*, 2903–2913.

52 T. Wang, H. Cölfen, M. Antonietti, *J. Am. Chem. Soc.* **2005**, *127*, 3246–3247.

53 Y. Politi, T. Arad, K. Eugenia, S. Weiner, L. Addadi, *Science* **2004**, *306*, 1161–1164.

54 L. Addadi, S. Raz, S. Weiner, *Adv. Mater.* **2003**, *15*, 959–970, and references therein.

55 A.W. Xu, Q. Yu, W.F. Dong, M. Antonietti, H. Cölfen, *Adv. Mater.* **2005**, *17*, 2217–2221.

6
Avian Eggshell as a Template for Biomimetic Synthesis of New Materials

José Luis Arias, José Ignacio Arias, and María Soledad Fernandez

Abstract

Living organisms have evolved a vast diversity of mineralized structures through mechanisms that often are significantly different from those used by the materials engineer. These biomaterials often exhibit a fine-scale microstructure with variable porosity, unusual crystal habits and morphologies, and remarkable mechanical, optical, or other interesting physical properties. Among other biomineralizing systems, the avian eggshell is one of the most interesting models. Its distinctive features, relative to the characteristics of other mineralized systems, are the nature of the mineral deposit, the absence of cells intermixed with the mineralized structure, and the rapidity of the mineralization of the avian eggshell, this being the most rapid biomineralization process known. Due to an absence of cells in the eggshell, the interaction between organic matrices and inorganic crystals can be studied in this system without interference by cells that normally populate other biomineralizing systems. In addition, the different components of the eggshell can easily be separated, so that their properties can be exploited for a variety of purposes. This chapter emphasizes the use of avian eggshell components as support materials for immobilization and adsorption, as a template for crystal growth, as a reinforcement in composite materials, and in biomedical applications.

Key words: eggshell, template, membranes, composites, biomimetic, biosensors, materials.

6.1
Introduction

Living organisms have evolved a vast diversity of structures through mechanisms that often are significantly different from those used by the materials engineer. For example, mineralized tissues, such as eggshells and seashells, spicules and

Handbook of Biomineralization. Edited by P. Behrens and E. Bäuerlein

ISBN: 978-3-527-31805-6

spines, exoskeletons, pearl and coral, and bones and teeth, among others, are naturally fabricated bioceramic–biopolymer composites that are assembled from readily available materials, usually in aqueous media, at ambient conditions, and in specific shapes [1]. These biomaterials often exhibit a fine-scale microstructure with variable porosity, unusual crystal habits and morphologies, and remarkable mechanical, optical, or other interesting physical properties. Natural bioceramics are normally produced very slowly and present a limited range of compositions, dominated by calcium carbonate or phosphate, silica, and iron oxide; nevertheless, more than 60 biominerals are now known [2], and more have been discovered recently, such as copper- or zinc-based structures in marine worms [3]. The processing strategies used by biological systems should be mimicked to fabricate materials that could provide novel physical properties not currently available by conventional technologies [1]. Such biological concepts, mechanisms and design features have inspired numerous investigations during the past two decades, when a growing number of interdisciplinary approaches have emerged at the frontier of biology, physics, chemistry, and materials science [1, 4].

Biocomposites are produced via cell-mediated bottom-up processes, and their production involves an exquisite level of control both of the spatial and temporal regulation of the nucleation and growth of mineral, and of the development of microarchitecture [1]. An analysis of a variety of biomineralizing systems has led to general principles that have important implications for materials science [1]:

- That biomineralization occurs within specific confined microenvironments, which implies crystal production at certain functional sites and inhibition or prevention of the process at other sites.
- That a specific mineral is produced with a defined crystal size and orientation.
- That macroscopic growth is accomplished by packing many incremental units together; this results in unique composites with layered microarchitectures that impart exceptional material properties.

The fabrication of these structures consists of a precise spatiotemporal arrangement of sequentially deposited macromolecules and ions in a four-step mechanism [5]:

1. The fabrication of an inert laminar substrate or framework, which compartmentalizes the microenvironment where mineralization will take place.
2. The fabrication of specific macromolecules, which are deposited on the previously formed inert scaffolding, and where nucleation of the first inorganic crystals takes place.
3. The fabrication of a gel structure where polymorphism, diffusion-controlled growth, face-growing rates, and crystal habit are controlled.

4. The arrest of crystal formation both by a decrease in ionic availability and by the fabrication of a new inert scaffolding or the deposition of specific inhibitory macromolecules.

Among other biomineralizing systems, the avian eggshell is one of the most interesting. Its distinctive features as compared with other systems are the nature of the mineral deposit, the absence of cells intermixed with the mineralized structure, and the rapidity of mineralization, this being the most rapid biomineralization process known to mankind. Due to an absence of cells in the eggshell, the interaction between organic matrices and inorganic crystals can be studied in this system without interference by the cells that normally populate other biomineralizing systems. In addition, the different components of the eggshell can easily be separated, and the properties of each component exploited for a variety of purposes. Another major incentive to use eggshells is the huge amount of eggshell waste material which is produced by egg-processing plants worldwide and which, eventually, is deposited in dump sites. This chapter emphasizes the main materials and devices that have been inspired by, or derived from, the avian eggshell, other than food supplements.

6.2 Eggshell Organization and General Composition

This subject is treated extensively elsewhere (for a review, see [6]), and will be dealt with only briefly at this point. The avian eggshell is a highly ordered structure composed of multiple layers of fibrillar membranes and calcified matrix (Fig. 6.1). The innermost layer of the eggshell is composed of two non-mineralized fibrillar sublayers referred to as the inner and outer *shell membranes*. The shell membranes adhere to each other around almost the entire inner surface of the

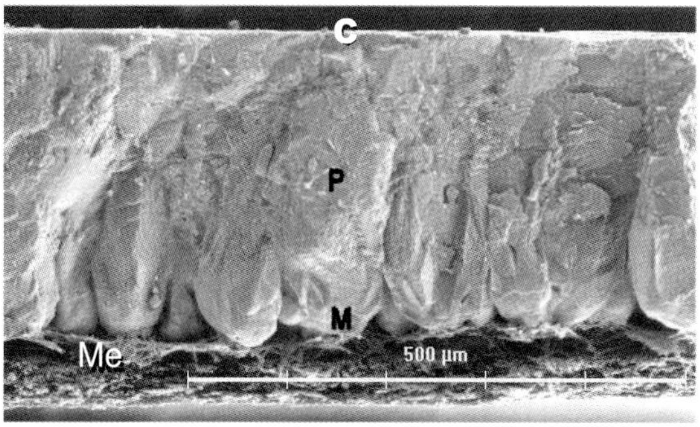

Fig. 6.1 Scanning electron micrograph of eggshell, transverse section. Me = eggshell membranes; M = mammilla; P = palisade; C = cuticle.

shell, but are separated at the obtuse side of the egg, thus defining the boundaries of the air chamber. Fibers in both sublayers have a central core with a smooth outer sheath, and their main composition is a highly cross-linked type X collagen. On the outer side of the shell membranes, discrete aggregations of organic matter, called *mammillae*, are formed. These structures are nucleation sites where the initiation of calcium carbonate (calcite) crystal formation occurs. Subsequently, the growth of calcite columns takes place to form the calcified layer, known as the eggshell proper or *palisades*. This crystal growth is controlled not only by physical mechanisms but also by a spatiotemporally regulated secretion of numerous macromolecules during eggshell formation. As a biological material, the eggshell possesses excellent gas and water permeability, which is essential for proper development of the avian embryo. As materials, both the shell membranes and the palisades are slightly porous, and the shell membranes act as a semi-permeable membrane [7].

6.3 The Eggshell Membrane as an Immobilization Support and Adsorbent

Eggshell membranes have been employed as substrates for the immobilization of enzymes, where their lifetimes are much extended [8]. With this information, Choi et al. [9] produced an optical glucose biosensor based on glucose oxidase immobilized on eggshell membranes. The system displayed good stability and high sensitivity to glucose, and subsequently several other enzyme-immobilized eggshell membrane biosensors have been successfully fabricated, either by glutaraldehyde cross-linking of the enzyme to the eggshell membrane matrix [10] or by using chitosan as an intermediary [11].

Eggshell membranes have been shown capable of binding various metal ions from aqueous solutions [12]. Eggshell membranes adsorb different types of metal ions, the affinity of adsorption being of the order $Au > Ag > Co > Cu > Pb > Ni > Zn$ [13]. In addition, eggshell membranes can remove selenium and, less effectively, arsenic ions from aqueous solutions [14]. Although the specific mechanisms of sorption are not known, an electrostatic interaction with the positively charged eggshell membrane ligands has been postulated [13].

6.4 The Eggshell Membrane or Matrix as a Template for Crystal Growth

Under physiological conditions, eggshell membranes do not mineralize but rather support mineralization that occurs on the mammillae (Fig. 6.2) [6, 15]. Eggshell membranes can be experimentally mineralized with calcium phosphate crystals after enzymatic treatment, however (Fig. 6.3) [16]. Self-assembled hydroxyapatite nanoribbon spherulites have also been successfully synthesized on eggshell membranes in the presence of ethylenediamine, under mild conditions [17].

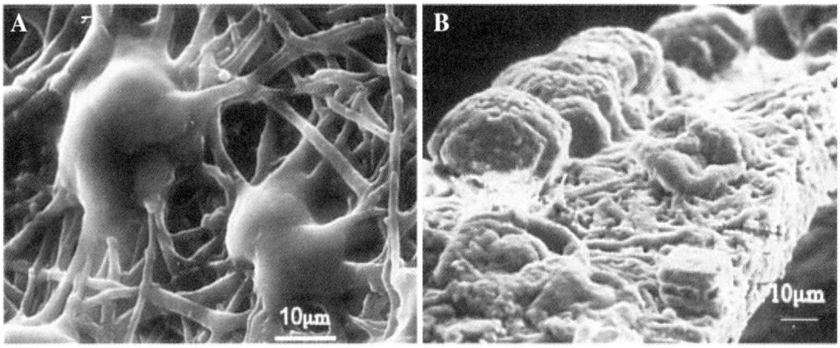

Fig. 6.2 Scanning electron micrographs of mammillae during eggshell formation. (A) Two mammillae before mineralization. (B) Columns of calcium carbonate grown on each mammilla.

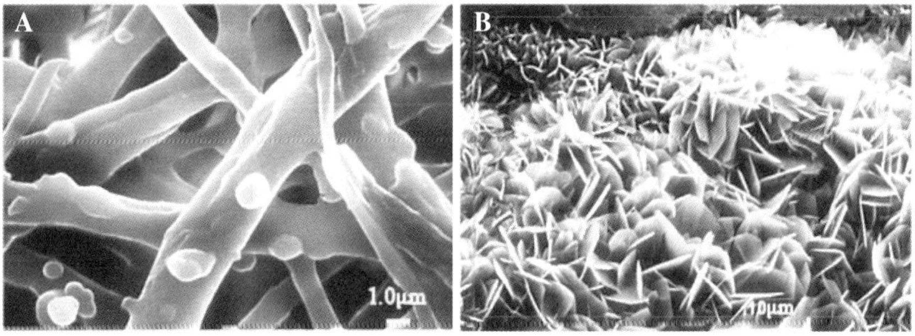

Fig. 6.3 Scanning electron micrographs of eggshell membranes (ESM). (A) An overview of the outer surface of the ESM. (B) Growth of hydroxyapatite crystals on the ESM surface after experimental mineralization.

Derivatization of eggshell membranes with acidic polymers shows remarkable effects on the growth, morphology, and polymorph selectivity of calcium carbonate crystals [18]. Designer peptides which mimic the function of eggshell matrix proteins may induce noticeable changes in calcite crystal morphology and produce polycrystalline crystal aggregates [19].

The synthesis of inorganic crystals of specific size and morphology (other than calcium salts) has attracted much interest due to the potential for designing new materials and devices for applications in catalysis, electronics, ceramics, biomedicine, and cosmetics [20]. Eggshell membranes have been shown to act as a supramolecular template by controlling the transport and growth of barium sulfate and tungstate crystals under mild conditions [21]. Thermogravimetric analysis shows that eggshell membranes begin to pyrolyze at 270 °C, and are completely pyrolyzed by 500 °C to leave ~24% inorganic solid. This property, when combined with a sol-gel approach, has been used to develop an eggshell membrane templating procedure for the synthesis of hierarchically ordered oxides, such as TiO_2, ZrO_2, and SnO_2 [22].

6.5
Composite Reinforcement with Eggshell

Recently, polypropylene composites using natural reinforcements from eggshells have been shown to have improved mechanical and thermal behaviors compared to such composites reinforced with "traditional" materials such as mineral calcium carbonate or zinc oxide (Fig. 6.4) [23].

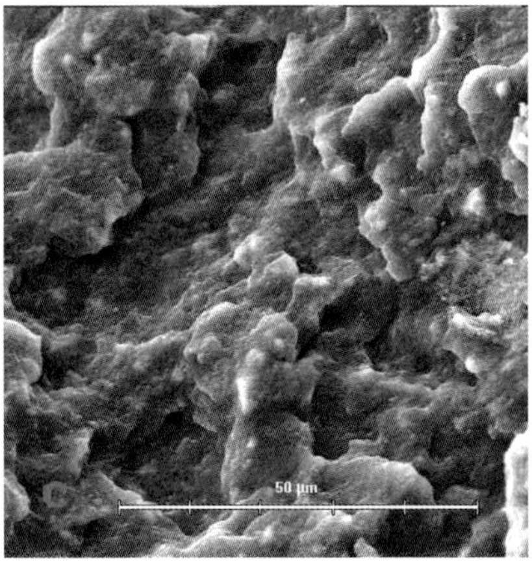

Fig. 6.4 Scanning electron micrograph of the fracture surface of polypropylene-eggshell filler composites with 40% (w/w) filler content.

6.6
Biomedical Applications of Eggshell

In the popular medicine of southern Chile, one folk remedy for wound healing is to use a piece of eggshell membrane as a wound dressing. In fact, such membranes have been shown to be biocompatible not only as wound dressings but also as a cell or tissue engineering scaffold (Fig. 6.5) [24]. The biomedical properties of eggshell membranes have also been improved by conjugating their constituents with collagen [25].

The calcitic eggshell has been used to obtain different compounds of calcium phosphate, such as hydroxyapatite, beta-tricalcium phosphate, and wollastonite ($CaSiO_3$) [26]. Although the direct use of eggshell as a bone substitute has been controversial, some promising effects are expected [27].

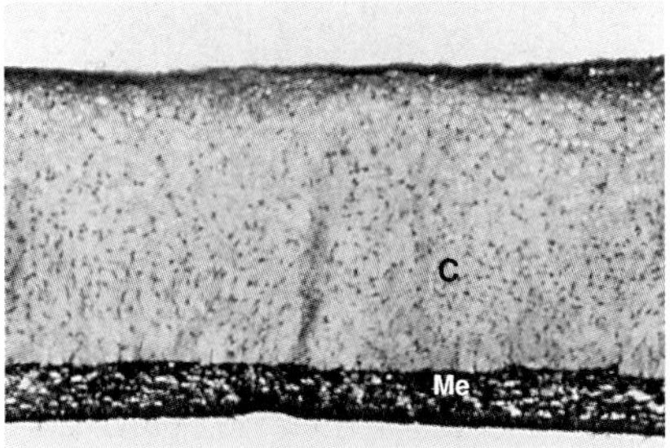

Fig. 6.5 Histologic section of chondrocytes derived from chicken periosteal cells cultured on eggshell membranes (original magnification, ×400). Me = eggshell membranes; C = chondrocytes.

6.7 Summary and Future Prospects

The relationship between eggshell biomineralization and materials science offers a new approach to producing materials with beneficial and desirable properties, but using only mild conditions. A closer inspection of the mineralized structures produced by Nature will, in time, lead us to mimic naturally occurring processes in order to create a range of superior materials that cannot be prepared using "conventional" technologies.

Acknowledgments

These studies were supported by FONDAP 11980002, granted by the Chilean Council for Science and Technology (CONICYT) through CIMAT. The authors wish to thank Dr. David Carrino for his careful proofreading of this manuscript.

References

1 A.H. Heuer, D.J. Fink, V.J. Laraia, J.L. Arias, P.D. Calvert, K. Kendall, G.L. Messing, J. Blackwell, P.C. Rieke, D.H. Thompson, A.P. Wheeler, A. Veis, A.I. Caplan, *Science* **1992**, *255*, 1098–1105.

2 (a) K. Simkiss, K.M. Wilbur (Eds.), *Biomineralization: Cell Biology and Mineral Deposition*. Academic Press, San Diego, USA, **1989**; (b) H.A. Lowenstam, S. Weiner (Eds.), *On Biomineralization*. Oxford University Press, Oxford, UK, **1989**; (c) S. Mann, J. Webb, R.J.P. Williams (Eds.), *Biomineralization: Chemical and Biochemical Perspectives*. VCH Publishers, Weinheim, Germany, **1989**.

3 (a) H.C. Lichtenegger, T. Schobert, M.H. Bartl, J.H. Waite, G.D. Stucky, *Science* **2002**, *298*, 389–392; (b) H.C. Lichtenegger, T. Schobert, J.T. Ruokolainen, J.O. Cross, S.M. Heald, H. Birkedal, J.H. Waite, G.D. Stucky, *Proc. Natl. Acad. Sci. USA* **2003**, *100*, 9144–9149; (c) H.C.

Lichtenegger, H. Birkedal, D.M. Casa, J.O. Cross, S.M. Heald, J.H. Waite, G.D. Stucky, *Chem. Mater.* **2005**, *17*, 2927–2931.

4 (a) M. Alper, P. Calvert, R. Frankel, P. Rieke, D. Tirrell (Eds.), *Materials Synthesis Based on Biological Processes*, Vol. 218. Materials Research Society, Pittsburgh, USA, **1991**; (b) E. Baeuerlein, *Biomineralization: From Biology to Biotechnology and Medical Application*. Wiley-VCH, Weinheim, Germany, **2000**; (c) S. Mann. *Biomineralization: Principles and Concepts in Bioinorganic Materials Chemistry*. Oxford University Press, London, UK, **2001**; (d) J. Aizenberg, J.M. McKittrich, C.A. Orme (Eds.), *Biological and Biomimetic Materials–Properties to Function*. Vol. 724. Materials Research Society, Pittsburgh, USA, **2002**; (e) F.C. Meldrum, *Int. Mater. Rev.* **2003**, *48*, 187–224; (f) J.L. Thomas, K.L. Kiick, L.A. Gower (Eds.), *Materials Inspired by Biology*, Vol. 774. Materials Research Society, Pittsburgh, USA, **2003**; (g) J. Aizenberg, W.J. Landis, C. Orme, R. Wang (Eds.), *Biological and Bio-inspired Materials and Devices*, Vol. 823. Materials Research Society, Pittsburgh, USA, **2004**; (h) E. Bäuerlein, *Biomineralization: Progress in Biology, Molecular Biology and Applications*. Wiley-VCH, Weinheim, Germany, **2004**; (i) J. Aizenberg, J. Livage, S. Mann (Eds.), *New Developments in Bio-related Materials. J. Mater. Chem.* **2004**, *14*, 2059–2354; (j) J.L. Arias, M.S. Fernandez (Eds.), *Biomineralization: From Paleontology to Materials Science*. Ed. Universitaria, Santiago, Chile, **2006**.

5 J.L. Arias, M.S. Fernandez. *Mater. Character.* **2003**, *50*, 189–195.

6 (a) J.L. Arias, D.J. Fink, S.-Q. Xiao, A.H. Heuer, A.I. Caplan, *Int. Rev. Cytol.* **1993**, *145*, 217–250; (b) J.L. Arias, K. Mann, Y. Nys, J.M. Garcia-Ruiz, M.S. Fernandez, in: E. Bäuerlein, P. Behrens, M. Epple (Eds.), *Handbook of Biomineralization, Vol. 1*. Wiley-VCH, Weinheim, Germany, **2007**.

7 W.T. Tsai, J.M. Yang, C.W. Lai, Y.H. Cheng, C.C. Lin, C.W. Yeh, *Biores. Technol.* **2006**, *97*, 488–493.

8 (a) P. Hu, Y. Fang, T. Zhou, M. Zhu, *Fenxi Ceshi Xuebao* **1995**, *14*, 51–53; (b) J. Deng, Y. Yuan, J. Xu, D. Xiao, K. Wang, *Fénix Huaxue* **1998**, *10*, 1257–1259; (c) M. Akagawa, Y. Wako, K. Suyama, *Biochim. Biophys. Acta* **1999**, *1434*, 151–160.

9 M.M.F. Choi, W.S.H. Pang, D. Xiao, X. Wu, *Analyst* **2001**, *126*, 1558–1563.

10 (a) J. Deng, L. Liao, Y. Yuan, D. Xiao, *Chin. J. Anal. Lab.* **2002**, *21*, 64–66; (b) D. Xiao, M.M.F. Choi, *Anal. Chem.* **2002**, *74*, 863–870; (c) M.M.F Choi, T.P. Yu, *Enzyme Microbiol. Technol.* **2004**, *34*, 41–47; (d) B. Wu, G. Zhang, S. Shuang, M.M.F. Choi, *Talanta* **2004**, *64*, 546–553; (e) B. Wu, G. Zhang, S. Shuang, M.M.F. Choi, *Anal. Biochem.* **2005**, *340*, 181–183; (f) G. Wen, Y. Zhang, Y. Zhou, S. Shuang, C. Dong, M.M.F. Choi, *Anal. Lett.* **2005**, *38*, 1519–1529; (g) M.M.F. Choi, M.M.K. Liang, A.W.N. Lee, *Enzyme Microbiol. Technol.* **2005**, *36*, 91–99; (h) B. Wu, G. Zhang, S. Shuang, C. Dong, M.M.F. Choi, A.W.N. Lee, *Sensors Actuators B* **2005**, *106*, 700–707; (i) G. Zhang, D. Liu, S. Shuang, M.M.F. Choi, *Sensors Actuators B* **2006**, *146*, 936–942.

11 M.M.F. Choi, *Food Chem.* **2005**, *92*, 575–581.

12 (a) K. Suyama, Y. Fukazawa, Y. Umetsu, *Appl. Biochem. Biotechnol.* **1994**, *45/46*, 871–879; (b) S. Ishikawa, K. Suyama, *Appl. Biochem. Biotechnol.* **1998**, *70/72*, 719–728.

13 S. Ishikawa, K. Suyama, K. Arihara, M. Itoh, *Biores. Technol.* **2002**, *81*, 201–206.

14 S. Ishikawa, S. Sekine, N. Miura, K. Suyama, K. Arihara, M. Itoh, *Biol. Trace Elem. Res.* **2004**, *102*, 113–127.

15 (a) M.S. Fernandez, M. Araya, J.L. Arias, *Matrix Biol.* **1997**, *16*, 13–20; (b) M.S. Fernandez, K. Passalacqua, J.I. Arias, J.L. Arias, *J. Struct. Biol.* **2004**, *148*, 1–10.

16 (a) J.L. Arias, M.S. Fernández, V.J. Laraia, J. Janicki, A.H. Heuer, A.I. Caplan, *Mater. Res. Soc. Symp. Proc.*

1991, *218*, 193–201; (b) T.-M. Wu, D.J. Fink, J.L. Arias, J.P. Rodriguez, A.H. Heuer, A.I. Caplan, in: H.C. Slavkin, P. Price (Eds.), *Chemistry and Biology of Mineralized Tissues*. Elsevier, Amsterdam, **1992**; (c) J.L. Arias, O. Nakamura, M.S. Fernandez, J.J. Wu, P. Knigge, D.R. Eyre, A.I. Caplan, *Connect. Tissue Res.* **1997**, *36*, 21–33.

17 J.K. Liu, Q.S. Wu, Y.P. Ding, *Eur. J. Inorg. Chem.* **2005**, *20*, 4145–4149.

18 P.K. Ajikumar, B.J.M. Low, S. Valiyaveettil, *Surf. Coat. Technol.* **2005**, *198*, 227–230.

19 (a) P.K. Ajikumar, R. Lakshminaratanan, B.T. Ong, S. Valiyaveettil, R.M. Kini, *Biomacromolecules* **2003**, *4*, 1312–1326; (b) P.K. Ajikumar, S. Vivekanandan, R. Lakshminaratanan, S.D.S. Jois, R.M. Kini, S. Valiyaveettil, *Angew. Chem Int. Ed.* **2005**, *44*, 2–5.

20 (a) M. Li, H. Schnablegger, S. Mann, *Nature* **1999**, *402*, 393–395; (b) S.H. Yu, H. Cölfen, M. Antonietti, *Chemistry: A European Journal* **2002**, *8*, 2937–2945; (c) S.H. Yu, A. Markus, H. Cölfen, *Nano Lett.* **2003**, *3*, 379–382.

21 (a) J.K. Liu, Q.S. Wu, Y.P. Ding, S.Y. Wang, *J. Mater. Res.* **2004**, *19*, 2803–2806; (b) J.K. Liu, Q.S. Wu, Y.P. Ding, *Chem. Res. Chin. Univ.* **2005**, *21*, 243–245; (c) J.K. Liu, Q.S. Wu, Y.P. Ding, *Cryst. Growth Des.* **2005**, *5*, 445–449.

22 (a) D. Yang, L. Qi, J. Ma, *Adv. Mater.* **2002**, *14*, 1543–1546; (b) D. Yang, L. Qi, J. Ma, *J. Mater. Chem.* **2003**, *13*, 1119–1123; (c) Q. Dong, H. Su, D. Zhang, F.Y. Zhang, *Nanotechnology* **2006**, *17*, 3986–3972.

23 (a) J.L. Arias, R. Quijada, P. Toro, M. Yazdani-Pedram. Chilean Patent Appl. No. 2542-2004, **2004**; (b) J.L. Arias, R. Quijada, P. Toro, M. Yazdani-Pedram. USA Patent Appl. No. 428292100 USPTO, **2006**.

24 (a) M. Tavassoli, *Experientia* **1983**, *39*, 411–412; (b) K. Maeda, Y. Sasaaki, *Burns* **1984**, *8*, 313–316; (c); F. Yi, Z.X. Guo, L.X. Zhang, J. Yu, Q. Li, *Biomaterials* **2004**, *25*, 4591–4599; (d) H.P. Lee, H.C. Chen, C.W. Lai, S.F. Chiang, C.J. Liao, Y.C. Hu, *J. Chin. Inst. Chem. Eng.* **2005**, *36*, 321–330; (e) F. Yi, Q. Li, Z.X. Guo, J. Yu, *J. Appl. Polym. Sci.* **2006**, *99*, 1340–1345.

25 (a) K. Takahashi, K. Shirai, M. Kitamura, M. Hattori, *Biosci. Biotechnol. Biochem.* **1996**, *60*, 1299–1302; (b) T. Ino, M. Hattori, T. Yoshida, S. Hattori, K. Yoshimura, K. Takahashi, *Biosci. Biotechnol. Biochem.* **2006**, *70*, 865–873.

26 (a) E.M. Rivera, M. Araiza, W. Brostow, V.M. Castaño, J.R. Diaz-Estrada, R. Hernandez, J.R. Rodriguez, *Mater Lett.* **1999**, *41*, 128–134; (b) E.M. Rivera, R. Curiel, R. Rodríguez, *Mater. Res. Innov.* **2003**, *7*, 85–90; (c) S.J. Lee, S.H. Oh, *Mater. Lett.* **2003**, *57*, 4570–4574; (d) K. Prabakaran, A. Balamurugan, S. Rajeswari, *Bull. Mater. Sci.* **2005**, *28*, 115–119; (e) K. Prabakaran, U. Vijayalakshmi, S. Rajeswari, *Surf. Eng.* **2005**, *21*, 225–228; (f) S.J. Lee, S.Y. Chun, *Eco-Mater. Proc. Des. VI Mater. Sci. Forum* **2005**, *486/487*, 293–296; (g) S.J. Lee, Y.S. Yoon, M.H. Lee, N.S. Oh, *Mater. Lett.* **2006**.

27 (a) L. Dupoirieux, *Br. J. Oral Maxilofac. Surg.* **1999**, *37*, 467–471; (b) L. Dupoirieux, D. Pourquier, M.C. Picot, M. Neves, *Int. J. Oral Maxilofac. Surg.* **2001**, *30*, 58–62; (c) L. Dupoirieux, D. Pourquier, M. Neves, L. Teot, *J. Craniofac. Surg.* **2001**, *12*, 53–58; (d) E. Durmus, I. Celik, A. Ozturk, Y. Ozkan, M.F. Aydin, *J. Int. Med. Res.* **2003**, *31*, 223–230; (e) J. Park, C. Lee, B. Choi, J. Suh, *Bioceramics* **2006**, *309/311*, 183–186.

7
Biomimetic Mineralization and Shear Modulation Force Microscopy of Self-Assembled Protein Fibers

Elaine DiMasi, Seo-Young Kwak, Nadine Pernodet, Xiaolan Ba, Yizhi Meng, Vladimir Zeitsev, Karthikeyan Subburaman, and Miriam Rafailovich

Abstract

Biological mineralization relies upon proteins which preferentially nucleate minerals and control their growth. This process is often referred to as "templating", but the term is inclusive of a variety of mineral–organic interactions demonstrated in diverse model systems. In this chapter, details are presented of studies designed to differentiate between structured and unstructured proteins, which can be assembled together on submicron length scales and probed simultaneously at early stages of biomimetic mineralization. The approach utilizes extracellular matrix proteins, which self-assemble into fiber networks when induced onto negatively charged sulfonated polystyrene surfaces. A novel technique, based on atomic force microscopy is introduced; this is used to measure the elastic modulus of both structured and disorganized protein, prior to and during calcium carbonate mineralization. Mineral induced thickening and stiffening of the protein fibers occurs during the early stages of mineralization, well before discrete mineral crystals are large enough to image. Calcium carbonate stiffens the protein fibers selectively, without affecting the regions of disorganized protein between them. Secondary ion mass spectroscopy reveals calcium to be concentrated along protein fibers. In this unique model system, organized versus unstructured proteins can be assembled only nanometers apart and probed in identical environments, demonstrating a mineralization process which requires the structural organization imposed by fibrillogenesis of the extracellular matrix.

Key words: biomimetic, self-assembling, protein fibers, mineralization, scanning modulation force microscopy, polysaccharides, elastin, fibronectin.

7.1
Introduction

Many reports have described how biomaterials motivate the creation of new synthetic materials, and likewise have pointed out the value of *in-vitro* experiments,

Handbook of Biomineralization. Edited by P. Behrens and E. Bäuerlein

ISBN: 978-3-527-31805-6

based on materials science methods, in elucidating the mechanisms by which biomineralization may occur. The question persists, however, as to whether the biomineral can motivate the design of the *in-vitro* experiment itself. In the case of the present experiments, it has, and in this chapter we take the opportunity to describe a new model system for biomineralization, based on self-assembled extracellular matrix (ECM) protein fibrils. The process by which this model system was conceived is also reviewed. These studies have been inspired in equal measure by biological mineralization in the avian eggshell and by some simple – but very memorable – mineral–organic interfaces which have been studied *in vitro*. Both of these sources of inspiration presented ideas which have evolved during the course of the studies, and also during the course of recent developments in the field.

The basic question to be addressed was: given that the presence of organic molecules and matrices will nearly always affect mineralization in some way, how can one determine which of many possible mechanisms is acting? Minerals might nucleate homogeneously from solution and migrate to an insoluble organic interface, or nucleate heterogeneously at an organic interface (Fig. 7.1a). A soluble protein might poison or catalyze a crystal face, thus affecting growth (Fig. 7.1b), or cations might bind to an ordered organic, creating a template which is epitaxially or stereochemically matched to a select crystal face (Fig. 7.1c). Such an assembly – if it acted as a catalyst – would alter the mineralization kinetics, and this could in turn affect polytype selection. Soluble organics might bind ions in solution, affecting local concentration gradients and thereby, the kinetics of mineralization again (Fig. 7.1d). And finally, in any of these scenarios a fluidic or amorphous precursor mineral – perhaps containing water or macromolecules – could form and give rise to crystalline material in a later process, itself influenced by any of the above mechanisms (Fig. 7.1e).

As with many other research groups, we were drawn to the concept of structural templating – a mechanism which is apparently able to select and orient calcium carbonate polytypes, based on the functionalization and structure of two-dimensional (2-D) films of fatty acid-based molecules. Templating has enor-

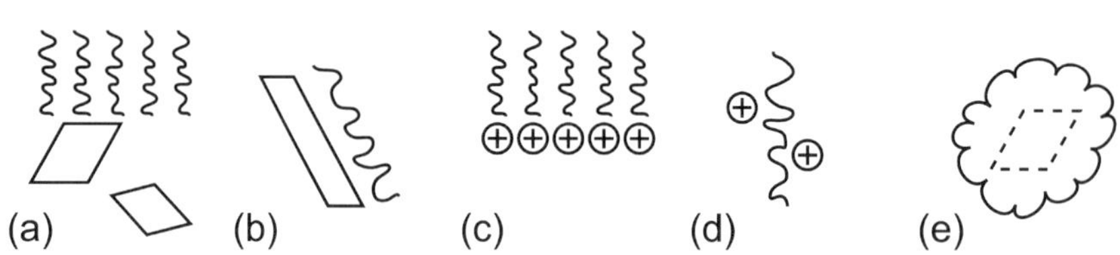

Fig. 7.1 Illustrations of potential mineral-organic interactions during mineralization. Rhombuses represent mineral crystals, curved lines represent soluble or insoluble organics, ⊕ symbols represent cations, and the cloud shape is an amorphous mineral precursor. (a) Heterogeneous nucleation at an insoluble organic interface, or homogeneous nucleation from solution followed by drift towards the interface. (b) Poisoning or catalysis of growth at a crystal face by a chemisorbed macromolecule. (c) Epitaxial or stereochemical templating process with mineral cations bound at the organic interface. (d) Macromolecule sequestering cations in solution. (e) Formation of an amorphous precursor phase.

mous appeal from an engineering standpoint: if the interactions between a structured organic and a mineral face are strong enough to overcome other mechanisms, there is a great potential for the design of synthetic nanocomposites. In this context, two main bodies of investigation have been cited extensively. In 1988, a series of reports by Heywood, Mann, Rajam, and Birchall highlighted the formation of oriented vaterite crystals beneath fatty acid monolayers assembled on supersaturated calcium bicarbonate solution [1–3]. A stereochemical match was depicted between the presumed stearic acid unit cell and the vaterite *a–b* plane, and the illustration of this interface became an icon for the concept of structural templating in many subsequent discussions. However, Heywood's publications acknowledged the difficulty of describing the vaterite results by stereochemical models alone, and emphasized the importance of considering the kinetics of mineralization and dynamics of the film structure and cation binding [2, 3]. These studies were followed by a larger body of experimental results that explored the limits of templating among other surfactant and mineral types [4]. Ultimately, *in-situ* synchrotron X-ray scattering measurements made by two independent groups were performed on this same fatty acid/$CaCO_3$ system. It was found that vaterite crystals beneath fatty acid monolayers were not well oriented [5], and that by changing the solution saturation, either calcite or vaterite polymorphs could be kinetically selected against the same monolayer film structure [6]. For crystallization under Langmuir monolayers, the epitaxial and stereochemical models have been supplanted by discussions of charged interfaces and other mineral–organic interactions.

Another significant body of work on the templating of $CaCO_3$ was produced by Aizenberg, Black, Han, Travaille, Whitesides, and co-workers. These investigations utilized self-assembled monolayers (SAMs) of functionalized alkanethiols on metal film substrates [7, 8]. Here, the metal surface orients the functional group of the SAM, creating a very tunable system in which calcite nucleates with a dramatic demonstration of preferred orientation. But even here, parameters related to the solution conditions were apparently able to overcome the effect of the oriented template. In some cases, samples were immersed in calcium chloride solution and exposed to the decomposition products of ammonium carbonate salts, while in others supersaturated calcium bicarbonate was used, yielding different crystal orientations [9]. Attempts to mineralize the substrates by exposing the films to the $CO_2{}^{2-}$ counterions first and the Ca^{2+} ions subsequently, were less successful at producing oriented crystals [10]. Experiments using sulfate-functionalized alkanethiols did not always yield the same proportion of variously oriented calcite crystals [11, 12]. Previously, Travaille and co-workers had attempted to use a flow cell method, supplying a constant composition of carbonate ions and a known constant driving force, to nucleate calcite on these SAMs; however, too little nucleation occurred and the group reverted to the ammonium carbonate method for nucleation and used the flow cell only for crystal growth [9]. These latter examples showed that the structure of the organic template did not always exert enough control to override other influences on mineralization. Furthermore, the homogeneous, "infinite" 2-D template model somewhat begged the question of how and where nucleation of the discrete crystals occurred.

Discussions of nucleation at faults and disordered domain boundaries were already under way within this same SAM literature [13], and also in the Langmuir monolayer literature where groups were using Brewster Angle microscopy to image crystals forming at domain boundaries in the films [14, 15]. Although the influence of environmental gradients and interfacial disorder on crystal nucleation – in other words, nucleation under *heterogeneous* conditions – has been a significant part of the literature all along, it has never been adequately reviewed in one place; meanwhile, the discussion of templating at a *homogeneous* interface continues to attract interest.

Heterogeneous conditions would be combined with structural templating in biological mineralization, but how could these aspects be combined in model experiments? One demonstration of these dual influences was achieved by soft lithography patterning of the functionalized alkanethiol SAMs [16]. In these experiments, micron-scale islands of acid-terminated molecules were stamped into a methyl-terminated SAM. The polar acid-terminated regions preferentially nucleated calcite crystals. When deposited as a single island, the polar nucleation site was observed to be surrounded by a depletion region, beyond which crystals also grew on the methyl-terminated SAM. This was explained by the local decrease in solution saturation at the nucleating point, where ions are depleted relative to the bulk. This report led us to consider whether such a patterned template might be used to *quantify* the effects of soluble macromolecules. For example, suppose that a soluble, acidic protein sequesters cations, removing them from solution; this lowers the concentration curves (as shown in Fig. 7.2a) and effectively widens the depletion region around a single nucleating site. As described elsewhere [16], in a solution without macromolecules, a given spacing of nucleators results in a particular maximum effective concentration between nucleation sites, controlling crystal growth; this is illustrated in Figure 7.2b. With macromolecules present, the same maximum effective concentration would be achieved with a different nucleator spacing (Fig. 7.2c). This raises the really novel possibility of using the patterned template as a tool to quantify the effects of the soluble proteins, balancing protein concentration against nucleator spacing to probe the limits of saturation conditions and the crystal growth rates. The question remained, however, that even if it were to be an elegant physical chemistry experiment, would it be "biomimetic"?

Enter the avian eggshell. As noted in Chapter 6 of this Volume, the eggshell is a marvel of *extracellular* mineralization: a collagen membrane, planar on the scale of millimeters, which is exposed within the oviduct to fluids the composition of which varies comparatively slowly, over a 12-h time scale. The first step in eggshell formation is the deposition of protein-rich material (mammillae), which are localized on approximately 50-μm spacings on the membrane. These deposits are rich in polysaccharides and will act as nucleation points for calcium carbonate mineral. As cations and counterions are introduced from the fluid, crystals nucleate at the mammillae, and their subsequent orientation and growth are controlled by other combinations of ions and macromolecules (Fig. 7.3) [17]. When we became aware of this process, we were motivated to assemble a synthetic system

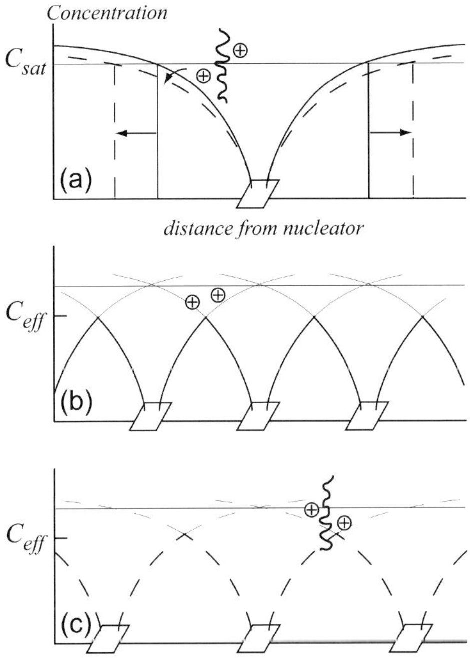

Fig. 7.2 Concentration profiles around crystal nucleators, schematically drawn after Figure 4 of Ref. [16]. (a) When macromolecules bind to cations and remove them from solution, the effective saturation curves are different, affecting the depletion region around a nucleation site where the concentration falls below the saturation concentration C_{sat}. (b) For a given nucleator spacing, an effective maximum cation concentration C_{eff} is achieved between nucleators. (c) The same C_{eff} value applies to a larger nucleator spacing when soluble macromolecules sequester cations. A series of patterned templates might be used as a tool to quantify the cation binding of a soluble inhibitor in a mineralizing system.

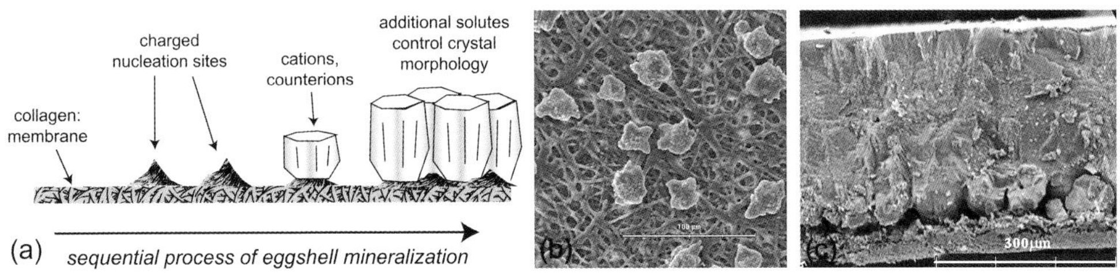

Fig. 7.3 (a) Simplified stages of mineralization for the avian eggshell. The collagen membrane is exposed first to proteins which form nucleation sites. Next, mineral nuclei form, and finally solution conditions change to control oriented crystal growth and eventually to halt mineralization. (b) Scanning electron micrograph of a chicken eggshell membrane, showing the localized nucleation sites. (c) Scanning electron micrograph of eggshell cross-section showing the fully mineralized structure. Micrographs (b) and (c) appear by courtesy of J.L. Arias, Universidad de Chile, Santiago, Chile.

that would share key features of this structure – that is, having 50-μm-scale protein assemblies which could be mineralized *in vitro* and monitored at very early stages.

7.2
Self-Assembled ECM Protein Networks

Until recently, it has been thought that ECM proteins such as fibronectin and elastin require the presence of cells to undergo fibrillogenesis and unfold from their globular forms in solution. However, our group has recently shown that these proteins can be induced onto sufficiently charged surfaces as self-assembled networks having 10 to 50-μm mesh size [18]. We proceeded to study these self-assembled networks to determine their capacity for nucleating calcium carbonate, as a model of extracellular biomineralization at a heterogeneous interface.

Our method utilizes sulfonated polystyrene (SPS) films spin-coated onto silicon wafers, as shown schematically in Figure 7.4a. On an arbitrary surface, a monolayer of globular protein is formed first [19]. The degree of sulfonation is tuned to produce a sufficiently negatively charged surface such that, when incubated in a buffer solution containing the protein, the protein subsequently unfolds onto the substrate, making the appropriate domains available to undergo fibrillogenesis [18]. AFM topography images of the elastin and fibronectin networks show that they have dimensions similar to those of the natural ECM (Fig. 7.4b,c). These networks form the starting point for our model of mineralization at heterogeneous, and also biomimetic, templates [20].

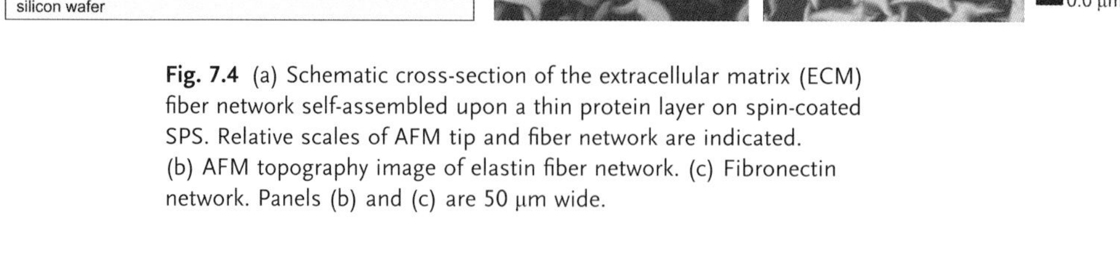

Fig. 7.4 (a) Schematic cross-section of the extracellular matrix (ECM) fiber network self-assembled upon a thin protein layer on spin-coated SPS. Relative scales of AFM tip and fiber network are indicated. (b) AFM topography image of elastin fiber network. (c) Fibronectin network. Panels (b) and (c) are 50 μm wide.

7.3
Shear Modulation Force Microscopy

An important feature of our model system is that the films incorporate disorganized proteins (the thin layer) and fibrils side-by-side on micron-length scales.

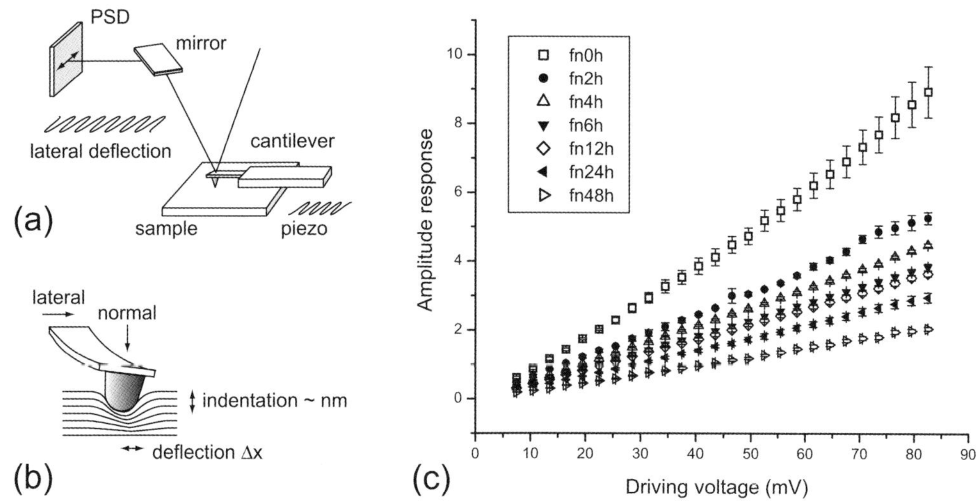

Fig. 7.5 (a) For shear modulation force microscopy (SMFM), a 25 nN normal force is applied to the tip while simultaneously driving the piezo scanner laterally at 1400 Hz with a small amplitude. The lateral deflection is measured by the position-sensitive detector (PSD). (b) The tip will indent several nanometers into soft materials. The combination of indentation and deflection which twists the tip is related to the Young's modulus. (c) Series of raw amplitude response data obtained in one SMFM relative modulus measurement (fibronectin fibers mineralized in a flow cell, described in Refs. [20, 23]). Each curve represents a series of measurements of the amplitude response as a function of driving voltage applied to oscillate the tip laterally. The slope of these curves drops after each time step, indicating that the material is stiffer after being exposed to $CaCO_3$ mineralizing conditions.

As shown in Figure 7.4a, the 20-nm AFM tip can independently probe the fibers versus the flat regions between them. Along with the morphology which can be monitored by AFM imaging, a new technique is applied, termed shear modulation force microscopy (SMFM), which measures the mechanical response of materials with 10-nm spatial resolution [21]. The modification made to the AFM method for this measurement is shown in Figure 7.5a. With a small normal force applied to keep the tip imbedded a few nanometers into the sample, the scanner piezo is also oscillated laterally with a 1400-Hz sinusoidal frequency. Lateral deflection of the tip is detected at a position-sensitive detector and recorded as an amplitude response. The amplitude response of the tip is a function of the normal force and the physical properties of the sample in which the tip is partially buried (Fig. 7.5b). By modeling the system as a set of coupled springs, it can be shown that the ratio of the response amplitude to the driving force is proportional to $E^{-2/3}$, where E is Young's modulus [21, 22]. By measuring a series of response curves on the sample over the linear range of driving voltage applied, a relative measure of the modulus is obtained which can be compared over different sample regions at chosen times during the mineralization process (Fig. 7.5c). The great advantage of this approach is that it provides with the means to probe the very early mineralization stage, well before mineral crystals can be imaged by any type of microscopy. With SMFM, it is possible to measure the mechanical response of the *organized protein fibrils* after mineralization, for comparison with the results obtained from the *unstructured* protein regions between fibers. Thus, we can distinguish between mineral adsorption induced by simple *chemical* properties and mineralization resulting from tertiary *structure* of the ECM proteins.

7.4 Comparative $CaCO_3$ Mineralization of Elastin and Fibronectin Networks

Calcium carbonate is an abundant biomineral with a complicated phase diagram that includes hydrous and non-hydrous forms, each with several possible crystal structures including amorphous phases. As discussed in the Section 7.1, it is known that kinetics plays a role in crystal habit and polytype selection even when the $CaCO_3$ phase is being modified by the presence of organics. For this reason, our group has surveyed several *in-vitro* mineralization methods. The use of $CaCl_2$ solution exposed to the decomposition products of ammonium carbonate was rejected, based on the rapid increase of pH to values above 9.5, which would most likely denature the proteins. A free drift method, in which calcium carbonate is dissolved into pure water by bubbling carbon dioxide gas, produces supersaturated calcium bicarbonate solution with $[Ca^{2+}] \approx 9$ mM. Allowing this solution to outgas causes the $CaCO_3$ to precipitate rather quickly, but under poorly controlled conditions; this method was used as a survey method of mineralization. In a third method, we prepared a mixture of $NaHCO_3$, $CaCl_2$, and NaCl bubbled with N_2/CO_2 gas to maintain a constant driving force and pH of 1.92 and 8.57, respectively [23]. Samples were placed in a closed cell and flowed with this solution using a peristaltic pump, thus creating slow – but reproducible – mineralization conditions. In the following sections, the data shown are principally from the free drift method of $CaCO_3$ mineralization. More details and comparisons to the flow cell have been published elsewhere [20].

The first sign that mineralization affects the proteins is found from section analyses of the AFM images. The elastin network morphology is superficially the same after 120 min of immersion in the calcium bicarbonate solution (free drift method) (see Fig. 7.6a and b). Although no evidence of discrete mineral crystals is obtained, by averaging many measurements of the fiber sections it is found that fiber thickness increases monotonically during exposure to the mineralizing solution (Fig. 7.6c). From this it is evident that mineral is accumulating in or on the fibers, but not collecting on the flat substrate between them.

By using SMFM, it is possible to probe the effects of mineralization along the fibers. The results of SMFM response curves probed both on the fibers (■ symbols in all panels) and on the flat monolayer of globular protein between the fibers (○), are shown in Figure 7.7. On the elastin and fibronectin fibers (Fig. 7.7A,B), the relative modulus follows the same linear trend as the fiber heights, increasing monotonically by about a factor of two as mineralization progresses

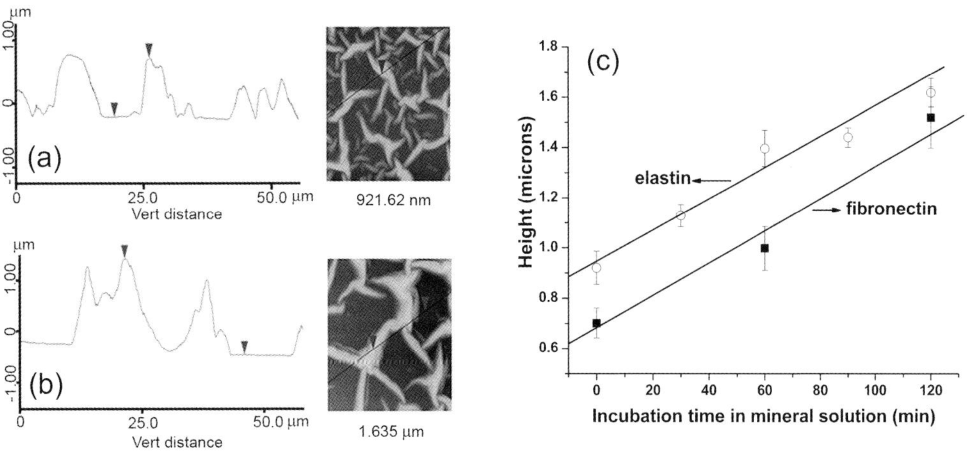

Fig. 7.6 Height profiles obtained by atomic force microscopy along indicated lines of elastin network (a) prior to mineralization, and (b) after $CaCO_3$ mineralization by free drift method for 120 min. (c) Average heights of elastin and fibronectin fibers as a function of mineralization time.

during a 2-h interval. By contrast, the results obtained from the flat regions between the fibers showed no difference in modulus after 2 h, and this modulus is much smaller than that of the fibers. This indicates that hardening mainly occurred due to interactions between the mineral and the proteins *in the form of fibers.*

These observations may be attributed to the fact that the protein adsorbed in the thin layer has a different conformation from that self-assembled into fibers. Those domains of the fibrillar protein which are exposed are apparently the most conducive to mineralization. The results also confirm that the SMFM measurement is truly localized and sensitive for surface study. Otherwise, the regions between the fibers, due to the proximity to the Si substrate, would have appeared much stiffer.

In these experiments, both elastin and fibronectin follow the same kinetics – that is, doubling their fiber height and stiffness in the same 2-h interval. Evidently, the solution parameters dominate this behavior in the system, although after mineralization for 24 h the elastin and fibronectin behave differently. No matter which method of mineralization is used, elastin networks always become covered with a high density of crystals, the majority phase being calcite with a typical rhombohedral morphology observed by SEM (Fig. 7.8a), optical microscopy, and X-ray diffraction (Fig. 7.8c). There is no evidence that elastin networks can affect the morphology or orientation of these crystals, but regions of the hydrophilic fiber crosslinks appear to be especially conducive to crystal growth [20]. By contrast, fibronectin networks are able to inhibit calcite formation, and observations by SEM show a thin coating of very small particles (Fig. 7.8b). Synchro-

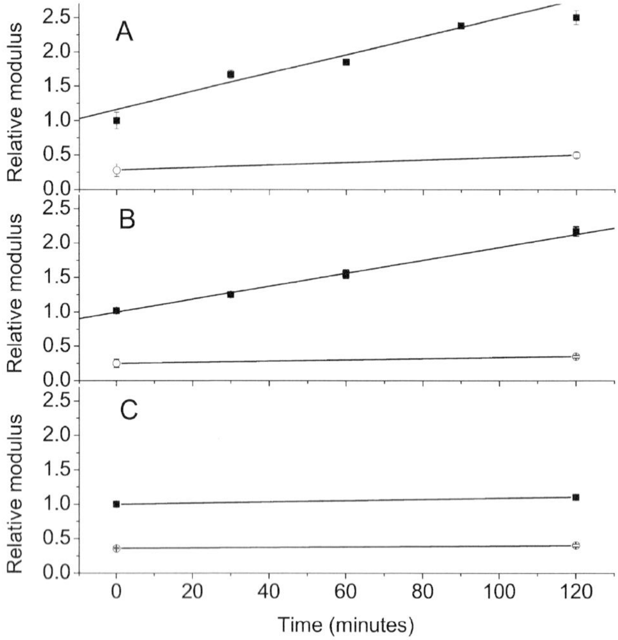

Fig. 7.7 Relative modulus measured by SMFM as a function of immersion time in solution. ■, Measurements taken on fibers (average of 10–20 response curves at different points). ○, Measurements taken on flat regions between fibers. (A) Elastin film in calcium bicarbonate solution. (B) Fibronectin film in calcium bicarbonate solution. (C) Elastin film in 9 mM $CaCl_2$ solution as non-mineralizing control. Data are normalized to the zero time, on-fiber modulus for each of the sample conditions. The lines serve as guides for the eye.

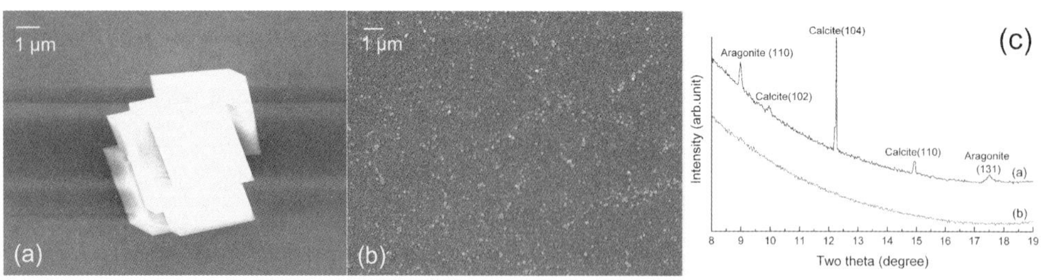

Fig. 7.8 Scanning electron micrographs of samples mineralized for 24 h by the free drift method. (a) Elastin, showing typical calcite morphology. (b) Fibronectin usually exhibits a thin coating of small particles. (c) Synchrotron X-ray diffraction ($\lambda = 0.65$ Å) from the mineralized (a) elastin and (b) fibronectin films.

tron X-ray diffraction measurements of these films are featureless. However, the presence of Ca in the fibers has been confirmed by secondary ion mass spectrometry [20]. Thus, our results suggest that an amorphous mineral phase has been inducted into the fiber network, and the tiny particles imaged by SEM may be the dried remnants of this protein–mineral composite. It is concluded that whilst various ECM proteins may be conducive to mineral filling, only the biomineral-associated proteins control growth at late stages; this may explain, for example, why fibronectin is implicated only in pathological biomineralization (as in arterial calcification) [24].

7.5 Mineralization of ECM Produced by Cells

A natural way to follow this line of thought is to apply our techniques to the natural ECM produced by cells such as osteoblasts. Using cultured cells, we can perform the same *in-vitro* mineralization measurements and compare the behavior of the complete ECM to that of the purified ECM proteins. This might lead to a more precise understanding of the roles of different proteins. However, we are also motivated by the prospect of developing these experimental techniques to study biomineralization of whole-cell systems. Specifically, the study of mineral formation in the ECM of bone cells can yield valuable information about the environmental conditions that are needed to facilitate its production, and the cellular activities that respond to extracellular stimuli. As the details of how bone mineral is formed are not yet well understood, the development of new experimental tools is clearly worthwhile.

For these experiments, MC3T3-E1 osteoblast-like cells were cultured in media for 4 days [25], after which they were incubated either in a supersaturated $CaCO_3$ solution or in phosphate-buffered saline (PBS) as control at room temperature for 2 or 24 h. A fluorescence microscopy image of the cells in culture is shown in Figure 7.9a. Using the AFM, the ECM proteins can be imaged, and are found to resemble the pure protein networks very closely (Fig. 7.9b). When SMFM measurements were performed on these fibers to probe the modulus change after a 2-h immersion in calcium bicarbonate solution, the modulus was found to have increased by 57% – somewhat less than the factor of two observed for the pure elastin and fibronectin networks. This may mean that mineral induction is not occurring to such a large extent in the natural ECM. Conversely, as the natural ECM contains a large proportion of collagen (which is a stiffer protein), its mechanical properties may not change as much with mineralization. A third possibility is that other components of the ECM, aside from the fibers, are taking up the mineral.

It is possible to distinguish between these scenarios by using secondary ion mass spectrometry (SIMS), an ion-scattering technique which provides element-specific information at very high lateral resolution [26]. Previously, it has been shown that for pure fibronectin networks, protein fibers and Ca are co-localized

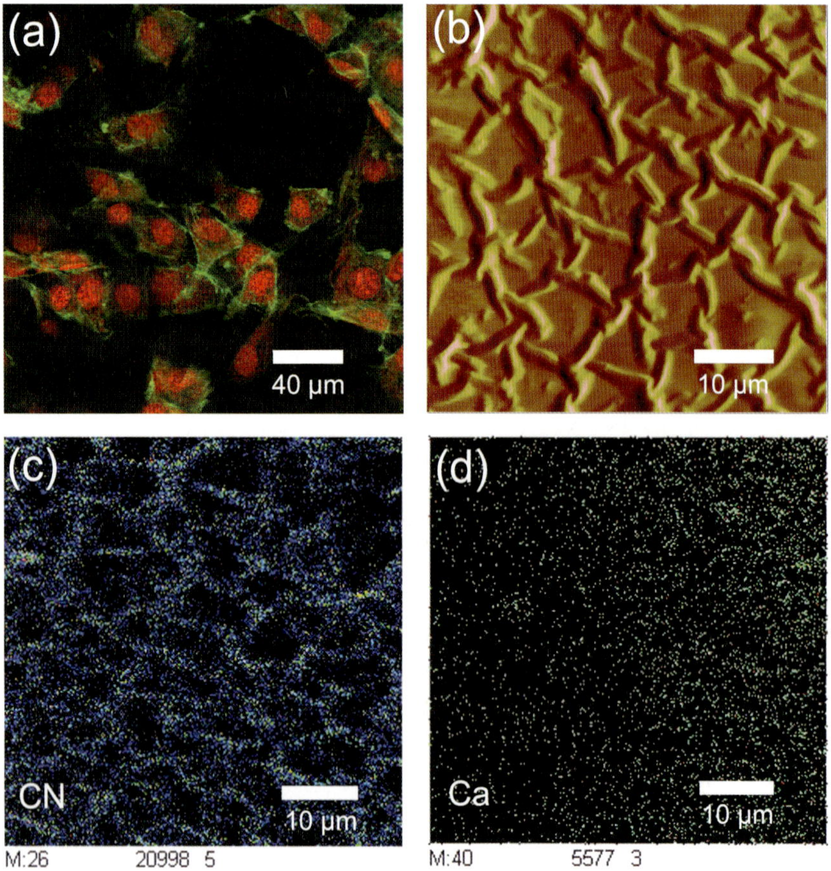

Fig. 7.9 (a) Fluorescence micrograph of unmineralized osteoblasts in culture. The F-actin fibers and the nuclei are stained green and red, respectively. (b) AFM image of protein fibers in the osteoblast extracellular matrix. (c) CN signal from SIMS showing distribution of the proteins in the ECM network [26]. (d) Ca signal from SIMS shows no evidence for co-localization of Ca with the protein fibers.

after mineralization [20]. Here, the same ion scattering technique was applied to image the Ca distribution on the mineralized osteoblast ECM. Figure 7.9c shows the CN signal, which has the same topology as the fiber network imaged by AFM in Figure 7.9b. This is also very similar to the SIMS CN signal obtained from the pure fibronectin networks [20]. The Ca signal (see Fig. 7.9d) is not co-localized with the fibers, as shown by an absence of any network pattern. Thus, it is concluded that the induction of calcium carbonate into the fibrous ECM of the osteoblasts is reduced compared to pure fibronectin.

7.6 Outlook

Inspired by "templating" at model mineral/organic interfaces, and also by biomineralization of the eggshell, we have discussed several new approaches for *in-vitro* mineralization experiments. We have developed a model system based on self-assembled ECM protein fiber networks which have two advantages: (i) that they present a heterogeneous landscape for mineralization, unlike the unrealistic 2-D organic monolayers studied in the past; and (ii) that the proteins spontaneously undergo fibrillogenesis, assembling into structures very similar to the natural ECM, yet at the same time incorporating an unstructured monolayer of globular protein. This has enabled us to show how mineralization preferentially affects organized protein structures while leaving the unstructured protein unaffected. This model system has been probed by a novel technique, SMFM, an AFM-based technique that allows the detection of very early mineralization stages, well before macroscopic crystals are visible by either AFM or SEM imaging. SMFM measurements also have the high spatial resolution required to distinguish between the protein fibers and the flat regions between them, on a time-resolved basis. Finally, we have shown how these techniques can be applied to a whole-cell system, by imaging, probing, and mineralizing the ECM produced by cells in culture.

As we review these accomplishments, it should be noted that there has been a departure from the "physical chemistry" experiments originally envisioned. The original concept would have utilized patterned templates of different uniform spacings, to test how cation depletion around a nucleator is affected by soluble macromolecules. The protein networks are neither tunable enough in their mesh size, nor uniform enough, to support this type of experiment. A return to soft lithography or other methods of chemical patterning on 1 to 50-μm length scales would be necessary to pursue this idea. The ECM model system and SMFM probe are ideally suited to determine which properties of the matrix proteins are important for advances in mineralized biomaterials and other tissue engineering applications. We look forward to new developments in these directions.

Acknowledgments

The authors thank L.B. Gower of the University of Florida, C. Orme, J. De Yoreo, and Y.-J. Han of LLNL, J. Aizenberg of Lucent Technologies, A.M. Travaille of Radboud University Nijmegen, and J.L. Arias of University Chile for their fruitful discussions of biological and biomimetic mineralization. They also acknowledge S. Ge of SUNY-Stony Brook and N.-L. Yang of CUNY-Staten Island for their contributions to the SPS and SMFM experimental methods. Brookhaven National Laboratory is supported under DOE Contract DE-AC02-98CH10886. These studies were also supported by the NSF MRSEC and the BNL-SBU Seed Grant Program.

References

1 S. Mann, B.R. Heywood, S. Rajam, J.D. Birchall, *Nature* 1988, *334*, 692–695.

2 S. Mann, B.R. Heywood, S. Rajam, J.D. Birchall, *Proc. R. Soc. London* 1989, *A425*, 457–471.

3 B.R. Heywood, S. Rajam, S. Mann, *J. Chem. Soc. Faraday Trans.* 1991, *87*, 735–743.

4 P.J.J.A. Buijnsters, J.J.J.M. Donners, S.J. Hill, B.R. Heywood, R.J.M. Nolte, B. Zwanenburg, N.A.J.M. Sommerdijk, *Langmuir* 2001, *17*, 3623–3628.

5 J. Kmetko, C. Yu, G. Evmenenko, S. Kewalramani, P. Dutta, *Phys. Rev. B* 2003, *68*, 085415–085420.

6 E. DiMasi, M.J. Olszta, V.M. Patel, L.B. Gower, *Cryst. Eng. Commun.* 2003, *5*, 346–349.

7 Y.-J. Han, J. Aizenberg, *Angew. Chem. Int. Ed.* 2003, *42*, 3668–3670.

8 A.M. Travaille, L. Kaptijn, P. Verwer, B. Hulsken, J.A.A.W. Elemans, R.J.M. Nolte, H. van Kempen, *J. Am. Chem. Soc.* 2003, *125*, 11571–11577.

9 A.M. Travaille, PhD Thesis, Radboud University Nijmegen, published by PrintPartners Ipskamp, 2005.

10 J. Aizenberg, personal communication.

11 S.-Y. Kwak, E. DiMasi, Y.-J. Han, J. Aizenberg, I. Kuzmenko, *Crystal Growth Des.* 2005, *5*, 2139–2145.

12 Y.-J. Han, unpublished data.

13 J. Aizenberg, A.J. Black, G.M. Whitesides, *Nature* 1998, *394*, 868–869.

14 E. Loste, E. Díaz-Martí, A. Zarbakhsh, F.C. Meldrum, *Langmuir* 2003, *19*, 2830–2837.

15 D. Volkmer, M. Fricke, D. Vollhardt, S. Siegel, *J. Chem. Soc. Dalton Trans.* 2002, 4547–4554.

16 J. Aizenberg, A.J. Black, G.M. Whitesides, *Nature* 1999, *398*, 495–498.

17 J.L. Arias, D.J. Fink, S.-Q. Xiao, A.H. Heuer, A.I. Caplan, *Int. Rev. Cytol.* 1993, *145*, 217–250.

18 N. Pernodet, M. Rafailovich, J. Sokolov, D. Xu, N.L. Yang, K. McLeod, *J. Biomed. Mater. Res.* 2003, *642*, 684–692.

19 Polished Si wafers were treated with a modified Shiraki technique to create a hydrophobic surface. Sulfonated polystyrene (SPS: $M_w \sim 15$ K, polydispersity < 1.1) was dissolved in dimethylformamide (10 mg mL^{-1}), spun-cast onto the wafers forming a 20 nm-thick film, and dried under vacuum at 150 °C. Elastin from bovine neck ligament and fibronectin from bovine plasma were dissolved in physiological phosphate-buffered saline. Prepared SPS substrates were incubated in protein solution at 37 °C, 100% humidity for 3 days. For further details, see Ref. [20].

20 K. Subburaman, N. Pernodet, S.-Y. Kwak, E. DiMasi, S. Ge, V. Zeitsev, X. Ba, N.L. Yang, M. Rafailovich, *Proc. Natl. Acad. Sci. USA* 2006, *103*, 14672–14677.

21 Y. Zhang, S. Ge, M. Rafailovich, J. Sokolov, R.H. Colby, *Polymer* 2003, *44*, 3327–3332.

22 S. Ge, Y. Pu, W. Zhang, M. Rafailovich, J. Sokolov, C. Buenviaje, R. Buckmaster, R.M. Overney, *Phys. Rev. Lett.* 2000, *85*, 2340–2343.

23 Flow cell solution: 2.0 mM $NaHCO_3$ and 2.2 mM $CaCl_2$ were mixed and bubbled with a premixed N_2/CO_2 gas mixture ($pCO_2 = 10^{-3.5}$) to maintain a constant driving force and pH of 1.92 and 8.57, respectively. NaCl was used to fix the ionic strength at 0.08 M. For details, see Ref. [20].

24 K.E. Watson, F. Parhami, V. Shin, L.L. Demer, *Arterioscler. Thromb. Vasc. Biol.* 1998, *18*, 1964–1971.

25 Cells were maintained at 37 °C (5% CO_2, humidified) in α-MEM culture medium supplemented with 10% fetal bovine serum, 100 U mL^{-1} penicillin and 100 μg mL^{-1} streptomycin, 50 μg mL^{-1} L-ascorbic acid and 4 mM β-glycerophosphate.

26 Static and dynamic imaging SIMS experiments utilized an Ion ToF-IV instrument equipped with a liquid metal analyzing gun and a dual sputtering gun. Negative and positive ion dynamic imaging was performed with $^{133}Cs^+$ and Ar^+ beams respectively. Twenty scans of each secondary ion have been combined in the images shown to enhance contrast. For more details, see Ref. [20].

8
Model Systems for Formation and Dissolution of Calcium Phosphate Minerals

Christine A. Orme and Jennifer L. Giocondi

Abstract

Calcium phosphates are the mineral component of bones and teeth. As such, there is great interest in understanding the physical mechanisms that underlie their growth, dissolution, and phase stability. Control is often achieved at the cellular level by the manipulation of solution states and the use of crystal growth modulators such as peptides or other organic molecules. This chapter first discusses solution speciation in body fluids and relates this to important crystal growth parameters such as the supersaturation, pH, ionic strength and the ratio of calcium to phosphate activities. The use of scanning probe microscopy as a tool to measure surface kinetics of mineral surfaces evolving in simplified solutions is then discussed. The two primary themes of the chapter are: (i) the use of microenvironments that temporally evolve the solution state to control growth and dissolution; and (ii) the use of various growth modifiers that interact with the solution species or with mineral surfaces to shift growth away from the lowest energy facetted forms. The study of synthetic minerals in simplified solution lays the foundation for an understanding of the mineralization process in more complex environments found in the body.

Key words: calcium phosphates, scanning probe microscopy, solution speciation, kinetics, crystal growth, nucleation, impurities, brushite.

8.1
Introduction

Organisms use complex, biologically regulated processes with many feedback loops to create microenvironments that induce *nucleation* and *growth*. This is often achieved at the cellular level by generating organic templates and solution states that, in turn, control crystallization. Similarly, *dissolution* processes, such as found in healthy bone remodeling, or in diseases such as caries and osteopo-

Handbook of Biomineralization. Edited by P. Behrens and E. Bäuerlein

ISBN: 978-3-527-31805-6

rosis, reflect an underlying chemistry that has been biologically manipulated to favor resorption. In general, solutions, proteins, and templates act together to control growth and dissolution. In this chapter, attention is focused on the role of the solution as a model environment for modifying the growth kinetics and thermodynamics of calcium phosphate minerals.

Model solutions represent aspects of body fluids such as serum, urine, or saliva, but do not contain the full complement of ions and proteins found in biological systems. There are many virtues of model solutions, chief among them reproducibility and the ability to explore systematic variations about a norm. The use of simplified environments is not meant to mimic biosystems so much as to address how changes in the solution chemistry impact upon mineralization. This simplification has the advantage of allowing one to ask specific questions such as: "How does citrate impact brushite crystallization?" – a query that complements biological or medical-motivated questions such as: "Does citrate act as a therapeutic for kidney stone formation?" The study of synthetic minerals in simplified solution lays the foundation for an understanding of mineralization processes in more complex environments found in the body.

First, we will discuss solution speciation in body fluids and relate this to important crystal growth parameters such as the supersaturation, pH, ionic strength and the concentration ratio of calcium to phosphate activities. We then discuss how scanning probe microscopy can be used to investigate surface kinetics of mineral surfaces. The goal of this style of experiment is to develop an understanding of the mechanisms by which organic and inorganic growth modifiers alter crystal growth.

8.2
Calcium Phosphate Phases Found in Biology

Organisms have adopted a variety of strategies for creating hard tissues suitable for bearing loads. Marine organisms such as mollusks and algae have adapted to take advantage of materials in their external environment by creating exoskeletons and grinding appendages composed of calcium carbonates or silica. For example, the layered structure of the abalone shell is composed of alternating sheets of crystalline calcite and aragonite intertwined with organic binders; the intricate frustule of the diatom is a glass composed of very pure silica; and the grinding teeth of the sea urchin are composed of calcite but hardened with "impurities" of magnesium [1]. These examples serve as archetypes of the complex structures found in nature due to the bioavailability of carbonates and silicates in natural waters. By contrast, the *interiors* of organisms are bathed in solutions containing phosphate, as part of the adenosine triphosphate (ATP)-driven energy cycle. As a result, the organism's body fluids are consistently supersaturated with respect to calcium phosphates, and the endoskeletons and other load-bearing mineralized tissues such as teeth have evolved to reflect this chemistry.

Whether present as carbonate, silicate or phosphate, biominerals adopt interesting strategies to expand the versatility of the relatively simple base materials.

In the examples above, the abalone shell gains strength and limits crack propagation because it is an organic–inorganic composite composed of two polymorphs; the diatom makes use of amorphous instead of crystalline materials; and the sea urchin uses impurities to improve mechanical properties. Organisms use similar strategies in calcium phosphate structures. For example, bones tune elasticity and hardness with a collagen–mineral composite [2]; matrix vesicles, which mineralize bone, are thought to store stabilized amorphous calcium phosphate [3]; and teeth are made less soluble by the incorporation of small amounts of fluoride into their outer layers [4]. Beyond the current volume, a number of books and reviews have discussed the general strategies that lead to complex biomaterials [5–8], and these will not be reviewed at this point. Hence, the two primary themes that will be encountered are: (i) the use of microenvironments that temporally evolve the solution state to control growth and dissolution; and (ii) the use of various growth modifiers that interact with the solution species or with mineral surfaces to shift growth away from the lowest energy facetted forms that are defined by the crystal Wulff plots.

The calcium phosphate phases of interest in biology include amorphous calcium phosphate, as well as several crystalline forms the names, abbreviations and some physical properties of which are summarized in Table 8.1. These materials have been extensively reviewed by several authors [9, 10]. Typically, biogenic calcium phosphate is not pure, but rather is substituted, as shown for the apatites: bone [9], enamel [11], and dentin [11].

For the most part, healthy mineralized tissues are hierarchical composites composed principally of carbonate-substituted hydroxyapatite, closely associated with a collagen matrix. Dentin and cementum in the tooth, and the various forms of bone [12] all contain approximately 70 wt% apatite with 20% collagenous matrix and 10% water. Enamel which, unlike the other tissues, contains no cells or pores, is almost 95 wt% mineral [9]. Enamel also differs in that it is associated with the proteins enamelin and ameliogenin rather than collagen. While these biomaterials have broad similarities, they differ in important details including their minor elements, water composition, and the degree of crystallinity [9].

Biogenic apatite, enamel, dentin, and bone, are impure and non-stoichiometric. The major impurities (in wt%) include carbonate (3.5–7.4%), which has been shown to substitute for phosphate, as well as sodium (0.5–1%), magnesium (0.4–1.2%), potassium (0.03–0.08%), chloride (0.01–0.3%) and fluoride (0.01–0.06%) [13, 14]. The impurities typically make biogenic apatite considerably more soluble than synthetic hydroxyapatite (HAP), as shown by the solubility products listed in Table 8.1. Despite the impurities, there is a need only to examine the skeleton to realize that the bones and teeth are sparingly soluble. Yet, although skeletons can last for thousands of years in the environment, within the body the bones are constantly being dissolved from one location and deposited in another as a response to mechanical stress and hormonal environment. On average, bones are dissolved and rebuilt every 30 years (based on 3% per year rate for cortical bone in healthy adults). To accomplish this, the body uses specialized cells and vesicles, to create microenvironments that regulate Ca^{2+}, $HPO_4{}^{2-}$, and H^+ to favor crystal growth or dissolution.

Table 8.1 Calcium phosphate minerals and biominerals.

Calcium phosphate [Reference]	Mineral name	Chemical formula	$-\text{Log } K_{sp}$ [37 °C]	Ca/P[a]
Dicalcium phosphate (DCPA)	Monetite	$CaHPO_4$	7.04	1
Dicalcium phosphate dihydrate (DCPD)	Brushite	$CaHPO_4 \cdot 2H_2O$	6.63	1
β-Tricalcium phosphate (TCP)	Whitlockite	$Ca_3(PO_4)_2$	29.55	1.5
Octacalcium phosphate (OCP)		$Ca_8H_2(PO_4)_6 \cdot 5H_2O$	97.4	1.33
Hydroxyapatite (HAP)	Apatite	$Ca_{10}(OH)_2(PO_4)_6$	117.3	1.67
Fluoroapaptite (FAP)		$Ca_{10}F_2(PO_4)_6$	115.8–120.2	1.67
Carbonated apatite (CAP)		$Ca_{10}(OH)_2(PO_4,CO_3)_6$	111.5–115.6	
Human bone [9]		$(Ca,Z)_{10}(PO_4,Y)_6(OH,X)_2$[b]		v
Human enamel [11]		$(Ca,Z)_{9.4}(PO_4,Y)_{5.98}(OH,X)_{1.3}$[b,c]	96.1–117.5	v
Human dentine [11]		$(Ca,Z)_{8.96}(PO_4,Y)_{5.96}(OH,X)_{0.78}$[b,d]	88.8–104.0	v

[a] Molar ratio, v = varies.
[b] Z = Na, Mg, K, Sr, etc; Y = CO_3, HPO_4; X = Cl, F.
[c] $(Ca)_{9.12}(Mg)_{0.06}(Na)_{0.22}(HPO_4)_{0.20}(CO_3)_{0.46}(PO_4)_{5.32}(OH,F)_{1.3}$.
[d] $(Ca)_{8.44}(Mg)_{0.28}(Na)_{0.24}(HPO_4)_{0.26}(CO_3)_{0.72}(PO_4)_{4.98}(OH,F)_{0.78}$.

Table 8.2 Pathological and normal calcium phosphate minerals found in the body.

Tissue	Mineral	Disease	Model solution	Reference(s)
Loops of Henle	Apatite, DCPD	Kidney stone formation	Urine	15
Teeth	HAP, DCPD, TCP, OCP	Calculus/caries	Plaque/saliva	16, 17
Salivary glands	DCPD	Sialolith	Saliva	18
Joint	DCPD, HAP	Rheumatoid and osteoarthritis	Synovial (joint) fluid	19

For abbreviations, see Table 8.1.

Apatite is the dominant calcium phosphate phase found in mineralized tissue; however, other transitory phases are postulated in healthy mineralization and are found stabilized for longer durations in pathological mineralization. A few examples (which by no means are exhaustive) [15–19] are indicated in Table 8.2. Model solutions that may serve as a starting point for investigating the dynamics of mineralization are also indicated.

8.3
Solution Chemistry in the Body

8.3.1
Solution Speciation

Solution speciation is the starting point for developing a quantitative model of crystal growth. In order to model the thermodynamic properties of a solution, it is first necessary to know the solubility product (K_{sp}) and the association constants (K_a) for all possible solid and solution species. Due to the many solid calcium phosphate phases (see Table 8.1) and solution complexes (not shown), calcium phosphate chemistry is extremely complicated even in simplified, model environments. Moreover, despite the maturity of this field the databases continue to evolve as new methods improve the accuracy of solubility measurements and identify the existence of new aqueous species [20].

Solution speciation techniques, which have been made routine with commercial and shareware programs, calculate the concentrations and activities of all so-

Table 8.3 Composition ranges (in mM) of human fluids and associated mineral saturation states of phosphates.

	Serum	Saliva [26, 27]	Enamel fluid [28]	Plaque fluid [29]	Urine [30]
Na^+	130–150	10	140	6.8–50	50–250
Cl^-	99–110	23	150	14–52	64–380
Ca^{2+}	2.1–2.9	0.40–2.1	0.5	0.8–8.6	0.81–7.8
P_i	0.74–1.5	2.9–11	3.9	7.8–29	7.2–45
HCO_3^-	8.2–9.3	2.1–25	10		
K^+	3.6–5.6	23	21	38–90	20–96
Mg^{2+}	0.74–1.5	0.21	0.8	1.3–5.5	0.70–7.8
F^-	0.01–0.02	0.005	0.005	0.0013–0.018	
SO_4^{2-}	0.08–0.12				0.50–50
NH_4^+		4		19–64	10–56
Growth parameters					
pH	7.4	5.5–7.5	7.2–7.3	5.69–7.08	4.8–8.0
$\{Ca\}/\{P_i\}$	1.31–2.54	0.02–0.69	0.145[a]	0.01–1.40	0.05–1.89
I (mM)	152–153	39–46	165[a]	152–155	258–274
σ_{gu} DCPD	−0.9–−0.5	−1.6–1.0	−0.50[a]	−1.1–1.7	−2.1–1.6
σ_{gu} TCP	0.8–1.2	−1.6–2.6	0.81[a]	−1.2–3.0	−2.7–3.5
σ_{gu} OCP	0.4–0.8	−1.3–2.0	0.49[a]	−0.9–2.4	−2.1–2.7
σ_{gu} HAP	2.3–2.7	−0.2–3.8	2.22[a]	0.2–4.2	−1.3–4.8

[a] Average values.

lution ions and complexes from a list of initial reactants and a database of potential reactions with their respective association constants and solubility constants [21]. Typically, a solution is described in terms of ion or salt concentrations (as in Table 8.3), although it is the activities that determine the full speciation. Speciation shown in this chapter uses the GeoChemists' WorkBench [22] with an extended Debye–Huckle formula [23] to define the activity coefficients.

8.3.2
Crystal Growth Parameters

Once speciation has been accomplished, the activities can be recast in terms of parameters that affect crystallization (lower half of Table 8.3). In general, these parameters can affect both the solution speciation (which is a thermodynamic consideration) as well as the surface of the mineral (which can be either a thermodynamic consideration such as shifting the hydrated state of the surface with pH, or a kinetic consideration affecting the activation barriers associated with incorporation). In this section we will briefly discuss how the precipitation–dissolution process is affected by the composition of the solution, including the effects of supersaturation, pH, ionic strength and ratio of cations to anions. The effects of each of these parameters are summarized in Table 8.4. The effects of additives will be described in Section 8.5.

Table 8.4 Crystal growth controls and their effect on the bulk solution and the crystal surface.

Parameter (symbol)	Effect on bulk solution	Effect on surface
Supersaturation (S)	Stability of solid phases	Net flux to surface; determines mode of growth (island nucleation versus incorporation at existing steps).
pH	Solution speciation (and subsequently supersaturation)	Net charge of surface due to degree of protonation
Ionic strength (I)	Screening length within the solution – activity coefficients	Debye length of the double layer
Temperature (T)	Solution speciation through temperature-dependence of association constants	Kinetics of adsorption, desorption, diffusion
Ratio of calcium to phosphate ions $\{Ca^{2+}\}/\{P_i\}$	Solution speciation	Kinetics of incorporation – in principle, activation barriers differ for calcium and phosphate ions
Additive concentration ([X])	Can change solution speciation (and subsequently supersaturation)	Various: step-pinning, surfactant, blocking layer, incorporation, etc.

8.3.2.1 Supersaturation

The most important crystallization parameter is the thermodynamic driving force or the *supersaturation*. The supersaturation, $\sigma = \Delta\mu/kT$, is a unitless number which is proportional to the chemical potential difference associated with molecules transferring from the bulk solution to the bulk solid phase, and can be determined from speciation calculations. Three related representations for the driving force are found in the literature: the *supersaturation*; the *supersaturation ratio*; and the *relative supersaturation*. It is instructive to spend a moment to define the relationship between these terms.

The supersaturation ratio can be computed by using the solution speciation results to calculate the ion activity products (IP) for minerals of interest. The *supersaturation ratio*, S is then given by:

$$S = \frac{IP}{K_{sp}}, \quad S_{HAP} = \frac{\{Ca^{2+}\}^5\{PO_4{}^{3-}\}^3\{OH^-\}^1}{K_{sp,\,HAP,\,half}},$$
$$S_{DCPD} = \frac{\{Ca^{2+}\}\{HPO_4{}^{2-}\}}{K_{sp,\,DCPD}}, \tag{1}$$

where the supersaturation ratio for hydroxyapatite (half unit cell) and brushite are shown explicitly. For $S = 1$, the mineral and solution are in equilibrium, for $S < 1$ the solution is undersaturated and the mineral will dissolve, and for $S > 1$ the solution is supersaturated and the mineral will grow. Usually, the half unit cell notation is used for octacalcium phosphate and the various apatites. In this representation, $K_{\text{sp, HAP, half}} = 10^{-58.65}$ and nine growth units are denoted in the activity product. In the full unit cell representation (as shown in Table 8.1), 18 growth units appear in the activity product and the solubility product is correspondingly squared, $K_{\text{sp, HAP, full}} = 10^{-117.3}$. From this example, it can be seen that $S_{HAP,\,full} = S^2_{HAP,\,half}$, which highlights the necessity of normalizing by the number of growth units (n) when making comparisons between minerals with different numbers of growth units. Traditionally, the supersaturation is written per molecule rather than per growth unit, and is related to S through $\sigma = \Delta\mu/kT = \ln S$. To allow comparisons between phosphate phases, it is useful to define a driving force that is normalized for the number of growth units in the unit cell (n),

$$\sigma_{gu} = \frac{\Delta\mu_{gu}}{kT} = \ln S^{1/n} \tag{2}$$

where "gu" denotes growth unit. Note that for $S \sim 1$ (i.e., near saturation) the logarithm can be expanded ($\ln x \sim 1 - x$ for $x \sim 1$), leading to what is commonly termed, the relative supersaturation, defined as:

$$\sigma_{rel} = S^{1/n} - 1. \tag{3}$$

In this chapter we will use Eq. (2) because of the broad range of supersaturation ratios found in body fluids. For σ, σ_{gu}, and σ_{rel}, positive values imply crystal

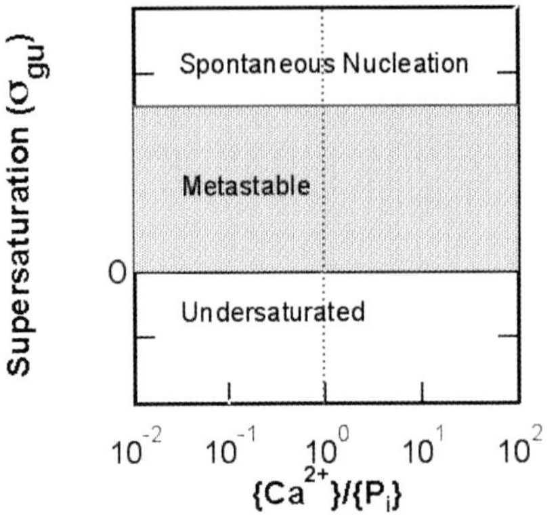

Fig. 8.1 Schematic graph of supersaturation normalized per growth unit [as defined in Eq. (2)] versus the ratio of the activities of calcium and total inorganic phosphate. For $\sigma < 0$, solutions are undersaturated and the mineral will dissolve; for $\sigma = 0$, the solution is in equilibrium with a solid mineral phase; and for $\sigma > 0$, the solution is supersaturated with respect to the mineral phase. The slope of the line bounding the metastable region from unstable precipitation can have either a positive or negative slope, depending on the kinetics associated with calcium versus phosphate incorporation.

growth. However, most solutions have a metastable region, where the solution is supersaturated but not enough to overcome the energy barrier that prevents crystals from precipitating spontaneously from solution phase (on reasonable timescales). In this region, crystal growth occurs on existing crystal surfaces, without nucleating new crystals. Above a supersaturation threshold the solution becomes unstable and both nucleation and growth occur; here, crystals are said to "crash" out of solution. These regimes are represented schematically in Figure 8.1, plotted against the calcium to phosphate activity ratio in order to highlight the fact that a solution with a fixed supersaturation can be either phosphate-rich (left side of the graph) or calcium-rich (right side of the graph). Thermodynamically, these solutions will have the same driving force for growth, but the kinetics can differ substantially.

8.3.2.2 pH

The pH affects both the solution as well as the mineral surface. In the solution, a shift to lower pH will lower the saturation state by shifting the balance of phosphate species from PO_4^{3-} to HPO_4^{2-} to $H_2PO_4^-$. At the mineral surface, the pH can shift the surface charge by changing the distribution of proton and hydroxyl groups hydrating the interface. For hydroxyapatite, the point of zero charge (in solutions without calcium) occurs at pH = 7.3. For more alkaline solutions the surface is negatively charged, whereas for more acidic solutions it is positively charged. However, ions other than H^+ and OH^- can adjust surface charge, and in calcium-containing solutions the Ca^{2+} ions bind to the surface for pH > 7, leaving the surface neutral rather than negative (provided that calcium is present) [24].

8.3.2.3 Ionic Strength

Ionic strength plays a role in screening both ion–ion electrostatic interactions in solution (which is accounted for by the activity coefficient) and electrostatic interactions between ions in solution and the surface. The ionic strength of a solution, I, is defined as:

$$I = 1/2\sum_i [i]{z_i}^2,$$

where $[i]$ is the concentration and the z_i the charge, of each ionic species, i. The Debye length sets the screening range at the mineral surface, and thus sets a length scale for the electric field generated at the charged surface. At a distance greater than the Debye length, the electric field is effectively shielded and therefore does not affect charged species. In most biological systems the ionic strength is near 0.15 M, corresponding to a Debye length of approximately 1 nm.

8.3.2.4 Temperature

Temperature is generally an important crystal growth parameter, although it is not a variable in a regulated environment such as the body. Solution speciation changes with temperature because solubility products and association constant are temperature-dependent. Temperature also affects the kinetics of adsorption, desorption, and diffusion. Within transition state theory, these motions are typically modeled as activated hopping processes where the probability of making the jump can be written as $P = \nu(T)e^{-E_a/kT}$. The attempt frequency, υ, is weakly temperature-dependent and is typically treated as independent of temperature. Thus, the primary temperature dependence is the exponent. However, in biological systems temperature is regulated and remains nearly constant (37 °C for humans); thus, while temperature can be used as a tool *in vitro* to measure activation barriers (E_a), it is not a control used *in vivo*.

8.3.2.5 Cation to Anion Ratios

Parameters such as the ratio of calcium and phosphate activities acknowledge that growth rates may not be dictated simply by the supersaturation and the surface energies but rather that kinetics may play a role. In principle, ion ratios can affect growth rates, growth shape, and the transformation of metastable phases. The growth of a multi-species crystal relies on the relative rates of adsorption and desorption of the various ions or growth units that make up the unit cell. For a simple salt such as NaCl, the growth units are the Na^+ and Cl^- ions. However, in general the growth units represent the pathway with the lowest activation barrier that allows an ion to move from the solution state to the solid state, and vice-versa. In a binary ionic compound such as brushite, it is tempting to think of the growth units as Ca^{2+} and ${HPO_4}^{2-}$, but it is possible that in the process of shedding waters of hydration and incorporating into the solid that the activation barrier is lower for a multi-step process wherein one of the other phosphate complexes (e.g., H_2PO^{4-} or $PO_3{}^{3-}$) adsorbs and then adds or sheds a hydrogen. The graphs shown in this chapter use the ratio of the calcium ion activity to the total phosphate ion activity ($\{P_i\} = \{H_2PO_4{}^-\} + \{HPO_4{}^{2-}\} + \{PO_4{}^{3-}\}$) to accommodate the uncertainty in rate-limiting ions. Many groups have shown that ion ratios play a role in kinetics, and some progress has been made at modeling these effects [25]. However, most crystal growth models assume a single species and further studies are required to fully describe multi-component crystals.

8.3.3 The Speciation of Body Fluids

The compositions and resultant crystal growth parameters for common body fluids including serum, saliva [26, 27], enamel fluid [28], plaque [29], and urine [30] are summarized in Table 8.3. It should be noted that the calcium, phosphate and pH values have broad ranges reflecting temporal and population variability. Saliva, plaque fluid and urine are especially affected by external factors such as diet, and this is reflected in their wide range of values. The exceptions are the pH of serum (which is regulated) and enamel fluids, for which few data are available. Table 8.3 also lists the range of values of the important crystal growth parameters, as determined by speciating the average solution chemistries over all combinations of maximum and minimum calcium and phosphate concentrations and pH.

One way of summarizing the variability of parameters in these solutions is to construct a map. The solution maps in Figure 8.2 express the supersaturation [per growth unit; Eq. (2)] as a function of the calcium to phosphate activity and the pH for HAP (which has low solubility) and dicalcium phosphate dihydrate (DCPD) (which has high solubility). These are shown for the body fluids listed in Table 8.3. The bounding areas of Figure 8.2a were determined by speciating all combinations of the maximum and minimum calcium and phosphate concentrations and pH values in the presence of the average concentrations of the other ions listed. The curves in Figure 8.2b were generated by speciating the average concentrations of all listed ions over the entire normal pH range. Note that HAP and DCPD were chosen to display as bounding cases and that the minerals with intermediate solubilities will fall between these plotted range of values.

From these maps we can make several observations. First, most body fluids are richer in phosphate than calcium, with serum being the exception. In serum, the Ca/Pi ratio is near that found in bone (1.66), and likely plays a role in creating crystals with this stoichiometry during the body's cycle of bone reformation. Second, HAP is supersaturated in all fluids over almost their entire range, with the result that under most conditions our bones and teeth are stable. In fact, solution conditions need to be explicitly modified to dissolve biological apatites, as will been seen in the example of the osteoclast. DCPD, on the other hand, lies nearer to the saturation border and is both under- and super-saturated. In healthy conditions, as represented by these solutions, HAP is always more supersaturated than DCPD; however, at lower pH values one can see that the HAP and DCPD stability

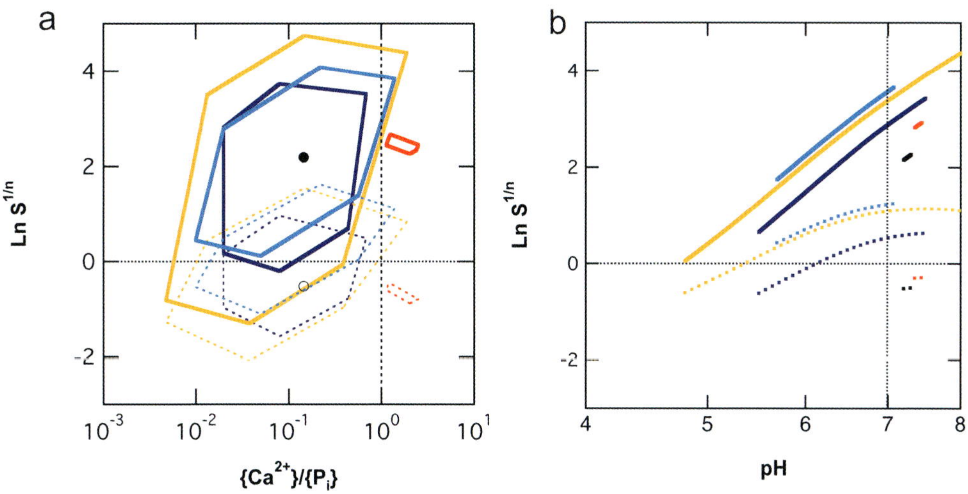

Fig. 8.2 A solution map that plots the supersaturation [as defined in Eq. (2)] versus (a) the ratio of activities of Ca^{2+} and total inorganic phosphate (P_i) and (b) the pH for various biological fluids. For each fluid: serum (red), saliva (dark blue), plaque fluid (light blue), enamel (black), and urine (yellow), the supersaturation normalized per growth unit is shown with respect to hydroxyapatite (HAP) and dicalcium phosphate dihydrate (DCPD) in solid and dotted lines, respectively. The boundaries are approximate based on the range of fluid compositions cited in Table 8.3 and solubility products cited in Table 8.1.

lines will cross due to shifts in the relative concentrations of $PO_4{}^{3-}$ and $HPO_4{}^{2-}$. It is thus possible under pathological conditions for DCPD to be more stable than HAP, although this requires both higher acidity and higher calcium and phosphate concentrations than is normal. The data in Table 8.2 suggest some instances where DCPD forms; however, it is not clear whether these are cases where DCPD is thermodynamically stable or a transitory phase.

Another interesting point is that crystals do not normally nucleate in fluids such as serum, despite the fact that they are highly supersaturated. This can be attributed in part to the metastable region where the probability of homogeneous nucleation is small, and in part to the presence of nucleation inhibitors (such as proteins) that are not accounted for in a thermodynamic map. In some ways the solution maps of normal, easily accessible body fluids give the regions over which nothing much happens – apatites neither dissolve, nor do they nucleate. One of the more difficult aspects of biomineralization is determining the composition and time-evolution of specialized local environments where crystals are forming or resorbing. Several possible pathways have been postulated that illustrate how the solution conditions might be manipulated to favor nucleation and dissolution; some of these are shown in Table 8.5. The remainder of this section will describe, as an example, the possible mechanisms at play when bone minerals are formed and resorbed.

Table 8.5 Examples of biological regulation that change the mineralizing fluid.

Cell activity	Biological regulation	Potential effect on mineralization	Base fluid
Bone growth	Osteoblasts produce enzyme alkaline phosphatase	Increase P_i	Matrix vesicle
Bone growth	Osteoblasts produce enzyme alkaline phosphatase	Reduce inhibitor concentration	Matrix vesicle
Bone resorption	Osteoclasts activate H^+-ATPase pump	Decrease pH	Resorption lacuna/serum
Tooth dissolution (caries)	Bacteria produce lactic acid	Decrease pH	Saliva or plaque

It is currently thought that the growth of individual crystals for bone mineralization takes place in two stages – nucleation and subsequent growth – and that these processes occur in different solution environments [31, 32]. The nucleation phase occurs in the regulated environment of small (~100-nm), fluid-filled, phospholipid "containers" called matrix vesicles (MV) that are released by osteoblast cells into the surrounding tissue, termed the *osteoid*. In general, more favorable nucleation environments can be created by increasing calcium or phosphate concentrations, creating suitable templates that reduce interfacial energy, or reducing inhibitor concentration. Within the interior of the MVs one process (or several of these processes) must occur. Amongst several potential pathways for initiating nucleation [32–34], two involve the enzyme alkaline phosphatase (ALP), which is produced by osteoblasts and can react with phosphorylated proteins to produce free phosphate. In principle, this would be a mechanism that could be used to regulate the total extracellular phosphate concentration [35]. On the schematic solution map (see Fig. 8.1), the initiation of nucleation implies increasing the supersaturation above the metastable line into the labile region. Alternatively, or in parallel, ALP hydrolyzes pyrophosphates, which are known inhibitors of mineralization, thereby lowering the barrier to nucleation. Inside the MV, the first phase to form is amorphous calcium phosphate, which then transforms into apatite with a needle-like morphology. At some point the MV membrane is disrupted and the apatite needles finish their growth in the surrounding serum-like solution. This example highlights one route for separating nucleation from growth when the two processes have different solution requirements.

Controlled dissolution also occurs in an occluded environment, as seen under an osteoclast cell during mineral resorption. Osteoclasts actively dissolve bone through a complex set of linked processes that ultimately produce high concentrations of HCl in an occluded region abutting the bone. In order to generate these high local concentrations of acid, the cell seals off a region known as the

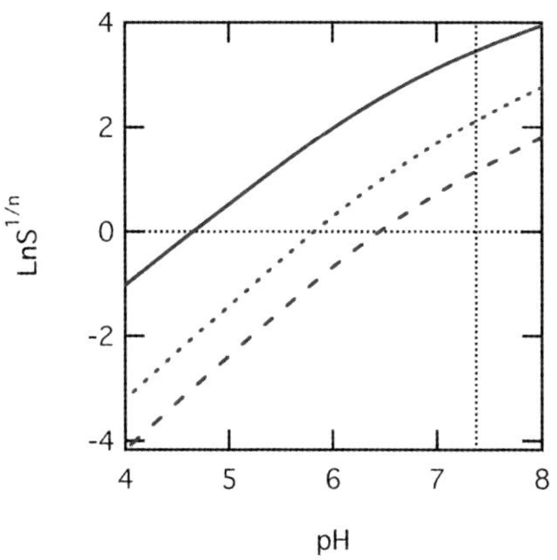

Fig. 8.3 Plot of supersaturation [Eq. (2)] versus pH for a range of apatites from the least soluble fluoroapaptite (solid line) to the most soluble dentine (dashed line). The apatites are modeled with the stoichiometry of hydroxyapatite (dotted line) but with solubility product as cited in Table 8.2. This is not strictly correct, but provides a reasonable approximation for determining a range.

resorption lacuna and use a vacuolar-like proton pump (H^+-ATPase) to transport H^+ against a gradient. Chloride ion channels maintain charge balance [36–38]. Within the resorption lacuna, the pH falls to 4.5 [39]. The change in saturation state associated with reducing the pH of a serum-like solution is depicted in Figure 8.3. The band spans a range of solubilities from the least-soluble synthetic fluoroapaptite (FAP) to the most soluble dentine; the actual solubility will depend upon the bone composition. A more rigorous simulation of this process would need to account for changes to the solution composition caused by dissolving mineral. However, from this simple titration it can be seen that that the solution moves from being supersaturated with respect to apatite to being undersaturated, thus favoring dissolution. A similar pH-mediation is found in caries, where bacterial colonies on the teeth produce lactic and acetic acid as part of their metabolic cycle.

8.3.4
Limitations of Speciation Modeling

A number of limitations must be borne in mind particularly with respect to modeling mineralization within the body. First, this is a thermodynamic approach, and if the reactions are slow the solution may not reflect its thermodynamic values. Similarly, mineralization inhibitors typically slow kinetics and are not accounted for in a thermodynamic model. Also, inherent to any speciation modeling, the calculations reflect the quality and completeness of the databases that contain the solubility products and the association constants. Speciation becomes more difficult as one includes proteins and organic surfaces because the association constants are rarely known for these materials. Similarly, we have imperfect knowledge of the *local* solution environments where mineralization occurs. Typical compositions and ranges of bodily fluids such as blood serum and urine are accessible and well documented, but the time evolution of the solution composition in an occluded environment, such as within a matrix vesicle or underneath an osteoclast, is less well known. Nevertheless, quantitative crystal growth relies on speciation, and a great number of biomineralization studies could be more easily compared and made more rigorous by a more widespread use of this approach.

8.4 Measuring Crystal Growth

There are a limited number of *in-situ* techniques that are viable in a fluid environment and that quantify crystallization dynamics. These include optical spectroscopies such as Raman and Fourier-transform infrared spectroscopy; diffraction techniques, such as X-ray or neutron (although these typically require a dedicated facility), interferometry, and imaging techniques such as video microscopy and scanning probe microscopy (SPM) or atomic force microscopy (AFM). In addition, a number of solution probes have been developed to monitor aspects of solution chemistry, such as pH probes and ion-selective electrodes. Most kinetic measurements have been obtained on bulk powders that reflect a distribution of facets and surface morphologies.

8.4.1 Bulk Crystallization

Bulk crystallization experiments can be classified either as *free drift*, where the solution composition varies due to nucleation and growth, or as *constant composition* (CC) [40, 41], where changes in the solution composition are monitored and compensated as the crystallization process occurs. In a seeded CC experiment, crystals (with known surface area) are placed in a temperature-controlled growth chamber with an automated titration system to maintain a constant pH or other ion-selective electrode reading. As the crystals grow, consuming calcium and phosphate from solution, the pH (or $\{Ca^{2+}\}$) shifts. Based on these changes (and knowledge of the seed materials), calcium and phosphate ions are automatically added to the solution to balance the material grown. The crystal growth rate is obtained from the additions after normalizing by the crystal surface area, and noting that the crystal surface area changes throughout the experiment and thus must be calculated or measured. Variations on the seeded CC method allow the measurement of dissolution rates, nucleation rates, and growth rates for multiple species.

Constant composition experiments have clear advantages for quantitative crystal growth. They allow growth rates to be measured systematically as a function of the parameters delineated in Section 8.3.2; they test and help direct development of the speciation database; and they determine which additives act as inhibitors. Although CC has arguably done more than any other single method to develop the quantitative physical chemistry of calcium phosphates, it is a bulk method with a limited ability to deduce how and where molecular changes occur.

8.4.2
Scanning Probe/Atomic Force Microscopy

Scanning probe microscopy has emerged as a complementary technique that, like CC, quantifies kinetics in a fluid environment but measures the crystal surface rather than the solution state. The technique measures surface morphology by rastering a cantilever over the surface and detecting its deflection. The lateral resolution is limited by the probe used to scan the surface and is typically 10 nm; thus, the individual growth units (such as Ca^{2+}, $HPO_4{}^{2-}$ or $PO_4{}^{3-}$) of a growing crystal are not resolvable, except when the crystal is composed of large building blocks such as proteins [42]. What makes SPM valuable for crystal growth is its sub-Angstrom z-resolution, making it capable of measuring atomic steps. For *in-situ* studies, temperature- and pH-controlled solutions are pumped through a fluid cell (Fig. 8.4) while the crystal is being imaged. The flow rate is adjusted until the surface dynamics are independent of flow, ensuring that the system is not mass-transport-limited. This also ensures that the crystal is responding to a solution with the same (constant) composition as the reservoir.

Figure 8.5 shows a typical AFM image of atomic steps emanating from a dislocation source. In this case the surface is the [010] face of a DCPD crystal and the 0.38-nm steps are readily imaged, even when scanning relatively fast (~12 s per image). Sequential images make a movie that captures the trajectory of the growing crystal steps and allows direct measurement of the step velocities in the different crystallographic directions. Similarly, when the solution is undersaturated, etch pits form (Fig. 8.6) and their dissolution can be monitored. Both growth hillocks and etch pits reflect the underlying symmetry of the crystal (Fig. 8.5c) which, for the case of DCPD, has three steps forming a triangle.

The ability to monitor step dynamics also enables several other fundamental measurements beyond step velocity and morphology. The reader is referred to

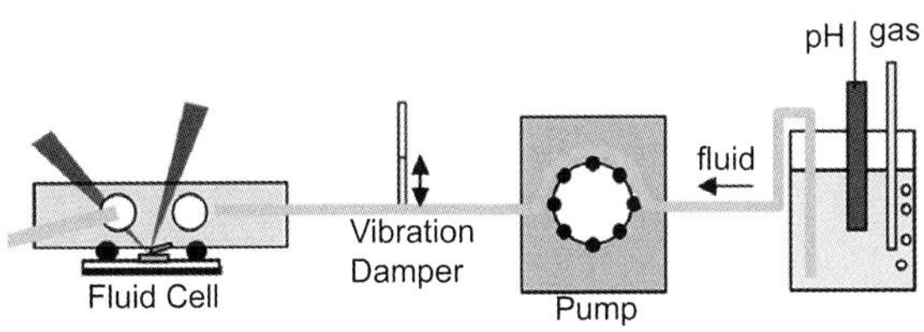

Fig. 8.4 Flow set-up for *in-situ* atomic force microscopy (AFM) experiments. The crystal is imaged while solution flows over the surface.

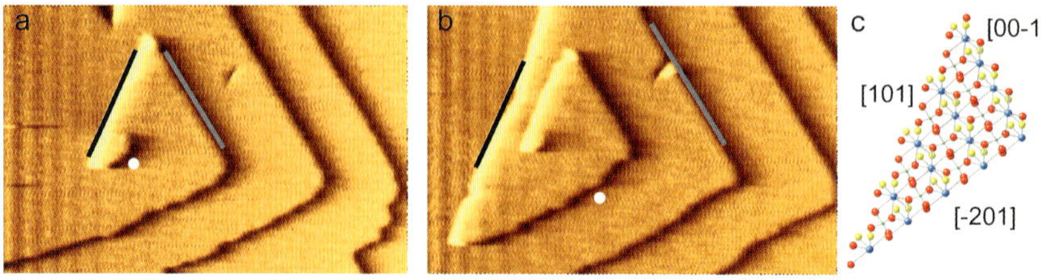

Fig. 8.5 (a,b) Two sequential images showing a growing hillock on the (010) facet of a dicalcium phosphate dihydrate (DCPD) crystal. Each image is 2 μm × 1.5 μm, and took 12 s to capture. Lines and a dot mark the same step in the two images. The step marked with a black line has a lower velocity than the other two steps, and does not travel as far. The ratio of terrace widths is proportional to the ratio of step velocities. (c) Atomic model of step geometry with Ca in red, P in blue, and O in yellow. The directions on the model are given for space group Ia.

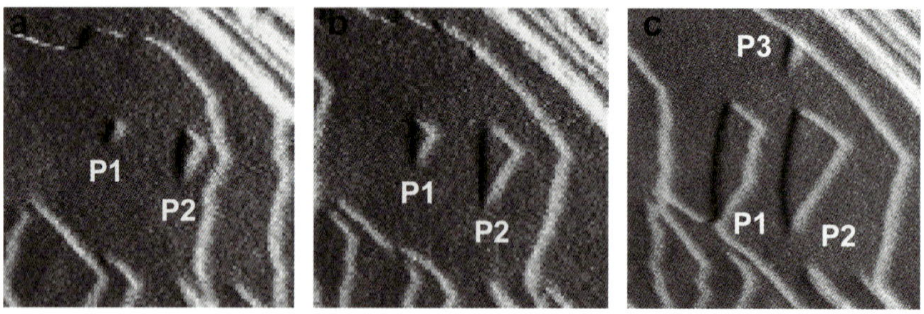

Fig. 8.6 Three sequential images showing the growth of etch pits. Pits are labeled P1, P2, and P3.

several reviews [43, 44] and books [45] that cover these topics. In brief, one important parameter that can be measured is the length needed for a step to propagate. When the step is in equilibrium, this length is termed the *critical length* and reflects a balance between the reduction of energy associated with the chemical potential difference between solution and solid, and the increase in energy associated with creating a step-edge. In this case, $L_c \propto \gamma/\Delta\mu$, where γ is the step-edge free energy; from this relationship, the step free energy can be obtained by measuring the critical length as a function of supersaturation [46]. If the step is not in equilibrium, but rather has a low probability of moving due to kinetics, then the step length is instead related to the probability of nucleating a kink [47]. There is growing evidence that this is a common case for sparingly soluble materials such as calcium phosphates [48]. However, the theoretical description of this case is still under development.

In the next section we will discuss how SPM can be used to augment bulk rate measurements by helping to pinpoint the mechanisms by which impurities or additives interaction with the growing crystal surface.

8.5 Impurity Interactions

Thus far, we have discussed the speciation and growth parameters associated with base solutions without inhibitors or growth modifiers such as proteins. The growth and stability of calcium phosphates is influenced by a great many inorganic and organic species. Some of the major modifiers include: (i) the ions that

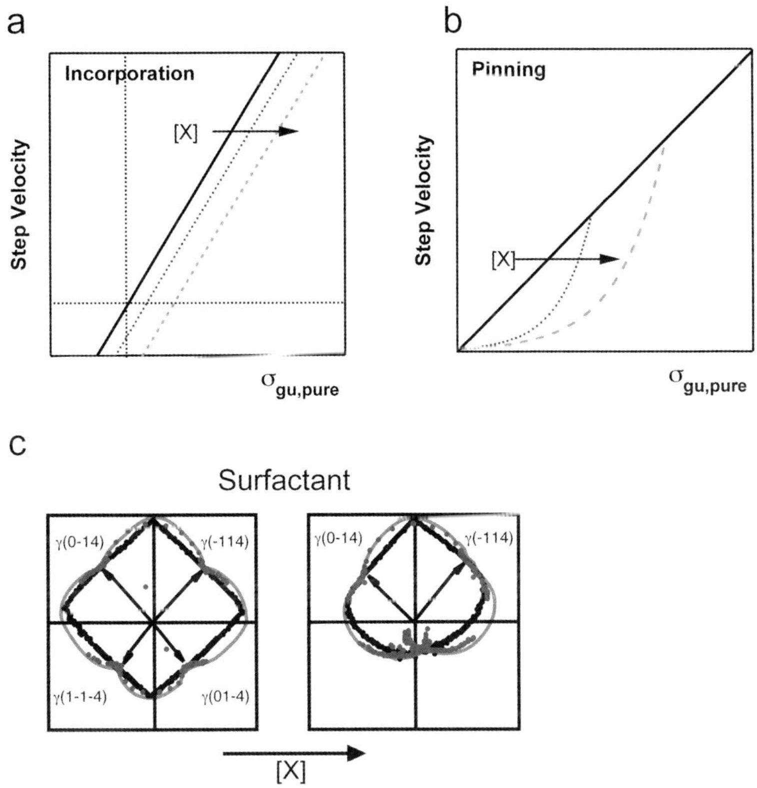

Fig. 8.7 Schematic illustration of how different mechanisms of impurity interactions change the step velocity. As the impurity concentration ([X]) is increased, the velocity versus supersaturation (a,b) or velocity versus orientation (c) changes in characteristically different ways. (a) Strain caused by the substitution of impurity ions changes the equilibrium solubility (σ_0), which is the concentration where the step velocity goes to zero. This causes the velocity curves to shift over, but not to change shape. (b) Impurities that adsorb to steps can prevent the steps from moving due to the high local curvature of the step between the blocked points. This causes a "dead zone" where the velocity of the steps is slow compared to the clean solution (solid line). (c) Surfactants change the step free energy. Two-dimensional slices of pseudo-Wulff plots of calcite before and after the addition of aspartic acid, show that the step-edge free energy, $\gamma(\theta)$, changes due to adsorbed molecules at the steps.

are found incorporated into biogenic apatites, such as Mg, Cl, Na, CO_3, K, and F; (ii) proteins, such as collagen, amelogenin, and albumin; (iii) molecules with phosphate moieties, such as pyrophosphate; and (iv) molecules with carboxyl moieties, such as carbonate and citrate.

There are several generic ways in which an adsorbate can affect growth: (i) it can substitute for similar ions within the crystal which may induce strain [49]; (ii) it can bind to the surface hindering or pinning step motion; and (iii) it can adsorb to steps such that the composite has new equilibrium facet directions (i.e., by acting as a surfactant) [50]. In addition, some adsorbates may act as local ion sources, thereby effectively increasing the local supersaturation; this is particularly true for proteins, many of which are known to have a high capacity for binding calcium and phosphate. It also possible for adsorbates to block or reduce the efficacy of step sources; this has the effect of reducing the step density and thereby the facet growth rate. In general, when impurities are added to the system, the step velocities will change in characteristically different ways (some of which are shown in Fig. 8.7), and different mechanisms can be distinguished by measuring the step velocity as a function of adsorbate concentration and mineral supersaturation. In addition to kinetics, AFM images also provide detailed information about which steps are affected by the adsorbate, and whether new facets or step morphologies are formed.

We will close with two examples that illustrate different mechanisms of inhibiting crystal growth. The first example looks at a classic step pinning process as peptides interact with a calcium carbonate (calcite) crystal, first slowing and finally stopping step propagation. The second example examines the interaction of citrate with DCPD, and shows that step kinetics are unaltered but that step density is reduced. In both cases, bulk measurements of the crystal growth rate would show inhibition but would not be able unambiguously to pinpoint the mechanism by which this occurred.

8.5.1
Inhibition Through Step Pinning

The classic description of pinning mechanisms is based on impurity adsorption at surfaces, steps, or kinks [51–54]. In this case, impurities act as blockers at the sites where they adsorb, preventing the crystal step from propagating locally and thus causing a straight step to become scalloped. As steps curve, their velocity is reduced until they are eventually stopped when their radius of curvature reaches the critical radius (as defined in Section 8.4). In this model, the degree of inhibition depends on the supersaturation and the "blocker" concentration ([X]) on the surface (shown schematically in Fig. 8.7b). Higher concentrations of adsorbate cause a greater reduction in velocity; these effects are more pronounced at lower supersaturations. Figure 8.8 illustrates a surface (calcite) where the normally straight steps are being pinned due to adsorbed peptides. At this particular supersaturation and impurity concentration, the steps continue to propagate but are slowed compared to a clean solution at the same supersaturation. A higher-

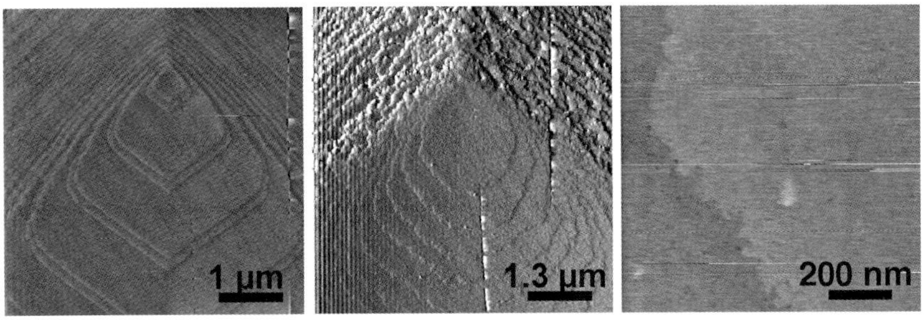

Fig. 8.8 Atomic force microscopy images of the (104) face of calcite. The crystal is growing in solutions with and without peptide additions. Image (a) is in pure solution, whereas images (b) and (c) are in solutions with 6 μM of peptide. The steps become pinned (b) when the peptide is added, reducing the velocity of the growth steps. The image in (c) shows the detail of the torturous path of the step edge.

resolution image (Fig. 8.8c) shows the tortuous path of the step edge, one of the classic signatures of step-pinning.

8.5.2
Inhibition by Reduction of Step Density

The interaction of citrate with brushite presents a non-intuitive example of how modifiers can be used to tune growth rate. Brushite or apatitic phosphates are often found at the center of kidney stones. It has been suggested that brushite may aid the nucleation of calcium oxalate, the majority component of many stones, and could play a role in the aggregation of crystals to form a stone. Citrate is a common therapy administered to recurrent stone formers, and is thought to inhibit crystal growth. In fact, constant composition experiments [40] have shown that citrate does inhibit the growth of brushite [56]. To investigate the mechanisms by which this occurs, AFM was used to directly image the atomic step kinetics and morphology.

Because citrate forms complexes with calcium in solution, there are two ways of performing this experiment, as shown schematically in a supersaturation versus citrate concentration graph (Fig. 8.9a). Citrate can either be added directly to a supersaturated calcium phosphate solution (Expt. 1) or it can be added while supplementing calcium to compensate for Ca-citrate complexes that form, thus keeping the supersaturaton constant (Expt. 2). While biological realism is always difficult to argue in *in-vitro* experiments of this type, Expt. 1 may seem a more natural experiment (after all, when you ingest citrate you do not compensate calcium). However, from a crystal growth point of view the experiment designated by Expt. 1 has two simultaneous affects: the change in supersaturation; and the effects of citrate, which makes interpretation more difficult.

The step velocity data obtained from AFM experiments bear this out. If one uses a titration method (Expt. 1) without compensating for complexation then,

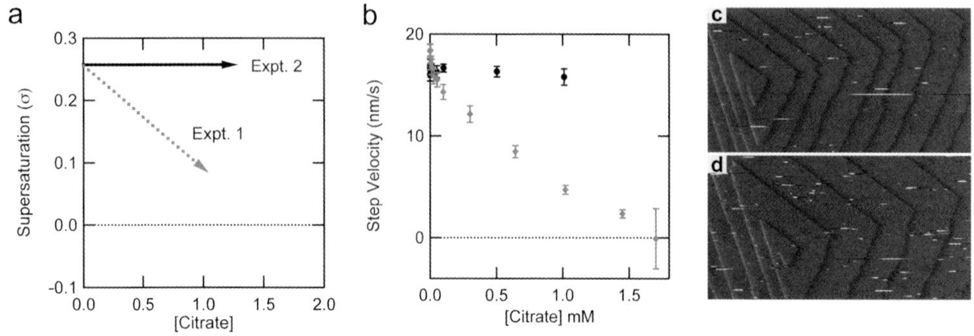

Fig. 8.9 Brushite-citrate experiments designed to study the influence of additives on crystal growth. (a) Schematic showing the addition of a complexing agent (in this case citrate) when added to a growth solution (in this case with respect to brushite). Expt. 1 (dotted gray) demonstrates the case when the complexing of citrate with solution calcium is not accounted for resulting in a decrease in the supersaturation of brushite. Expt. 2 (solid black) shows that a constant supersaturation can be maintained by compensating calcium to determine the affect of citrate independently. (b) Actual step velocity data showing a decrease in step velocity when the supersaturation is allowed to vary (gray), but is constant when the supersaturation is held constant (black). (c,d) AFM images of the same 3 μm × 1.5 μm area on the (010) face of brushite. The crystal is growing in 37 °C solutions with $\sigma_{gu} = 0.14$ before (c) and after (d) the addition of 2×10^{-6} M citrate. The density of steps decreases when citrate is added, reducing the growth rate of this facet.

as might be expected, the step velocities decrease when citrate is added (Fig. 8.9b, gray diamonds). However, by adjusting the calcium to assure a constant supersaturation this effect is removed and the step velocities are constant up to 1 mM citrate (Fig. 8.9b, black circles).

But, in the case where supersaturation is held constant, if citrate does not reduce step velocities and yet does reduce the bulk growth rate, we are left with a puzzle. One possibility is that the step kinetics on facets other than the [010] face are modified. Another possibility presents itself from the AFM images (Fig. 8.9); although step velocities and morphology remains unchanged, the *step density* decreases when citrate is added. Because facet growth rate is proportional to both the number and the speed of the steps, a lower density has a concomitant lower facet growth rate.

In spiral growth, the distance between parallel steps (or step density) is a function of the time it takes for new step to begin to propagate. As discussed in Section 8.4, this can be related either to the critical length or to the probability of nucleating a kink depending on whether thermodynamics or kinetics dominate. If the steps are in equilibrium, then the decrease in step density implies an increase in the critical length and a concomitant increase in step free energy. On the other hand, if the step motion is kinetically limited, the effect of citrate is to decrease the probability of nucleating kinks, which are necessary for step motion. In either case, the interaction with citrate has interesting implications for how organisms might use small amounts of additives to tune crystal growth rate.

This example demonstrates two points. The first point is well appreciated, but worth repeating – namely that additives which complex with other ions in the solution can be misidentified as inhibitors if changes in saturation state are not taken properly into account. The second point is that bulk crystal growth rates depend not only on the velocity of atomic steps but also on the density of steps. In this example, AFM images showed that citrate molecules inhibit crystallization not by changing the speed of the atomic steps (as is generally assumed from bulk rate experiments) but rather by modifying the rate at which steps are generated.

8.6 Outlook

Over the past decade, a greater degree of quantification has been made possible through the use of SPM and advanced diffraction techniques. This, coupled with a greater control of mineral interfaces through advances in synthesis and genetic engineering and the creation of templates with greater specificity and finer-scaled features, has led to our ability to measure and control biomineral growth to with unprecedented precision.

Whilst it is clear that the understanding and exploitation of self-assembling processes has made great advances, it is equally clear that this is an area of great depth that still has untapped potential. Most bio-inspired assembly processing occurs in aqueous environments, whereas most of the highest-resolution characterization tools are only viable in vacuum. Thus, new experimental tools enabling molecular-level characterization and the monitoring of dynamical processes in fluids would have a great impact on the field as a whole. Future advances will rely on yet faster and higher-resolution imaging tools, and there is a gaping need for sensitive spectroscopies capable of determining the bonding of surface species in fluid environments. Theoretical advances in crystal growth will need to embrace the complexity of multiple-component crystals and to examine the regime where the generation of kinks is rate-limiting. Overall, advances will continue to rely on fundamental understanding of the physical controls on materials assembly from intermolecular force, to activation barriers, to thermodynamics.

Acknowledgments

These studies were conducted under the auspices of the U.S. Department of Energy by the University of California, Lawrence Livermore National Laboratory under Contract No. W-7405-Eng-48. Portions of these investigations were supported by the National Institutes of Health (NIDCR grant number DE03223).

References

1 R.Z. Wang, L. Addadi, S. Weiner, *Philos. Trans. Roy. Soc. B* **1997**, *352*, 469.

2 J.F.V. Vincent, *Structural Biomaterials*. Princeton University Press, **1990**.

3 L.N.Y. Wu, B.R. Genge, D.G. Dunkelberger, R.Z. LeGeros, B. Concannon, R.E. Wuthier, *J. Biol. Chem.* **1997**, *272*, 4404.

4 N.H. de Leeuw, *Phys. Chem. Chem. Phys.* **2004**, *6*, 1860.

5 S. Mann, J. Webb, R.J.P. Williams, *Biomineralization: chemical and biochemical perspectives*. VCH, Weinheim, New York, **1989**, p. 541.

6 H.A. Lowenstam, S. Weiner, *On biomineralization*. Oxford University Press, New York, **1989**.

7 E. Bäuerlein, *Biomineralization: Progress in Biology, Molecular Biology and Application*. Wiley-VCH, Weinheim, **2000**.

8 S. Mann, *Principles and Concepts in Bioinorganic Materials Chemistry*. Oxford University Press, Oxford, **2001**.

9 R.Z. Legeros, *Calcium Phosphate in Oral Biology and Medicine*, Vol. 15. Karger, Basel, **1991**.

10 Z. Amjad, *Calcium phosphates in biological and industrial systems*. Kluwer Academic Publishers, Boston, **1998**, p. 515.

11 T. Aoba, *Oral Dis.* **2004**, *10*, 249.

12 S. Weiner, W. Traub, *FEBS Lett.* **1986**, *206*, 262.

13 G. Daculsi, J.M. Bouler, R.Z. LeGeros, *Int. Rev. Cytol.* **1997**, *172*, 129.

14 M. Finke, K.D. Jandt, D.M. Parker, *J. Colloid Interfac. Sci.* **2000**, *232*, 156.

15 A. Evan, J. Lingeman, F.L. Coe, E. Worcester, *Kidney Int.* **2006**, *69*, 1313.

16 R.Z. Legeros, *J. Dent. Res.* **1974**, *53*, 45.

17 J. Abraham, M. Grenon, H.J. Sanchez, C. Perez, R. Barrea, *J. Biomed. Mater. Res. A* **2005**, *75A*, 623.

18 T. Sakae, H. Yamamoto, G. Hirai, *J. Dent. Res.* **1981**, *60*, 842.

19 G. Faure, P. Netter, B. Malaman, J. Steinmetz, *Lancet* **1977**, *2*, 142.

20 Z.F. Chen, B.W. Darvell, V.W.H. Leung, *Arch. Oral Biol.* **2004**, *49*, 359.

21 C.M. Bethke, *Geochemical Reaction Modeling*. Oxford University Press, Oxford, **1996**.

22 C. Bethke, GeoChemist's Workbench, 5.0 ed., RockWare Inc., Golden, CO.

23 K.S. Pitzer, *J. Phys. Chem.* **1973**, *77*, 268.

24 I.S. Harding, N. Rashid, K.A. Hing, *Biomaterials* **2005**, *26*, 6818.

25 A.A. Chernov, *J. Cryst. Growth* **2004**, *264*, 499.

26 R.P. Shellis, *Arch. Oral Biol.* **1978**, *23*, 485.

27 M.J. Larsen, E.I.F. Pearce, *Arch. Oral Biol.* **2003**, *48*, 317.

28 T. Aoba, E.C. Moreno, in: R. Fearnheard, S. Suga (Eds.), *Tooth Enamel IV*. Elsevier Science Ltd, St. Louis, Missouri, USA, 1985, pp. 163–167.

29 H.C. Margolis, E.C. Moreno, *Crit. Rev. Oral Biol.* **1994**, *5*, 1.

30 D. White, *A constant composition study of the kinetics of crystallization and demineralization of calcium oxalate: applications to renal stone disease*. PhD thesis, State University of New York, Buffalo, **1982**.

31 H.C. Anderson, *Clin. Orthop. Relat. Res.* **1995**, 266.

32 H.C. Anderson, R. Garimella, S.E. Tague, *Front. Biosci.* **2005**, *10*, 822.

33 M. Balcerzak, E. Hamade, L. Zhang, S. Pikula, G. Azzar, J. Radisson, J. Bandorowicz-Pikula, R. Buchet, *Acta Biochim. Pol.* **2003**, *50*, 1019.

34 L. Zhang, M. Balcerzak, J. Radisson, C. Thouverey, S. Pikula, G. Azzar, R. Buchet, *J. Biol. Chem.* **2005**, *280*, 37289.

35 R. Robison, *Biochemistry* **1923**, *17*, 286.

36 H.C. Blair, *BioEssays* **1998**, *20*, 837.

37 A.V. Rousselle, D. Heymann, *Bone* **2002**, *30*, 533.

38 S.L. Teitelbaum, *Science* **2000**, *289*, 1504.

39 I.A. Silver, R.J. Murrills, D.J. Etherington, *Exp. Cell Res.* **1988**, *175*, 266.

40 G.H. Nancollas, M.B. Tomson, *J. Dent. Res.* **1978**, *57*, 87.

41 A. Ebrahimpour, J.W. Zhang, G.H. Nancollas, *J. Cryst. Growth* **1991**, *113*, 83.

42 P.G. Vekilov, *Prog. Cryst. Growth Ch.* **2002**, *45*, 175.

43 J.J. De Yoreo, C.A. Orme, T. Land, in: K. Sato, Y. Furukawa, K. Nakajima (Eds.), *Advances in Crystal Growth Research*. Elsevier Science, Amsterdam, **2001**, p. 419.

44 J.J. De Yoreo, P.G. Vekilov, in: P.M. Dove, J.J. De Yoreo, S. Weiner (Eds.), *Biomineralization, Vol. 54*. The Mineralogical Society of America, Washington, **2003**, p. 57.

45 I.V. Markov, *Crystal Growth for Beginners*, 2nd edn. World Scientific Publishing, Singapore, **2003**.

46 H.H. Teng, P.M. Dove, C.A. Orme, J.J. DeYoreo, *Science* **1998**, *282*, 724.

47 A.A. Chernov, L.N. Rashkovich, I.V. Yaminski, N.V. Gvozdev, *J. Phys.-Condens. Mater.* **1999**, *11*, 9969.

48 A.A. Chernov, J.J. De Yoreo, L.N. Rashkovich, P.G. Vekilov, *Mater. Res. Bull.* **2004**, *29*, 927.

49 K.J. Davis, P.M. Dove, J.J. De Yoreo, *Science* **2000**, *290*, 1134.

50 C.A. Orme, A. Noy, A. Wierzbicki, M.T. McBride, M. Grantham, H.H. Teng, P.M. Dove, J.J. DeYoreo, *Nature* **2001**, *411*, 775.

51 N. Cabrera, D.A. Vermilyea, in: R.H. Doremus, B.W. Roberts, D. Turnbull (Eds.), *Growth and perfection in crystals*. Wiley, New York, **1958**, p. 393.

52 N. Kubota, M. Yokota, J.W. Mullin, *J. Cryst. Growth* **2000**, *212*, 480.

53 G.W. Sears, *J. Chem. Phys.* **1958**, *29*, 1045.

54 K. Sangwal, *Cryst. Res. Technol.* **2005**, *40*, 635.

55 J.L. Giocondi, I. Kim, C.A. Orme, J.S. Evans, unpublished results.

56 R.K. Tang, M. Darragh, C.A. Orme, X.Y. Guan, J.R. Hoyer, G.H. Nancollas, *Angew. Chem. Int. Ed.* **2005**, *44*, 3698.

9
Biomimetic Formation of Magnetite Nanoparticles

Damien Faivre

Abstract

Magnetic nanoparticles have both fundamental and technological applications, ranging from environmental to life sciences, and from nanotechnology to mechanics. Magnetotactic bacteria produce magnetic nanoparticles called magnetosomes; these magnetite crystals are embedded in an organic matrix, and have tailored properties. The crystals have a permanent magnetization, though laboratory strains have been created which produce magnetosomes of superparamagnetic size. These magnetic properties, together with the lipidic membrane, confer a very high nanobiotechnological potential to the magnetosomes. The production of magnetosomes in high quantities is problematic, however, and as a consequence biomimetic approaches have been developed in an attempt to mimic the formation of materials observed in the living world. In this chapter, several such synthetic pathways are presented. The best biomimetic approach will be developed when it is realized how these bacteria biomineralize their magnetic inclusions. At that point, the process will be reproduced to permit the development of novel, abiotic routes of synthesis. Attempts to circumvent the problems of working with slow-growing magnetotactic bacteria, by using an "abiomimetic" approach to understand biomineralization, are presented. Finally, an approach coupling *in-vitro* and *in-vivo* experiments is described which should not only pave the way towards an understanding of the magnetite biomineralization process by magnetotactic bacteria, but also aid in the development of successful biomimetic synthetic routes.

Key words: magnetite, biomimetics, biomineralization, magnetosome, magnetotactic bacteria, nanoparticles, nanotechnology, biotechnology.

Handbook of Biomineralization. Edited by P. Behrens and E. Bäuerlein

ISBN: 978-3-527-31805-6

9.1 The Ubiquitous Interest for Magnetite Nanoparticles

Magnetite nanoparticles are of fundamental interest for applications in environmental sciences, in biomedicine, and/or in nanotechnology. This chapter presents an overview of the potential uses of magnetite [$Fe^{3+}{}_{Td}(Fe^{2+}Fe^{3+})_{Oh}O_4$; where Td and Oh are tetrahedral and octahedral lattice position] [1] nanoparticles, in attempt to illustrate the numerous possibilities of these particles. For example, the isotopic and/or magnetic properties of fossilized magnetite particles in sediments or in rocks can be used as paleoenvironmental tracers to reconstruct past climates and understand climatic changes [2–4]. The reactivity of iron oxyhydroxide nanoparticles can also be used in remediation strategies to repair environmental damage resulting from pollution by toxic metals, for example in ancient mines or by radionuclides [5–7]. Biomedical applications are also abundant: for example, nanomagnetite can be used for cellular therapy such as cell labeling, targeting, and also as a tool for cell-biology research to separate and purify cell populations. These materials can also be utilized for tissue repair, drug delivery, magnetic resonance imaging, hyperthermia, or magnetofection [8–11]. Today, research investigations in nanosciences and nanotechnology are aimed at using magnetic nanoparticles as nanomotors, nanogenerators, nanopumps, and other similar nanoscale devices [12]. In this respect, technical applications include the use of magnetite in films [13], in the form of ferrofluids as magnetic inks, in magnetic recording media, in liquid sealings, as dampers in motors and shock absorbers, and for heat transfer in loudspeakers [14].

Initially, we will outline the details of magnetotactic bacteria and their magnetosomes, and the tailored properties of the magnetosomes that enhance the value of these bacterial magnetites. We will then briefly discuss the possibility of using magnetotactic bacteria and their magnetosomes in bionanotechnological applications. As magnetotactic bacteria are unable to produce their magnetic inclusions with high yields, a biomimetic approach is required if nanobiotechnological applications requiring large amounts of particles are foreseen. Thus, we will suggest the types of biomimetic approach that might be undertaken to synthesize inorganic nanoparticles of magnetite that resemble their biogenic counterparts. In addition to obtaining such crystals, an "abiomimetic" approach may also provide suggestions as to the formation conditions of biogenic magnetite. The chapter will conclude by describing possible future developments for the biomimetic formation of magnetite nanoparticles.

9.2 Biogenic Magnetite Nanocrystals

One of the most fascinating examples of microbial synthesis of nanostructures is the biomineralization of magnetosomes [15] by magnetotactic bacteria [16] (Fig.

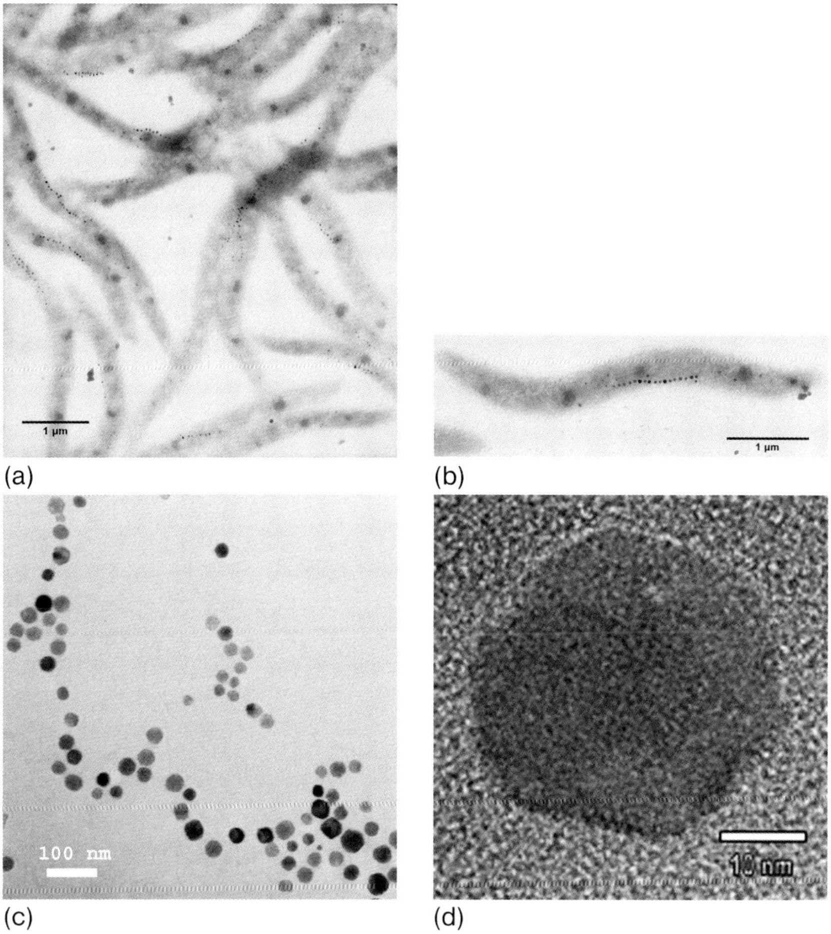

Fig. 9.1 Transmission electron microscopy image of: (a) a group of *Magnetospirillum gryphiswaldense*; (b) an isolated magnetotactic bacterium; (c) isolated magnetosomes; and (d) a magnetosome.

9.1). Magnetosomes consist of a magnetic mineral crystal, magnetite or greigite [17] embedded in a biological membrane that contains phospholipids and specific proteins [18]. Magnetosomes are formed intracellularly, aligned in chains, and serve as a navigational device for spatial orientation along chemical gradients in stratified aquatic habitats, by interaction with the Earth's magnetic field [19, 20].

Magnetosome crystals of magnetite typically range from 30 to about 140 nm in size [21–25]. This means that their size falls within the permanent single magnetic domain [26], maximizing the efficiency of the particle as magnetic carrier. Engineered mutants are available with an altered size, the magnetosome size fall-

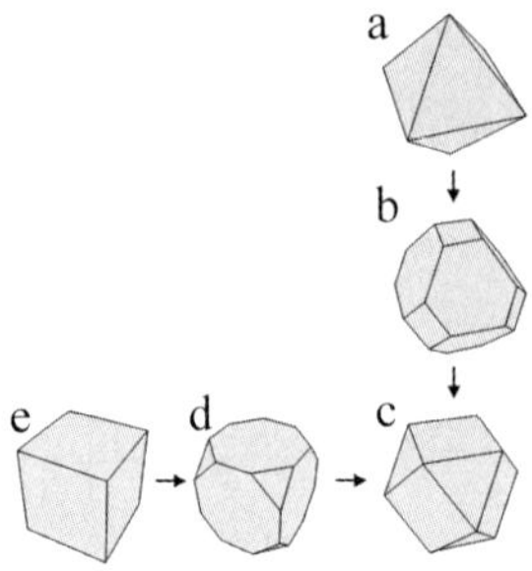

Fig. 9.2 Variation of the theoretical crystal morphology observed for magnetosomes of different strains based on the contribution of the {1 1 1} faces (octahedral) and the {1 0 0} faces (cube). (a) Octahedron; (b) truncated octahedron; (c) cuboctahedron; (d) truncated cube; (e) cube. The {1 1 0} faces can also be observed.

ing within the superparamagnetic domain [27, 28]. Statistical analyses of crystal size distributions (CSDs) of magnetosomes containing magnetite from cultured strains or enriched environmental samples show a narrow, asymmetrical distribution with sharp cut-offs towards a larger size, and with shape factor (or width-to-length ratio) consistent for a given strain or sample [21–25]. The morphology of magnetosomes crystals varies, also being consistent for a given bacterial strain [29], which shows that the process of magnetosome formation is under strict genetic control. The most widespread morphology seems to be the cuboctahedral one, as first discovered for *Magnetospirillum magnetotacticum* MS-1 [30]. Other magnetospirilla, such as *Magnetospirillum gryphiswaldense* MSR-1 or *Magnetospirillum magneticum* AMB-1, produce crystals with slightly different morphologies [21] (Fig. 9.2). Non-equidimensional (elongated) crystals have also been reported [25]. "Truncated hexa-octahedral" or crystals are observed for MV-1 bacteria, and bullet-shaped (or tooth- or arrowhead-shaped) particles have also been reported. All of the magnetosomes genes are arranged in the magnetosome island [31]. For example, it has been shown that formation of the magnetite chain is under genetic control [32, 33], although the process leading to the formation of magnetite is not clear [34].

The specific interest in magnetosomes, which has been constantly growing over the past few years [35], began in 1996 when McKay et al. proposed that nanoparticles of magnetite found in the Martian meteorite ALH84001 might have a biogenic origin [36]. In fact, although the ancient "life on Mars" hypothesis has been extensively challenged, it has inspired numerous studies in the field of geology, crystallography or mineralogy on these nanomagnetite crystals formed by magnetotactic bacteria (see, among others, the following, both contradictory, reports in [25, 37]). Nowadays, geologists use nanoparticles of iron oxides and iron oxyhydroxides as sensitive proxies of environmental change near the Earth's surface. Several studies have aimed to provide information about the environmental control variables (temperature, rainfall, pH, as well as microbial type and concentration) for the process of Fe(III) to Fe(II) reduction and solid-phase precipitation. Such solid phases of environmental change proxies (e.g., magnetite and maghemite) are being produced in the laboratory for the calibration of field data [4]. Of course, it is not only geologists and environmental scientists who are interested in the magnetosomes – interest has been much broader, reaching biology and medicine as magnetosomes may serve as models in our understanding of magnetic nanoparticle formation as magnetic field receptor by higher organisms [38–43] or even humans [44, 45]. Magnetosomes also represent the most complex subcellular structure in prokaryotes, and might inspire cell-biological research. Finally, a multidisciplinary interest has focused on the biogenic magnetite crystals due to possible applications in bio- and nanotechnology [28].

Briefly, magnetosomes are perfect for nanotechnological purposes, as they have all the required properties:

- they present an organic envelope;
- they can have either a single magnetic domain size or superparamagnetic (from mutant cells or non-mature crystals);
- they are nicely dispersed without the need of any organic solvent or surface coating;
- they have a narrow crystal size distribution;
- they have a given morphology, and a high surface-to-volume ratio [28].

The question, therefore, is why do we need to design highly sophisticated methods to produce magnetite crystals by biomimetics? There is one key factor that bacteria cannot presently satisfy – they are neither effective, nor rapid enough to compete with bio-inorganic syntheses. In fact, while simple inorganic coprecipitation experiments can transform grams of iron into magnetite with a near-100% yield in a few minutes, a maximum of only 6.3 mg magnetite per liter of culture and per day can be obtained for *Magnetospirillum gryphiswaldense* [46]. Jogler and Schüler [47] suggested that the heterologous expression of magnetosome-related genes in a faster-growing host bacterium might overcome these limitations, but very few attempts have been made to facilitate magnetosome formation through gene-technology approaches.

Magnetite can also be formed intracellularly or extracellularly by other bacteria or fungi [48–50]. For example, the bacterium *Actinobacter* spp. is capable of magnetite synthesis by reaction with suitable aqueous iron precursors under fully aerobic conditions [48]. This is of advantage because the cultivation of magnetotactic bacteria requires a specific system in order to achieve the micro-aerobic conditions that are not necessary for *Actinobacter* spp. However, as yet the quality of magnetite issued from magnetotactic bacteria has been neither matched nor replicated; neither has the bacterium been persuaded to grow more rapidly.

9.3 Biomimetics

Human beings are always amazed by the high degree of sophistication and miniaturization found in natural materials. Nature is a school for materials science, and its associated disciplines such as chemistry, biology, physics or engineering. Materials chemistry has long been known for its traditional methods of "heat and beat" to prepare compounds, but today nanotechnology requires the fabrication of hybrid organic-inorganic nanocomposites, for which an integrative material chemistry has been developed which attempts to mimic the formation of materials observed in the living world. Such attempts to identify synthetic pathways to obtain materials that have similar properties as those produced naturally is termed "biomimetics". In fact, in biomineralizing systems, whole organisms exert – via organic molecules such as proteins – a high level of control over nucleation and growth of inorganic materials such as carbonate, silicate, or iron oxides.

Several interconnected approaches may be related to biomimetics. First, biological concepts can be tested in inorganic approaches that attempt to create materials with properties that usually are characteristic of biogenic minerals. In the case of magnetite nanocrystals, for example, biomimetic approaches can be purely inorganic syntheses trying to generate nanoparticles of uniform size (a narrow size-distribution), and/or with tailored morphology. For such synthetic routes, the first step consists of magnetite precipitation, and for this methods such as pyrolysis, gas deposition, sol–gel, microemulsion or bulk solution can be used [11]. The most popular syntheses are conducted in bulk solution, as large quantities of products can be formed. In such cases, the co-precipitation of ferric and ferrous iron in alkaline environments leads to the formation of magnetite crystals [51] (Fig. 9.3). Precipitation can be also performed under more controlled conditions,

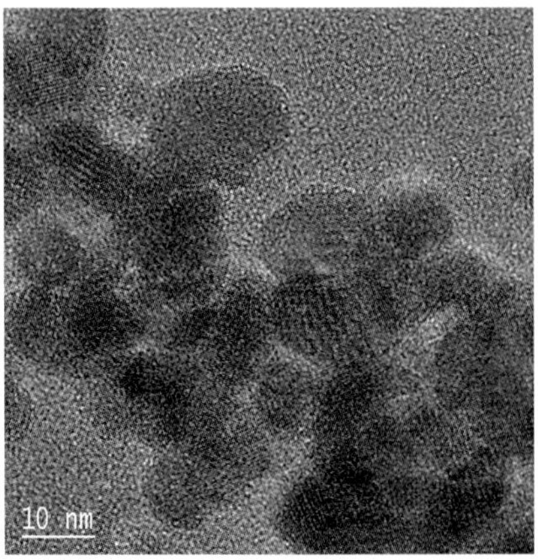

Fig. 9.3 Transmission electron micrograph of typical magnetite crystals obtained by inorganic co-precipitation of ferrous and ferric iron. Note the agglomeration of the crystals.

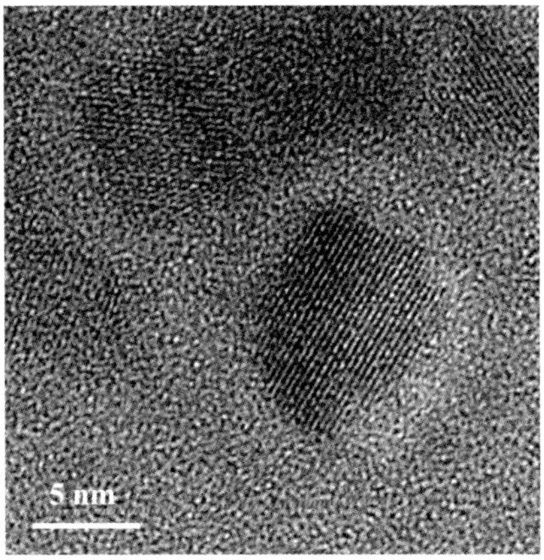

Fig. 9.4 Transmission electron micrograph of magnetite nanoparticles obtained by inorganic co-precipitation of ferrous and ferric iron in the presence of oleic acid to avoid agglomeration and allow later functionalization. Note that superparamagnetic crystals are obtained.

as inspired by biological processes, for example in the presence of a given salt and at a given pH, and this leads to narrower size distributions [52]. The next step is to coat the nanoparticle surfaces to prevent their agglomeration, and for this organic solvents such as oleic acid or alkyl phosphate or phosphonate [53], folic acid or polyethylene glycol [54, 55] have been shown to be powerful surfactants (Fig. 9.4).

The second approach to biomimetics consists of incorporating biological compounds into inorganic assays. For example, preformed magnetite inorganic nanoparticles can be incorporated into macroscopic threads of *Bacillus subtilis* by reversible swelling of the superstructure in colloidal sols. In this case, the organized bacterial superstructures is used as a three-dimensional (3-D) template for the fabrication of ordered inorganic–organic fibrous composites [56, 57]. The most promising approach, however, uses biological compounds such as proteins that are first extracted from the biomineralizing organism, and then used in an *in vitro* bio-inorganic assay. This type of approach has achieved great success in the case of carbonate, for example when phase-switching between calcite and aragonite [58], or for silicate biomineralization [59, 60]. At present, only one promising study has been reported for magnetite formation [61], though further control investigations must be performed.

9.4 Abiomimetics

The best biomimetic approach will be developed once it is known how the bacteria biomineralize their magnetic inclusions. In fact, we will then be able to reproduce the process and to develop novel abiotic routes of syntheses. Despite the fact that over 25 years have elapsed since their discovery, very few magnetotactic bacteria have been isolated, and all of these are difficult to grow. Thus, little is known of how such bacteria biomineralize their magnetic mineral inclusions at the biochemical/chemical and molecular levels. Consequently, the mechanism of physico-chemical control is largely unknown, and must be specified before developing optimal biomimetic conditions. Therefore, an indirect approach called "abiomimetics" can also be followed: improvements in knowledge on magnetite biomineralization are achieved by comparing the crystal's properties as obtained by inorganic syntheses (knowing the conditions of their formation) with the magnetosomes. Thus, the aim is to understand biomineralization by abiomimetics: Rather than simply trying to obtain inorganic crystals that are similar to their biogenic counterparts, the aim also was to deduce the conditions in which biogenic crystals are formed, and the reaction pathway leading to the formation of magnetite nanocrystals. For this purpose, it can be assumed that if inorganic crystals present similarities to their biogenic counterparts, it is implied that they were obtained in an analogous environment and conditions [62].

Abundant methods of synthesis which allow the formation of inorganic magnetite in aqueous media without any control have been reported [23, 51]. However, specific crystals properties cannot, in this case, be correlated to the formation conditions as the chemical affinity changes during the crystal growth. "Chemical affinity" represents the capacity of a reaction to happen, and a slight change in such affinity might have a drastic effect on the reaction products and their properties. Particles obtained by these synthetic pathways will have, for example, a variety of morphologies and dimensions that lead to broad crystal size distributions (CSDs) [23]. Thus, chemical affinity is the key parameter for controlling the synthesis and providing a better understanding of the conditions under which the magnetotactic bacteria form their magnetosomes.

The following section includes a brief (theoretical) explanation for this abiomimetic approach. The rate and mechanism of heterogeneous reactions are a function of the reaction chemical affinity *A*. Classically, mineral dissolution and precipitation reaction rates have most often been expressed in terms of a functional dependence. Using this formalism [63], the rate of reaction R_{net} can be linked to the temperature by:

$$R_{net} = R_{+}(1 - e^{-A/RT}) \tag{1}$$

where R is the perfect gas constant, R_{+} is the rate for the forward reaction in Eq. (2), and T is the temperature (in K).

A variation of the chemical affinity might have an important effect on the global rate of the reaction by affecting the reaction pathway and the mechanism of mineral formation. These changes in mechanism and rate will in turn affect the properties of the reaction product which, in the present case, is the formed mineral (magnetite).

A series of controlled experimental magnetite precipitations can be carried out by the co-precipitation of ferrous and ferric ions in aqueous solution under constant pH [62]. This methodology allows magnetite nanocrystals to be formed under constant and controlled chemical affinity conditions. At a given constant pH, the overall reaction of magnetite precipitation can be represented schematically by a simplified mass balance equation:

$$2\,Fe^{3+} + Fe^{2+} + 4\,H_2O \underset{R_-}{\overset{R_+}{\rightleftharpoons}} Fe_3O_4 + 8\,H^+ \qquad (2)$$

where R_+ is the rate of the forward reaction and R_- the rate of the reverse reaction. Equation (2) indicates that the reaction produces protons. For controlled experimental conditions, the pH can be held constant by the automatic addition of sodium hydroxide solution. These experimental conditions are suitable for forming magnetite at low temperature under constant chemical affinity. The possibility of forming magnetite crystal depends on the saturation state of the parent solution, Ω, defined classically by:

$$\Omega = \frac{a_{Fe^{2+}} a^2_{Fe^{3+}}}{a^8_{H^+} K_S} \qquad (3)$$

where a_X is the activity of the species X and K_s is the thermodynamic magnetite solubility. Typical values for K_s are in the range between $10^{9.61}$ and $10^{12.02}$ [64, 65]. Operationally it is difficult, however, to estimate the real saturation state under environmental low-temperature conditions, as the solubility constants of magnetite are known at high temperature [65, 66] and extrapolations to lower temperatures may be problematic. In addition, recent estimations of solubility on nanometer-size minerals [67–69] have shown that, at this scale, mineral solubility is several orders of magnitude different from that evaluated at the micro- and millimeter scales.

Using this technique, the kinetics of magnetite precipitation were shown to be related to the iron concentration by a first-order rate law [62]. This finding can be explained by a model involving a first step of Fe(III) oxide or hydroxide formation, a second step of reaction of Fe(II) with this solid, and then rapid evolution to magnetite [70]. Furthermore, the fractionation of oxygen isotopes between water and inorganic magnetite was found to be close to 0, which was similar to observations in bacterial magnetites [2] (Fig. 9.5). Oxygen isotopic effects of the same order thus occur during the bacterial precipitation of magnetites and during inorganic magnetite synthesis in aqueous medium at 25 °C. The important implications of these results are that the solution chemistry used here for inorganic precipitation is relevant to a better understanding of magnetite precipitation in bacterial magnetosomes, which might thus be characterized by high saturation states and pH. It would therefore be of great interest to measure actual pH-values and iron concentrations within bacterial magnetosomes. For example, it is known that bacteria can form crystals with as little as 1 μM of iron in their

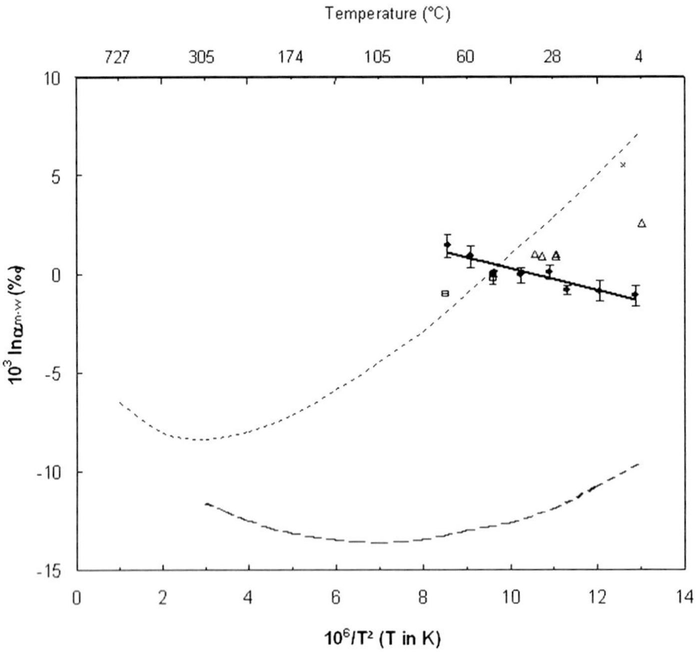

Fig. 9.5 Comparison of oxygen isotopic fractionation between magnetite and water. Data ($10^3 \ln \alpha_{m\text{-}w}$) are plotted as a function of temperature ($10^6/T^2$, T in K). Solid circles: pure inorganic magnetite [71]. Open triangles: magnetite formed by magnetotactic bacteria (intracellular crystals) [2]; open squares: magnetite formed by thermophilic iron-reducing bacteria [72]; crosses: magnetite from chiton tooth [73]. The dotted curve is from Ref. [3] with compilation of data from Refs. [72, 74–76]. The dashed curve E is also from Ref. [3], after Ref. [77] with higher-temperature extension calculated using data from Refs. [78, 79]. These two latter curves represent the extreme behaviors of the magnetite-water fractionation at low temperature (>ca. 15‰ differences between 0 and 50 °C).

environment, whereas magnetite cannot be formed inorganically with so little iron present [62]. Thus, it seems that the bacteria must use a very efficient "iron pump" to enrich the iron in their magnetosomes in order to synthesize their magnetic inclusions.

9.5 Future Considerations

In future, it is likely that biomimetics may achieve great success when our understanding of biomineralization pathways has been significantly advanced, whereupon it would surely be possible to create bioproducts *in vitro* using a biomimetic approach. So, where do we start? One approach which involves extracting proteins from biomineralizing bacteria and using them in a bioinorganic assay has proved very effective for other biomineralizing systems, and will clearly be pursued in the case of magnetite. The latter approach is problematic, however, as many possible candidate proteins exist and several micrograms of each are required in these assays. Therefore, the use of simpler biomolecules – for example, the peptide portion of the proteins, which is considered to possess a specific biomineralizing effect – might be helpful. Such a technique might also permit the identification of proteins, and identify not only the protein interactions required for magnetite precipitation but also those responsible for the given morphology.

Finally, coupling of the *in-vivo* deletion of the gene producing the protein of interest with a bio-inorganic *in-vitro* assay using the same proteins extracted from magnetotactic bacteria, would create the "ultimate" approach to identify the role played by any given protein, as well as improving our understanding of biomineralization and the development of biomimetics.

Acknowledgments

The author thanks Professors Baeuerlein and Behrens for inviting this contribution to the book, and N. Menguy and C. Lang for providing the electron microscopy images. The author's thanks are also expressed to D. Schüler for the opportunity to join his group, to J.-P. Jolivet, F. Guyot and P. Zuddas for much-appreciated discussions, and to C. Lang, C. Jogler and A. Scheffel for corrections to, and suggestions for, the text. These studies were supported by the Deutsche Forschungsgemeinschaft, the German BMBF and the Max Planck Society. D.F. is supported by a Marie Curie fellowship from the European Union.

References

1 W.A. Deer, R.A. Howie, J. Zussman, *An introduction to the rock forming minerals*. Longmans, Grenn and Co. Ltd, London, **1966**.

2 K.W. Mandernack, D.A. Bazylinski, W.C. Shanks, T.D. Bullen, *Science* **1999**, *285*, 1892.

3 R.L. Ripperdan, L.R. Riciputi, D.R. Cole, R.D. Elmore, S. Banerjee, M.H. Engel, *J. Geophys. Res.* **1998**, *103*, 21015.

4 S.K. Banerjee, *Phys. Earth Planet. Inter.* **2006**, *154*, 210.

5 J.R. Lloyd, *FEMS Microbiol. Rev.* **2003**, *27*, 411.

6 J. Hu, I.M.C. Lo, G. Chen, *Water Sci. Technol.* **2004**, *50*, 139.

7 A.S. Bahaj, I.W. Croudace, P.A.B. James, F.D. Moeschler, P.E. Warwick, *J. Magn. Magn. Mater.* **1998**, *184*, 241.

8 M. Sarikaya, C. Tamerler, A.K.-Y. Jen, K. Schulten, F. Baneyx, *Nat. Mater.* **2003**, *2*, 577.

9 I. Safarik, M. Safarikova, *Monatshefte fur Chemie* **2002**, *133*, 737.

10 P. Tartaj, M.D. Morales, S. Veintemillas-Verdaguer, T. Gonzalez-Carreno, C.J. Serna, *J. Phys. D: Appl. Phys.* **2003**, *36*, R182.

11 A.K. Gupta, M. Gupta, *Biomater.* **2005**, *26*, 3995.

12 F. Gazeau, C. Baravian, J.-C. Bacri, R. Perzynski, M.I. Shliomis, *Phys. Rev. E* **1997**, *56*, 614 LP.

13 L. An, Z. Li, W. Li, Y. Nie, Z. Chen, Y. Wang, B. Yang, *J. Magn. Magn. Mater.* **2006**, *303*, 127.
14 M. Zahn, *J. Nanoparticle Res.* **2001**, *3*, 73.
15 D. Balkwill, D. Maratea, R.P. Blakemore, *J. Bacteriol.* **1980**, *141*, 1399.
16 R.P. Blakemore, *Science* **1975**, *190*, 377.
17 R.B. Frankel, R. Blakemore, *Science* **1979**, *203*, 1355.
18 D. Schüler, *Arch. Microbiol.* **2004**, *181*, 1.
19 D.A. Bazylinski, R.B. Frankel, *Nat. Rev. Microbiol.* **2004**, *2*, 217.
20 C.B. Flies, H.M. Jonkers, D. de Beer, K. Bosselmann, M.E. Böttcher, D. Schüler, *FEMS Microbiol. Ecol.* **2005**, *52*, 185.
21 D. Faivre, N. Menguy, D. Schüler, **2006** (unpublished results).
22 B. Arató, Z. Szanyi, C.B. Flies, D. Schüler, R.B. Frankel, P.R. Buseck, M. Pósfai, *Am. Miner.* **2005**, *90*, 1233.
23 B. Devouard, M. Posfai, X. Hua, D.A. Bazylinski, R.B. Frankel, P.B. Buseck, *Am. Miner.* **1998**, *83*, 1387.
24 U. Lins, M. Farina, *Microsc. Res. Tech.* **1998**, *42*, 459.
25 K.L. Thomas-Keprta, D.A. Bazlinski, J.L. Kirschvink, S.J. Clemett, D.S. McKay, S.J. Wentworth, H. Vali, E.K. Gibson, Jr., C.S. Romanek, *Geochim. Cosmochim. Acta* **2000**, *64*, 4049.
26 R.F. Butler, S. Banerjee, *J. Geophys. Res.* **1975**, *80*, 4049.
27 A. Hoell, A. Wiedenmann, U. Heyen, D. Schüler, *Phys. B* **2004**, *350*, e309.
28 C. Lang, D. Schüler, in: B. Rehm (Ed.), *Microbial Bionanotechnology: Biological Self-assembly Systems and Biopolymer-based Nanostructures.* Horizon Bioscience, Wymondham, **2006**, p. 107.
29 E. Baeuerlein, *Angew. Chem. Int. Ed.* **2003**, *42*, 614.
30 S. Mann, R.B. Frankel, R.P. Blakemore, *Nature* **1984**, *310*, 405.
31 S. Ullrich, M. Kube, S. Schübbe, R. Reinhardt, D. Schüler, *J. Bacteriol.* **2005**, *187*, 7176.
32 A. Scheffel, M. Gruska, D. Faivre, A. Linaroudis, J.M. Plitzko, D. Schüler, *Nature* **2006**, *440*, 110.
33 A. Komeili, Z. Li, D.K. Newman, G.J. Jensen, *Science* **2006**, *311*, 242.
34 D. Faivre, L. Boettger, B. Matzanke, D. Schüler, **2006** (unpublished results).
35 E. Baeuerlein, *Biomineralization.* Wiley-VCH, Weinheim, **2000**.
36 D.S. McKay, E.K. Gibson, Jr., K.L. Thomas-Keprta, H. Vali, C.S. Romanek, S.J. Clemett, X.D.F. Chilier, C.R. Maechling, R.N. Zare, *Science* **1996**, *273*, 924.
37 D.C. Golden, D.W. Ming, R.V. Morris, A.J. Brearley, H.V. Lauer, Jr., A.H. Treiman, M.E. Zolensky, C.S. Schwandt, G.E. Lofgren, G.A. McKay, *Am. Miner.* **2004**, *89*, 681.
38 M.E. Deutschlander, S.C. Borland, J.B. Phillips, *Nature* **1999**, *400*, 324.
39 M. Winklhofer, E. Holtkamp-Rötzler, M. Hanzlik, G. Fleissner, N. Petersen, *Eur. J. Miner.* **2001**, *13*, 659.
40 W. Wiltschko, U. Munro, R. Wiltschko, J.L. Kirschvink, *J. Exp. Biol.* **2002**, *205*, 3031.
41 B.A. Maher, *Proc. R. Soc. Lond. B* **1998**, *265*, 733.
42 K.J. Lohmann, S. Johnsen, *Trends Neurosci.* **2000**, *23*, 153.
43 V. Courtillot, G. Hulot, M. Alexandrescu, J.-L. le Mouël, J.L. Kirschvink, *Terr. Nova* **1997**, *9*, 203.
44 J. Dobson, P. Grassi, *Brain Res. Bull.* **1996**, *39*, 255.
45 J.R. Dunn, M. Fuller, J. Zoeger, J. Dobson, F. Heller, J. Hammann, E. Caine, B.M. Moskowitz, *Brain Res. Bull.* **1995**, *36*, 149.
46 U. Heyen, D. Schüler, *Appl. Microbiol. Biotechnol.* **2003**, *61*, 536.
47 C. Jogler, D. Schüler, in: D. Schüler (Ed.), *Magnetoreception and magnetosomes in bacteria.* Springer, Heidelberg, **2006**.
48 A. Bharde, A. Wani, Y. Shouche, P.A. Joy, B.L.V. Prasad, M. Sastry, *J. Am. Chem. Soc.* **2005**, *127*, 9326.
49 A. Bharde, D. Rautaray, V. Bansal, A. Ahmad, I. Sarkar, S.M. Yusuf, M. Sanyal, M. Sastry, *Small* **2006**, *2*, 135.
50 D. Schüler, *Magnetoreception and magnetosomes in bacteria.* Springer, Heidelberg, **2006**.

51 R.M. Cornell, U. Schwertmann, *The Iron Oxides: Structure, Properties, Reactions, Occurrences and Uses*. Wiley-VCH, Weinheim, **2003**.

52 L. Vayssières, C. Chanéac, E. Tronc, J.P. Jolivet, *J. Colloid Interfac. Sci.* **1998**, *205*, 205.

53 Y. Sahoo, H. Pizern, T. Fried, D. Golodnitsky, L. Burstein, C.N. Sukenik, G. Markovich, *Langmuir* **2001**, *17*, 7907.

54 Y. Zhang, J. Zhang, *J. Colloid Interfac. Sci.* **2005**, *283*, 352.

55 Y. Zhang, N. Kohler, M. Zhang, *Biomater.* **2002**, *23*, 1553.

56 S.A. Davis, H.M. Patel, E.L. Mayes, N.H. Mendelson, G. Franco, S. Mann, *Chem. Mater.* **1998**, *10*, 2516.

57 N.H. Mendelson, *Science* **1992**, *258*, 1633.

58 A.M. Belcher, X.H. Wu, R.J. Christensen, P.K. Hansma, G.D. Stucky, D.E. Morse, *Nature* **1996**, *381*, 56.

59 N. Kröger, R. Deutzmann, M. Sumper, *Science* **1999**, *286*, 1129.

60 H. Menzel, S. Horstmann, P. Behrens, P. Bärnreuter, I. Krueger, M. Jahns, *Chem. Commun.* **2003**, *2003*, 2994.

61 A. Arakaki, J. Webbs, T. Matsunaga, *J. Biol. Chem.* **2003**, *278*, 8745.

62 D. Faivre, P. Agrinier, N. Menguy, P. Zuddas, K. Pachana, A. Gloter, J.-Y. Laval, F. Guyot, *Geochim. Cosmochim. Acta* **2004**, *68*, 4395.

63 A.C. Lasaga, *Kinetic theory in the earth sciences*. Princeton University Press, Princeton, **1998**.

64 R.M. Cornell, U. Schwertmann, *The Iron Oxides: Structure, Properties, Reactions, Occurrence and Uses*. VCH, Weinheim, **1996**.

65 F.H. Sweeton, C.F. Baes, Jr., *J. Chem. Thermodyn.* **1970**, *2*, 479.

66 S.E. Ziemniak, M.E. Jones, K.E.S. Combs, *J. Solution Chem.* **1995**, *24*, 837.

67 M.F. Hochella, Jr., *Earth Planet. Sci. Lett.* **2002**, *203*, 593.

68 M.F. Hochella, Jr., *Geochim. Cosmochim. Acta* **2002**, *66*, 735.

69 Y. Wang, C. Bryan, H. Xu, H. Gao, *Geology* **2003**, *31*, 387.

70 J.P. Jolivet, P. Belleville, E. Tronc, J. Livage, *Clays Clay Miner.* **1992**, *40*, 531.

71 D. Faivre, P. Zuddas, *Earth Planet. Sci. Lett.* **2006**, *243*, 53.

72 C. Zhang, S. Liu, T.J. Phelps, D.R. Cole, J. Horita, S.M. Fortier, M. Elless, J.W. Valley, *Geochim. Cosmochim. Acta* **1997**, *61*, 4621.

73 J.R. O'Neil, R.N. Clayton, in: H. Craig (Ed.), *Isotopic and Cosmic Chemistry*. North Holland, **1964**, p. 157.

74 R.H. Becker, R.N. Clayton, *Geochim. Cosmochim. Acta* **1976**, *40*, 1153.

75 P. Blattner, W.R. Braithwaite, R.B. Glover, *Isotope Geosci.* **1983**, *1*, 195.

76 S.M. Fortier, D.R. Cole, D.J. Wesolowski, L.R. Riciputi, B.A. Paterson, J.W. Valley, J. Horita, *Geochim. Cosmochim. Acta* **1995**, *59*, 3871.

77 M.W. Rowe, R.N. Clayton, T.K. Mayeda, *Geochim. Cosmochim. Acta* **1994**, *58*, 5341.

78 R.N. Clayton, S.W. Kiefer, in: H.P. Taylor, Jr. et al. (Eds.), *Stable Isotope Geochemistry: A Tribute to Samuel Epstein, Vol. 3*. Geochemical Society Special Publication, **1991**, p. 3.

79 P. Richet, Y. Bottinga, M. Javoy, *Annu. Rev. Earth Planet. Sci.* **1977**, *5*, 65.

Part II
Bio-Inspired Materials Synthesis

Handbook of Biomineralization. Edited by P. Behrens and E. Bäuerlein

ISBN: 978-3-527-31805-6

10
Using Ice to Mimic Nacre: From Structural Applications to Artificial Bone

Sylvain Deville, Eduardo Saiz, and Antoni P. Tomsia

Abstract

Materials that are strong, ultra-lightweight and tough are in demand for a range of applications from automotive to medical. These requirements will best be met by new composite materials, the components and interfaces of which are engineered at the molecular level, and the architectures of which are carefully designed from the meso-scale down to nano-scale dimensions while combining the favorable characteristics of several components. Nacre (seashells) and bone are frequently used as examples for how Nature achieves this through biomineralized, hybrid organic–inorganic composites that are highly optimized for specific functions. Unfortunately, it has proven extremely difficult to transcribe nacre-like clever designs into synthetic materials, in part because their intricate structures need to be replicated at several length-scales. In this chapter we describe how nacre-like materials can be obtained by controlling the freezing of ceramic slurries followed by subsequent ice sublimation and sintering, leading to multilayered porous ceramic structures with homogeneous and well-defined architecture, which can be subsequently filled with a selected second phase to obtain dense, complex composites.

Key words: biomineralized structure, nacre, biomimetics, ice, freezing, ceramic, composites, particles segregation, multilayer.

10.1
Nacre as a Blueprint

10.1.1
Biomineralized Natural Structures

There exist innumerable examples where biomineralized natural structures encountered in living organisms are far more sophisticated and efficient than

Handbook of Biomineralization. Edited by P. Behrens and E. Bäuerlein

ISBN: 978-3-527-31805-6

manufactured materials. The targeted properties in natural materials are often achieved through a complete integration of components from the nanometer scale to the millimeter, or even centimeter, scale. Nanocomposites – the new grail of materials scientists – have been encountered in biological systems for millions of years. The unique properties of natural layered materials and nanocomposites are achieved through a fine control of the layers thickness, selection of the correct components, and manipulation of the roughness and adhesion at the organic–inorganic interface. Although built at room temperature and from a narrow selection of intrinsically weak materials (phosphates, carbonates), the functional properties exhibited by biological composites often surpass those of synthetic materials, including those made with state-of-the-art techniques. Such biological structures have hence long attracted the attention of scientists and engineers as fine blueprints to guide the design of new advanced materials. In recent decades, one of the most investigated substances found in such organisms has been nacre.

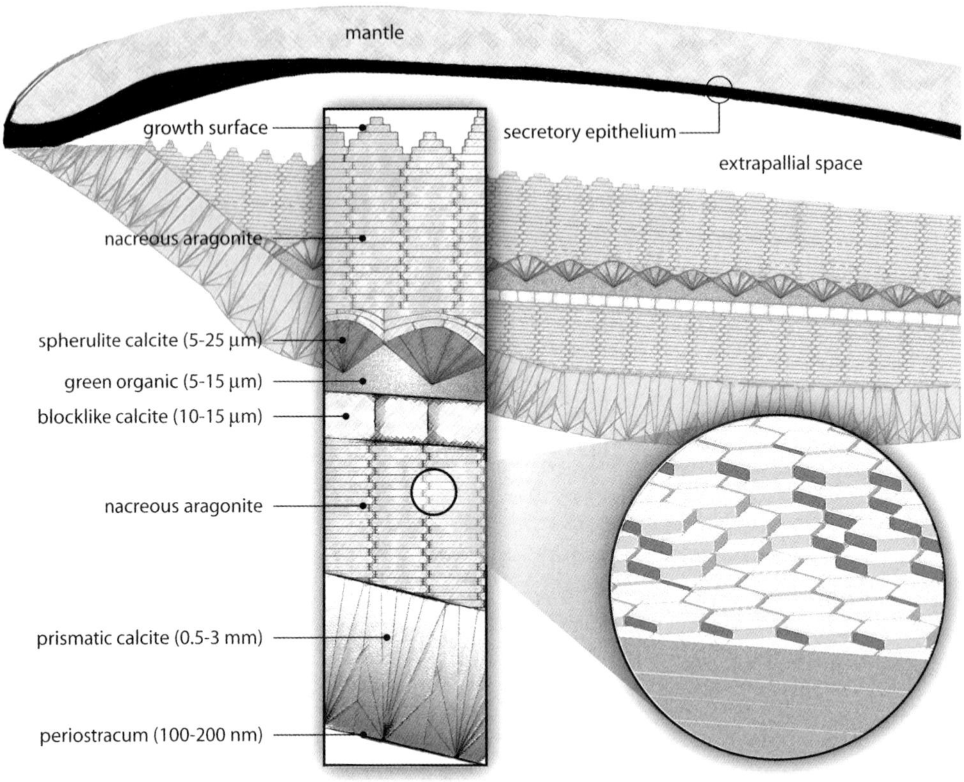

Fig. 10.1 Schematic of abalone shell and mantle (adapted from Ref. [32]).

10.1.2
Structure of Nacre

Nacre is the iridescent material that thickly coats many shells and mollusks, such as the abalone shell, and which has long attracted the attention of materials scientists. The structural part of interest, on the basis of its astonishing properties, is the layer of nacreous aragonite (Fig. 10.1) which covers the inside of the shell. Nacre is a nanocomposite composed of 95% calcium carbonate, one of the most abundant – but also one of the weakest – minerals on Earth. Despite the intrinsic weakness of calcium carbonate, this natural nanocomposite surpasses by far all manufactured ceramics with similar composition. With regard to its mechanical properties, nacre is very good, even compared with technical ceramics created by different chemistries. For abalone nacre, the work of fracture is between 1000 and 3000-fold greater than that of a single crystal of the pure mineral [1].

The underlying reasons for this strength arise from the complex hierarchical architecture of nacre, which is carefully defined across numerous length scales, from the nanometer to the millimeter. When examined under a microscope, nacre appears as a marvelously arranged layered material (Fig. 10.2a), with the

Fig. 10.2 The structure of nacre. (a) Multilayered structure. (b) Protein layer encountered between the mineral platelets. (c) Mineral platelets and their roughness and interlocks. (d) Nanograins of calcium carbonates observed within the mineral platelets.

layers being composed of individual mineral platelets each of approximately 5 μm in diameter and 0.5 μm in thickness (Fig. 10.2c). Platelets exhibit nano-asperities that provide a specific surface roughness, and a few nanoscale-sized columns are found in the organic matrix layers [2]. The aragonite columns, which in biomineralization are traditionally referred to as *mineral bridges*, pass through the mortar layers from one platelet to another and appear to be almost circular. Interlocks are also observed (Fig. 10.2c) between the platelets of nacre [3]; when two platelets are stacked one above another, the upper platelet with organic material penetrates into the lower platelet. The depth of the interlocking feature was found, on average, to be 20% of the thickness of the platelets.

The tablets of nacre are constructed from a continuous organic matrix, which breaks the mineral up into coherent nanograins (20- to 50-nm size; Fig. 10.2d) which share the same crystallographic orientation [4]. The mineral platelets are separated by thin, 10- to 30-nm-thick layers of biological organic adhesive (Fig. 10.2b) composed of polysaccharide and protein fibers; this represents the remaining 5 vol% of the material, and contributes extensively to the unique physical and mechanical properties of nacre [5].

10.1.3
Toughening Mechanisms in Nacre

Numerous toughening mechanisms have been identified in nacre, each of which is related to one particular or several features of the architecture.

The first level is the multilayered structure. Multilayering provides properties of toughness to composite materials by forcing the crack to take a convoluted path, and so require more energy. The layers also provide stiffening to the structure and weaken the crack-tip stress concentration [6]. The platelets of calcium carbonate play a key role in stress redistribution. Upon loading in tension parallel to the platelets, the interface at the platelets' edges rupture, whereby slip can occur, leading to the formation of dilatation bands (arising from the climbing of opposite asperities). These dilatation bands provide stress redistribution, and their development prevents the formation of a single brittle crack.

Progressive failure of the interlocks during loading guides the fracture path and provides an increase of yield stress [3].

The surface of the platelets is covered by inorganic asperities. The sliding of platelets involves the climbing of opposite asperities, which in turn creates a transverse compressive stress which is responsible for hardening. Stress redistribution also occurs because of the inelastic strain. The asperities cannot be too large, however, or failure of the platelets will occur before sliding and the formation of dilatation bands [7]. The opposite asperities are topographically matched before loading, but otherwise have dispersed wavelengths and aspect ratios. The strain hardening allows the formation of multiple dilatation bands, ensuring a ductile behavior rather than catastrophic failure at stress concentration sites.

Mineral bridges that connect adjacent platelets have been observed. Such bridges are found mostly on the center of aragonite bricks, but none is found at the platelets' junctions. Crack resistance and fracture toughness increase as the crack moves towards the center of the platelet, but when the crack is arrested in the central region it has to renucleate in the gap of a neighboring platelet layer, hence causing a periodic deflection of the crack.

Nanograins constituting the platelets are stacked along the same crystallographic direction in order to achieve extremely high adhesion [8]. The large number of nanograins within individual platelets seems to provide ductility to the platelet, allowing them to deform plastically before failure. As a result, stress redistribution occurs around the stress concentration sites, blunting the crack tip. In addition, the nanograins have a size just below the identified critical size, below which the Griffith criterion is no longer operative. As a result, the material becomes insensitive to flaws, in the sense that pre-existing cracks no longer propagate in the structure [9].

Finally, the thin organic layer plays a key role in energy dissipation and bridging of the cracks; this is provided by a gradual unfolding of the domain structure of the proteins. The organic layer must be weak enough to separate before the plates fail, ensuring multiple dilatation bands and a large shear strain before fracture. In addition, it also determines the crystallography of underlying crystals. Finally, it provides lubrication of the interface, leading to a friction coefficient close to zero [10], allowing stress redistribution through easier sliding of the platelets.

Thus, the various toughening mechanisms can be summarized as follows: crack blunting/branching, microcrack formation, plate pull-out, crack bridging, sliding of platelets, crack bridging by protein fibrils, weakened tip stress concentration and stress redistribution, and strain hardening.

The outer prismatic layer of the abalone shell (see Fig. 10.1) does not show these crack-diversion mechanisms, and serves mainly as a brittle outer shield.

10.1.4 Why Mimic Nacre?

For materials scientists, nacre has long been one of the ultimate natural structures to copy. With mechanical properties comparable to those of classical technical ceramics, yet being processed from intrinsically weak materials, the unique hierarchical architecture of nacre represents the optimum of how to overcome materials weakness by hierarchical design to strengthen and toughen structures. From an engineering point of view, two aspects of nacre have long been envied:

- This natural structure is probably one of the best known in regards of the armor properties – that is, the resistance to impact. The numerous toughening mechanisms of nacre lead to a remarkable fracture resistance in the range of 1000 to 3000 J m^{-2}, which is a three-orders of magnitude increase compared to that of calcium carbonate (which represents 95% of the structure). Composites made from technical ceramics achieve a maximum increase in properties of one order of magnitude compared to the bulk materials.

- This composite is processed under very mild chemical conditions and at room temperature. This should be compared to the densification temperatures routinely used in the ceramics industry (1300–2000 °C).

Thus, synthetic nacre processed from technical ceramics might lead to an entirely new class of highly optimized materials with unique properties.

10.1.5
Currently Available Techniques for Mimicking Nacre

The nacre structure can be considered from two different points of view: (i) as a multilayered structure; or (ii) as a self-assembled nanocomposite. Strategies for mimicking nacre have hence been developed following this analysis.

The first strategy utilizes the top-down approach – that is, the processing of bulk multilayered materials. While the potential of layered materials has long been recognized, their creation requires solving a two-fold problem, namely the design of optimum microstructures, and the development of fabrication procedures to implement these designs. The ideal fabrication process must be not only simple but also adaptable enough to fabricate layers with a large number of material combinations and a wide range of layer dimensions. Currently available layering techniques offer only a coarse control of the layer thickness or material range limitations. Practical limitations regarding the number of layers that can be fabricated are still encountered, in particular when small thicknesses are desired (<50 μm). In addition, multilayered materials typically offer control over only two degrees of hierarchy: (i) thickness and orientation of the layers; and (ii) control of the interfacial properties between layers (roughness, residual stresses).

The second approach is the bottom-up one, gathering techniques like crystallization beneath Langmuir monolayers, crystallization on self-assembled monolayers, supramolecular self-assembly, and sequential deposition [11]. All these techniques have in common a control of the microstructure at the nanometer scale. However, they are intrinsically limited to a narrow range of materials exhibiting the proper interfacial reactions and compatibility. Practical limitations regarding the number of layers that can be fabricated are still always encountered with these techniques, which makes their scaling-up for practical applications very difficult. Great hopes are attached to self-assembly techniques with promising preliminary results [12, 13], with concepts similar to those encountered in Nature.

10.2
A Natural Segregation Principle

Among the ceramics and polymer community the technique introduced here has long been referred to as *freeze-casting* or *freeze-drying*; it is a simple technique that produces porous, complex-shaped polymeric or ceramic parts. In freeze-casting, a ceramic slurry (a suspension of ceramic particles, usually in water) is poured into a mold and then frozen. The frozen solvent acts temporarily as a binder to hold the part together for demolding. Subsequently, the part is subject to freeze-drying to sublimate the solvent under vacuum, thus avoiding the drying stresses and shrinkage that may lead to cracks and warping during normal drying. After drying, the compacts are sintered in order to fabricate a porous material with improved strength, stiffness, and desired porosity. The result is a scaffold with a complex and often anisotropic porous microstructure that is generated during freezing. By controlling the growth direction of the ice crystals, it is possible to impose a preferential orientation for the porosity in the final material [14].

10.2.1 Basics of the Technique

The processing steps (Fig. 10.3) take advantage of the water–ice phase diagram. Freezing of the ceramic slurry is performed to build an interpenetrating scaffold of ice and ceramic particles. The ice is then removed by sublimation, such that a ceramic scaffold, the microstructure of which is a negative replica of the ice, is produced. The resultant ceramic part is then taken back to room temperature be-

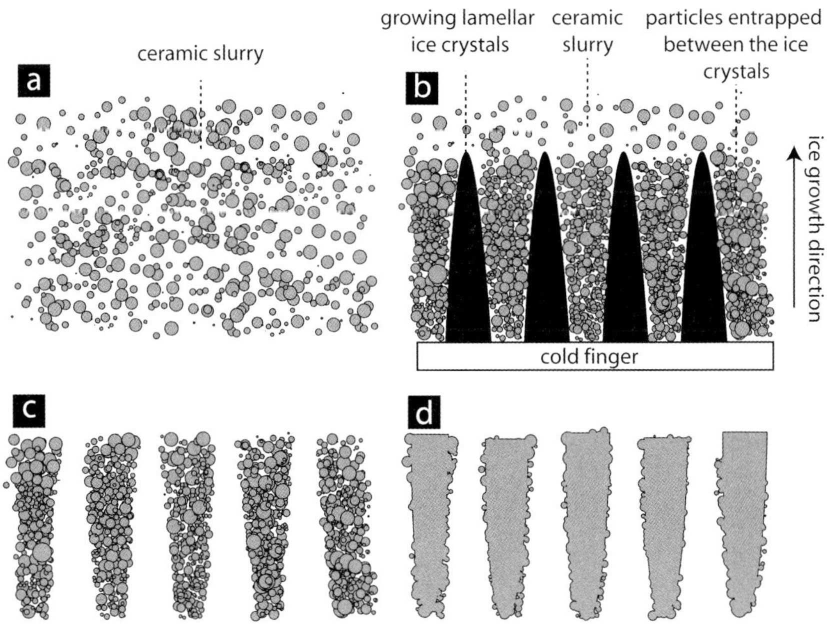

Fig. 10.3 Freeze-casting processing steps. (a) Suspension of ceramic particles in the slurry. (b) Directional freezing and segregation of particles between lamellar ice. (c) Sublimation of ice. (d) Binding of particles (densification) at high temperature.

fore the usual densification stage at high temperature to consolidate the structure. The porosity of the sintered materials is a replica of the ice structure before drying. This ceramic scaffold can then be used as a basis for a dense composite, by infiltrating it with a suitable second phase, whether organic (polymer) or inorganic (metal).

10.2.2 Previous Achievements

To date, freeze-casting (also known as freeze-drying or freeze-gelation) has been applied to a variety of different materials, including ceramics, polymers, composites made of both, and hydrogels.

10.2.2.1 Ceramics

With regards to ceramics processed by freezing-casting, two features of the process have attracted the attention of ceramists, with different applications in mind. Different types of ceramics have hence been freeze-casted, including alumina [15], silicon nitride [16], and NiO-YSZ [17]. The first idea is to take advantage of the specific porosity templated by ice, thereby processing materials that could be used for gas filtration, separation filters, catalyst supports, etc. The requirements are very specific to each application, and the main interest of the technique in such cases is the control of the total porosity, orientation of the porosity, and the control of its characteristics (shape, size). In addition, the simplicity of the process (limited shrinkage, variety of materials that might be used, absence of solvent, etc.) provides additional advantages. The porous structure is based on the interaction between the ice front and the incorporated ceramic particles. Optimal conditions should be identified in terms of porous structure and strength.

A second approach was developed, with interest not in the porous structure but rather in the possibility of processing complex-shaped ceramic parts, by obtaining net-shape green bodies [18]. In such cases, dense materials are desired, so that the suspensions are highly loaded and no porous structures developed. The use of ice as a binder allows minimization of the additives concentration and the use of organic binders, which must later be removed by burning. Sublimation of the binder also reduces the amount of shrinkage, as well as subsequent cracking and warping. Since the formation of continuous ice crystals turning later into pores is not desired, additives such as glycerol must be used to modify the water crystallization behavior.

With all ceramic materials, a densification stage at high temperature (sintering) is necessary after ice sublimation in order to consolidate the structure and bind the particles together.

10.2.2.2 Polymers

Polymer scientists have already taken advantage of the technique for a wide variety of materials, including collagen [19], chitosan [20], agarose [21], and alginate

[22]. Sponges made from biodegradable materials have found many applications, including bone or soft connective tissue implantation, nerve regeneration, wound dressing, and tissue engineering. The pores are elongated and oriented along the freezing direction. Several strategies have been identified to modify the pore structure, including modification of the freezing regime, with faster freezing rates yielding smaller pores and the use of additives, such as acetic acid or ethanol [19], to increase the amount of constitutional supercooling due to solute rejection. The technique is of particular interest since the porogen (ice) can be removed totally by sublimation, and washing steps are not necessary. Hence, drugs, proteins, or any other active substance can be efficiently incorporated into the scaffolds during the processing, without affecting their biological activity. Finally, extremely high porosity (>90%) can be easily achieved. The process is somewhat simpler than with ceramics, where a high-temperature densification stage is necessary to consolidate the scaffolds. In the case of collagen, for example, water removal causes cross-linking between the collagen aggregates, and the scaffolds are ready to use after the ice is sublimated.

10.2.2.3 **Composites**

Following the examples of ceramics and polymers, composites made of both materials have been processed, opening up a new class of functional properties. Such composites include hydroxyapatite (HAP)/collagen [23] or cerium oxide/poly(vinyl alcohol) (PVA) or PVA/silica [24, 25]. The addition of micro- or nanoparticles can be justified for several reasons, such as improvement of mechanical properties or activation of catalytic properties. The structure is logically very similar to that of polymers and ceramics.

10.2.2.4 **Hydrogels (Silica)**

It is also worth mentioning the application of freeze-casting to hydrogels, such as silica gel [26]. The phase separation occurring between the water and the freshly gelled hydrogels is used to tailor the pore structure. This results in a unique structure with macro-, meso-, and micro-pores in advanced materials that might find numerous applications in separation and reaction processes.

10.2.3 Underlying Physical Principles

The process described here is self-assembly in nature, and inspired by a naturally occurring phenomenon – the freezing of sea ice – which occurs at the surface of the Earth's polar oceans. In sea ice, pure hexagonal ice platelets with randomly oriented horizontal *c*-axes are formed, and the various impurities originally present in seawater (salt, biological organisms, etc.) are expelled from the forming ice and entrapped within the channels between the ice crystals.

The physics of water freezing has long attracted the attention of scientists, and with few exceptions most of these studies have concentrated on the freezing of pure water or very diluted suspensions. This phenomenon is critical for various applications, such as the cryopreservation of biological cell suspensions and the purification of pollutants.

In order to obtain ceramic samples with a lamellar porous structure, two requirements must be satisfied:

- The ceramic particles, in suspension in the slurry, must be rejected from the advancing solidification front and piled up between the growing columnar or lamellar ice crystals. An important observation in fundamental studies of such model systems [27, 28] is that, during the freezing of such suspensions, there is a critical particle size above which the suspended particles will be trapped by the moving water-ice front. Size requirements for the ceramic particles must be fulfilled for the segregation to occur.
- The ice front must have a columnar or lamellar morphology. The crystal structure of ice is such that it does not allow the inclusion of impurities, except within defects in the crystal structure. Consequently, once ice crystals are formed, any solute initially present in the liquid will be excluded from these pure ice crystals. The rejection and building of a concentration gradient of any solute initially present in the slurry will eventually provoke morphological instability of the interface. The interface morphology will hence undergo a transition to a cellular or lamellar morphology. In addition, the ice front velocity parallel to the *c*-axis is 10^2 to 10^3 times slower than perpendicular to this axis. Ice platelets with a very large anisotropy can then be formed very rapidly, with ice growing along the *a*-axes, while the thickness (along the *c*-axis) remains low. The freezing process is easier for crystals with horizontal *c*-axes, such that upward growth can occur along an *a*- or *b*-direction. The crystals with *c*-axes horizontal will therefore grow at the expense of the others and continue to grow upward, in an architecture composed of long vertical lamellar crystals with horizontal *c*-axes. In the final scaffolds, the direction perpendicular to the layers corresponds thus to the original *c*-axis of ice crystals.

10.3
Type of Materials Processed and Mechanical Properties

The segregation phenomenon occurring during freezing is based on physical interactions (not chemical); thus, any type of ceramic particles can be used. We will show here only two examples of materials that can be processed this way, namely alumina (a model technical ceramic for the ceramist's community) and HAP (a bioceramic with high potential for implant and tissue engineering applications).

10.3.1
Scaffolds and Composites

The microstructure of the lamellar zone is to some extent very similar to that obtained for polymeric materials processed using the same technique. The multilayered structure (Fig. 10.4a) consists of ceramic plates (Fig. 10.4a,b), with flat interconnected macropores between them, aligned along the ice-growth direction. On the internal walls of the lamellae, a dendritic, branch-like structure can be observed (Fig. 10.4c) following the microscopic ice formation. The individual plates are made of individual grains (Fig. 10.4d).

The porous ceramic scaffolds can be infiltrated in a second step with any suitable second phase, depending on the targeted functional properties. To date, we have developed two simple different approaches to obtain two classes of materials: (i) polymer/ceramic composites, which were obtained after epoxy infiltration; and (ii) metal/ceramic composites, which were obtained after aluminum alloy infiltration. Epoxy infiltration was carried out by embedding the porous ceramic

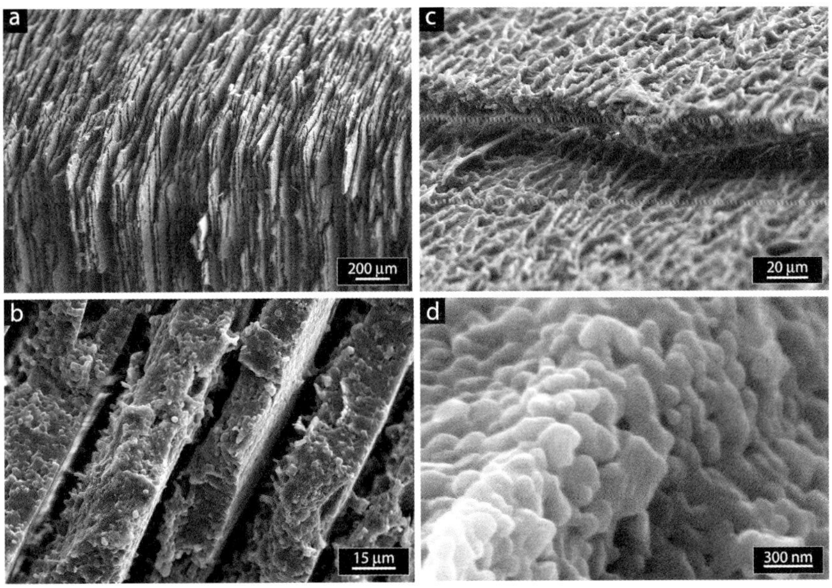

Fig. 10.4 The different degree of structure in the ice-templated ceramic scaffolds (shown here for alumina). (a) Multilayer aspect. (b) Individual lamellae. (c) Roughness covering the lamellae. (d) Submicronic grains constituting the lamellae.

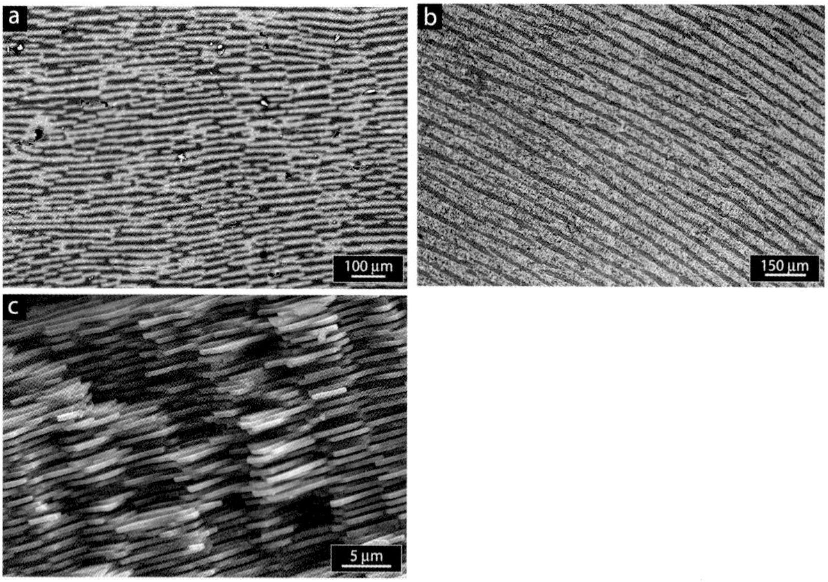

Fig. 10.5 Dense composites obtained after infiltration of the scaffolds. (a) Alumina-aluminum alloy. (b) Alumina-epoxy and comparison with nacre from abalone (c).

scaffolds in a standard epoxy resin under vacuum. Aluminum infiltration was performed by dipping the porous alumina scaffold into aluminum alloy melted in vacuum, and applying a gas pressure of 40 kPa to force the alloy into the ceramic scaffold. The obtained composites (Fig. 10.5) are completely dense, with no visible evidence of residual porosity.

The porous scaffolds and dense composites obtained by this process exhibit striking similarities to the meso- and micro-structures of the inorganic component of nacre, replicating its multilayer structure (see Fig. 10.2a). The inorganic lamellae are parallel to each other and very homogeneous throughout the entire sample. Particles trapped in between the ice dendrites lead to a dendritic surface roughness of the walls, just as in nacre. Some dendrites span the channels between the lamellae, mimicking the tiny inorganic bridges that link the inorganic platelets of nacre, which are believed to increase the fracture resistance. Finally, the ceramic plates are composed of individual grains.

10.3.2
Preliminary Reports of Properties of Ice-Templated Materials

Compressive strength versus total porosity for the HAP scaffolds is plotted in Figure 10.6. Although the strength for high-porosity content (typically >60 vol%) is comparable to that reported in the literature, it increases rapidly when the poros-

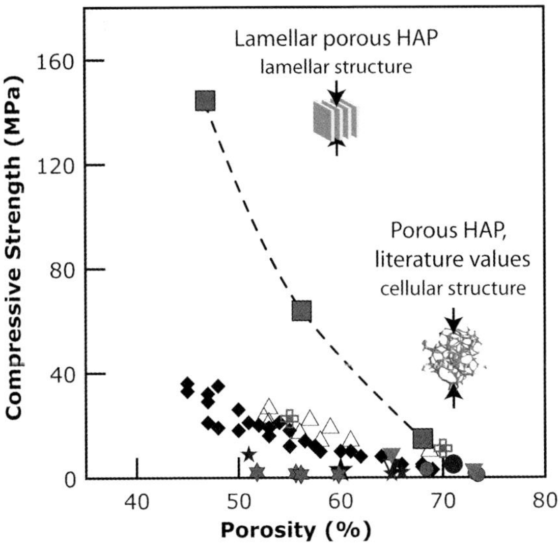

Fig. 10.6 Compressive strength of the porous hydroxyapatite scaffolds and comparison with results from the literature (for precise details, see Ref. [30]). Each style of point corresponds to a different literature source.

ity decreases. Values obtained for these samples are well above those reported to date. The presence of inorganic bridges between the ceramic lamellae (a feature that parallels the microstructure of nacre) prevents Euler buckling of the ceramic lamellae and contributes to the high strength. In fact, the strength of the porous lamellar HAP is similar to that of compact bone. Load-bearing biological applications, requiring high strength, might now be considered with such materials.

The composites obtained after epoxy or aluminum infiltration were tested in three-point bending to investigate the crack-propagation mechanisms. Extensive crack deflection at the interface between lamellae was observed (Fig. 10.7a). As in nacre, this delamination creates tortuous cracks (Fig. 10.7b) that propagate in a stable manner and increase the toughness of the materials. Multiple cracking, and bridging of the main crack, were also observed (Fig. 10.7b) in the metal/ceramic composites.

The load-displacement curves of the tested samples (Fig. 10.7c) revealed a gradually decreasing load, very similar to that observed in the testing of nacre. This behavior can be directly related to a stable crack propagation perpendicular to the layer with active toughening. Even if the toughness value of the HAP composite (220 J m^{-2}) is still much lower than that of nacre (1000–3000 J m^{-2}) – partly because epoxy has very low toughness in comparison with the tough protein layer found in nacre – the crack-propagation mechanisms seem very similar. This type of microstructure can therefore offer promising perspectives for increasing toughness in ceramic composites.

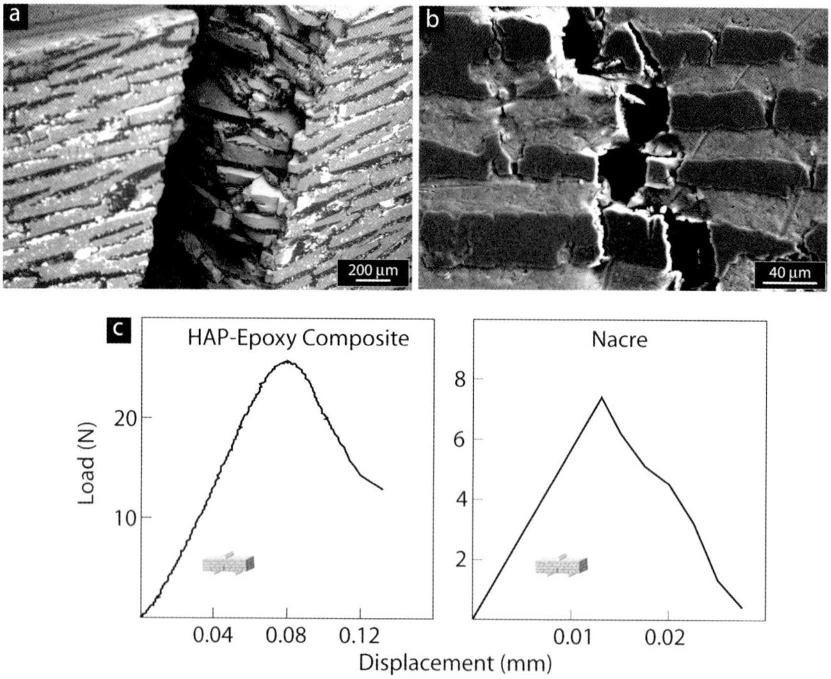

Fig. 10.7 Crack propagation in the composites. (a) Delamination in the alumina-epoxy composite. (b) Crack deflection and bridging at the interface in metal/ceramic composites. (c) Crack propagation in hydroxyapatite-epoxy composite (three-point bending) and comparison with nacre. The three-point bending load-displacement data are qualitatively very similar, with a gradually decreasing load after the elastic limit (characteristic of a stable crack propagation and active toughening) for cracks propagating in the direction perpendicular to the inorganic lamellae.

10.4 Control of the Structure: Influence of Processing Parameters

Many microstructural features can be controlled by applying a few principles of the physics of ice formation.

10.4.1 Mesostructural Gradients

The mesostructure often determines the mechanical response of natural or synthetic structures [29]. The optimization of the material response in the present case might be expected by a control of the lamellae orientation. Preliminary results [30] indicate that such control might be obtained by controlling the orientation of the first-formed ice crystals. Engraving patterns at the surface of the cold finger on which the ice crystals grow might promote patterns and gradient formation.

10.4.2 Porosity or Relative Importance of the Two Phases

The scaffolds obtained by the freezing technique can be used either where porous scaffolds are desired (e.g., tissue engineering, catalysts support), or as a template for dense composites after infiltration with the second phase of choice. In the first case, the global amount of porosity will be critical with regard to the functional properties. In the second case, the relative importance of the two phases of the composites will be of interest to optimize the response for the targeted application. Such control can be achieved by adjusting the initial content of ceramic particles in the slurry. For low particle content (e.g., 40 wt%), highly porous samples (porosity > 65 vol%) will be obtained. Ceramics will be the minor phase in the composite after infiltration. For a high particle content (e.g., 65 wt%), much lower porosities (35 vol%) are obtained. Less space is available for infiltration with a second phase; composites with a higher Young's modulus can hence be obtained as the ceramic becomes the major phase. The relationship between total porosity and particle content in the initial slurry is linear (see Fig. 4 in [31]).

10.4.3 Lamellae Characteristics

The lamellae obtained by the freezing technique can be characterized by three main parameters: their thickness, their surface roughness, and the ceramic bridges linking adjacent lamellae. These three parameters may be controlled to some extent by controlling the characteristics of the growing ice crystals during

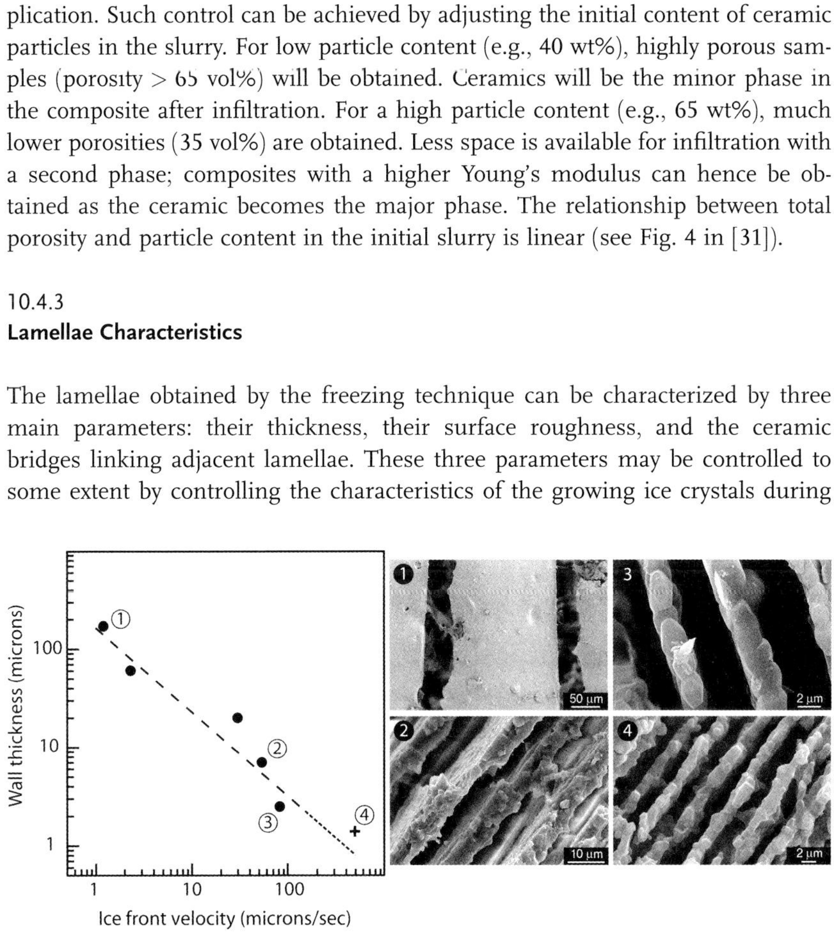

Fig. 10.8 Effect of the speed of the solidification front on the thickness of the lamellae for alumina samples fabricated from powders with an average grain size of 0.3 µm. The scanning electron micrographs numbers shown in the graph correspond to cross-sections parallel to the direction of movement of the ice front. Sample (4) was obtained with ultrafast freezing to estimate the thickness limit achievable by this technique. The approximate ice-front velocity for this extreme case is in agreement with the extrapolation of the controlled freezing results.

freezing. It has been demonstrated, for example, that increasing the ice front velocity leads to ice crystals with a smaller tip radius, which in turn results in lamellae of smaller thickness. The thickness can be adjusted over almost two orders of magnitude (Fig. 10.8), from 2 to 200 μm.

The surface roughness covering the lamellae of the ceramics is linked to the dendritic shape of the ice crystals. The scaffold is a negative of the ice structure before sublimation; hence, a modification of the shape of the ice crystals will be directly reflected into the scaffolds' morphology and surface characteristics of the ceramic lamellae. Particles trapped in between the ice dendrites (see Fig. 1 in [30]) led to a dendritic surface roughness of the walls. The ceramic particles' size will therefore be critical for replicating the ice structure. If large particles are used, then only a rough replicate of the crystals will be obtained. Smaller particles will yield better replicas of the ice crystals.

10.4.4 Grain Size

The grain size has been proven to have a critical influence over numerous properties of technical ceramics, such as strength, toughness, or creep. The same influence could be expected here, and control of the grain size can be achieved by usual techniques: control of sintering conditions, sintering aids, and additives that restrict or favor grain growth. The characteristics will depend heavily on the desired properties; for example, toughness usually arises from elongated grains, while high strength is bound to small grains.

10.4.5 Interface

Although the inorganic portion represents 95 vol% of the nacre composition, the highly specific properties of nacre are due to the remaining 5%, the organic phase found between the calcium carbonate platelets. Nature shows that the optimum fracture properties are encountered not only when the organic/inorganic interface is strong, but also when delamination at the organic/inorganic interface occurs before the crack goes across the stiff, brittle layer. It is believed that nature manipulates adhesion in two ways – mechanical and chemical. In nacre, this is effected by controlling both the roughness and the highly specific properties of the polymer adhesive phase. Our process allows us also to control the chemistry of the interface. For example, the mechanical response of alumina/aluminum layered composites can be manipulated by controlling the interfacial bonding. Specifically, by adding as little as 0.5 wt% Ti to the aluminum eutectic (which is known to segregate at the metal/ceramic interfaces), the strength increases by 50% and fracture toughness by 80% (Fig. 10.9), respectively, from 400 to 600 MPa and from 5.5 to 10 MPa $m^{-1/2}$. A similar control in the polymer/ceramic composites will be pursued in the future to optimize the mechanical response of the processed materials.

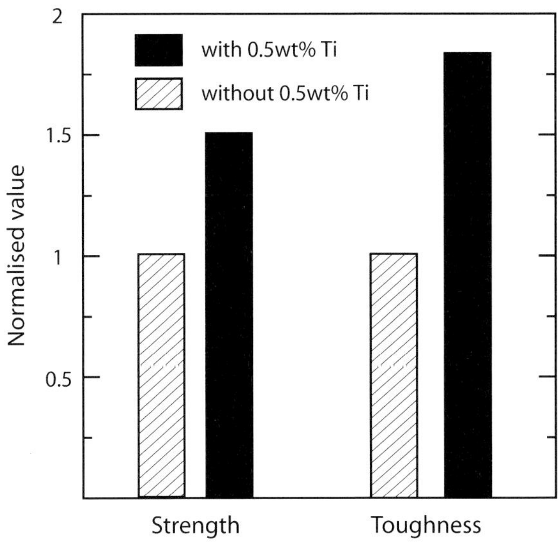

Fig. 10.9 Effect of Ti doping on the mechanical response of the alumina-aluminum composites. Strength and toughness are largely increased due to modification of the interface properties.

10.5 Conclusions

In the pursuit of processing routes for advanced materials designs following biomimetic concepts, the natural particle segregation that occurs during the freezing of ceramic suspensions appears to be an appealing alternative. Materials obtained with the freezing concepts shown here exhibit striking similarities with the structure of nacre, not only visually, but also with regard to their functional properties. By using a natural, self-organizing phenomenon, we allow Nature to guide the design and processing. In comparison to alternative strategies, such as sequential deposition or multilayering, the physical basis of the technique makes it very versatile, with any type of particle capable of being used. The processing of large samples (1 cm to 1 m) can be achieved in a short time, and industrial transfer might be straightforward as the technologies are readily available. The vast knowledge of the underlying physical principles – that is, the physics of ice and particle interaction with a solidification front – provides a large variety of strategies for controlling the structure of the materials at all length scales.

Although still at the experimental stage, variations of these materials could also be used in a myriad of applications in which strength and lightness are imperative, such as dental implants, airplane manufacturing, and microelectronic packaging.

This work was supported by the NiH under grant SR01 DE015633 and by the Director, Office of Science, Office of Basic Energy Sciences, Division of Materials

Sciences and Engineering of the US Department of Energy under contract DE-AC03-76SF00098.

References

1 J. Currey, *Proc. R. Soc. London B* **1977**, *196*, 443–463.
2 F. Song, A.K. Soh, Y.L. Bai, *Biomaterials* **2003**, *24*, 3623–3631.
3 K.S. Katti, D.R. Katti, S.M. Pradhan, A. Bhosle, *J. Mater. Res.* **2005**, *20*, 1097–1100.
4 M. Rousseau, E. Lopez, P. Stempfle, M. Brendle, L. Franke, A. Guette, R. Naslain, X. Bourrat, *Biomaterials* **2005**, *26*, 6254–6262.
5 B.L. Smith, T.E. Schaffer, M. Viani, J.B. Thompson, N.A. Frederick, J. Kindt, A. Belcher, G.D. Stucky, D.E. Morse, P.K. Hansma, *Nature* **1999**, *399*, 761–763.
6 K. Okumura, P.G. de Gennes, *Eur. Phys. J. E.* **2001**, *4*, 121–127.
7 A.G. Evans, Z. Suo, R.Z. Wang, I.A. Aksay, M.Y. He, J.W. Hutchinson, *J. Mater. Res.* **2001**, *16*, 2475–2484.
8 X.D. Li, W.C. Chang, Y.J. Chao, R.Z. Wang, M. Chang, *Nano Lett.* **2004**, *4*, 613–617.
9 H.J. Gao, B.H. Ji, I.L. Jager, E. Arzt, P. Fratzl, *Proc. Natl. Acad. Sci. USA* **2003**, *100*, 5597–5600.
10 R.Z. Wang, Z. Suo, A.G. Evans, N. Yao, I.A. Aksay, *J. Mater. Res.* **2001**, *16*, 2485–2493.
11 Z.Y. Tang, N.A. Kotov, S. Magonov, B. Ozturk, *Nat. Mater.* **2003**, *2*, 413–418.
12 Y. Oaki, H. Imai, *Angew. Chem.* **2005**, *44*, 6571–6575.
13 H.-W. Kang, Y. Tabata, Y. Ikada, *Biomaterials* **1999**, *20*, 1339–1344.
14 T. Fukasawa, M. Ando, T. Ohji, S. Kanzaki, *J. Am. Ceram. Soc.* **2001**, *84*, 230–232.
15 T. Fukasawa, Z.Y. Deng, M. Ando, T. Ohji, S. Kanzaki, *J. Am. Ceram. Soc.* **2002**, *85*, 2151–2155.
16 J.W. Moon, H.J. Hwang, M. Awano, K. Maeda, *Mater. Lett.* **2003**, *57*, 1428–1434.
17 S.W. Sofie, F. Dogan, *J. Am. Ceram. Soc.* **2001**, *84*, 1459–1464.
18 H. Schoof, J. Apel, I. Heschel, G. Rau, *J. Biomed. Mater. Res.* **2001**, *58*, 352–357.
19 S.V. Madihally, H.W.T. Matthew, *Biomaterials* **1999**, *20*, 1133–1142.
20 S. Stokols, M.H. Tuszynski, *Biomaterials* **2004**, *25*, 5839–5846.
21 S. Zmora, R. Glicklis, S. Cohen, *Biomaterials* **2002**, *23*, 4087–4094.
22 S. Yunoki, T. Ikoma, A. Monkawa, K. Ohta, M. Kikuchi, S. Sotome, K. Shinomiya, J. Tanaka, *Mater. Lett.* **2006**, *60*, 999–1002.
23 M.C. Gutiérrez, M. Jobbágy, N. Rapún, M.L. Ferrer, F. del Monte, *Adv. Mater.* **2006**, *18*, 1137–1140.
24 H.F. Zhang, I. Hussain, M. Brust, M.F. Butler, S.P. Rannard, A.I. Cooper, *Nat. Mater.* **2005**, *4*, 787–793.
25 S.R. Mukai, H. Nishihara, H. Tamon, *Chem. Commun.* **2004**, *874–875*.
26 H. Ishiguro, B. Rubinsky, *Cryobiology* **1994**, *31*, 483–500.
27 M.G. Worster, J.S. Wettlaufer, *J. Phys. Chem. B* **1997**, *101*, 6132–6136.
28 V.F. Petrenko, R.W. Whitworth, *Physics of Ice*. Oxford University Press, Oxford, New York, **2002**.
29 J. Aizenberg, J.C. Weaver, M.S. Thanawala, V.C. Sundar, D.E. Morse, P. Fratzl, *Science* **2005**, *309*, 275–278.
30 S. Deville, E. Saiz, R.K. Nalla, A. Tomsia, *Science* **2006**, *311*, 515–518.
31 S. Deville, E. Saiz, A. Tomsia, *Biomaterials* **2006**, *27*, 5480–5489.
32 C.M. Zaremba, A.M. Belcher, M. Fritz, Y. Li, S. Mann, P.K. Hansma, D.E. Morse, J.S. Speck, G.D. Stucky, *Chem. Mater.* **1996**, *8*, 679–690.

11
Bio-Inspired Construction of Silica Surface Patterns

Olaf Helmecke, Peter Behrens, and Henning Menzel

Abstract

The organic macromolecules isolated from diatom shells are able to influence silica condensation *in vitro*. It has been shown that aggregation and phase separation of these macromolecules are important, and different models have been suggested to explain the structure formation process leading to the highly organized diatom shells. Among these models, phase separation – which involves the formation of silica at the membrane of the silica deposition vesicle – is of particular interest and has prompted experiments with patterned surfaces. Some examples of the preparation of surfaces with well-defined delineated organically modified areas are presented in this chapter. The deposition of silica on these surfaces results in interesting silica surface arrays (pillars or lenses). However, in some experiments, which more closely mimic the natural system, silica structures are formed which resemble some of the features occurring in diatom shells. The structure formation in these model systems can be explained taking into account phase separation, silica sol formation, droplet formation, and wetting and drying phenomena.

Key words: surface patterning, photochemical grafting, polyamines, silica condensation.

11.1
Bioorganic Molecules and their Influence on Silica Condensation

Diatoms have highly organized exoskeletons – the frustules – with structures of striking complexity and beauty [1]. The patterns of the frustules are species-specific and are composed of silica and organic macromolecules. While the frustules are highly organized on the micrometer scale, they are completely amorphous on the molecular level, as can be shown by X-ray diffraction [2]. In diatoms, silica formation takes place inside the diatom cell within a specialized membrane-bound compartment termed the silica deposition vesicle (SDV) [3].

Handbook of Biomineralization. Edited by P. Behrens and E. Bäuerlein

ISBN: 978-3-527-31805-6

Polysaccharides, as well as proteins, have been found to be present as organic components [1, 4]. Some of the proteins have been identified and their structure determined [1, 2, 5, 6]. Perhaps the best-investigated protein is silaffin 1A, which was isolated from the diatom *Cylindrotheca fusiformis* [5, 7–9]. Silaffin 1A is a short peptide that is rich in phosphorylated serine; of particular interest are the modifications of the lysines by methylation and by polyamine side chains. The interaction of the anionic phosphoryl groups and the cationic polyamine groups are important for the influence of silaffin on silica deposition, although the effect of the phosphorylation can be replaced by phosphate anions present in the solution [8, 10].

Polyamines similar to the side chains in silaffin 1A have been isolated from diatom shells, and these are also able to influence the silica mineralization and induce precipitation of silica particles [11–13].

In addition to the cationic peptides and polyamines, peptides which are negatively charged under physiological conditions have also been found. These do not induce silica precipitation in their own right, but rather in combination with polyamines or silaffin 1A [6, 14]. Thus, it was hypothesized that larger supramolecular aggregates of the polyamines with anionic entities (either simple inorganic phosphate ions or large protein molecules) are involved in the pattern formation.

The chemistry of silica and the process of silicic acid condensation from aqueous solutions have been reviewed in detail elsewhere [15]. Briefly, at concentrations of silicic acid in water higher than ~100 ppm condensation occurs which involves three distinct steps:

1. condensation to form stable nuclei;
2. growth of nuclei leading to fundamental particles; and
3. particle aggregation.

The naturally occurring molecules involved in silica formation are able to control this process at each of these steps. Both, polypeptides [16, 17] and polyamines [18–20] have repeatedly been suggested to have a catalytic effect on the silica condensation, thus accelerating step 1. However, recent results have indicated that this effect may be small and mostly related to a shift in the pH [13, 19]. Furthermore, the growth of the nuclei (i.e., step 2) has also been suggested to be accelerated by polyamine species [19]. However, the major influence can be expected in step 3; as already highlighted by Iler [15], cationic polymers are possible candidates to act as flocculating agents in silica precipitation.

Poly(L-lysine) [21, 22], poly(L-arginine) [23, 24] poly(allylamine) [20, 23–27], linear oligo- [19] and poly(ethylene imines) [28], as well as linear poly(propylene imines) [13, 28], were investigated as models for the naturally occurring polyamines. These were found to form aggregates when phosphate ions are present in the solution, with the size of the aggregate depending on the concentration of phosphate and the pH-value of the solution. The formation of silica nanospheres from a solution of mono- and disilicic acid was found to depend on microscopic phase separation [25, 29]. The same parameters influence the size of the silica spheres formed. From these results is was postulated that the polyamines are involved in the pattern formation by bridging primary particles and forming templates by aggregation [30].

11.2 Structure Formation Models

During the shell formation process in diatoms, two events are recognized:

- Macromorphogenesis, or membrane-mediated morphogenesis, in which the cellular and cytoskeleton activity is involved in molding of the SDV to create the well-known large honeycomb structures and large pores of the valves [31, 32].
- Micromorphogenesis, in which the formation of smaller structures and delicate details (including mesopores) is controlled by processes inside the SDV [31, 33].

Although several bioorganic macromolecules have been identified as being involved (see above), the exact mechanism of morphogenesis in the SDV remains speculative to some extent, and several models have been proposed. For example, Gordon and Drum have proposed the structures to be the result of instabilities in the diffusion-limited aggregation of silica particles within the SDV. Following their release inside the SDV, particles diffuse until they encounter growing aggregates, to which they bind. After the sintering process, the aggregates tend to reorganize into a dense, thermodynamically stable packing [34]. Vrieling and coworkers have suggested that, in the presence of silica precursors, short-chain silaffins and polyamines induce rapid precipitation of silica. In this model, larger peptides contribute to the aggregation process by interacting not only with the silica particles, but also with each other, and this results in a hybrid mesophase [35]. Alternatively, Sumper has postulated the existence of repeated phase-separation processes within the SDV which produce emulsions of micro- and nanodroplets consisting of a polyamine-containing phase. According to this model, silica formation takes place at the border between a polyamine droplet, the SDV membrane and the interior of the SDV. The repeated phase separation in the mixture of silica precursors and surface active polyamines in ever-smaller compartments at the SDV membrane results in the formation of self-similar patterns. This model was used to explain the pattern formation observed by electron microscopic investigation of *Coscinodiscus wailesii* cell walls during growth [30].

11.3 Silica Deposition on Patterned Surfaces

In-vitro experiments may help in our understanding of the principles of silica biomineralization, and most likely could also be used for the fabrication of

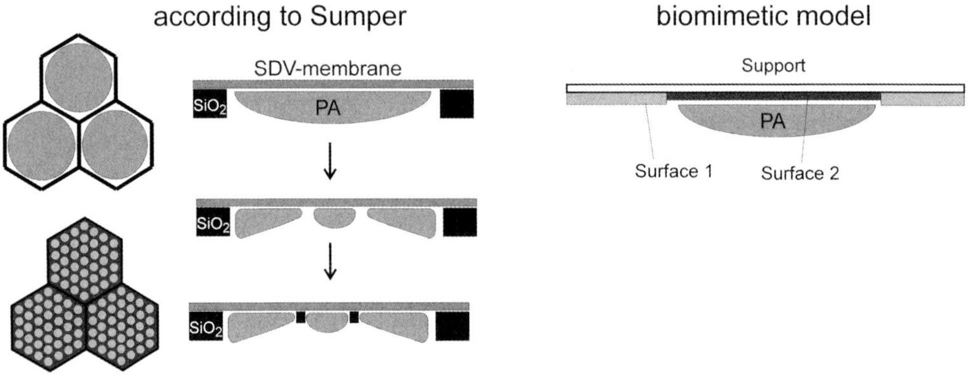

Fig. 11.1 Schematic representation of Sumper's phase-separation model (drawn according to Ref. [30]) for the structure formation of diatom shells and a possible biomimetic model for this geometry by delineated organically modified areas. PA = polyamine-containing droplets. SDV = silica deposition vesicle.

bio-inspired materials. One important question in connection with the phase-separation model is whether the restricted geometry at the SDV membrane has a decisive influence on the self-organization processes (see Fig. 11.1) [30, 34, 35] and can possibly already promote the formation of self similar structures.

In order to test the influence of a restricted geometry, it is essential to use spatially delineated areas. The necessary patterning can be achieved by creating surfaces with different properties on a substrate in such a way that, for example wettable areas, are separated from each other by non-wettable areas (Fig. 11.1). This can be accomplished in several different ways.

Coffman et al. studied the silica deposition on poly-L-lysine-coated surfaces [36]. The poly(L-lysine) was patterned onto the surface of a silicon wafer either by reagent jetting (this resulted in spots of ca. 500 μm diameter) or by adsorption in areas of 0.7 μm diameter, which had been created by a photolithographic process. The polymer layers prepared with the latter method were thin (3–5 nm) and fairly uniform. The patterned surfaces were brought into contact with a solution of sodium orthosilicate in borate buffer (pH 8.5) or sodium phosphate buffer (pH 7.0). By using this procedure, silica particles of approximately 30 nm diameter were deposited on the poly-L-lysine-coated areas (Fig. 11.2). With longer exposure times a rim was observed to form around the spot and, in addition to the 30-nm particles, smaller (5-nm) particles were also seen throughout the whole surface. It was speculated, that these smaller particles had formed in the solution and settled on the surface.

Kim et al. applied micro-contact printing and backfilling to install a patterned self-assembled monolayer of a polymerization initiator on a gold surface. Using this initiator, poly(2-(dimethylamino)ethyl methacrylate (pDMAEMA) was grafted onto the surface by atom transfer radical polymerization, and in this way uniform polymer pillars of up to 70 nm thickness and 10 μm diameter were created (Fig. 11.3) [37]. The patterned polymer film was brought into contact with a freshly

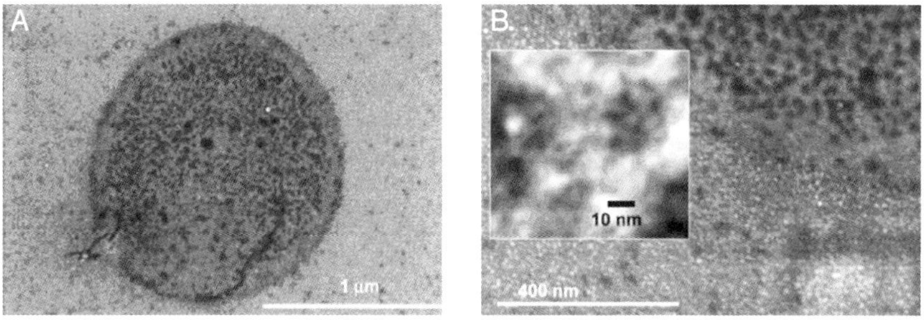

Fig. 11.2 Scanning electron microscopy images of silica structures derived from photolithographically patterned poly(L-lysine) (PLL) [36]. A pattern of ~700 nm diameter spots of PLL was exposed to a dilute solution of silicic acid for 1 h. (A) Magnification shows highly interconnected silica nanostructures (B). Note the 5-nm particles overlaying larger 30-nm structures in the inset in (B). These smaller particles are seen covering the entire surface of the silicon wafer. (Reprinted with permission from Langmuir. Copyright 2005 American Chemical Society.)

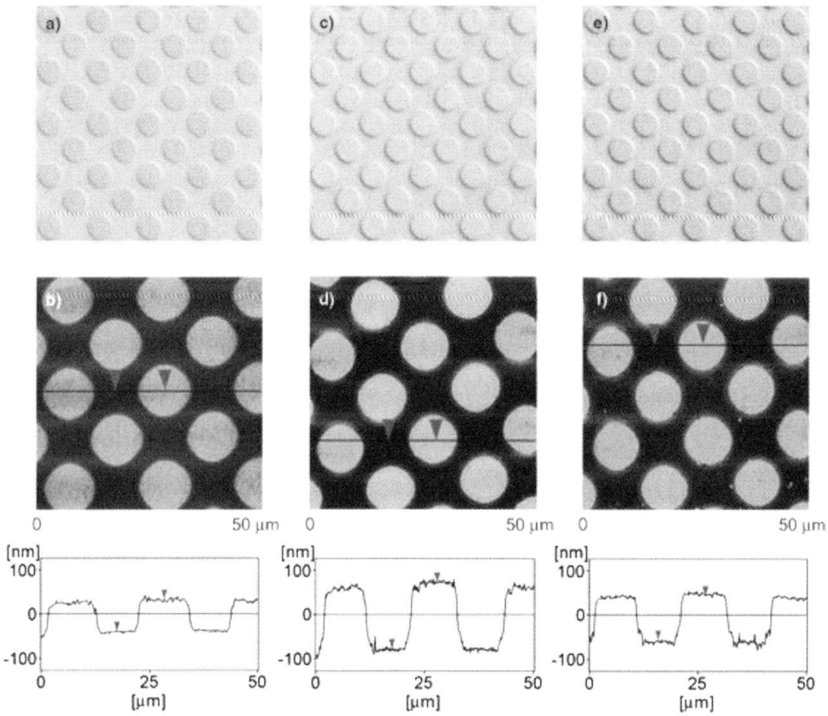

Fig. 11.3 Optical and atomic force microscopy (AFM) images of the substrates. (a,b) Patterned with poly(2-(dimethylamino)ethyl methacrylate (pDMAEMA) [37]; (c,d) after silicification; and (e,f) after heating at 300 °C for 20 min at 5×10^{-6} Torr. (Reprinted with permission from SMALL. Copyright 2005 Wiley-VCH.)

prepared solution of silicic acid (from the hydrolysis of tetramethoxysilane; TMOS) in phosphate buffer at pH 5.5, and silica was precipitated inside the polymer pillars. After silicification, the thickness of the patterned film was increased to 150 nm and the integrity of the circular shapes was maintained. The silicification required the presence of phosphates (or other multivalent anions).

Brott et al. used holographic two-photon-induced photopolymerization to create peptide-enriched lines on a polymer surface [38]. This process is feasible because certain areas of the sample cure more rapidly than others due to the inhomogeneous holographic irradiation, and the smaller molecules in the mixture (namely water and peptide) phase separate from the areas of higher crosslink density and migrate into areas of lower density. When these peptide nanopatterned holographic structures are exposed to a silicic acid solution, an array of silica nanospheres is deposited onto the polymer substrate (Fig. 11.4).

Well-defined, spatially delineated, organically modified areas can also be installed on silicon wafers by employing a photochemical grafting method [39–41] to bind different polymers in a laterally structured manner [42]. The principle of this method is illustrated in Figure 11.5. A monolayer of a benzophenone-containing anchor is deposited on the native SiO_2-layer present on silicon wafers.

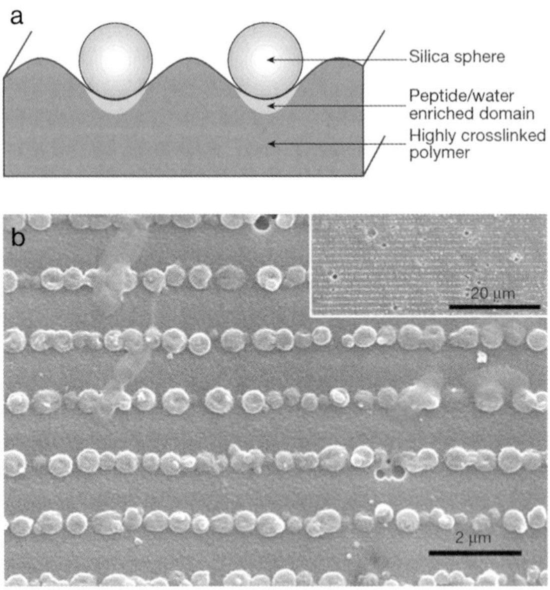

Fig. 11.4 (a) Schematic cross-section of a hologram inscribed by two-photon photopolymerization into a mixture of a monomer, water, and peptide [37]. (b) SEM images of the silica nanostructure created by reacting the silane with a peptide-embedded hologram [38]. (Reprinted from Ref. [38], with permission from Macmillan Publishers Ltd.)

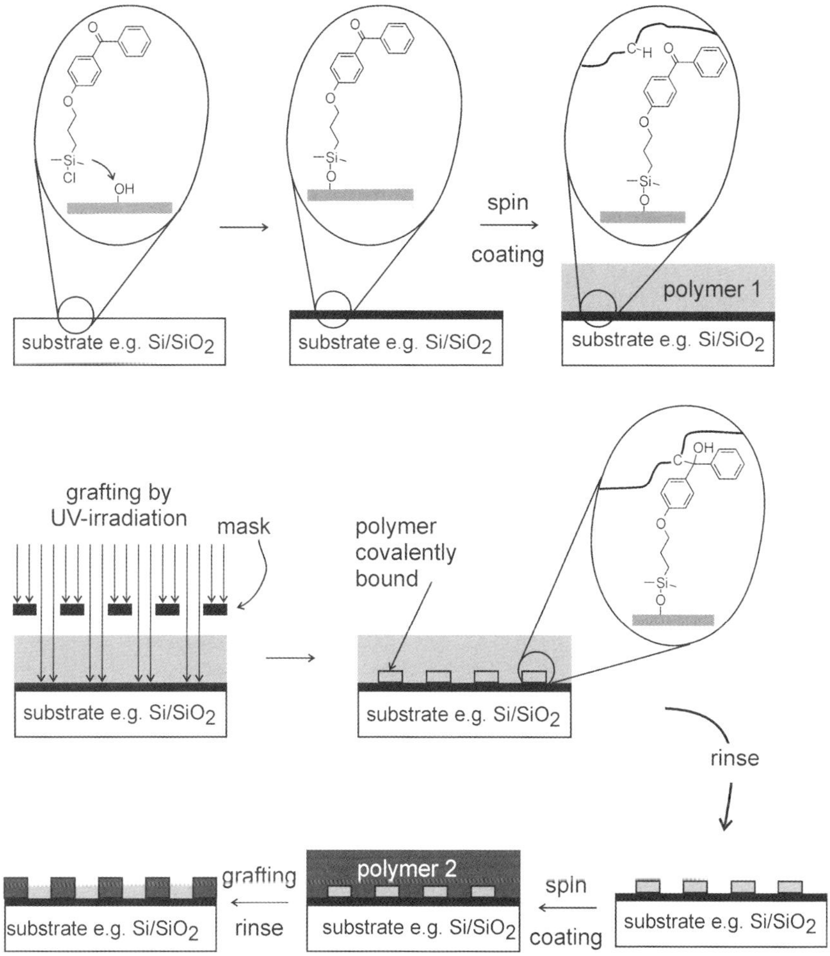

Fig. 11.5 Schematic representation of the preparation of laterally structured polymer films on surfaces by photochemical grafting.

Subsequently, a polymer can be spin-coated onto the anchor layer and photochemically bound to the surface. By using appropriate masks, this can be carried out in a laterally structured manner. Photo-masks with hexagons (and other geometries) of different sizes were used. The combination of the polymers used for structuring is also very important. Successful structuring was possible with hydrophobic, low glass-transition polymers (e.g., poly-2-ethyl-hexyl acrylic acid ester; PEHAA) and branched poly(ethylene imine) (PEI) [43]. The structures can easily be observed with scanning electron microscopy (SEM), even when the polymer layers are in the range of 7 nm and 5 nm for PEHAA and PEI, respectively. The contrast mechanism in the SEM image is not absolutely clear, but is most likely

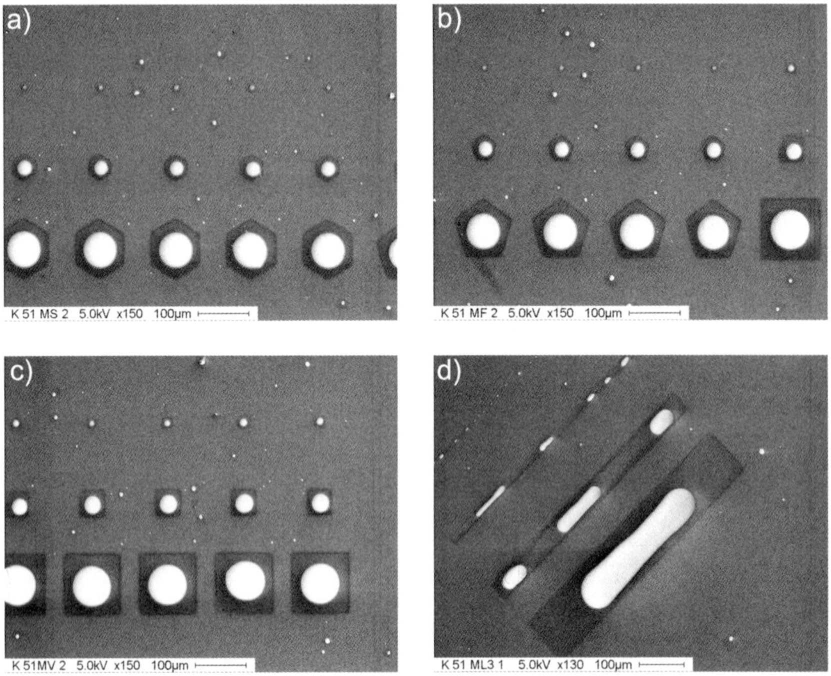

Fig. 11.6 Scanning electron microscopy images of silica deposits in areas of different geometry: (a) hexagons; (b) pentagons; (c) squares; (d) rectangles [44].

due to a higher conductivity of the PEI. This technique can easily be transferred to other polymer combinations [44].

The substrates with laterally structured polymer surfaces were exposed to silicic acid solutions (either by dipping into it for a certain time or by spin-coating the solution onto the substrate). This results in a silica deposition almost exclusively in the PEI-coated areas (see Fig. 11.6a). The deposited silica forms regular spherical particles on the previously inscribed pattern.

Investigations using atomic force microscopy (AFM) have indicated that the particles are very smooth and have a height which is in the range of one-tenth of their diameter (Fig. 11.7). Thus, by using a simple dipping procedure very homogeneous and regular arrays of small glass lenses have been created, and these may find applications in micro optical devices, although the transparency of the deposited silica has not yet been investigated.

It should be noted that the conditions used during silica deposition (dipping or spin coating, dipping time, pH of the solution) can influence the structure of the deposited silica, and further investigations are required in order to elucidate these effects in more detail.

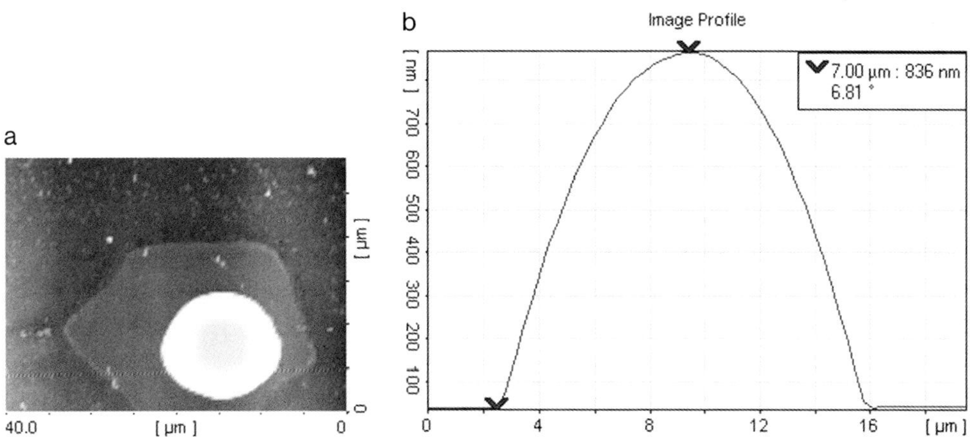

Fig. 11.7 Atomic force microscopy image and height profile for a silica particle formed in a hexagon with 32 μm diameter. Height scale in nm; distance scale in μm [44].

11.3.1 Influence of the Geometry

Structures with different geometries were created on the surface by employing modified masks [44]. These structures were also subjected to a freshly prepared silicic acid solution (by hydrolysis of TMOS) by spin-coating. As all structures are on the same support, the deposition conditions are identical for all different geometries. As can be seen from Figure 11.6, the resultant silica particles are very similar, and their shapes do not depend significantly on the geometry of the reaction area, although for the larger rectangular areas the silica particles were not round but rather elongated. Thus, the geometry does not appear to have any major influence on silica deposition. Rather, the silica structures in rectangular areas suggest that droplet formation and contraction may play an important role in the development of these structures.

11.3.2 Influence of the Polymer at the Reaction Area

The photochemical grafting procedure also allows the binding of other hydrophilic polymers in the reaction area [44]. This has been achieved: (i) with low-molecular-weight linear PEI (linPEI), which mimics the polyamines found in diatoms [28]; (ii) with polyacrylic acid (PAA), which is an acidic polymer with reduced hydrophilicity at low pH values; and (iii) with poly(ethyleneglycol) (PEG), which is a very hydrophilic neutral polymer the hydrophilicity of which is not influenced by the pH of the environment. The patterned surfaces were prepared as

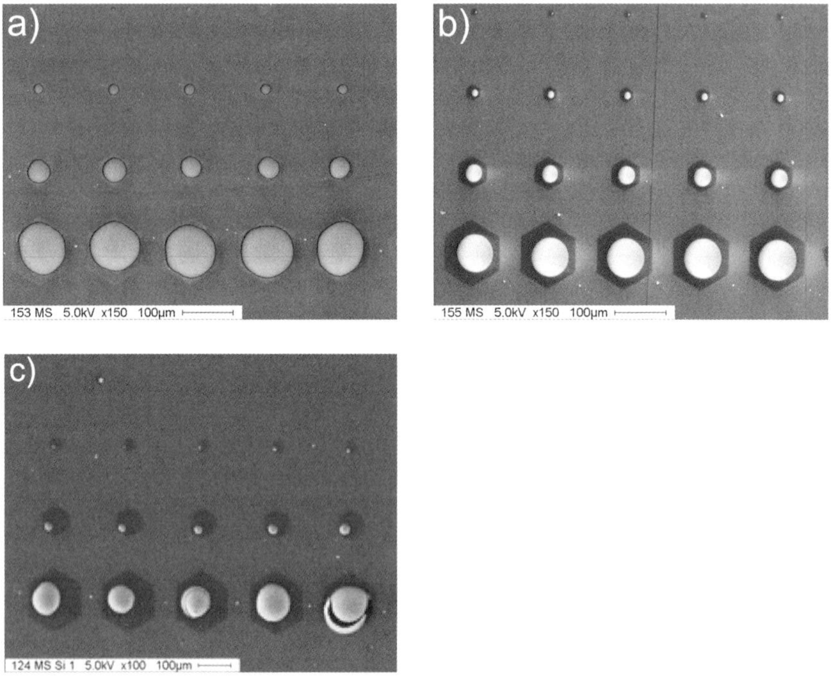

Fig. 11.8 Scanning electron microscopy images of silica deposits on structures filled with: (a) poly(ethylene glycol); (b) poly(acrylic acid); and (c) linear poly(ethylene imine) with an average degree of polymerization of 14.

described before for the branched PEI [44], and the substrates were again spin-coated with a freshly prepared silicic acid solution (from hydrolysis of TMOS).

The resultant silica deposits are depicted in Figure 11.8, and exhibit in all cases a spherical shape similar to the structures found for PEI-coated reaction areas (cf. Fig. 11.6). However, there were some differences. The deposits on PEG were larger, and filled most of the reaction area, but were less regular in shape. On PAA, however, almost perfectly round shaped structures were formed, which covered the reaction area only partially. This difference might be explained by the wetting behavior of the surfaces, and would support the hypothesis that formation of the lens-shaped deposits is due to droplet formation and wetting phenomena – that is, the silicic acid solution wets only the reaction areas and forms droplets within them. Such a droplet has a low contact angle with PEG surfaces, and upon silica condensation and drying a flat particle is formed. However, on PAA – which has a higher contact angle with water at low pH-values – smaller droplets are formed and thus more spherical particles are created. The slight difference in the deposits on branched high-molecular-weight PEI (see Fig. 11.6) and those on linear low-molecular-weight PEI (Fig. 11.8c) might then due to the less-homogeneous coating obtained with a short-chain polymer. As the photochemical grafting binds only those molecules directly in contact with the surface, the layer thickness is much smaller in the case of the oligomeric linear PEI. Consequently, the coating will be less homogeneous, and the droplet formation less regular.

The results obtained by changing the shape of the reaction area – and particularly those obtained for different polymers in the reaction areas – indicate that the structure of the silica deposits is mainly governed by wetting behavior and droplet formation, when only silicic acid solution is brought into contact with the reaction areas.

11.3.3
Influence of Additives in the Silicic Acid Solution

Experiments with different natural and synthetic macromolecules have shown that their presence can influence the silicic acid condensation, and eventually result in different silica structures. Thus, experiments with the patterned surfaces were carried out in which the polyamine was present not only at the surface but also in the silicic acid solution.

Preliminary experiments were carried out with poly(allyl amine hydrochloride), which has been used previously as a model for the naturally occurring polyamines [20, 24–26]. The aggregation of this polyamine is observed only in the presence of multivalent anions such as phosphate; thus, poly(allyl amine hydrochloride) and phosphate were added to the silicic acid solution (for concentrations, details in Fig. 11.9) [44]. Again, silica was deposited only onto the polyamine-coated areas, and very regular spherical silica particles were found (Fig. 11.9a). Thus, the influence of the poly(allylamine)/phosphate system on the structure of the deposited silica is also rather limited.

A completely different situation is found when the poly(allyl amine) in the solution is replaced by linPEI with a chain length ($P_n = 14$) which is comparable to the naturally occurring polyamines [28]. SEM images in this case show a more or less homogeneous coverage of the reaction areas (see Fig. 11.9b). Upon magnification (Fig. 11.9c), a granular structure of the silica coating can be observed, which suggests that the coating is formed by deposition and coalescence of spherical particles. On the other hand, when no phosphate is added the surface appears to be smooth (not shown).

A closer examination of the silica coating by AFM (Fig. 11.10) reveals indeed a granular structure inside the coated area. The rather sharp spikes in the profile indicate that the silica particles are relatively small (the lateral resolution may be insufficient to judge their size accurately) but, based on the height of their features, a particle diameter in the range of 50 nm can be estimated. It can be assumed, that the linPEI and phosphate in the solution form aggregates which serve as a template for the formation of spherical silica particles in the solution [25, 29, 45]. These silica particles are then deposited onto the PEI-coated reaction area and subsequently fuse to a certain degree.

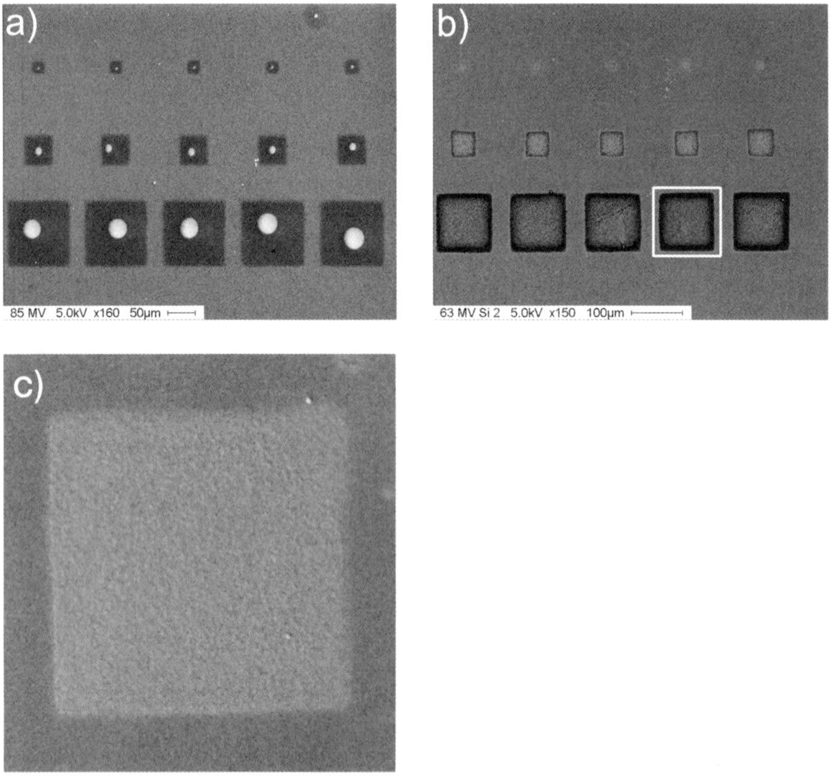

Fig. 11.9 Silica deposits on poly(ethylene imine) (PEI)-coated reaction areas in the presence of a mixture of 0.05 mL HCl (0.01 M), 0.1 mL TMOS, and 1 mL phosphate solution (0.007 M) containing: (a) poly(allylamine hydrochloride) (PAH) (ratio P/N = 0.5) or (b) linear PEI (P_n = 14) (P/N = 0.5). (c) Magnification of an area as marked in (b).

The most striking feature of the deposited silica, however, is an accumulation at the border between the polyamine-coated area and the surrounding hydrophobic polymer. The AFM line scan shows a wall of approximately 130 nm height compared to the coating inside the reaction area (Fig. 11.10). Patterns in which there is accumulation of material at the periphery of a wetted area are often observed when a drop of a suspension is drying while being pinned to the surface [46, 47]. Three phase-contact lines with contact angles less than 90° are sites of rapid evaporation. If the contact line is pinned, then continuity demands a flow of liquid towards the contact line, which carries suspended particles with it. Thus, the particles are collected near the contact line. A similar accumulation at the periphery of the PEI-coated area was observed when a silica particles were brought into contact with the patterned surface [44].

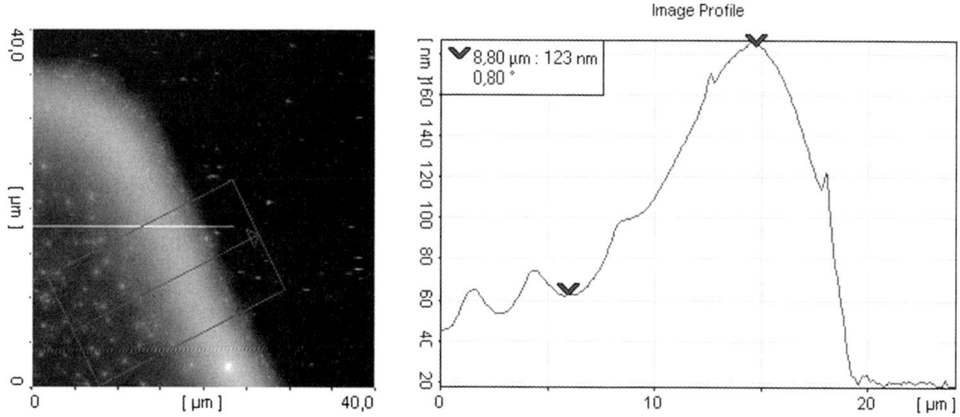

Fig. 11.10 Atomic force microscopy image and height profile for silica deposits at the border between the reaction area and surrounding hydrophobic polymer, indicating a strong accumulation of silica in this region. The arrowhead in the image indicates the line for the profile scan.

11.4
Summary

Diatoms build their frustules by assembling silica in the SDV. During this building process it has been shown that cationic peptides and polyamines are involved, and that their aggregation and phase separation in the presence of multivalent anionic species is of particular importance for structure formation. Several mechanisms have been suggested, most of which include physico-chemical processes such as aggregation and phase separation as important structuring elements. In particular, those processes which occur at the membrane of the SDV are believed to induce the formation of highly organized, self-similar structures. Therefore, patterned surfaces are valuable model systems for testing hypotheses of silica biomineralization.

A number of different ways exist by which to pattern the surface and to imitate the situation in the SDV. Some approaches result in interesting silica patterns on the surface (pillars or lenses), while others – which more closely mimic the natural system – show silica structures which resemble some of the features in natural silica (spherical particles arranged in structures). Structure formation in these model systems may result from phase separation, silica sol formation, droplet formation, and wetting and drying phenomena.

References

1 N. Kröger, M. Sumper, in: E. Baeuerlein (Ed.), *Biomineralization, Progress in Biology, Molecular Biology and Application*, 2nd edn. Wiley-VCH, Weinheim, **2004**, p. 137.

2 C.C. Perry, T. Keeling-Tucker, *J. Biol. Inorg. Chem.* **2000**, *5*, 537.

3 B.E. Volcani, in: T.L. Simpson, B.E. Volcani (Eds.), *Silicon and Siliceous Structures in Biological Systems.* Springer, New York, **1991**, p. 157.

4 R.E. Hecky, K. Mopper, P. Kilham, E.T. Degens, *Mar. Biol.* **1973**, *19*, 323.

5 N. Kröger, R. Deutzmann, M. Sumper, *Science* **1999**, *286*, 1129.

6 N. Poulsen, N. Kröger, *J. Biol. Chem.* **2004**, *279*, 42993.

7 M. Sumper, N. Kröger, *J. Mater. Chem.* **2004**, *14*, 2059.

8 N. Kröger, S. Lorenz, E. Brunner, M. Sumper, *Science* **2002**, *298*, 584.

9 N. Kröger, R. Deutzmann, M. Sumper, *J. Biol. Chem.* **2001**, *276*, 26066.

10 M. Sumper, S. Lorenz, E. Brunner, *Angew. Chem. Int. Ed.* **2003**, *115*, 53500.

11 N. Kröger, R. Deutzmann, C. Bergsdorf, M. Sumper, *Proc. Natl. Acad. Sci. USA* **2000**, *97*, 14133.

12 M. Sumper, E. Brunner, G. Lehmann, *FEBS Lett.* **2005**, *579*, 3765.

13 P. Behrens, H. Menzel, in: P. Behrens, E. Baeuerlein (Eds.), *Handbook of Biomineralization Vol. 2. Biomimetic and Bio-Inspired Materials Chemistry.* Wiley-VCH, Weinheim, **2006**, Chapter 1.

14 N. Poulsen, M. Sumper, N. Kröger, *Proc. Natl. Acad. Sci. USA* **2003**, *100*, 12075.

15 R.K. Iler, *The Chemistry of Silica.* Wiley, New York, **1979**.

16 J.N. Cha, G.D. Stucky, D.E. Morse, T.J. Deming, *Nature* **2000**, *403*, 289.

17 Y. Zhou, K. Shimizu, J.N. Cha, G.D. Stucky, D.E. Morse, *Angew. Chem. Int. Ed.* **1999**, *38*, 780.

18 N. Kröger, M. Sumper, in: E. Baeuerlein (Ed.), *Biomineralization: from Biology to Biotechnology and Medical Application*, 1st edn. Wiley-VCH, Weinheim, **2000**, p. 138.

19 D.J. Belton, S.V. Patwardhan, C.C. Perry, *J. Mater. Chem.* **2005**, *15*, 4629.

20 T. Mizutani, H. Nagase, N. Fujiwara, H. Ogoshi, *Bull. Chem. Soc. Jpn.* **1998**, *71*, 2017.

21 D. Belton, G. Paine, S.V. Patwardhan, C.C. Perry, *J. Mater. Chem.* **2004**, *14*, 2231.

22 S.V. Patwardhan, M. Mukherjee, N. Steinitz-Kannan, S.J. Clarson, *Chem. Commun.* **2003**, 1122.

23 T. Coradin, O. Durupthy, J. Livage, *Langmuir* **2002**, *18*, 2331.

24 S.V. Patwardhan, S.J. Clarson, *Silicon Chem.* **2002**, *1*, 207.

25 E. Brunner, K. Lutz, M. Sumper, *Phys. Chem. Chem. Phys.* **2004**, *6*, 854.

26 K. Lutz, C. Gröger, M. Sumper, E. Brunner, *Phys. Chem. Chem. Phys.* **2005**, *7*, 2812.

27 S.V. Patwardhan, S.J. Clarson, *Polym. Bull.* **2002**, *48*, 367.

28 H. Menzel, S. Horstmann, P. Behrens, P. Bärnreuther, I. Krueger, M. Jahns, *Chem. Commun.* **2003**, 2994.

29 M. Sumper, *Angew. Chem. Int. Ed.* **2004**, *43*, 2251.

30 M. Sumper, *Science* **2002**, *295*, 2430.

31 E.G. Vrieling, Q. Sun, T.P.M. Beelen, S. Hazelaar, W.W.C. Gieskes, R.A. Van Santen, N.A.J.M. Sommerdijk, *J. Nanosci. Nanotechnol.* **2005**, *5*, 68.

32 A.M. Schmidt, *Protoplasma* **1994**, *181*, 43.

33 S.A. Crawford, M.J. Higgins, P. Mulvaney, R. Wetherbee, *J. Phycol.* **2001**, *37*, 543.

34 R. Gordon, R.W. Drum, *Int. Rev. Cytol.* **1994**, *150*, 243.

35 E.G. Vrieling, T.P.M. Beelen, R.A. van Santen, W.W.C. Gieskes, *Angew. Chem. Int. Ed.* **2002**, *41*, 1543.

36 E.A. Coffman, A.V. Melechko, D.P. Allison, M.L. Simpson, M.J. Doktycz, *Langmuir* **2004**, *20*, 8431.

37 D.J. Kim, K.B. Lee, T.G. Lee, H.K. Shon, W.J. Kim, H.J. Paik, I.S. Choi, *Small* **2005**, *1*, 992.

38 L.L. Brott, R.R. Naik, D.J. Pilkas, S.M. Kirkpatrick, D.W. Tomlin, P.W. Whitlock, S.J. Clarson, M.O. Stone, *Nature* **2001**, *413*, 291.

39 O. Prucker, C.A. Naumann, J. Rühe, W. Knoll, C.W. Franck, *J. Am. Chem. Soc.* **1999**, *121*, 8766.

40 M.A. Bartlett, M. Yan, *Adv. Mater.* **2001**, *13*, 1449.

41 N. Adden, L.J. Gamble, D.G. Castner, A. Hoffmann, G. Gross, H. Menzel, *Langmuir* **2006**.

42 J.D. Jeyaprakash, S. Samuel, J. Rühe, *Langmuir* **2004**, *20*, 10080.

43 O. Helmecke, A. Hirsch, P. Behrens, H. Menzel, *Small* **2007** (submitted).

44 O. Helmecke, P. Behrens, H. Menzel, Presented at BASF Symposium on Bio-inspired Materials for the Chemical Industry, Strasbourg August 7th–9th, 2006. Unpublished results.

45 M. Sumper, E. Brunner, *Adv. Funct. Mater.* **2006**, *16*, 17.

46 R.D. Deegan, O. Bakjin, T.F. Dupont, G. Huber, S.R. Nagel, T.A. Witten, *Nature* **1997**, *389*, 827.

47 F. Fan, K.J. Stebe, *Langmuir* **2004**, *20*, 3062.

12
Template Surfaces for the Formation of Calcium Carbonate

Wolfgang Tremel, Jörg Küther, Mathias Balz, Niklas Loges, and Stephan E. Wolf

Abstract

Calcium carbonate ($CaCO_3$), one of the most abundant biominerals on Earth, exists in three main crystalline polymorphs: aragonite, calcite, and vaterite. These polymorphs have a wide range of naturally occurring crystal habits, and they are often found assembled into hierarchical structures that result in a variety of intriguing properties in organisms. As the process of biomineral formation (which involves additives such as amphiphiles, proteins, nucleic acids, a structure directing insoluble matrix, and the action of specialized cells) is too complex to be understood at the molecular level, one must resort to simplified models which allow an understanding of certain key factors of the biomineralization process. Two such models – Langmuir monolayers and self-assembled monolayers (SAMs) – are reviewed in this chapter. Mineral formation at organic surfaces in natural systems is affected by physical, chemical, and molecular interactions, and molecular interactions at the organic aqueous interface, which can be controlled with molecular precision in Langmuir layers and SAMs. Phase selection and crystal orientation (i.e., the nucleating plane) of the growing crystal are determined by: (i) surface polarity; (ii) surface ordering/roughness; (iii) surface geometry/symmetry; and (iv) head group orientation due to even or odd chains. The concepts of template-induced crystallization on SAMs, and the use of polymer additives, can finally be combined to a new strategy where, through the cooperative interaction of a matrix involved in the nucleation process, an additive in solution and the dissolved ions, hierarchically ordered mineral structures are formed.

Key words: calcium carbonate, self assembled monolayers, Langmuir monolayers, template-induced crystallization.

Handbook of Biomineralization. Edited by P. Behrens and E. Bäuerlein

ISBN: 978-3-527-31805-6

12.1 Introduction

Organisms form their mineralized tissue for a number of functions from different minerals, mainly calcium carbonate, apatite or silica, in a variety of ways. In some cases the mineral is amorphous, whereas in others it is crystalline. In the latter case, the mineral may be deposited as large, specifically shaped single crystals, or alternatively as an array of smaller organized crystallites, again each with a well-defined shape. Organisms are able to control the mineralization very precisely, and examples abound in Nature of systems that are able to achieve control over all levels of organization of inorganic materials: control over the crystalline phase, over the size and morphology of the individual crystallites, and over the manner in which they are organized within a single matrix. Exemplary in this aspect is the formation of pearls [1, 2]. Implanting flat inorganic surfaces between the mantle and shell of the red abalone switches the deposition of $CaCO_3$ from the usual nacreous aragonite to calcite [2, 3] In the later stages of deposition, there is another switch back to aragonite. Both, the calcite and the aragonite phases, grow in a highly oriented manner, with the initial deposition being controlled by an organic complex protein machinery. In flat pearls deposited on inorganic substrates implanted near abalone shell-forming tissues, the transition from calcite to aragonite was observed to occur abruptly, without the apparent intervention of a nucleating organic matrix [2, 3]. Previous studies had shown that the mixture of soluble proteins found in abalone nacre causes the nucleation and growth of aragonite needles on the (104) faces of calcite seed crystals in supersaturated calcium carbonate solutions [4]. Similar control of calcium carbonate polymorphs has also been reported for soluble proteins extracted from other mollusk species [5–7]. Furthermore, soluble proteins associated with a variety of calcium carbonate biominerals have been shown to interact with calcite growth, become occluded within calcite crystals, and to affect a variety of crystal properties [4, 5, 8, 9].

Although the biochemical machinery has been characterized in some detail, the biological processes relevant to the functional and structural variety of biominerals are still not well understood. A central tenet in this field of research is the "template hypothesis", which assumes that nucleation, growth and the final morphology of the inorganic species (e.g., $CaCO_3$, apatite, silica) are determined by pre-organized assemblies of organic molecules where the organic matrix of calcium-containing biominerals (e.g., bone, mollusk shells) contains specific macromolecules that can initiate or inhibit the crystal nucleation space in a selective manner [10–12a].

12.2 *In-Vitro* Models

The process of biomineralization is still far from being understood, due mainly to the complexity expressed in different levels of hierarchy, the action of additives (e.g., amphiphiles, proteins, nucleic acids), a structure-directing insoluble matrix, and the action of specialized cells, which are difficult to take into account by biomineralization model systems. Therefore, during recent years a common approach in the field was to apply simplified model systems to understand certain key aspects of the biomineralization process, or to study biominerals themselves in order to reveal characteristic features, which might help to understand the actual biomineral formation [13]. Most such studies focused almost exclusively on the nucleating matrix. One strategy is to isolate the matrix proteins, which are involved in the nucleating process, from the biological material and to use them afterwards in *in-vitro* experiments [14, 15]. An alternative approach is to employ synthetic models of the matrix proteins. In this context, Langmuir monolayers [16], protein-covered substrates [12b], liquid crystalline systems [17, 18], self-assembled monolayers (SAMs) [19], or colloids and dendrimers [20a–c, 21–23] were used as templates or substrates for the crystallization of inorganic compounds. From the results of these and other [24, 25] studies it has become increasingly clear, that the original idea of an "epitaxial" crystal growth on a two-dimensional protein layer in biomineralization does not hold in a strict sense, but must be replaced in a qualitative sense by a global charge density which takes some influence on the crystal growth process.

Model systems also facilitate the study of chemical and physical properties at the interfaces between cells and the biomaterial, or between organic and inorganic components. This information can subsequently be used in the design of new bio-inspired materials with superior properties as scaffolds for tissue engineering.

12.3 Control of Polymorphism in Homogeneous Crystallization

Any crystalline process must cross the nucleation barrier. It is generally assumed that, as the ions or molecules begin to associate in supersaturated solutions, they form embryos with structures that resemble those of the crystals to be grown. For polymorphous systems, embryos of all phases may be formed, with each type of embryo resembling structurally the crystal into which it will eventually develop. If this hypothesis is correct, one may use the structural information on a mature crystal to design inhibitors of one particular crystalline phase. The overall consequence of this stereospecific inhibition process can be that the unaffected phase, if less stable, will grow by kinetic control.

For simplicity, we consider one distinct and simple case, where the logic of the above arguments applies. Calcium carbonate forms three crystalline polymorphs – vaterite, aragonite, and calcite – of which vaterite is the least stable and calcite the most stable. The additive molecules, some amino acids, are chiral. A racemic mixture of the amino acids produces a conglomerate of two crystalline phases. The individual enantiomers bind enantioselectively and stereoselectively; hence, only one of the polymorphs is formed, while growth of the other one is suppressed [25]. The use of tailor-made additives to control crystal shape has been demonstrated for many systems [26], although a strict control of phase selection through the enantioselective binding of additive molecules has not yet been observed.

Crystal nucleation is generally a heterogeneous process. This means that the activation barrier for nucleation is lowered by interaction with foreign surfaces. This process can occur at different levels of specificity, ranging from non-specific adsorption to epitaxial growth. One can therefore envisage an induced nucleation of desired crystalline structures, even with a specific crystal orientation, by designing appropriate nucleation promoters which match the structure of the crystal on a specific plane.

12.4 Control of Nucleation and Structure Formation Processes at Interfaces: Langmuir Monolayers

Langmuir monolayers were among the first controllable and versatile model systems to be used for approximating the bilayer structure of a biomembrane. When placed in aqueous media, water-insoluble amphiphiles arrange themselves at the gas–liquid interface in an ordered fashion. The resulting monolayers have an electrostatic pattern that mimics the charge distribution at the surfaces of three-dimensional (3-D) crystals in such a way that an oriented nucleation event is initiated. Lahav, Leiserowitz and collaborators have shown that Langmuir monolayers can initiate the formation of crystals of which the polymorphic phase, orientation and morphology is determined by the head group size and polarity, the packing and the conditions of the subphase [27]. The crystal systems studied include examples such as α-glycine [28] or calcium carbonate. The philosophy behind this approach relies on the structural fit and electrostatic attraction between the monolayer structure and a specific face of the growing crystal. Based on the results of these studies, it was possible to understand the structure-directing processes to such a degree that the design of new templates for the oriented mineralization of several organic and inorganic compounds could be achieved.

In order to understand this phenomenon at the molecular level, Mann and coworkers [29–31] and later Volkmer [32] and Sommerdijk and co-workers [33], studied the template-induced crystallization under Langmuir monolayers in the presence of other amphiphiles. The appeal of the Langmuir monolayer approach to control crystallization relies on the ease by which the interfacial properties of the model membranes can be varied in order to induce and regulate crystal growth. This approach was exploited subsequently using oxoanion head groups (e.g., –OH, $-CO_2H$, $-SO_3H$, $-PO_3H_2$) that complemented in terms of size, stereochemistry, or charge the crystal faces of other simple inorganic compounds such as $BaSO_4$ and $CaSO_4$.

For the sake of simplicity, we focus again on calcium carbonate. The crystallization of $CaCO_3$ in the absence of a monolayer of acid or amine leads to the forma-

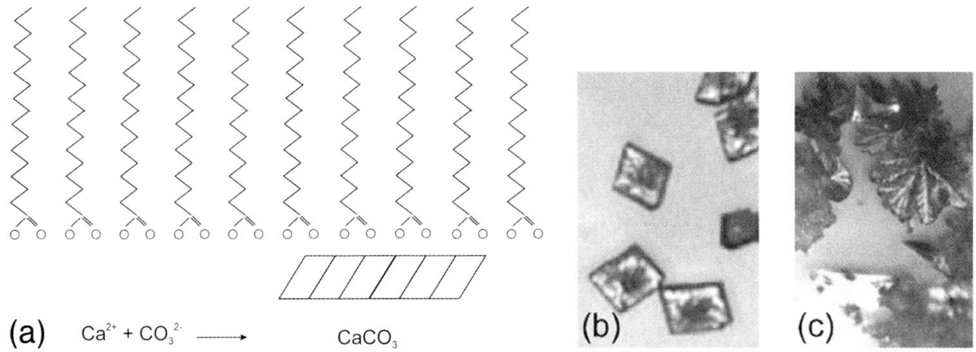

Fig. 12.1 (a) Schematic drawing of a fatty acid template on the supersaturated calcium bicarbonate solution. Calcium carbonate nucleates preferentially at the monolayer. (b) Rhombohedral calcite favored during fast mineralization. (c) Vaterite florets obtained from rate-inhibited mineralization.

tion of rhombohedral calcite crystals, whereas vaterite was formed under monolayers of long-chain carboxylic acids (Fig. 12.1). The structure of the monolayers corresponds to pseudohexagonally packed amphiphiles with an average head group spacing of about 5 Å for the carboxylate and amine terminated monomers. The (001) face of calcite contains the trigonal-planar carbonate groups – that is, a trigonal cation pattern with a cation–cation separation of approximately 5 Å should match the anion distribution of this face. Because the stereochemical arrangement of the head groups is unknown, it is difficult to show a clear structural relationship between monolayer structure and mineral orientation. The apparent lack of influence of the monolayer compression indicates that an exact structural relationship is not of vital importance. Nonetheless, the phase selection was found to depend on the Ca^{2+} concentration. The growth of oriented calcite requires a negatively charged surface, and the favorable electrostatic interaction with calcium seems to induce formation of the thermodynamically favored calcite. Calcite nucleated with the (1-10) axis perpendicular to the layer for higher Ca^{2+} concentration. The inspection of structural models showed that the (1-10) face of calcite did indeed contain linear arrangements of carbonate anions, the twofold axis of which was oriented perpendicular to the crystal face. In addition, the anion–anion separations of 4.96 Å were in registry with the head group separations in the monolayer. Therefore, it might be assumed that the nucleation barrier is lowered by the chemical complementarity and the resulting charge pattern of the constituents.

According to Ostwald's rule, vaterite should be favored kinetically over calcite, and the negatively charged surfaces should enhance the kinetics of calcite formation. Interestingly, lower Ca^{2+} concentrations under stearic acid monolayers led to the formation of thermodynamically less-favored vaterite polymorph on the (001) face. Models of the solution-crystal interface could be devised from two-dimensional (2-D) crystallographic studies using X-ray and neutron diffraction techniques [34], and from structural data of the calcium arachidate monolayers [35]. One current model assumes that charge balance requires the repeat per Ca^{2+} to be twice that of the fatty acid. As the cation density is low, there must be complexating water molecules to fill any empty space, which serves the purpose of screening the cation charge and forming hydrogen bonds with the carboxylate groups of the monolayer, the orientation of which is similar to that of the $CO_3{}^{2-}$ groups in $CaCO_3$ and the carbonate anion neighbors of the nascent vaterite. The different arrangement of the $CO_3{}^{2-}$ groups in the calcite structure may explain the preferred nucleation of vaterite. The complexity of this model exceeds that of a simple epitaxial effect, where the preferred association of cations to the template molecules induces nucleation and crystal growth, and the polytype selection may be dominated by kinetics [25].

12.5
Control of Nucleation and Structure Formation Processes at Interfaces: Self-Assembled Monolayers

An alternative approach to model the (bio)mineralization process is based on the templating features of solid, submerged organic substrates. In contrast to Langmuir films, solid substrates can be easily manipulated and characterized by using *ex-situ* methods. Furthermore, practical applications such as thin-film deposition on metal parts for biomedical purposes involve in some sense solid substrates.

Organic films have been prepared on supports such as silicon wafers and aluminum oxide. Tarasevich and Rieke studied the nucleation and growth of iron oxyhydroxides on polymers such as functionalized polyethylene [36], sulfonated polystyrene [37], or sulfonated SAMs on oxidized silicon [38]. The lack of structural in-depth information of the functionalized polymer surfaces did not allow an understanding of the structural and chemical parameters involved in the mineralization process. As might be expected from the complexating behavior of phosphate and sulfate anions, sulfonate-functionalized surfaces are effective promoters of calcium phosphate (or apatite) mineralization [38]. In addition, a variety of metal oxides (TiO_2, FeOOH, MnO_2) have been deposited on polymer, metal or ceramic substrates [39, 40a,b, 41]. The general result of these studies was that thin-film deposition is dominated by the coordination properties of the anionic surface group; that is, the sulfonate and phosphonate groups are efficient binding groups for transition metals and alkaline earth elements. The first examples of oriented crystal growth on ordered SAMs were reported by Bein and co-workers, who showed that a zinc phosphate zeolite could be grown in an oriented manner on zirconium phosphate multilayer films on gold [42].

The organic molecules can be synthesized and arranged, with atomic level precision, making them the templates of choice. Additionally, relevant physical properties such as polarity, chain length – and thereby the surface order or head group symmetry – can be varied by the choice of functional groups. In addition, the SAMs are chemically stable and their structure can be studied using atomic force microscopy (AFM), which allows the establishment of a lattice match between the substrate and the templated crystal. A quantitative analysis of the phase distribution can be made using X-ray powder diffraction, whilst X-ray microdiffraction allows an analysis to be made of the nucleating planes. Surface plasmon spectroscopy (SPS) or quartz crystal micro-balance (QCM) are convenient and powerful tools with which to study the assembly and kinetics of the crystal deposition. Furthermore, simple lithographic methods may be used to pattern the surface at the microscopic or submicroscopic level. This has also been observed in experiments involving the deposition of iron oxide [40a] and titania [41]. By using dithiol SAMs, it is possible to attach gold colloids to the gold surface, thereby achieving a microscopic roughening of the surface and increasing the number of nucleation sites [20b].

To summarize, phase selection and crystal orientation (i.e., the nucleating plane) are determined by: (i) surface polarity; (ii) surface ordering/roughness; (iii) surface geometry/symmetry; and (iv) head group orientation due to even or odd chains.

12.5.1 Surface Polarity

Surface polarity is reflected most clearly by the crystal nucleation density. Polar head groups have a larger complexation ability and thus display a stronger binding to Ca^{2+}. The resulting Ca^{2+} cation accumulation at the surface leads in turn to an increased $CO_3{}^{2-}$ anion binding and eventually, through a higher nucleation rate, to a gross mass transport and faster crystallization (Fig. 12.2). The lower nu-

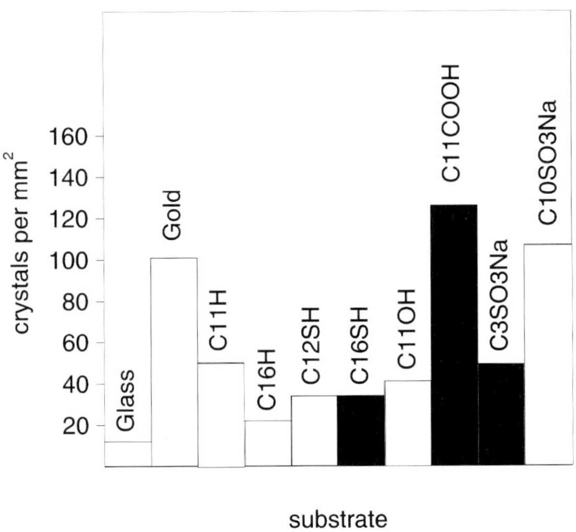

Fig. 12.2 Nucleation densities for $CaCO_3$ crystals on different substrates at room temperature [20a].

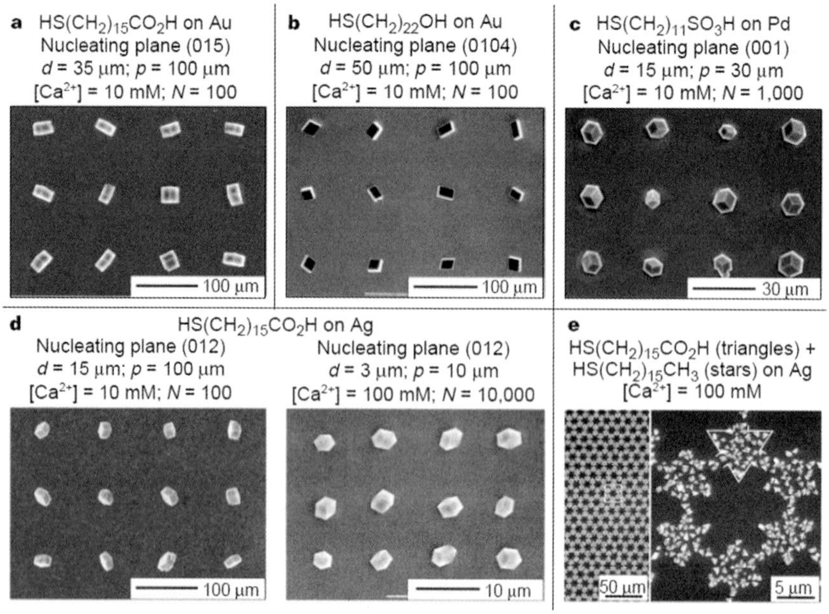

Fig. 12.3 Ordered two-dimensional arrays of single calcite crystals. The densities of nucleation, uniform sizes and crystallographic orientation are controlled by the micropatterned self-assembled monolayers (SAMs) consisting of regions of $HS(CH_2)_nX$ and $HS(CH_2)_{15}CH_3$. The density and sizes of features in the stamp and the concentration of the crystallizing solution were chosen to ensure the formation of one crystal per printed site. (a) Arrays of crystals with the density of nucleation $N \approx 100$ crystals mm^{-2} grew selectively from the (015) plane on SAMs of $HS(CH_2)_{15}CO_2H$ supported on Au(111). (b) Arrays of crystals with the density of nucleation $N \approx 100$ crystals mm^{-2} grew selectively from the (104) plane on SAMs of $HS(CH_2)_{22}OH$ supported on Au(111). (c) Arrays of crystals with the density of nucleation $N \approx 1000$ crystals mm^{-2} grew selectively from the (001) plane on SAMs of $HS(CH_2)_{11}SO_3H$ supported on Pd. (d) Arrays of crystals nucleated selectively from the (012) plane on SAMs of $HS(CH_2)_{15}CO_2H$ supported on Ag(111) with various densities of nucleation: $N \approx 100$ crystals mm^{-2} (left) and $N \approx 100$ crystals mm^{-2} (right). (e) An example of the fabrication of another complex crystalline pattern: a continuous, polycrystalline structure formed on SAMs consisting of a hexagonal array of "stars" of $HS(CH_2)_{15}CH_3$ ($d = 12$ μm; $p = 15$ μm) in a field of $HS(CH_2)_{15}CO_2H$ on Ag(111). The low-magnification SEM (left) illustrates the high fidelity of the procedure, and the high-magnification fragment (right) shows the formation of uniform crystals of sub-micrometer sizes [43].

cleation density due to reduced cation complexation also leads to a higher degree of supersaturation in solution. The $CaCO_3$ phase diagram suggests that aragonite is only stable at high pressure and high temperature, but kinetic factors – such as the degree of supersaturation in combination with a slight increase of temperature and absence of a nucleating plane for calcite or vaterite – may favor formation of the polymorph. Therefore, non-polar SAMs have a higher propensity to nucleate aragonite.

Microcontact printing of SAM patterns on gold resulted in further advances for controlling crystal nucleation and growth process in various systems. A recent example deals with calcite formation on SAMs having carboxyl-terminated regions from methyl-terminated regions [43]. It is well known from studies of growth on SAMs that the nucleation of specific crystallographic planes and the orientation are controlled by charge, stereochemistry, and geometric matching of the organic–inorganic interface. Therefore, the presence of a patterned SAM with different terminal groups allows control to be exerted over the nucleation of specific crystallographic planes, as well as growth in different crystallographic directions at a specific position on the substrate. As a result, ordered 2-D arrays of calcite crystals of uniform size and shape were obtained (Fig. 12.3). As demonstrated previously by Küther et al. [20b], the nucleation density is controlled by the binding constants of Ca^{2+} to the carboxylate and methyl SAM head groups, respectively. The complexation of Ca^{2+} by the carboxylate group is stronger than by the alkyl group, and this induces a higher nucleation rate at the polar area. As a result, mass transport to the growing crystals depletes ions from the slowly nucleating sites to the point of undersaturation.

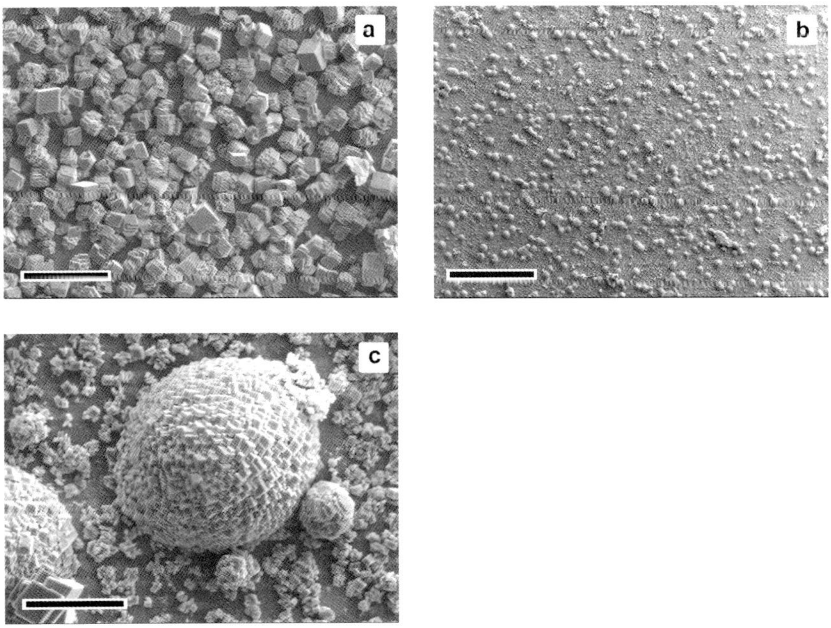

Fig. 12.4 (a) Scanning electron micrograph of the crystals formed by precipitation of $CaCO_3$ from solution at 22 °C. The rhombohedra are crystals of calcite displaying {104} faces. (b) Scanning electron micrograph of the crystals formed on crystallization in the presence of the seeds (100 mg L^{-1} at 22 °C). A typical crystallite assembly seen at higher magnification is shown in (c); the rhombohedra arrange spherically around some central point. Scale bars in (a), (b), and (c) = 50, 500, and 20 mm, respectively [20c].

Templating effects are not limited to organic monolayers on gold (111) surfaces; rather, the chemistry of thiol SAMs on gold surfaces has been extended to 3-D structures with the aid of gold colloids [20c]. Protecting gold colloids with thiols yields provides the possibility of forming stable nanoparticles which, in many ways, behave like molecules. This near-molecular behavior of the thiol-protected colloids should not detract from the fact that they possess a surface that is mostly close-packed (111). Passing from traditional SAMs to thiol-covered colloids introduces some interesting new aspects: the crystallization is carried out heterogeneously (at an interface; Fig. 12.4) from a homogeneous solution – that is, the seeds are perfectly wetted. Classical theory predicts that the seeds should act very much like homogeneous nuclei, but this is not observed. The colloidal crystallization templates testify to the ease with which nanoscopic seeds can be designed and used for controlled crystallizations.

12.5.2
Surface Ordering

Surface ordering is controlled by non-polar interactions between the alkyl chains of the SAM, with long chains leading to improved crystallinity by enhanced van der Waals interactions between the chains. Recently, we have carried out extensive model biomineralization studies using SAMs of substituted alkylthiols on gold to template the growth of the three forms of $CaCO_3$, namely calcite, vaterite, and aragonite. These studies point to aragonite being precipitated out of solution onto poorly crystallized surfaces (such as SAMs formed from short-chain thiols), while vaterite and calcite prefer ordered surfaces (such as SAMs formed from hexadecane thiol) [20a]. Crystalline surfaces (including clean gold) seem to inhibit the formation of aragonite. Furthermore, a high degree of preferred orientation is found for well-ordered surfaces (e.g., vaterite nucleated on the (100) plane), whereas no preferred orientation is found on less-ordered SAMs. Calcite and vaterite are nucleated preferentially on polar SAMs. However, in the absence of any highly crystalline surface pattern compatible with one of the calcite lattice planes, the crystals nucleate preferentially on the (001) plane, and tend to cluster due to secondary nucleation on the faces of pre-existing crystals.

Surface ordering can be suppressed intentionally by attaching gold colloids in the 8- to 10-nm size range to "sticky" gold surfaces (carrying SAM layers based on an α,ω-dithiol). Atomic force microscopy imaging of these gold colloid-coated SAM surfaces (Fig. 12.5) indicates that surfaces of controlled roughness have been obtained for use as substrates. The present authors have used dithiol-tethered colloid substrates for the crystallization of $CaCO_3$ and compared results with those obtained using clean gold surfaces as substrates. Figure 12.6 shows scanning electron microscopy (SEM) micrographs of $CaCO_3$ crystals grown on pure gold (Fig. 12.6a) and colloid-coated gold (Fig. 12.6b) surfaces. The very high nucleation densities on the modified surface are of interest. The morphology and habits of the crystals in the two cases are distinct; the needles on the gold surface are aragonite crystals, while the florets on the colloid-coated gold surface are vaterite.

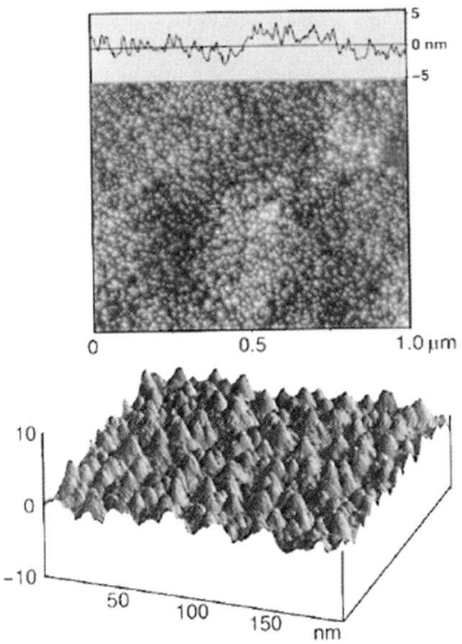

Fig. 12.5 Contact mode atomic force microscopy. Top view (top) and contour plot (bottom) of the gold surface covered by a dithiol self-assembled monolayer, and then by >10-nm colloids. The inset indicates a typical trace of the roughness of the surface [20b].

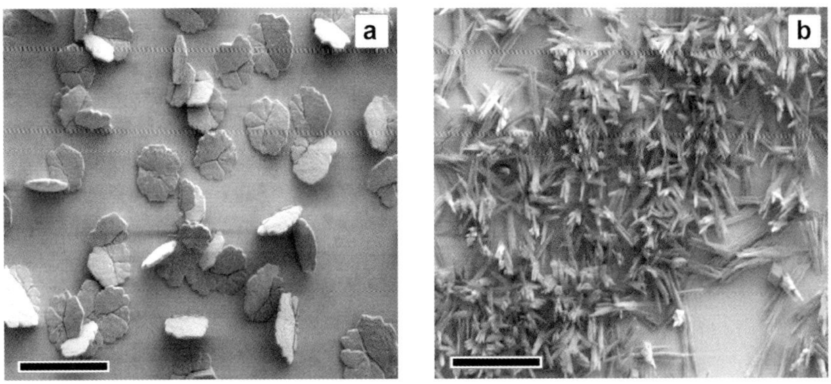

Fig. 12.6 Scanning electron micrographs of $CaCO_3$ crystals grown on (a) a clean gold surface and (b) on a colloid-modified surface. Scale bar = 200 mm [20b].

Quantitative estimates from X-ray profile analyses suggest that approximately equal amounts of calcite and vaterite are formed on clean gold under these conditions, but on the colloid-modified surface aragonite is the dominant phase. As the formation of aragonite under these conditions is kinetically favored [44], the effect observed here is equally one of aragonite inhibition by clean gold as one of aragonite promotion by the colloid-covered surface. Similar results could be derived for surfaces roughened by the attachment of non-polar polyglycerols [45].

12.5.3 Surface Geometry/Symmetry

The calcite and vaterite modifications of $CaCO_3$ possess threefold symmetry axes in their crystal structures, while the aragonite modification does not. As many SAMs of simple thiols on gold organize in a hexagonal close-packed manner, patterns for the templated crystallization of the first two polymorphs on SAMs are simple to achieve. This is not true in the case of aragonite. A thiol derived from anthracene-2-carboxylic acid (ANTH), which has been established as forming centered rectangular lattices when assembled on a gold (111) surface, has therefore been employed to form substrates for the crystallization of $CaCO_3$ with the specific intention of preferentially inducing the growth of aragonite [46].

The anthracene-derived disulfide ANTH displays a non-hexagonal pattern of thiol organization. AFM studies show a centered rectangular lattice structure with lattice parameters of 6.48 Å and 8.31 Å. The reason why ANTH organizes in this rectangular manner concerns the optimal packing of the oval-shaped anthracene group. A sketch of the two different schemes of thiol organization on the Au (111) surface is shown in Figure 12.7.

Figure 12.8 show the results obtained for the templated crystallization on the hexagonal and rectangular SAMs. At 22 °C all the surfaces show almost equal amounts of calcite and vaterite except for ANTH, which also shows a small amount of aragonite. At 45 °C, the high-pressure orthorhombic aragonite phase is formed in greater quantities. The organized hexagonal templates clearly disfavor aragonite. On the rectangular ANTH surface, about 15% aragonite is found

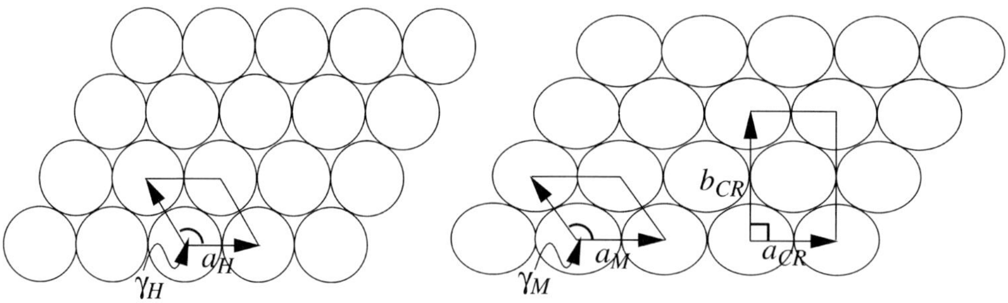

Fig. 12.7 (a) Scheme for the 2-D hexagonal close packing of simple thiols such as C16H on Au(111) surfaces. In the case of C16H, $a_H = 5.00(1)$ Å and $\gamma_H = 120°$. (b) Scheme for the close packing of thiols (for disulfides the shape of the thiolates is considered) that possess an oval rather than circular profile when viewed from above. The primitive monoclinic and centered rectangular lattices are outlined. For ANTH, $\gamma_M = 104.78°$; $a_M = 5.30(1)$ Å; $a_{CR} = 6.48(1)$ Å and $b_{CR} = 8.31(1)$ Å [46].

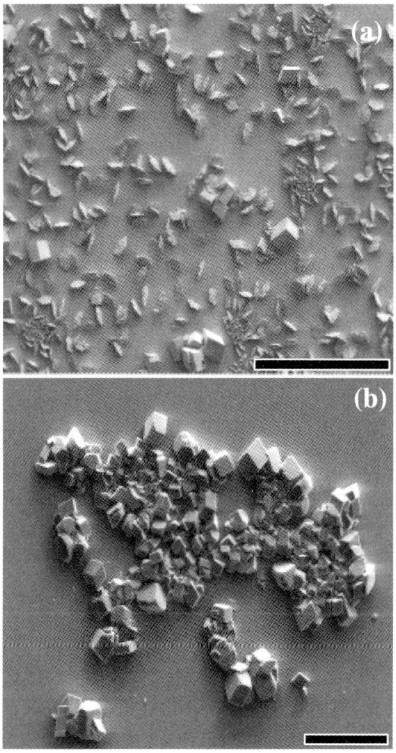

Fig. 12.8 (a) Scanning electron micrographs showing crystals of calcite (rhombi) and vaterite (florets standing on edge) obtained from $CaCO_3$ crystallization at 22 °C on C16H substrates (scale bar = 200 μm). (b) Scanning electron micrographs of calcite crystals formed at 45 °C on C16H substrates (scale bar = 50 μm) [46].

when the crystallization is carried out at 22 °C, and almost 90% when it is performed 45 °C.

The mode of templating of the different crystal polymorphs on the ordered, well-characterized surfaces can be rationalized by a comparison of the crystal habits and orientations, and the 2-D structure of the template surface. Figures 12.8a and b display the hexadecanethiol (C16H) surface with the crystals of calcite (rhombi) and vaterite (florets) grown at a temperature of 22 °C. The rhombic calcite crystals are often seen as standing on a vertex. This mode of templating of the calcite crystals is explained by recognizing the (001) plane of the calcite crystal as sitting flat on the SAM surface, as the dimensions of the (001) plane are in near-perfect registry with the 2-D hexagonal SAM structure. In the case of vaterite, such templating is more difficult to identify. The fact that the florets stand edge-on suggests that the planes in contact with the thiol substructure could be parallel to either (100) or (110). In Figure 12.8c and d, which show crystallizations on C16H performed at 45 °C, the clustering of the calcite rhombs on this low-polarity surface is more clearly seen, although the vaterite florets are less easily distinguished. From the isolated calcite crystals it can be seen that the templating pattern is maintained.

Figure 12.9a illustrates all three crystal polymorphs of $CaCO_3$ formed on the ANTH surface at 22 °C. The calcite rhombi are now oriented in an arbitrary fashion, and the vaterite florets are now seen to sit flat on their {001} crystal face. A comparison of this with the micrographs in Figure 12.7 indicates the effect of the different thiol structures on the crystals which they template. On raising the crystallization temperature to 45 °C, the vaterite florets are no longer formed (Fig. 12.9b), and now mostly bundles of aragonite needles and a small amount of calcite rhombi are observed. Whilst the calcite rhombi do not display any evidence for templating, the bundles of aragonite needles indicate that the substrate plays a specific role. Figure 12.9c shows, at high magnification, an efflorescent bundle of aragonite needles that fortuitously was turned over on the substrate during the sample preparation. The small hole in the center is due to growth of the bundle preventing the uniform accretion of material. The base of the bundle is about 5 μm in diameter. It is known from the AFM images that the substrates are not crystalline over such a large extent.

On the C16H surface, correlation of the (001) plane of the growing calcite crystal to the structure of the substrate is trivial (see Fig. 12.10a). Similar templating modes for vaterite do not seem as simple, and indeed on the surfaces presented here there is no clear evidence of templating. Of the three polymorphs, vaterite is the least stable, and its formation is possibly a manifestation of Ostwald's rule of successive crystallization, wherein ions precipitate from solution to yield the least stable phase first. Indeed, vaterite crystals redissolve over a period of time and reprecipitate as the other modifications. Figure 12.10b suggest that a plane parallel to (100) provides a reasonable lattice matching with the thiol substrate. This plane results in an absolute lattice mismatches of 4% and 11% along the two sides. The choice of this initial growth direction, namely [100], also allows us to reconcile the sudden fanning of the initial needles into the efflorescent bundle that is finally obtained. It would seem that secondary nucleation on the initial aragonite crystals takes place, and that this then permits the crystal to grow along the [001] direction. Figure 12.10b illustrates the plausible schemes for epitaxy between a plane parallel to the (100) and (001) planes of $CaCO_3$ aragonite and the centered rectangular thiol lattice. In the case of the (001) plane the mismatch is larger, being 4% and 23% along the two sides. The main reason for assuming (100) as the nucleating plane, rather than (001), is not the larger mismatch of the latter but rather the observation that the needles fan out after growing for a few micrometers.

At this point a word of caution is, however, in order. It should be noted that the planar extent of the domains of single-crystalline Au (111) is much smaller than the $CaCO_3$ crystals that we examine. This is particularly true in the case of the ANTH substrate, in which the crystalline domains extend only over 10 to 50 nm. The correlation between template and crystal structure is therefore justified only

Fig. 12.9 (a) Scanning electron micrograph of $CaCO_3$ crystals formed on anthracene-2-carboxylic acid (ANTH) substrates at 22 °C. All three modifications – calcite rhombi, vaterite florets sitting flat and aragonite needles – are seen. (b) Scanning electron micrograph of $CaCO_3$ crystals formed on the ANTH surface at 45 °C. The crystals are mostly aragonite. (c) Bundle of aragonite needles at higher magnification. The bundle is overturned and presents the flat surface on which it was nucleated. Scale bars: (a) = 200 mm; (b) = 200 mm; (c) = 10 mm [46].

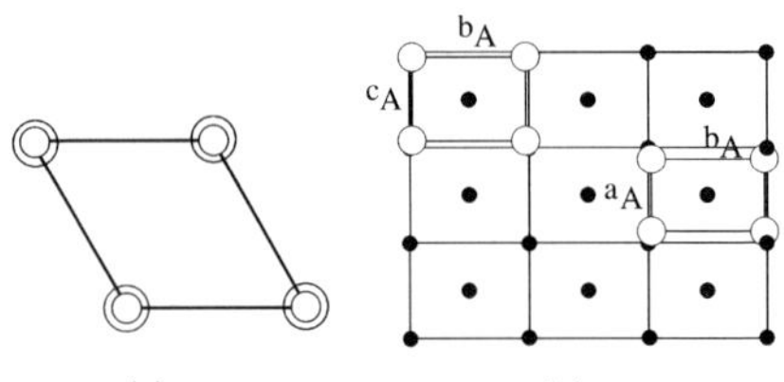

Fig. 12.10 (a) Scheme for epitaxy between the (001) plane of calcite and the underlying hexagonal (as in C16H) SAM substructure. The Ca atoms are depicted with hatched circles. (b) Plausible modes of epitaxy between aragonite and the anthracene-2-carboxylic acid (ANTH) SAM lattice. The Ca atoms are depicted with hatched circles. At the upper left corner is a scheme showing the (100) plane of aragonite; the middle right section shows a scheme of the (001) plane of aragonite; both are overlayed on the centered rectangular SAM lattice. The sides of the aragonite cell are $a_A = 4.96$ Å; $b_A = 7.98$ Å; and $c_A = 5.75$ Å [46].

by assuming that the template needs only to act in the very early stages of crystallization, when the nucleating crystal and the crystalline domains in the SAM substrate are similar in extent.

12.5.4 Head Group Orientation Due to Even/Odd Chains

Aizenberg et al. suggested that a match may exist between the direction of the SAM terminal groups and that of anions in the nucleated crystal [43]. This situation would imply that the mechanism of the face-selective nucleation involves translation of the orientation of the terminal groups on the SAM into the nucleating crystals. This mechanism could be verified by studying the oriented growth of calcite on SAMs in which only one parameter – the orientation of the functional group – was varied [47, 48]. For this purpose, the so-called "odd-even effect" in the monolayers was used [49, 50]. Detailed structural studies of SAMs [51] had shown that the orientation of long-chain alkanethiols ($HS(CH_2)_nCH_3$) adsorbed from solution onto metal surfaces is determined by the cant (α, Fig. 12.11a) and twist (β) angles which the thiol molecules adopt in relation to the metal film during formation of the monolayers. It had also been demonstrated that when gold films are used to support SAMs, α and β had the same value and sign for all alkanethiol chains, while alkanethiol molecules which are assembled on silver show cant angles of opposite signs for alkyl chains of different parity. Therefore, the orientation of the terminal group X in SAMs on Ag is constant for both odd and even chains, and the terminal group X in SAMs on Au forms two different angles with the interface for odd and even chain lengths (Fig. 12.11). SAMs with alkyl chains of different parity assembled on Au and Ag were studied in order to show the effect of terminal groups on the oriented growth of crystals. In fact, it could be shown that due to the face-selective nucleation by orientation of the functional

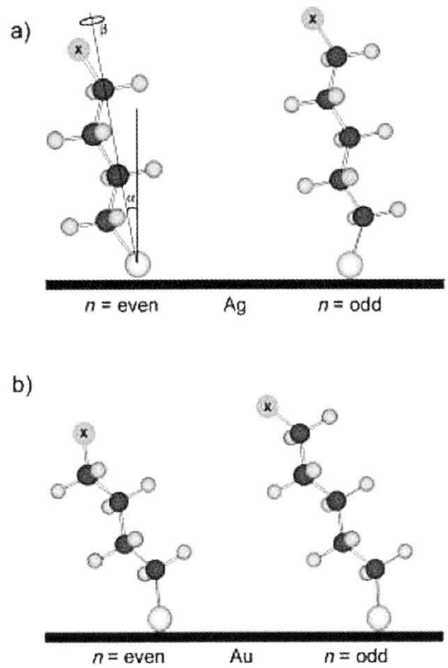

Fig. 12.11 Schematic representation of even and odd chain length ω-terminated alkylthiols adsorbed on (a) silver and (b) gold. The differences in the orientation of the functional group X with respect to the interface as well as cant (a) and twist (b) angles determine the nucleating plane of the crystals [47].

groups of the templating surface all SAMs on Ag induced the oriented crystal growth from the same crystal plane, while odd- and even-length SAMs on Au should induced nucleation in two different crystallographic directions.

12.6 Mechanistic Studies of the Crystallization on SAMs

Crystallization on SAM surfaces is usually monitored *ex situ*, by allowing the process to commence and to evolve for some time, removing the substrate from the mother solution, and then examining it using microscopy or diffraction, for example. We have studied the process also *in situ*, using high-energy monochromatic synchrotron X-radiation in conjunction with a 2-D detector to monitor, in a time-resolved fashion, the growth of $SrCO_3$ (strontianite) crystals on a SAM substrate. These experiments demonstrated that a process as complicated as the crystallization from solution of a mineral on a substrate can be monitored in an *in-situ*, time-resolved fashion. The nucleation and growth follow Avrami-type kinetics, and are indicative of the growth of crystals being auto-catalytic [52].

12.7 Studies of Cooperative Interactions in Template-Induced Crystallization Processes

In biological systems, insoluble macromolecules (e.g., collagen, polysaccharides) form a rigid matrix to which soluble macromolecules are tethered. These processes have been mimicked in a simplified model system consisting of a SAM template matrix and a growth modifier – for example, a dissolved polyelectrolyte (or a protein) and the ions in solution. Our model system utilizes macromolecules that form nanosize structures at an interface that can exert control during the nucleation and growth of an inorganic phase [45, 53a,b].

12.7.1 Mineralization of $CaCO_3$ on SAMs in the Presence of Polyacrylate

The crystallization of $CaCO_3$ or $SrCO_3$ in the presence of polyacrylate with a molecular weight of 30 000 or 240 000 Da, respectively, leads to the formation of long wires with a diameter of about 250 nm and a length of more than 100 μm (Fig. 12.12). Transmission electron micrographs of the vaterite wires indicated the growth of nanometer-sized crystallites around a common backbone, an interpretation that was backed by high-resolution (HR) SEM measurements. The phase selection may be rationalized, as it is well known that the crystallization of $CaCO_3$ in the presence of glutamate or aspartate additives leads to a preferred formation of vaterite as the kinetically stabilized polymorph of $CaCO_3$ [44].

One possible explanation for the formation of wire bundles is based on the unfolded polymer strand as a matrix for the crystallization process. In this case, the polyacrylate would act as organic backbone for the vaterite fibers; the carboxylate groups of the polyacrylate chain then act as complexating agents for the Ca^{2+} cations which, in turn, bind carboxylate groups from the solution. As a result, the polymer chain acts as a nucleation center. It is assume that the polymer is grafted as a coil of strands to the surface; the individual strands of the coil interpenetrate and exhibit a spaghetti-type appearance, which is reflected in the morphology of the product.

Attachment of the polymer strands to the surface is not due to direct interaction of the carboxylic groups of the polymer and the carboxylic groups of the SAM, but rather is mediated by the Ca^{2+} ions, and this can be demonstrated using the QCM. In addition, it is known from light-scattering experiments on polymethacrylic acid that high concentrations of bivalent cations can lead to an unfolding of the polymer chain [54]. This may be attributed to a complexation of the M^{2+} cations and a concomitant loss of hydrogen bridges within the polymer. Similarly, polymer unfolding may be expected at basic pH values. By applying these ideas to our model system we should expect – in contrast to the experimental results – wire-like polymer/$CaCO_3$-composites, irrespective of the underlying thiol monolayer.

An interpretation of the experimental results can be obtained only when an interaction between the polymer and the template surface is considered. It is as-

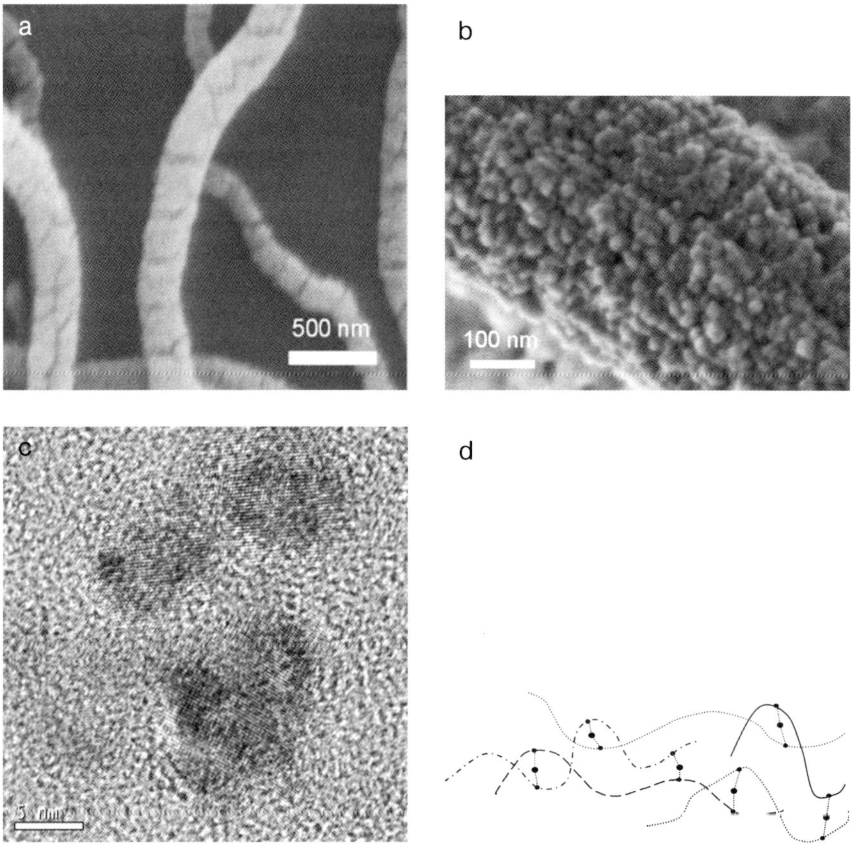

Fig. 12.12 (a) Scanning electron micrographs of $CaCO_3$-aggregates grown on a COOH-terminated gold slide in the presence of polyacrylate. (b) HR-SEM graph of the fiber-like aggregates. (c) TEM-image demonstrating the presence of nanometer-sized $CaCO_3$-particles about 1 h after starting the mineralization process. (d) Schematic linkage of polymer strands by complexation of Ca^{2+}-ion [52b].

sumed that the polymer can unfold when the Ca^{2+} concentration is large enough. However, if the polarity of the SAM surface layer is sufficiently high to ensure the attachment and unfolding of the polymer strands by Ca^{2+} complexation, the surface and/or the polymer may serve as a mineralization template. This condition is fulfilled only, if a carboxylate-terminated monolayer is present. This assumption is confirmed by the results of QCM-measurements on COOH- and CH_3-terminated SAMs.

Based on these data, it is impossible to differentiate whether $CaCO_3$ nucleates on the substrate with attached polymer directly, or by the aggregation of polyacrylate and $CaCO_3$ particles formed in solution, though it seems likely that nanometer-sized particles pre-formed in solution are attached to the polymer template. It was possible, however, to demonstrate the presence of these particles in a polyacrylate solution by using TEM measurements. The samples, which were taken at 1 h after the start of the mineralization process, contained small $CaCO_3$ particles that were about 10 nm in size and which were shown by X-ray diffraction to be crystalline in nature.

Samples taken about 20 min after starting the mineralization process seemed to be amorphous. During the examination of these particles by TEM, the transformation from amorphous to crystalline phase was seen to occur within only a few minutes. Clearly, amorphous particles are present at the start of the mineralization process, but these rapidly transform (within <60 min) to a crystalline material.

On replacing polyacrylic acid by polyaspartate, the amorphous hydrated polymer-stabilized $CaCO_3$ particles transform into spherical particles, the spherical shape being dictated by the interfacial energy. The subsequent transformation from a voluminous hydrated to a more dense crystalline material leads to the formation of hollow spheres (Fig. 12.13a,b), reminiscent of a 3-D analogue of the

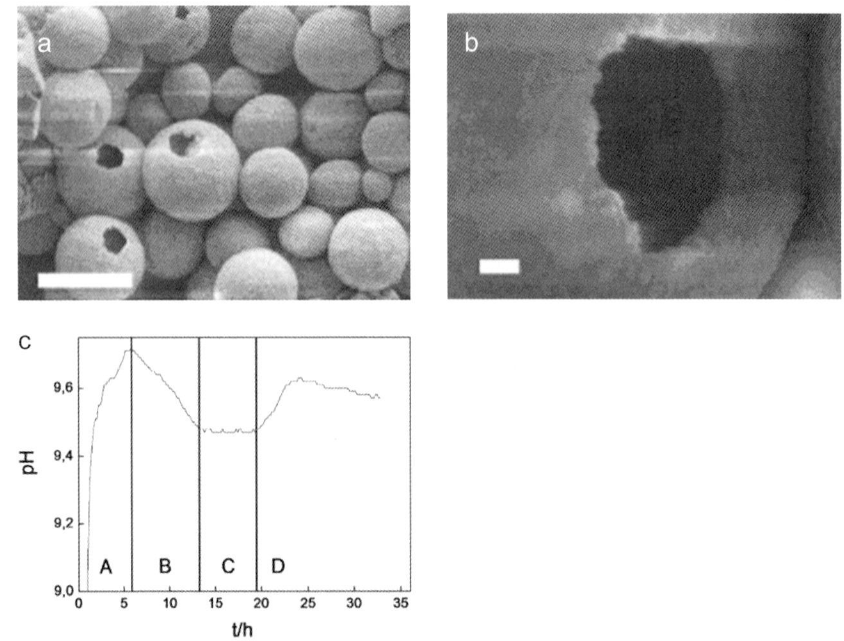

Fig. 12.13 Scanning electron micrographs of vaterite hollow spheres obtained after 2 days of crystallization on OH-terminated SAMs in the presence of poly(aspartate). (a) Scale bar = 5 μm; (b) scale bar = 200 nm [54]. (c) pH profile of $CaCO_3$ crystallization. At point A, pH equilibrium is dominated by NH_3 uptake. At B, the pH equilibrium is controlled by CO_2 uptake. At C, the CO_2 uptake and depletion by $CaCO_3$ precipitation have similar orders of magnitude. At D, the pH value is dominated by the release of aspartate.

so-called "coffee-stain effect". An outward flow of material in a crystallizing spherical particle is produced when the outer particle boundary is pinned, so that material compacting at the edge of the sphere must be replenished by mass flow from the interior. This flow is capable of transferring 100% of the amorphous calcium carbonate (ACC) precursor particles to the sphere boundary, and thus accounts for the formation of a hollow sphere [55].

One possible explanation for the formation of hollow spheres composed of vaterite nanocrystals may be provided on the basis of the development of solution pH during the course of the crystallization. The corresponding pH profile (Fig. 12.13c) showed a characteristic multistep process. When the reaction was started, the pH of the solution was seen to rise, within an induction period of 5 h, from pH 7 to pH 9.7 (this was due to the dissolution of ammonia formed by the decomposition of $(NH_4)_2CO_3$). The better solubility of NH_3 compared to that of CO_2 leads to the observed sudden pH change. This equilibrium adjusts until the NH_3 vapor pressure in the gas phase matches the NH_3 partial pressure of the solution. Then, CO_2 is absorbed by the solution over a period of about 6 h, during which time the pH value remains constant. This is the limit of an equilibrium between uptake and CO_2 in the solution and CO_2 depletion of the solution due to the precipitation of calcium carbonate. According to the Ostwald rule of stages, ACC is formed in the presence of polyaspartate by a liquid–liquid phase separation of droplets of the mineral precursor and the solution [56].

Acknowledgments

These studies were supported by grants from the Deutsche Forschungsgemeinschaft in the Priority Program "Principles of Biomineralization".

References

1 H.A. Heuer, D.J. Fink, V.J. Laraia, J.L. Arias, P.D. Calvert, K. Kendall, G.L. Messing, J. Blackwell, P.C. Rieke, D.H. Thompson, A.P. Wheeler, A. Veis, A.I. Caplan, Innovative materials processing strategies: a biomimetic approach. *Science* **1992**, *255*, 1098–1105.

2 C. Zaremba, A.M. Belcher, M. Fritz, Y. Li, S. Mann, P.K. Hansma, D.E. Morse, J.S. Speck, G.D. Stucky, Critical transitions in the biofabrication of abalone shells and flat pearls. *Chem. Mater.* **1996**, *8*, 679–690.

3 M. Fritz, A.M. Belcher, M. Radmacher, D.A. Walters, P.K. Hansma, G.D. Stucky, D.E. Morse, S. Mann, Flat pearls from biofabrication of organized composites on inorganic substrates. *Nature* **1994**, *371*, 49–51.

4 A.M. Belcher, X.H. Wu, R.J. Christensen, P.K. Hansma, G.D. Stucky, D.E. Morse, Control of crystal phase switching and orientation by soluble mollusc-shell proteins. *Nature* **1996**, *381*, 56–58.

5 G. Falini, G.S. Albeck, S. Weiner, L. Addadi, Control of aragonite polymorphism by mollusk shell macromolecules. *Science* **1996**, *271*, 67–69.

6 T. Samata, N. Hayashi, M. Kono, K. Hasegawa, C. Horita, S. Akera, A new matrix protein family related to the nacreous layer formation of *Pinctada fucata*. *FEBS Lett.* **1999**, *462*, 225–229.

7 M. Kono, N. Hayashi, T. Samata, Molecular mechanism of the nacreous layer formation in *Pinctada maxima*. *Biochem. Biophys. Res. Commun.* **2000**, *269*, 213–218.

8 (a) A. Berman, L. Addadi, S. Weiner, Interactions of sea-urchin skeleton macromolecules with growing calcite crystals: a study of intracrystalline proteins. *Nature* **1988**, *331*, 546–548; (b) A. Berman, H. Hanson, L. Leiserowitz, T.F. Koetzle, S. Weiner, L. Addadi, Biological control of crystal texture: a widespread strategy for adapting crystal properties to function. *Science* **1993**, *259*, 776–779.

9 J.M. Didymus, P. Oliver, S. Mann, A.L. Devries, P.V. Hauschka, P. Westbroek, Influence of low-molecular-weight and macromolecular organic additives on the morphology of calcium carbonate. *J. Chem. Soc. Faraday Trans.* **1993**, *89*, 2891–2900.

10 S. Mann, Mineralization in biological systems. *Struct. Bonding* **1983**, *54*, 125–174.

11 R. Traub, S. Weiner, Macromolecules in mollusc shells and their functions in biomineralization. *Philos. Trans. R. Soc. Lond. B* **1984**, *304*, 425–434.

12 (a) L. Addadi, S. Weiner, Interactions between acidic proteins and crystals: stereochemical requirements in biomineralization. *Proc. Natl. Acad. Sci. USA* **1985**, *82*, 4110–4114; (b) S. Weiner, L. Addadi, Design strategies in mineralized biological materials. *J. Mater. Chem.* **1997**, *7*, 689–702.

13 E. Bäuerlein, *Biomineralization, Progress in Biology, Molecular Biology and Application*. Wiley-VCH, 2004.

14 Y. Levi, S. Albeck, A. Brack, S. Weiner, L. Addadi, Control over aragonite crystal nucleation and growth: an in vitro study on biomineralization. *Chem. Eur. J.* **1998**, *4*, 389–396.

15 N. Gehrke, N. Nassif, N. Pinna, M. Antonietti, H.S. Gupta, H. Cölfen, Retrosynthesis of nacre via amorphous precursor particles. *Chem. Mater.* **2005**, *17*, 6514–6516.

16 D. Jaquemain, S.G. Wolf, F. Leveiller, M. Deutsch, K. Kjaer, J. Als-Nielsen, M. Lahav, L. Leiserowitz, Two-dimensional crystallography of amphiphilic molecules at the air-water interface. *Angew. Chem. Int. Ed.* **1992**, *31*, 130–152.

17 C.T. Kresge, M.E. Leonowicz, W.J. Roth, J.C. Vartuli, J.S. Beck, Ordered mesoporous molecular sieves synthesized by a liquid-crystal template mechanism, *Nature* **1992**, *359*, 710–712.

18 A. Sugawara, T. Ishii, T. Kato, Self-organized calcium carbonate with regular surface-relief structures. *Angew. Chem. Int. Ed.* **2003**, *42*, 5299–5303.

19 D.D. Archibald, S.B. Quadri, B.P. Gaber, Modified calcite deposition due to ultrathin organic films on silicon substrates. *Langmuir* **1996**, *12*, 538–546.

20 (a) J. Küther, R. Seshadri, W. Knoll, W. Tremel, Templated growth of calcite, vaterite and aragonite crystals on self-assembled monolayers of substituted alkylthiols on gold. *J. Mater. Chem.* **1998**, *8*, 641–650; (b) J. Küther, R. Seshadri, G. Nelles, H.-J. Butt, W. Knoll, W. Tremel, Rough surfaces by design: gold colloids tethered to gold surfaces as substrates for $CaCO_3$ crystallization. *Adv. Mater.* **1998**, *10*, 401–404; (c) J. Küther, R. Seshadri, W. Tremel, Crystallization of calcite spherules around designer nuclei. *Angew. Chem. Int. Ed.* **1998**, *37*, 3044–3047.

21 M. Bartz, N. Weber, J. Küther, R. Seshadri, W. Tremel, Sticky gold colloids through protection–deprotection and their use in complex metal-organic-inorganic architectures. *Chem. Commun.* **1999**, 2085–2086.

22 M. Bartz, J. Küther, G. Nelles, N. Weber, R. Seshadri, W. Tremel, Monothiols derived from glycols as agents for stabilizing gold colloids in water: synthesis, self-assembly and

use as crystallization templates. *J. Mater. Chem.* **1999**, *9*, 1121–1125.

23 J.J.J.M. Donners, B.R. Heywood, E.W. Meijer, R.J.M. Nolte, N.A.J.M. Sommerdijk, Control over calcium carbonate phase formation by dendrimer/surfactant templates. *Chem. Eur. J.* **2002**, *8*, 2561–2573.

24 M.J. Lochhead, S.R. Letellier, V. Vogel, Assessing the role of interfacial electrostatics in oriented mineral nucleation at charged organic monolayers. *J. Phys. Chem. B* **1997**, *101*, 10821–10827.

25 E. DiMasi, M.J. Olszta, V.M. Patel, L.B. Gower, When is a template directed mineralization really template directed? *Cryst. Eng. Commun.* **2003**, *5*, 346–350.

25 S. Wolf, N. Loges, M. Panthöfer, B. Mathiasch, W. Tremel, Phase selection of calcium carbonate through the chirality of adsorbed amino acids – a path to homochirality? *Angew. Chem.* **2007**, in press.

26 I. Weissbuch, L. Addadi, M. Lahav, L. Leiserowitz, Molecular recognition at crystal interfaces. *Science* **1991**, *253*, 637–645.

27 I. Weissbuch, M. Lahav, L. Leiserowitz, Toward stereochemical control, monitoring, and understanding of crystal nucleation. *Crystal Growth Des.* **2003**, *3*, 125–150.

28 E.M. Landau, M. Levanon, L. Leiserowitz, M. Lahav, J. Sagiv, Transfer of structural information from Langmuir monolayers to three-dimensional growing crystals. *Nature* **1985**, *318*, 353.

29 S. Mann, B.R. Heywood, S. Rajam, J.D. Birchall, Controlled crystallization of $CaCO_3$ under stearic acid monolayers. *Nature* **1988**, *334*, 692–695.

30 S. Mann, Molecular recognition in biomineralization. *Nature* **1988**, *332*, 119–124.

31 B. Heywood, S. Mann, Molecular construction of oriented inorganic materials: controlled nucleation of calcite and aragonite under compressed Langmuir monolayers. *Chem. Mater.* **1994**, *6*, 311–318.

32 D. Volkmer, M. Fricke, C. Agena, J. Mattay, Interfacial electrostatics guiding the crystallization of $CaCO_3$ underneath monolayers of calixarenes and resorcarenes. *J. Mater. Chem.* **2004**, *14*, 2249–2259.

33 P.J.J.A. Buijnsters, J.J.J.M. Donners, S.J. Hill, B.R. Heywood, R.J.M. Nolte, B. Zwanenburg, N.A.J.M. Sommerdijk, Oriented crystallization of calcium carbonate under self-organized monolayers of amide-containing phospholipids. *Langmuir* **2001**, *17*, 3623–3628.

34 J.M. Bloch, W. Yun, Condensation of monovalent and divalent metal ions on a Langmuir monolayer. *Phys. Rev. A* **1990**, *41*, 844–862.

35 D.A. Outka, J. Stöhr, J. Rabe, J.D. Swalen, H.H. Rotermund, Orientation of arachidate chains in Langmuir-Blodgett monolayers on Si(111). *Phys. Rev. Lett.* **1987**, *59*, 1321–1324.

36 P. Alvert, P. Rieke, Biomimetic mineralization in and on polymers. *Chem. Mater.* **1996**, *8*, 1715–1727.

37 B.J. Tarasevich, P.C. Rieke, J. Liu, Nucleation and growth of oriented ceramic films onto organic interfaces. *Chem. Mater.* **1996**, *8*, 292–300.

38 P.C. Rieke, R. Wiecek, B.D. Marsh, L.L. Wood, J. Liu, L. Song, G.E. Fryxell, B.J. Tarasevich, Interfacial free energy of nucleation for iron oxyhydroxide on mixed self-assembled monolayers. *Langmuir* **1996**, *12*, 4266–4271.

39 B.C. Bunker, P.C. Rieke, B.J. Tarasevich, A.A. Campbell, G.E. Fryxell, G.L. Graff, L. Song, J. Liu, J.W. Virden, Ceramic thin-film formation on functionalized interfaces through biomimetic processing. *Science* **1994**, *264*, 48–55.

40 (a) M. Nagtegaal, J. Küther, J. Ensling, P. Gütlich, W. Tremel, Hydrothermal deposition of small a-Fe_2O_3 (hematite) particles on ordered zirconium phosphonate multilayer SAMs on gold. *J. Mater. Chem.* **1999**, 9, 1115–1120; (b) M. Nagtegaal, P. Stroeve, J. Ensling, P. Gütlich, M. Schurrer, H. Voit, J. Flath, J.

Käshammer, W. Knoll, W. Tremel, Soft-Chemical Growth of g-FeO(OH) films on self-assembled monolayers of substituted alkylthiols on gold(111). *Chem. Eur. J.* **1999**, *5*, 1331–1337.

41 M. Bartz, A. Terfort, W. Knoll, W. Tremel, Stamping of monomeric SAMs as a route to crystallization templates: patterned titania films. *Chem. Eur. J.* **2000**, *6*, 4149–4153.

42 S. Feng, T. Bein, Growth of oriented molecular sieve crystals on organophosphonate films. *Nature* **1994**, *368*, 834–836.

43 J. Aizenberg, A.J. Black, G.M. Whitesides, Control of crystal nucleation by patterned self-assembled monolayers. *Nature* **1999**, *398*, 495–498.

44 J.L. Wray, F. Daniels, Precipitation of calcite and aragonite. *J. Am. Chem. Soc.* **1957**, *79*, 2031–2034.

45 M. Balz, E. Barriau, V. Istratov, H. Frey, W. Tremel, Controlled crystallization of $CaCO_3$ on hyperbranched polyglycerol adsorbed to self-assembled monolayers. *Langmuir* **2005**, *21*, 3987–3991.

46 J. Küther, G. Nelles, R. Seshadri, M. Schaub, H.-J. Butt, W. Tremel, Templated crystallization of calcium and strontium carbonates on centred rectangular self-assembled monolayer substrates. *Chem. Eur. J.* **1998**, *4*, 1834–1841.

47 A.M. Travaille, J.J.J.M. Donners, J.W. Gerritsen, N.A.J.M. Sommerdijk, R.J.M. Nolte, H. van Kempen, Aligned growth of calcite crystals on a self-assembled monolayer. *Adv. Mater.* **2002**, *14*, 492–495.

48 Y. Han, J. Aizenberg, Face-selective nucleation of calcite on self-assembled monolayers of alkanethiols: effect of the parity of the alkyl chain. *Angew. Chem. Int. Ed.* **2003**, *42*, 3668–3670.

49 R. Popovitz-Biro, J.L. Wang, J. Majewski, E. Shavit, L. Leiserowitz, M. Lahav, Induced nucleation of supercooled water into ice by self-assembled crystalline monolayers of amphiphilic alcohols at the air-water interfaces. *J. Am. Chem. Soc.* **1994**, *116*, 1179.

50 V.R. Thalladi, M. Nüsse, R. Boese, The melting point alternation in α,ω-alkanedicarboxylic acids. *J. Am. Chem. Soc.* **2000**, *122*, 9227–9236.

51 P.E. Laibinis, G.M. Whitesides, D.L. Allara, Y.T. Tao, A.N. Parikh, R.G. Nuzzo, A comparison of the structures and wetting properties of self-assembled monolayers of n-alkanethiols on the coinage metal surfaces Cu, Ag and Au. *J. Am. Chem. Soc.* **1991**, *113*, 7152.

52 J. Küther, M. Bartz, R. Seshadri, G.B.M. Vaughan, W. Tremel, Crystallization of $SrCO_3$ on a self-assembled monolayer substrate: an in-situ synchrotron X-ray study. *J. Mater. Chem.* **2001**, *11*, 503–506.

53 (a) M. Balz, H.A. Therese, J. Li, J.S. Gutmann, M. Kappl, L. Nasdala, W. Hofmeister, H.-J. Butt, W. Tremel, Crystallization of vaterite nanowires by the cooperative interaction of tailor-made nucleation surfaces and polyelectrolytes. *Adv. Funct. Mater.* **2005**, *4*, 683–688; (b) M. Balz, H.A. Therese, M. Kappl, L. Nasdala, W. Hofmeister, H.-J. Butt, W. Tremel, Morphosynthesis of strontianite nanowires using polyacrylate templates tethered onto self-assembled monolayers. *Langmuir* **2005**, *21*, 3981–3986.

54 Y. Ikeda, M. Beer, M. Schmidt, K. Huber, Ca^{2+} and Cu^{2+} induced conformational changes of sodium polymethacrylate in dilute aqueous solution. *Macromolecules* **1998**, *31*, 728–733.

55 N. Loges, K. Graf, L. Nasdala, W. Tremel, Probing cooperative interactions of tailor-made nucleation surfaces and macromolecules: a bioinspired route to hollow micrometer-sized calcium carbonate particles. *Langmuir* **2006**, *22*, 3073–3080.

56 L.B. Gower, D.J. Odom, Deposition of calcium carbonate films by a polymer-induced liquid-precursor (PILP) process. *J. Cryst. Growth* **2000**, *210*, 719–734.

Part III
Bio-Supported Materials Chemistry

Handbook of Biomineralization. Edited by P. Behrens and E. Bäuerlein

ISBN: 978-3-527-31805-6

13
Inorganic Preforms of Biological Origin: Shape-Preserving Reactive Conversion of Biosilica Microshells (Diatoms)

Kenneth H. Sandhage, Shawn M. Allan, Matthew B. Dickerson, Eric M. Ernst, Christopher S. Gaddis, Samuel Shian, Michael R. Weatherspoon, Gul Ahmad, Ye Cai, Michael S. Haluska, Robert L. Snyder, Raymond R. Unocic, and Frank M. Zalar

Abstract

The attractive three-dimensional (3-D) self-assembly characteristics of certain microorganisms may be coupled with the chemical versatility of synthetic processing to yield a revolutionary biologically enabling fabrication paradigm known as Bioclastic and Shape-preserving Inorganic Conversion (BaSIC). Nature provides impressive examples of the precise and scalable assembly of 3-D mineral (bioclastic) structures. A stunning variety of nanostructured 3-D silica assemblies are generated by diatoms (unicellular algae). Diatoms form rigid cell walls (frustules) comprised of inter-connected networks of amorphous silica. Each diatom species forms a 3-D frustule with a particular hierarchical structure; that is, the microscale frustule shape and patterned nanoscale frustule features (pores, channels, protuberances, etc.) are specific to a given diatom species. Because the frustule morphology is replicated with a high degree of fidelity upon diatom reproduction, sustained culturing of a particular diatom species can yield enormous numbers of frustules of similar shape (e.g., >1 trillion replicas in 40 reproduction cycles). Such genetically precise and massively parallel 3-D self-assembly under ambient conditions exceeds the capabilities of current synthetic protocols. However, the SiO_2 chemistry of diatom frustules is not appropriate for a variety of devices. With BaSIC, diatom frustules (and other bioclastic structures) can be converted into a variety of new functional chemistries through shape-preserving displacement reactions, conformal coating approaches, or combinations thereof (although this chapter will focus on the use of displacement reactions). If continued research on the genetic manipulation of diatoms (see Chapter 3 in Volume 1) leads to tailorable frustule morphologies, then such genetic engineering may be coupled with the BaSIC process to enable the manufacturing of low-cost 3-D Genetically Engineered Micro/nano-devices (3-D GEMs).

Handbook of Biomineralization. Edited by P. Behrens and E. Bäuerlein

ISBN: 978-3-527-31805-6

Key words: diatoms, frustules, bioclastic, silica, self-assembly, chemical conversion, reaction processing, inorganic conversion, shape-preserving, three-dimensional, nanocrystals, microstructures, replicas, magnesia, titania, zirconia, functional, ceramics.

13.1 Attractive Characteristics and Limitations of Biological Self-Assembly

Because nanoparticles can exhibit significantly enhanced, or altogether new, properties for advanced devices, extensive efforts are underway worldwide to develop robust protocols for the assembly of nanoparticles into functional three-dimensional (3-D) structures. Such protocols must be capable of producing 3-D assemblies that are: (i) structurally precise down to the nanoscale; (ii) chemically versatile for tailorable functionalities; and (iii) reproducibly scalable for economic mass production. The precise, versatile, scalable, and low-cost manufacturing of 3-D nanoparticle assemblies remains a difficult challenge in nanotechnology.

Impressive examples of precise and scalable biomineral assembly can be found in Nature [1–3], and a particularly diverse range of intricate 3-D mineralized structures is generated by diatoms [4]. Diatoms (Bacillariophyceae) are unicellular algae that form rigid cell walls (frustules) of amorphous silica [4, 5]. Each of the tens of thousands of diatom species generates a microscale frustule with a particular 3-D shape that is arrayed with specific patterns of finer features (10^1 to 10^2 nm pores, channels, nodules, etc.) [4–6]. Because the frustule morphology is faithfully replicated upon diatom reproduction, the sustained culturing (repeated doubling) of a single diatom species can yield enormous numbers (e.g., 80 reproduction cycles would yield $2^{80} = 1.2 \times 10^{24}$, or about twice Avogadro's number) of daughter diatoms with frustules of similar shape [5, 7, 8]. Such massively parallel and precise hierarchical (nano-to-microscale) 3-D assembly under ambient conditions is very attractive from a manufacturing perspective. While diatom frustules have been used as filters, porous substrates, absorbents, or fillers (for wastewater treatment, beverage filtration, catalysis, chromatography, and polymer processing [9–14]), the SiO_2 frustule chemistry severely limits the range of applications for these bioclastic structures.

13.2 The Bioclastic and Shape-Preserving Inorganic Conversion (BaSIC) Process

The massive parallelism and precision with which certain biomineralizing organisms assemble their intricate 3-D frustules under ambient conditions lie well beyond the capabilities of current man-made micro/nanofabrication methods. However, synthetic processing can provide a far greater variety of inorganic chemistries than are available among the 70 or so known minerals formed by living organisms [15–17]. In this chapter, a hybrid biological/synthetic chemical paradigm that couples the attractive characteristics of biological self-assembly (massive parallelism, nano-to-microscale precision, complex 3-D structure formation under ambient conditions) with those of synthetic processing (chemical versatility through reaction processing and/or conformal coating) is described: Bioclastic and Shape-preserving Inorganic Conversion (BaSIC) [18]. With BaSIC, diatom frustules or other bioclastic structures can be converted into a variety of non-natural chemistries without loss of the biogenic 3-D morphology. Such shape-preserving chemical conversion has been accomplished via: (i) gas/solid displacement reactions [18–24]; (ii) conformal coating methods [25–29]; or (iii) combinations of reaction and conformal coating methods [30–34]. This chapter is focused on the use of gas/silica displacement reactions to generate chemically modified replicas of diatom frustules. Before describing the chemical conversion of such biogenic microstructures, prior investigations related to the use of displacement reactions to synthesize functional 3-D macrostructures will be discussed.

13.3
Shape-Preserving Reactive Conversion of 3-D Synthetic Ceramic Macrostructures

Several patented reaction-based processes have been developed for altering the chemistries of 3-D synthetic ceramic macroscale (10^0 to 10^1 cm) structures while preserving the shapes and macroscopic dimensions of such structures [35–38]. For example, Breslin, et al. [35, 39–41] have converted dense, shaped silica (SiO_2) preforms into alumina/aluminum alloy (Al_2O_3/Al alloy) bearing composites via the following net displacement reaction with molten aluminum:

$$4\{Al\} + 3SiO_2(s) \Rightarrow 2Al_2O_3(s) + 3\{Si\} \qquad (1)$$

where the brackets { } denote a species dissolved in a liquid solution. For this reaction, the volume of the oxide product (2 moles of alumina) is smaller than the volume of the oxide reactant (3 moles of silica). This volume difference provided space for the further infiltration of the molten aluminum alloy into the reacting preforms (so-called "reactive metal penetration"). After cooling and solidification of the infiltrated aluminum, the resulting composites possessed shapes and dimensions that were similar to those of the starting dense preforms; that is, dense, near-net-shaped interpenetrating Al_2O_3/Al alloy composites were produced from dense SiO_2 preforms [35, 39]. Loehman et al. [42–44] have also utilized displacement reactions between aluminum-bearing liquids and dense oxides (e.g., $Al_6Si_2O_{13}$, TiO_2, NiO, $NiAl_2O_4$) to synthesize composites of Al_2O_3 with Al-bearing metallic or intermetallic phases.

Shape-preserving displacement reactions have also been used to convert porous macroscopic ceramic structures into dense, ceramic-bearing products. Claussen et al. [36, 45–49] have infiltrated molten aluminum into porous oxide preforms, such as porous TiO_2, to synthesize alumina-aluminide-alloy (3A) composites via net displacement reactions of the following type:

$$13\{Al\} + 3TiO_2(s) \Rightarrow 2Al_2O_3(s) + 3TiAl_3(s) \quad (2)$$

With the 3A process, the porosity of the starting preform could be tailored so as to control the relative amounts and stoichiometries of the aluminide product phase or phases. Sandhage et al. [37, 38, 49–53] have also developed and patented a reactive infiltration process ("displacive compensation of porosity"; or DCP) for synthesizing dense, near-net-shaped ceramic/metal composites via net displacement reactions of the following types:

$$3\{Mg\} + Al_2O_3(s) \Rightarrow 3MgO(s) + 2\{Al\} \quad (3)$$

$$\{Zr\} + WC(s) \Rightarrow ZrC(s) + W(s) \quad (4)$$

DCP-type reactions generate solid products that possess a total volume that is larger than the volume of the solid reactant(s). As a result, porous preforms can be infiltrated with the reactive metallic liquid and then undergo a displacement reaction that fills the prior pore spaces with new solid products (i.e., reaction-induced densification). Images of a porous WC preform, and the dense ZrC/W-bearing composite generated from this preform via DCP reactive infiltration [i.e., using net displacement reaction (Eq. (4)], are shown in Figures 13.1A and B, re-

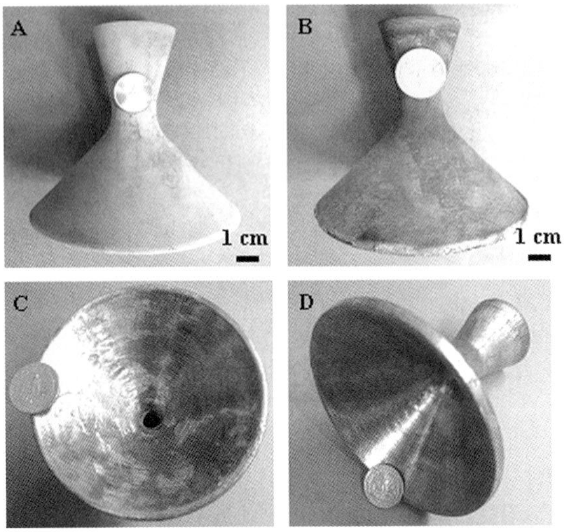

Fig. 13.1 Shape preservation upon chemical conversion of porous WC preforms into dense ZrC/W-bearing composites via the Displacive Compensation of Porosity process [53]. Optical photographs of: (A) a porous nozzle-shaped WC preform; (B) the same preform after conversion into a dense ZrC/W-bearing composite via reactive infiltration with a Zr-bearing liquid [see reaction in Eq. (4)]; and (C,D) after removal of excess solidified metal on the nozzle surfaces. (Images reproduced from Ref. [53] with permission of Springer.)

spectively. The 3-D shape and dimensions of the porous preform were retained in the dense, chemically converted composite product (e.g., the internal diameter of the larger end of the nozzle-shaped composite in Fig. 13.1B was within 0.2% of that for the starting nozzle-shaped preform in Fig. 13.1A). Images of the dense, near-net-shaped ZrC/W-bearing composite nozzle, obtained after removal of excess solidified metal on the nozzle surfaces, are shown in Figures 13.1C and D.

Such prior studies have demonstrated that fluid/solid displacement reactions can be used to alter the chemistries of 3-D ceramic-bearing macrostructures, without altering the shapes or dimensions (to within a few percent or less) of such macrostructures. This prior research led the present authors to consider the use of shape-preserving displacement reactions to alter the chemistries of microscale nanostructured bioclastic assemblies [18, 19].

13.4 Shape-Preserving Chemical Conversion of Diatom Frustules via Oxidation–Reduction Reactions

Previous studies have shown that oxidation-reduction (displacement) reactions between metallic fluids and macroscopic 3-D ceramic preforms may be used to generate ceramic-bearing products that retain the shapes of such preforms [35–53]. Oxidation–reduction reactions of the following type may also be used to alter the silica-based chemistry of diatom frustules:

$$(2/x)\{M\} + SiO_2(s) \Rightarrow (2/x)MO_x(s) + \{Si\} \tag{5}$$

where $\{M\}$ refers to a reactive element present within a gas or condensed phase, $MO_x(s)$ refers to a solid metal oxide, and $\{Si\}$ refers to silicon present in elemental form or as a component within a condensed phase. Several thermodynamically favored oxidation–reduction reactions may be utilized for such chemical conversion [18, 19]. The reaction of magnesium gas with silica diatom frustules will be discussed herein as an example of such reactions.

Sandhage et al. [18–22] have used the following net oxidation–reduction reaction to generate magnesia-bearing replicas of diatom frustules:

$$2Mg(g) + SiO_2(s) \Rightarrow 2MgO(s) + \{Si\} \tag{6}$$

For the case where the silicon product of this reaction is pure solid silicon, the critical thermodynamic value of the magnesium vapor pressure required for the conversion of silica into magnesia via this reaction at 900 °C is 1.1×10^{-7} atm (assuming ideal gas behavior for Mg(g), a pure Mg(l) reference state, pure crystalline reference states for Si and MgO, and a pure cristobalite SiO_2 reference state) [54, 55]. A similar thermodynamic calculation at 650 °C yields a critical magnesium vapor pressure of 5.4×10^{-10} atm (assuming pure Mg(l), pure crystalline Si and MgO, and pure quartz SiO_2 reference states) [54, 55]. For an amorphous (metastable) SiO_2 frustule reactant, the critical magnesium vapor pressure at 650 °C or 900 °C should be even lower. These critical values of the magnesium vapor pressure are more than six orders of magnitude lower than the equilibrium pressures of magnesium vapor over pure liquid magnesium at 650 °C or 900 °C (i.e., 4.1×10^{-3} atm at 650 °C; 0.16 atm at 900 °C) [54, 55]. Hence, by sealing silica diatom frustules and pure magnesium within steel tube ampoules, and then heating the ampoules to a temperature in the range of 650 to 900 °C, the magnesium vapor generated from the molten magnesium is more than sufficient to enable the reaction [Eq. (6)] to proceed to the right. By placing the frustules and magnesium at opposite ends of the steel tube ampoules, and then crimping and bending the middle of the sealed tubes (so as to form the tubes into an inverted V-shape), the direct contact of the frustules with magnesium liquid was avoided [22]. While magnesium liquid may also be used to react with silica frustules, the use of magnesium gas as the fluid reactant eliminated the need to extract the converted frustules from a solidified magnesium-rich matrix.

In order to ensure that sufficient magnesium is present to complete the conversion of silica frustules into magnesia-bearing products, an excess of magnesium should be sealed inside the steel ampoules; that is, the $Mg:SiO_2$ ratio within the ampoules should be kept well in excess of the 2:1 ratio associated with the reaction in Eq. (6) [18–22]. While such reactant ratios allow for complete consumption of the silica, the excess magnesium can undergo the following reaction with silicon to form the intermetallic compound, $Mg_2Si(s)$ [19, 20, 22, 56]:

$$2Mg(g) + Si(s) \Rightarrow Mg_2Si(s) \tag{7}$$

However, with a sufficient excess of magnesium in the ampoules, the Mg_2Si compound can continue to react with Mg(g) to form a Mg-rich liquid (note: a magnesium-rich liquid can form at ≥637.6 °C in the Mg–Si system for Mg:Si atomic ratios in excess of 2:1 [56]). The silicon in this Mg–Si liquid has been found to react with the iron in steel to form the solid compound, Fe_3Si [19]. This reaction, in turn, appeared to enhance the wetting of the Mg–Si liquid on steel, so that this liquid migrated away from the MgO-bearing frustule replicas [19, 21]. Hence, SiO_2-based diatom frustules have been converted into MgO-based replicas by conducting the oxidation–reduction reaction in Eq. (6) under the following conditions:

- $Mg:SiO_2$ reactant ratios in excess of 4:1 [to provide sufficient magnesium so as to enable the formation of a Mg–Si liquid after forming MgO and Mg_2Si via the reactions in Eqs. (6) and (7)].
- Reaction temperatures in excess of 637.6 °C (to enable the formation of a Mg-rich, Mg–Si liquid).
- Reaction within steel ampoules (to confine the Mg(g) in the vicinity of the SiO_2 frustules, and to allow for migration of the Mg–Si liquid away from the MgO-bearing frustule replicas via the wetting of steel).

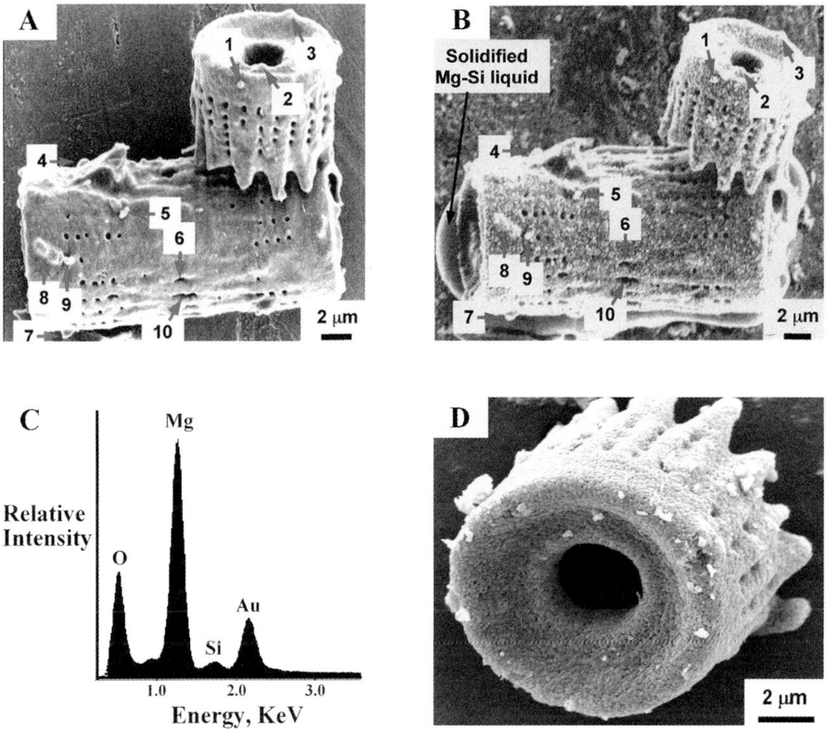

Fig. 13.2 Shape preservation upon chemical conversion of SiO_2-bearing diatom frustules into MgO-bearing frustule replicas via the Bioclastic and Shape-preserving Reactive Conversion (BaSIC) process [19, 21]. Secondary electron images of: (A) SiO_2-based *Aulacoseira* diatom frustules; (B) the same frustules after conversion into MgO-based replicas by exposure to Mg(g) within a sealed steel ampoule for 4 h at 900 °C; (C) EDX analysis of the MgO-converted frustule replica shown in (B); (D) Higher-magnification secondary electron image of a MgO-converted frustule. (Images (A) and (B) and EDX analysis in (C) reproduced from Ref. [21] with permission of The American Ceramic Society. Image (D) reproduced from Ref. [19] with permission of Wiley VCH Verlag GmbH.)

Secondary electron images of the same SiO_2-based *Aulacoseira* diatom frustules before and after conversion into MgO-based replicas are shown in Figure 13.2 [21]. The starting silica-based *Aulacoseira* frustules (Fig. 13.2A) were cylindrical in shape and comprised two halves joined end-to-end. One end of each half-frustule contained a circular hole that was surrounded by a protruding rim (a half-frustule can be seen next to a complete frustule in Fig. 13.2A). The cylindrical wall of each half-frustule contained rows of fine pores (several hundred nanometers in diameter). The other end of each half-frustule was closed and contained finger-like extensions. The protruding fingers of one half-frustule interlocked with those of the other half-frustule to form the complete frustule (note: gaps between the intercalating fingers of each half-frustule are observed as narrow channels at positions 6 and 10 in Fig. 13.2A). The reacted frustules shown in Figure 13.2B were generated by exposure to magnesium vapor within steel ampoules (with a $Mg:SiO_2$ molar reactant ratio $> 10:1$) at 900 °C for 4 h.

Although the surfaces were more granular in appearance, the reacted frustules retained the overall cylindrical morphology of the SiO_2 *Aulacoseira* diatom frustules. Finer features (at positions 1 to 10) were also preserved. The solidified Mg–Si liquid that formed and then poured out of the reacted frustules can also be seen in Figure 13.2B. Energy dispersive X-ray (EDX) analyses of reacted frustules (Fig. 13.2C) indicated the presence of magnesium and oxygen, with little silicon detected. EDX analyses of ion-milled cross-sections of such reacted frustules also revealed the presence of magnesium and oxygen, and little silicon [19]. These analyses indicated that the silicon had largely been removed from the frustule replicas through the formation and outward migration of Mg–Si liquid within 4 h at 900 °C [19, 21]. A higher-magnification image of a MgO-based frustule replica produced under these conditions is shown in Figure 13.2D.

The frustule replicas generated by conducting the oxidation–reduction reaction in Eq. (6) for 4 h at 900 °C were comprised of magnesia crystals that were several hundred nanometers in size [19]. By reducing the reaction temperature and time, magnesia-based frustule replicas with finer crystallite sizes have been produced [20]. A transmission electron microscopy (TEM) image of an ion-milled cross-section of a MgO-based *Aulacoseira* frustule replica generated via exposure to Mg(g) for 30 min at 700 °C ($Mg:SiO_2$ molar ratio $= 9.9:1$) is shown in Figure 13.3A. Electron diffraction and EDX analyses obtained from this cross-section are shown in Figures 13.3B and 13.3C, respectively. These latter analyses indicated that the reacted frustule was comprised of magnesium oxide, with very little silicon detected throughout the cross-section. Again, the loss of silicon was due to the formation of a Mg–Si liquid that migrated out of the converted frustule, as seen in the secondary electron image in Figure 13.3D. The TEM image in Figure 13.3A reveals that the MgO-based replica was comprised largely of crystallites with diameters well below 100 nm. Indeed, Williamson–Hall analyses of X-ray diffraction (XRD) patterns obtained from MgO frustule replicas generated within 1 h at 700 °C ($Mg:SiO_2$ molar ratio $= 8:1$) indicated that the average MgO crystallite size was below 20 nm [57].

Magnesia and magnesia-based compositions are widely used in agricultural (e.g., in fertilizers, in livestock feed, as a carrier for pesticides), environmental (e.g., for heavy-metal precipitation or neutralization of acidic wastewater streams, for $SO_2(g)$ removal from gaseous emissions), pharmaceutical (e.g., in ointments and cosmetics), chemical/petrochemical (e.g., as an acid acceptor, filler, or thickening catalyst in the production of plastics), and electrical (e.g., as insulation for heating elements) applications [58–64]. The shape-preserving conversion of diatom frustules into 3-D assemblages of magnesia nanocrystals (via use of an oxidation–reduction reaction) represents an attractive means of synthesizing magnesia powders with well-controlled morphologies for these and other applications [65–71]. Oxidation–reduction reactions may also be used to produce nanocrystalline assemblages of other functional oxides [18, 19].

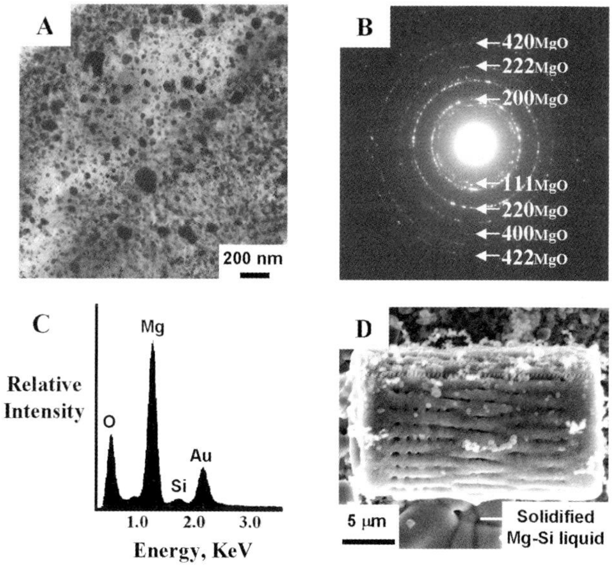

Fig. 13.3 Conversion of SiO_2-bearing diatom frustules into nanocrystalline MgO-bearing frustule replicas via an oxidation–reduction reaction [20]. (A) Transmission electron microscopy image of an ion-milled cross-section of a MgO-based replica of an *Aulacoseira* diatom frustule generated by exposure to Mg(g) within a sealed steel ampoule for 30 min at 700 °C. (B) Electron diffraction and (C) EDX analysis obtained from the cross-section shown in (A). (D) Secondary electron image of a MgO frustule replica generated by exposure to Mg(g) within a sealed steel ampoule for 30 min at 700 °C. (Images and analyses reproduced from Ref. [22] with permission of The American Ceramic Society.)

13.5 Shape-Preserving Chemical Conversion of Diatom Frustules via Metathetic Reactions

The shape-preserving conversion of silica diatom frustules into other functional oxides may also be accomplished with the use of net metathetic displacement reactions of the following type:

$$(2/y)\{MX_w\} + SiO_2(s) \Rightarrow (2/y)MO_y(s) + \{SiX_{2w/y}\} \quad (8)$$

where $\{MX_w\}$ refers to a reactive species present within a gas or condensed phase, $MO_y(s)$ refers to a solid metal oxide, and $\{SiX_{2w/y}\}$ refers to a silicon-bearing species within a gas or condensed phase. A number of thermodynamically favored metathetic reactions may be utilized for such chemical conversion [18, 19]. The reaction of titanium fluoride with silica diatom frustules will be described here to illustrate some of the issues associated with such reactive conversion.

The following net metathetic reaction was first proposed as a means of replacing silica in diatom frustules with titania [18, 19]:

$$TiF_4(g) + SiO_2(s) \Rightarrow TiO_2(s) + SiF_4(g) \qquad (9)$$

The critical thermodynamic ratio of $TiF_4(g)$:$SiF_4(g)$ partial pressures required for conversion of pure $SiO_2(s)$ into pure $TiO_2(s)$ via this reaction at 600 °C is 2.6×10^{-5} (assuming ideal gas behavior for $TiF_4(g)$ and $SiF_4(g)$, pure gaseous reference states for TiF_4 and SiF_4, and pure cristobalite SiO_2 and pure rutile TiO_2 reference states) [54]. A similar thermodynamic calculation at 350 °C yields a critical $TiF_4(g)$:$SiF_4(g)$ partial pressure ratio of 1.2×10^{-7}. For an amorphous (metastable) SiO_2 frustule reactant, the critical $TiF_4(g)$:$SiF_4(g)$ partial pressure ratio at 600 °C or 350 °C should be even lower. Given that the sublimation temperature of $TiF_4(s)$ is only 285 °C [54], such critical $TiF_4(g)$:$SiF_4(g)$ ratios may be readily achieved by sealing appropriate amounts of $TiF_4(s)$ and SiO_2 frustules within metal ampoules, and then heating the ampoules to $\geq$350 °C.

Unocic et al. [23] and Shian et al. [21] have evaluated the metathetic reaction between *Aulacoseira* diatom frustules and $TiF_4(g)$ within the confinement of titanium tubes. Initial experiments [23] were conducted by sealing solid TiF_4 powder and SiO_2 frustules in molar ratios of $\geq$4.9:1 within the titanium tubes, and then heating the sealed tubes at 5 °C min^{-1} to peak temperatures in the range of 500 to 700 °C. The solid products of such reaction at 600 °C for 2 h (with a TiF_4:SiO_2 molar reactant ratio = 4.9:1) are shown in Figure 13.4. The secondary electron images in Figures 13.4A and B indicated that the reaction product was largely comprised of relatively coarse plate-shaped crystals (with typical dimensions of 5–10 μm × 5–10 μm × 0.5–1 μm). EDX analyses (not shown) indicated that these crystals were composed of titania. While titania was produced, the hollow cylindrical shape of the starting *Aulacoseira* diatoms was not preserved under these reaction conditions. Upon careful inspection of the reaction products, a few partially disintegrated frustules were observed (see Fig. 13.4A). A sec-

Fig. 13.4 Reactive evaporation of silica diatom frustules. (A,B) Secondary electron images of the plate-shaped titania crystals generated upon exposure of *Aulacoseira* diatom frustules to $TiF_4(g)$ for 2 h at 600 °C within a sealed titanium ampoule (TiF_4:SiO_2 molar reactant ratio of 4.9:1). (C) Secondary electron image of a partially disintegrated *Aulacoseira* diatom frustule.

ondary electron image of such a partial frustule is shown in Figure 13.4C. Most of the length of this frustule was missing. A view through the circular hole at the end of this partial frustule also indicated that the frustule wall on the opposite side had vanished. Reaction at 600 °C for 2 h, with a TiF_4:SiO_2 molar reactant ratio = 4.9:1, resulted in reactive evaporation of the silica frustules. Such reactive evaporation must have proceeded through the formation of volatile Si–O-bearing species. Two reported volatile Si–O–F-bearing gas species are $SiOF_2$(g) and Si_2OF_6(g) [54, 72]. Thermodynamic analyses indicate that the following equilibrium reaction should strongly favor the formation of Si_2OF_6(g) over $SiOF_2$(g) at ≤827 °C. The equilibrium reaction constant for the reaction in Eq. (10) has been reported to range from 1.3×10^{-48} at room temperature to 1.3×10^{-7} at 827 °C [72]:

$$Si_2OF_6(g) = SiOF_2(g) + SiF_4(g) \tag{10}$$

If Si_2OF_6(g) was the predominant volatile Si–O-bearing gas species formed from the SiO_2 frustules within the sealed titanium ampoules at 600 °C (for TiF_4:SiO_2 = 4.9:1), then the formation of titania at locations removed from the frustule surfaces, and the reactive evaporation of the silica frustules, may have proceeded by the following cyclic reactions:

$$2Si_2OF_6(g) + TiF_4(g) \Rightarrow TiO_2(s) + 4SiF_4(g) \tag{11a}$$

$$3SiF_4(g) + SiO_2(s) \Rightarrow 2Si_2OF_6(g) \tag{11b}$$

$$\text{Net reaction:} \quad TiF_4(g) + SiO_2(s) \Rightarrow TiO_2(s) + SiF_4(g) \tag{11c}$$

TiO_2(s) could form at locations removed from the frustule surfaces by the reaction of TiF_4(g) (generated by sublimation of solid TiF_4) with Si_2OF_6(g) (generated from the frustules). The SiF_4(g) liberated by reaction in Eq. (11a) could, in turn, migrate back to the frustule surfaces to generate new Si_2OF_6(g) via the reaction in Eq. (11b). The net reaction [Eq. (11c)] would result in the reactive evaporation of silica, and the formation of titania platelets at locations removed from the silica, through the formation and consumption of the intermediate gas species, Si_2OF_6(g).

The reactive evaporation of silica diatom frustules was avoided by reducing the reaction temperature to ≤350 °C and by lowering the molar TiF_4:SiO_2 reactant ratio to values ≤ 2.4:1 [23]. Under these conditions, the reaction of TiF_4(g) with SiO_2(s) occurred on the frustule surfaces, so that the shape and fine features of the frustules were preserved during the course of reaction. A secondary electron image of an *Aulacoseira* frustule after reaction with TiF_4(g) for 2 h at 350 °C in a sealed titanium tube (with a molar TiF_4:SiO_2 reactant ratio = 2.4:1) is shown in Figure 13.5A. An associated EDX pattern is shown in Figure 13.5B. The overall shape and fine features (rows of fine pores, channels between finger-like extensions) of the starting frustule were well preserved in the reacted specimen. The absence of a silicon peak, and the presence of a strong titanium peak, in the

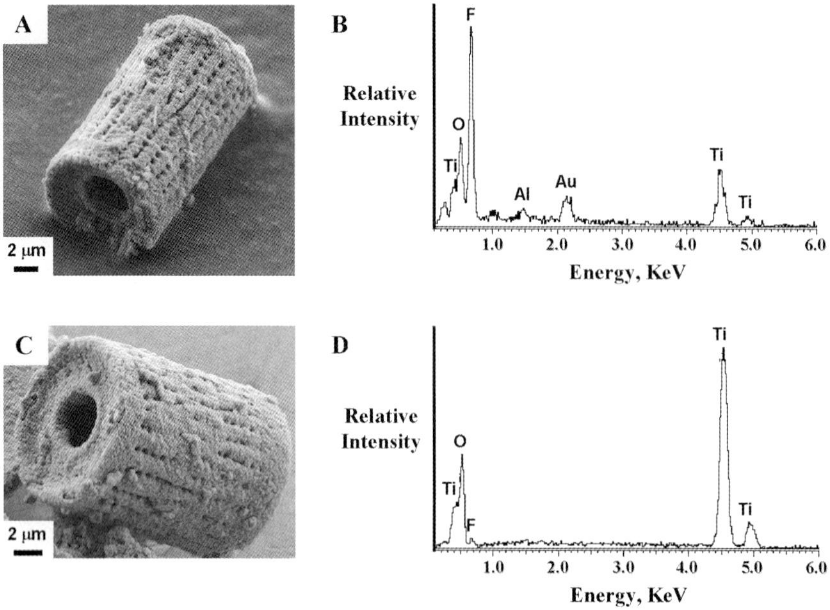

Fig. 13.5 Conversion of SiO_2-bearing diatom frustules into nanocrystalline TiO_2-bearing frustule replicas via use of a metathetic displacement reaction [21, 23]. (A) Secondary electron image of a $TiOF_2$-based replica of an *Aulacoseira* diatom frustule generated by exposure to TiF_4(g) within a sealed titanium ampoule for 3 h at 350 °C. (B) EDX analysis obtained from the replica shown in (A). (C) Secondary electron image of an anatase TiO_2-based frustule replica generated by exposure of a $TiOF_2$ replica to moist flowing oxygen for 5 h at 450 °C. (D) EDX analysis obtained from the replica shown in (C).

EDX pattern indicated that the silicon had been completely displaced by titanium within 2 h of reaction at 350 °C. However, the EDX pattern in Figure 13.5B also revealed the presence of appreciable fluorine. Subsequent XRD analysis indicated that the reacted frustules contained titanium oxyfluoride, $TiOF_2$ [23]. The formation of $TiOF_2$ may have occurred via metathetic reactions of the following type:

$$TiF_4(g) + 1/2\ SiO_2(s) \Rightarrow TiOF_2(s) + 1/2\ SiF_4(g) \tag{12}$$

$$TiF_4(g) + 2/3\ SiO_2(s) \Rightarrow TiOF_2(s) + 1/3\ Si_2OF_6(g) \tag{13}$$

In order to form the desired oxide product, TiO_2, the $TiOF_2$ replicas were exposed to flowing moist oxygen generated by passing pure oxygen (at 1 slpm) through a water bath heated to 45 °C. Titanium oxyfluoride may be converted into titania by the following reactions [23, 73, 74]:

$$TiOF_2(s) + 1/2O_2(g) \Rightarrow TiO_2(s) + F_2(g) \tag{14}$$

$$TiOF_2(s) + H_2O(g) \Rightarrow TiO_2(s) + 2HF(g) \tag{15}$$

Critical thermodynamic values of the partial pressure ratios, $[p_{O_2(g)}]^{1/2}:p_{F_2(g)}$ and $p_{H_2O(g)}:[(p_{HF(g)})^2]$, must be achieved in order for the reactions in Eqs. (14) and (15), respectively, to proceed spontaneously to the right. Unfortunately, values of the standard Gibbs free energy of formation of solid $TiOF_2$ at $\leq$600 °C are not available in standard thermodynamic tables (nor apparently in the literature) to allow for calculation of these critical thermodynamic ratios. Nonetheless, both reactions have been successfully conducted at modest temperatures (i.e., $\leq$350 °C) with the use of flowing oxygen or moist air [23, 73, 74]. A secondary electron image of a TiO_2-bearing frustule replica generated by passing moist oxygen past a $TiOF_2$-converted frustule for 5 h at 450 °C, and an associated EDX pattern obtained from this frustule, are shown in Figures 13.5C and D, respectively. Excellent preservation of the frustule shape and fine features was achieved in the TiO_2-based replica. Comparison of the EDX patterns in Figures 13.5B and D indicated that the fluorine was largely removed by this moist oxygen treatment. X-ray and EDX analyses (not shown) confirmed the conversion of the $TiOF_2$ frustule replicas into anatase TiO_2 by such treatments [21, 23]. TEM analyses [23] of ion-milled cross-sections of TiO_2-converted replicas revealed nanoporous networks of anatase crystals with sizes on the order of 50 nm. These studies showed that 3-D nanocrystalline anatase replicas of diatom frustules could be produced via use of the net metathetic displacement reaction [Eq. (9)] through proper selection of reaction conditions.

Owing to attractive chemical, optical, biological, and electrical properties, TiO_2-based compositions are among the most versatile of ceramics. The wide variety of applications for titania-based ceramics include use as photocatalysts (e.g., for the degradation of stable organic pollutants, such as detergents, dyes, and pesticides in water), gas sensors (e.g., for sensitive detection of CO(g) or H_2(g)), pigments (e.g., for paints, paper, plastics, ink, and cosmetics), photovoltaics (e.g., as an electrode in dye-sensitized solar cells), medical implants (e.g., as biocompatible coatings for bone implants), and antimicrobial agents (e.g., for killing *E. coli* bacteria) [75–80]. The present approach for converting diatom frustules into titania replicas provides a low-cost scalable route to intricate assemblages of titania nanocrystals with thousands of 3-D morphologies and with precisely controlled nanoscale features for such applications.

13.6
Shape-Preserving Chemical Conversion of Diatom Frustules via Sequential Displacement Reactions

The direct reactive conversion of silica diatom frustules into replicas composed of other functional materials may, in some cases, be inhibited by sluggish reaction kinetics or by poor preservation of the frustule shape and fine features upon complete reaction – for example, due to reactive evaporation of the frustules, or to coarsening of nanoscale features during reaction at an elevated temperature. In such cases, a series of displacement reactions may be used to sidestep such problems. The use of sequential displacement reactions also allows for the syntheses of functional composite replicas (i.e., if one of the reactions is not carried to completion).

To demonstrate that multiple, consecutive displacement reactions can be used to alter the chemistry of diatom frustules while preserving the 3-D frustule shape, a series of gas/solid displacement reactions has been utilized to synthesize zirconia frustule replicas [81]. This approach was developed after initial attempts to convert silica frustules into zirconia replicas by the following net metathetic reaction failed.

$$ZrF_4(g) + SiO_2(s) \Rightarrow ZrO_2(s) + SiF_4(g) \tag{16}$$

As was observed in initial experiments for metathetic titania conversion at 600 °C, the silica frustules underwent reactive evaporation during exposure to $ZrF_4(g)$ within sealed metal ampoules. However, for the $ZrF_4(g)/SiO_2(s)$ reactions, such reactive evaporation occurred over a wide range of temperatures (250–800 °C) and a wide range of ZrF_4:SiO_2 molar reactant ratios (from 0.36:1 to 3.6:1) [81]. In other words, conditions for conducting the reaction in Eq. (16) that avoided such reactive evaporation could not be found. Since such reactive evaporation required the formation of volatile Si–O-bearing gas species, an alternate reaction path that avoided such gas species was developed. The following series of displacement reactions was examined:

$$\text{oxidation–reduction:} \quad 2Mg(g) + SiO_2(s) \Rightarrow 2MgO(s) + \{Si\} \tag{6}$$

$$\text{metathetic:} \quad ZrCl_4(g) + 2MgO(s) \Rightarrow ZrO_2(s) + 2\{MgCl_2\} \tag{17}$$

The SiO_2 frustules were first converted into MgO-bearing replicas via the reaction in Eq. (6). These replicas were then reacted with zirconium chloride vapor to form zirconia and magnesium chloride. MgO tends not to form Mg–O–Cl-bearing gas species at the modest temperature (650 °C) used for the reaction in Eq. (17). Hence, by forming MgO as an intermediate product, the reactive evaporation of SiO_2 (due to formation of silicon–oxygen–halide gas species) was avoided.

The conversion of *Aulacoseira* frustules into magnesia was conducted within steel ampoules at 900 °C, as described above [18–22]. The MgO-bearing frustules were then exposed to a sodium hydroxide solution to allow for selective dissolution of any remaining silicon. The MgO replicas were sealed along with solid $ZrCl_4$ within Ni ampoules. The critical partial pressure of $ZrCl_4$ vapor required to enable the reaction in Eq. (17) to proceed to the right is only 5.7×10^{-5} atm at 650 °C (assuming ideal gas behavior for $ZrCl_4(g)$, a pure gaseous reference state for $ZrCl_4$, and pure solid $MgCl_2$, MgO, and monoclinic ZrO_2 reference

Fig. 13.6 Conversion of SiO_2-bearing diatom frustules into ZrO_2-bearing frustule replicas via use of a series of displacement reactions [81]. (A,B) Secondary electron images of a ZrO_2-based replicas of an *Aulacoseira* diatom frustule generated by exposure of MgO frustule replicas to $ZrCl_4$(g) within a sealed nickel ampoule for 2 h at 650 °C (molar $ZrCl_4$:MgO reactant ratio of 0.52:1). (C) EDX analysis of such a ZrO_2-based frustule replica.

states) [54, 55]. Given that $ZrCl_4$ sublimes at 336 °C [54], this critical partial pressure could be readily achieved by heating $ZrCl_4$(s) within sealed ampoules. The molar $ZrCl_4$:MgO ratios sealed within the ampoules were varied from 0.52:1 to 1.2:1 [i.e., near and above the stoichiometry of the reaction in Eq. (17)] [81]. The ampoules were then heated at 5 °C min^{-1} to 650 °C and held at this temperature for 2 h. After cooling and removal from the ampoules, the reacted frustules were immersed in pure water heated to 90 °C to allow for selective dissolution of the magnesium chloride product of the reaction in Eq. (17).

Secondary electron images of the resulting zirconia frustule replicas are shown in Figures 13.6A and B. The cylindrical frustule shape, and the protruding rim and open hole at the frustule end, were preserved in the reacted frustules. EDX analysis of such a reacted frustule is shown in Figure 13.6C. Predominant peaks for only Zr and O were detected. X-ray diffraction, electron diffraction, and TEM analyses (not shown) indicated that the reacted frustules were composed of nanocrystalline mixtures of the monoclinic and tetragonal polymorphs of zirconia [81]. These studies show that a series of displacement reactions may be used to alter the chemistry – but not the overall 3-D morphology – of diatom frustules. Such a sequential reaction approach may be used to generate nanostructured, microscale frustule replicas with a wide range of compositions for a variety of applications [18, 81].

13.7 Summary and Future Opportunities

Several gas/solid reaction-based methods for altering the SiO_2 composition of diatom frustules, but not the starting frustule morphologies, have been demonstrated. With proper control of reaction temperature, time, and reactant ratios, oxidation–reduction reactions (with elemental gas reactants), metathetic reactions (using halide gas reactants), or combinations of both, have yielded frustule replicas composed of new nanocrystalline oxides. While the syntheses of frustule replicas composed of nanocrystalline MgO, TiO_2, and ZrO_2 have been described, numerous other functional oxides may be produced using similar types of displacement reaction [18].

The synergistic combination of biological assembly with synthetic chemical functionalization provides a new biologically enabling paradigm (Bioclastic and Shape-preserving Inorganic Conversion or BaSIC) for fabricating large numbers of 3-D nanostructured micro-assemblies with chemistries that can be tailored for a host of devices [18]. Through the sustained reproduction of certain biomineralizing microorganisms – such as diatoms – enormous numbers of 3-D bioclastic structures may be generated under ambient conditions at low cost. Indeed, the technology for large-scale diatom culturing has already been developed for commercial aquaculture operations and for the syntheses of biotechnological compounds [7, 8]. Reaction-based methods [18–24], conformal coating-based approaches (not discussed here) [25–29], or combinations of coating and reaction processes [30–34] can then be used to tailor the chemistries of such biogenic 3-D micro/nanostructures for environmental, agricultural, chemical, biomedical, transportation, manufacturing, telecommunications, and other applications.

Although the frustules of diatoms provide a spectacular variety of nanostructured microtemplates for chemical conversion, the BaSIC process may also be applied to other bioclastic structures. The multifarious structures available among extant biomineralizing organisms may be selected for specific device applications. A tantalizing alternative is the use of genetic engineering to tailor the morphologies of bioclastic structures produced by diatoms and other biomineralizing organisms. The sequencing of the *Thalassiosira pseudonana* genome [82], and the development of transformation-based genetic manipulation methodologies for *T. pseudonana* (e.g., see Chapter 3 in Volume 1) and *Phaeodactylum tricornutum* [83, 84], are important initial strides in the controlled genetic manipulation of diatoms. If the genetic tailoring of diatoms (or other microorganisms) ultimately yields nanostructure microassemblies with precisely controlled 3-D morphologies, then such genetic engineering could be combined with large-scale cell culturing and with BaSIC chemical conversion processes to enable a revolutionary and powerful paradigm for large-scale and precise 3-D Micro/nano-fabrication: Three-Dimensional Genetically-Engineered Micro/nano-devices (3-D GEMs) [21].

Acknowledgments

The financial support of the Air Force Office of Scientific Research (Dr. Joan Fuller, Dr. Hugh C. De Long, program managers) is gratefully acknowledged.

References

1 E. Baeuerlein, *Angew. Chem. Int. Ed.* **2003**, *42*, 614–641.

2 S. Mann, G.A. Ozin, *Nature* **1996**, *382*, 313–318.

3 J.R. Young, S.A. Davis, P.R. Bown, S. Mann, *J. Struct. Biol.* **1999**, *126*, 195–215.

4 F.E. Round, R.M. Crawford, D.G. Mann, *The Diatoms: Biology & Morphology of the Genera*. Cambridge University Press, Cambridge, England, **1990**.

5 V. Martin-Jezequel, M. Hildebrand, M.A. Brzezinski, *J. Phycol.* **2000**, *36*, 821–840.

6 M. Hildebrand, *J. Nanosci. Nanotechnol.* **2005**, *5*, 146–157.

7 E.O. Duerr, A. Molnar, V. Sato, *J. Mar. Biotechnol.* **1998**, *7*, 65–70.

8 T. Lebeau, J.-M. Robert, *Appl. Microbiol. Biotechnol.* **2003**, *60*, 612–623.

9 J.M. Villora, C. Baudin, P. Callejas, M. Flora Barba, *Key Eng. Mater.* **2004**, *264–268*, 2437–2440.

10 J. Blanco, A. Bahamonde, E. Alvarez, P. Avila, *Catal. Today* **1998**, *42*, 85–92.

11 E.O. Obanijesu, O.O. Bello, F.A.O. Osinowo, S.R.A. Macaulay, *Int. J. Environ. Pollut.* **2004**, *22*, 701–709.

12 P.T. Flynn, Jr., *Adv. Filtr. Sep. Technol.* **2003**, *16*, 585–593.

13 A.B. Cummins, *Filtr. Sep.* **1973**, *10*, 215–216, 218–219.

14 S. Fustinoni, L. Campo, C. Colosio, S. Birindelli, V. Foa, *J. Chromatogr. B Biomed. Appl.* **2005**, *814*, 251–258.

15 H.A. Lowenstam, *Science* **1981**, *211*, 1126–1131.

16 H.A. Lowenstam, S. Weiner, in: P. Westbroek, E.W. de Jong (Eds.), *Biomineralization and Biological Metal Accumulation*. D. Reidel Publishing Co., Hingham, MA, **1983**, pp. 191–203.

17 S. Weiner, L. Addadi, *Science* **2002**, *298*, 375–376.

18 K.H. Sandhage, *U.S. Patent No. 7,067,104*, June 27, **2006**.

19 K.H. Sandhage, M.B. Dickerson, P.M. Huseman, M.A. Caranna, J.D. Clifton, T.A. Bull, T.J. Heibel, W.R. Overton, M.E.A. Schoenwaelder, *Adv. Mater.* **2002**, *14*, 429–433.

20 Y. Cai, S.M. Allan, F.M. Zalar, K.H. Sandhage, *J. Am. Ceram. Soc.* **2005**, *88*, 2005–2010.

21 K.H. Sandhage, R.L. Snyder, G. Ahmad, S.M. Allan, Y. Cai, M.B. Dickerson, C.S. Gaddis, M.S. Haluska, S. Shian, M.R. Weatherspoon, R.A. Rapp, R.R. Unocic, F.M. Zalar, Y. Zhang, M. Hildebrand, B.P. Palenik, *Int. J. Appl. Ceram. Technol.* **2005**, *2*, 317–326.

22 S.M. Allan, M.R. Weatherspoon, P.D. Graham, Y. Cai, M.S. Haluska, R.L. Snyder, K.H. Sandhage, *Ceram. Eng. Sci. Proc.* **2005**, *26*, 289–296.

23 R.R. Unocic, F.M. Zalar, P.M. Sarosi, Y. Cai, K.H. Sandhage, *Chem. Commun.* **2004**, *7*, 795–796.

24 Y. Cai, M.R. Weatherspoon, E. Ernst, M.S. Haluska, R.L. Snyder, K.H. Sandhage, *Ceram. Eng. Sci. Proc.* **2006**, *27*, 49–56.

25 C.S. Gaddis, K.H. Sandhage, *J. Mater. Res.* **2004**, *19*, 2541–2545.

26 N.L. Rosi, C.S. Thaxton, C.A. Mirkin, *Angew. Chem. Int. Ed.* **2004**, *43*, 5500–5503.

27 J. Zhao, C.S. Gaddis, Y. Cai, K.H. Sandhage, *J. Mater. Res.* **2005**, *20*, 282–287.

28 E.K. Payne, N.L. Rosi, C. Xue, C.A. Mirkin, *Angew. Chem. Int. Ed.* **2005**, *44*, 5064–5067.

29 D. Losic, J.G. Mitchell, N.H. Voelcker, *Chem. Commun.* **2005**, 4905–4907.

30 M.W. Anderson, S.M. Holmes, N. Hanif, C.S. Cundy, *Angew. Chem. Int. Ed.* **2000**, *39*, 2707–2710.

31 Y. Wang, Y. Tang, A. Dong, X. Wang, N. Ren, Z. Gao, *J. Mater. Chem.* **2002**, *12*, 1812–1818.

32 Y. Cai, K.H. Sandhage, *Phys. Stat. Sol. (A)*, **2005**, *202*, R105–R107.

33 M.R. Weatherspoon, S.M. Allan, E. Hunt, Y. Cai, K.H. Sandhage, *Chem. Commun.* **2005**, 651–653.

34 M.R. Weatherspoon, M.S. Haluska, Y. Cai, J.S. King, C.J. Summers, R.L. Snyder, K.H. Sandhage, *J. Electrochem. Soc.* **2006**, *153*, H34–H37.

35 M.C. Breslin, *U.S. Patent No. 5,214,011*, May 25, **1993**.

36 N. Claussen, F. Wagner, *U.S. Patent No. 6,051,277*, April 18, **2000**.

37 K.H. Sandhage, P. Kumar, *U.S. Patent No. 6,407,022*, June 18, **2002**.

38 K.H. Sandhage, R.R. Unocic, M.B. Dickerson, M. Timberlake, K. Guerra, *U.S. Patent 6,598,656*, July 29, **2003**.

39 M.C. Breslin, J. Ringnalda, L. Xu, M. Fuller, J. Seeger, G.S. Daehn, T. Otani, H.L. Fraser, *Mater. Sci. Eng. A* **1995**, *A195*, 113–119.

40 M.Y. Chen, M.C. Breslin, *Wear* **2001**, *249*, 868–876.

41 V. Imbeni, I.M. Hutchings, M.C. Breslin, *Wear* **1999**, *233–235*, 462–467.

42 W.G. Fahrenholtz, K.G. Ewsuk, R.E. Loehman, A.P. Tomsia, *Metall. Mater. Trans. A* **1996**, *27A*, 2100–2104.

43 W.G. Fahrenholtz, K.G. Ewsuk, D.T. Ellerby, R.E. Loehman, *J. Am. Ceram. Soc.* **1996**, *79*, 2497–2499.

44 R.E. Loehman, K.G. Ewsuk, *J. Am. Ceram. Soc.* **1996**, *79*, 27–32.

45 J. Bruhn, N. Claussen, R.H.J. Hannink, S. Lathabai, P.R. Miller, *Ceram. Trans.* **1998**, *99*, 65–79.

46 F. Wagner, D.E. Garcia, A. Krupp, N. Claussen, *J. Eur. Ceram. Soc.* **1999**, *19*, 2449–2453.

47 P. Rendtel, F. Wagner, R. Janssen, N. Claussen, *Mater. Sci. Forum* **1999**, *308–311*, 181–186.

48 N. Claussen, *Br. Ceram. Trans.* **1999**, *98*, 256–257.

49 P. Kumar, N. Travitsky, P. Beyer, K.H. Sandhage, R. Janssen, N. Claussen, *Scripta Metall.* **2001**, *44*, 751–757.

50 P. Kumar, K.H. Sandhage, *J. Mater. Sci.* **1999**, *34*, 5757–5769.

51 K.A. Rogers, P. Kumar, R. Citak, K.H. Sandhage, *J. Am. Ceram. Soc.* **1999**, *82*, 757–60.

52 M.B. Dickerson, R.L. Snyder, K.H. Sandhage, *J. Am. Ceram. Soc.* **2002**, *85*, 730–732.

53 M.B. Dickerson, P.J. Wurm, J.R. Schorr, W.P. Hoffman, E. Hunt, K.H. Sandhage, *J. Mater. Sci.* **2004**, *39*, 6005–6015.

54 I. Barin, *Thermochemical Data of Pure Substances*. VCH Verlagsgesellschaft, Weinheim, Germany, **1995**.

55 K.V. Gourishankar, M. Karaminezhad Ranjbar, G.R. St. Pierre, *J. Phase Equil.* **1993**, *14*, 601–611.

56 A.A. Nayeb-Hashemi, J.B. Clark, *Bull. Alloy Phase Diagr.* **1984**, *5*, 584–592.

57 M.S. Haluska, I. Dragomir, K.H. Sandhage, R.L. Snyder, *Powder Diff.* **2005**, *20*, 306–310.

58 A.N. Copp, *Bull. Am. Ceram. Soc.* **1995**, *74*, 135–137.

59 Z. Xin, W.B. Tucker, R.W. Hemken, *J. Dairy Sci.* **1989**, *72*, 462–470.

60 J.J. Mortvedt, J.J. Kelsoe, *Fert. Res.* **1988**, *15*, 155–161.

61 F. Giroud, C. Nocerino, *French Patent 2,822,684*, Oct. 4, **2002**.

62 S. Spychaj, F.J. Balta Calleja, *J. Mater. Sci. Lett.* **1993**, *12*, 1255–1257.

63 M.T. Frost, M.H. Jones, R.C. Flann, R.L. Hart, P.R. Strode, A.J. Urban, S. Tassios, *Trans. Inst. Mining Metall.* **1990**, *99*, C117–C124.

64 J. Palmer, *U.S. Patent No. 4,867,961*, Sept. 19, **1989**.

65 T. Karasuda, K. Aika, *J. Catal.* **1997**, *171*, 439–448.

66 V.B. Fenelonov, M.S. Mel'gunov, I.V. Mishakov, R.M. Richards, V.V. Chesnokov, A.M. Volodin, K.J. Klabunde, *J. Phys. Chem. B* **2001**, *105*, 3937–3941.

67 B.M. Choudary, M.L. Kantam, K.V.S. Ranganath, K. Mahendar, B. Sreedhar, *J. Am. Chem. Soc.* **2004**, *126*, 3396–3397.

68 J.V. Stark, D.G. Park, I. Lagadic, K.J. Klabunde, *Chem. Mater.* **1996**, *8*, 1904–1912.

69 R. Kon, M. Miyazaki, *Fragrance J.* **2003**, *31*, 63–68.

70 S.T. Hobson, E.H. Braue, Jr., E.K. Lehnert, K.J. Klabunde, O.P. Koper, S. Decker, *U.S. Patent 6,403,653*, June 11, **2002**.

71 O.B. Koper, J.S. Klabunde, G.L. Marchin, K.J. Klabunde, P. Stoimenov, L. Bohra, *Curr. Microbiol.* **2002**, *44*, 49–55.

72 M. Shinmei, T. Imai, T. Yokokawa, C.R. Masson, *J. Chem. Thermodynam.* **1986**, *18*, 241–246.

73 J.S. Machin, D.L. Deadmore, *Nature* **1961**, *189*, 223–224.

74 J.C. Lytle, H. Yan, R.T. Turgeon, A. Stein, *Chem. Mater.* **2004**, *16*, 3829–3837.

75 I. Justicia, P. Ordejon, G. Canto, J.L. Mozos, J. Fraxedas, G.A. Battiston, R. Gerbasi, A. Figueras, *Adv. Mater.* **2002**, *14*, 1399–1402.

76 T. Hyodo, G.S. Devi, C. Yu, Y. Shimizu, M. Egashira, *Chem. Sensors* **2002**, *18*, 178–180.

77 J.H. Braun, *J. Coatings Technol.* **1997**, *69*, 59–72.

78 C.J. Barbe, F. Arendse, P. Comte, M. Jirousek, F. Lenzmann, V. Shklover, M. Gratzel, *J. Am. Ceram. Soc.* **1997**, *80*, 3157–3171.

79 M. Manso, S. Ogueta, P. Garcia, J. Perez-Rigueiro, C. Jimenez, J.M. Martinez-Duart, M. Langlet, *Biomaterials* **2002**, *23*, 349–356.

80 G. Fu, P.S. Vary, C.-T. Lin, *J. Phys. Chem. B* **2005**, *109*, 8889–8898.

81 S. Shian, Y. Cai, M.R. Weatherspoon, S.M. Allan, K.H. Sandhage, *J. Am. Ceram. Soc.* **2006**, *89*, 694–698.

82 E.V. Armbrust, et al., *Science* **2004**, *306*, 79–86.

83 N. Poulsen, P.M. Chesley, N. Kröger, *J. Phycol.* **2006**, *42*, 1059–1065.

84 L.A. Zaslavskaia, J.C. Lippmeier, C. Shih, D. Ehrhardt, A.R. Grossman, K.E. Apt, *Science* **2001**, *292*, 2073–2075.

14
Organic Preforms of Biological Origin: Natural Plant Tissues as Templates for Inorganic and Zeolitic Macrostructures

Alessandro Zamperi, Wilhelm Schwieger, Cordt Zollfrank, and Peter Greil

Abstract

Nature displays a huge number of biological materials with complex hierarchically built anatomies and a large variety of cellular/porous structures and functional architectures, as the result of evolutionary processes. Most of these materials exhibit micro-/macro-structural features which are difficult, or impossible, to reproduce via technical processes. Nonetheless, reproducing such biostructures into engineering materials might be the key for the development of inorganic materials with novel structures and properties. In particular, inexpensive, abundant, and renewable biological structures such as plant tissues represent, from the technical point of view, interesting preforms/templates for the manufacturing of biomorphous ceramic substrates. The open cellular anatomy provides access for infiltration with liquid or gaseous reactants to convert the biological preform or the carbon template into carbide- or oxide-based ceramics. Such biomorphous cellular ceramics can be used as novel monolithic supports for zeolite layers, producing hierarchically porous composites with unique structures and properties. On the other hand, cellular biological preforms could be used directly as sacrificial templates, in order to obtain bio-inspired self-supporting zeolite macrostructures with complex morphologies and open-porous architectures. Owing to their unique structures and functionalities, zeolite-based biomorphous materials might open new scenarios in the development of catalysts, molecular sieves, adsorbents and membranes with hierarchical porosity and organization. In this chapter we present two approaches towards the realization of biomorphous cellular inorganic zeolite-based materials/replica with hierarchical porosity. On the one hand, cellular biological supports have been subjected to a three-step biotemplating process: (i) pyrolysis of the biotemplate; (ii) liquid silicon infiltration (LSI) for the development of biomorphous SiSiC cellular ceramic supports; and (iii) support self-transformation (SST) for the growth of zeolite crystals in the inner cavities of the above-mentioned porous substrate. On the other hand, biological structures such as *Luffa* sponges and cellulose fibers have been used as sacrificial templates for

Handbook of Biomineralization. Edited by P. Behrens and E. Bäuerlein

ISBN: 978-3-527-31805-6

the synthesis of self-supporting zeolite replicas, which mimic the morphology and microstructure of the organic materials.

Key words: zeolitic macrostructures, zeolite coatings, hierarchical porous materials, bio-inspired cellular ceramics, liquid silicon infiltration (LSI), biomorphic structured catalysts, cellulose-derived composites, biotemplating.

14.1 Introduction

Biological materials, for example wood, diatoms, and natural sponges, are inexpensive, abundant, environmentally benign and renewable resources [1–4], with

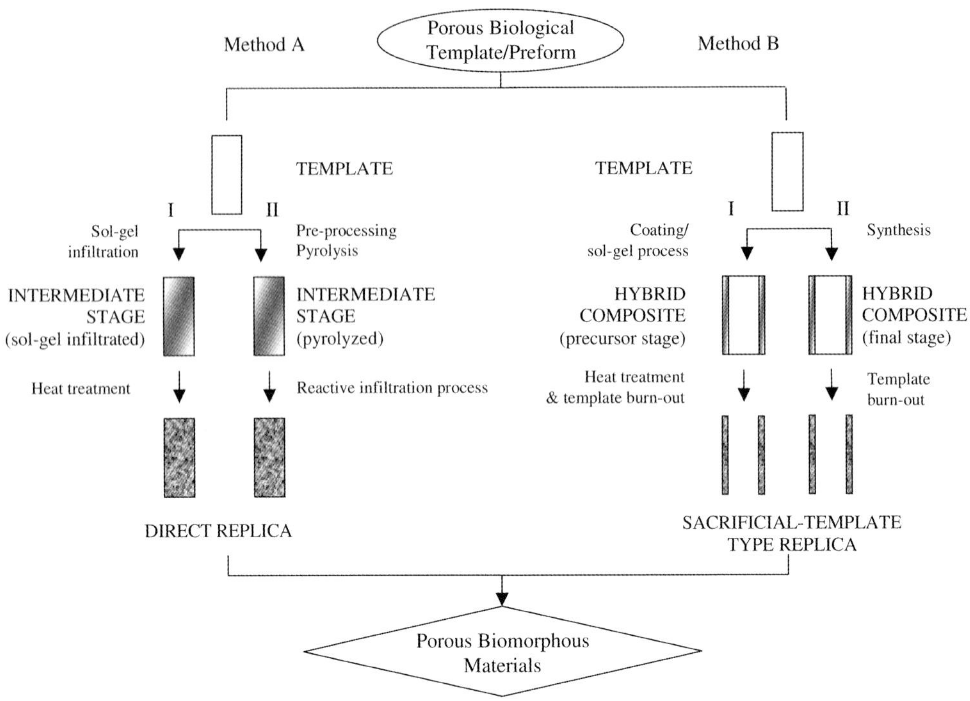

Method	Example	Reference
A-I	Sol-gel mineralization of biological structures	1
A-I	Morphosynthesis of ZnO replica using butterfly wings	2
A-II	Al_2O_3 and SiSiC replica of wood	3,4
B-I	Metal-oxide replica of natural cellulosic substances	5
B-II	TiC and TiO_2 replicas of cellulose paper by CVI-R	6
B-II	Zeolitic replicas of wood and zeolite replica of *Luffa cylindrica*	7,8

Fig. 14.1 Flow-chart: different approaches of the processing of cellular biomorphous materials.

complex hierarchically built anatomies (often not reproducible artificially) and a large variety of structures and functional architectures, as the result of their evolutionary process. Obtaining such features in technically processed materials represents a great challenge for researchers.

Biomimetic and biotemplating techniques open new ways for the preparation of bio-inspired or biomorphous inorganic materials with unique structures and properties, complex functional patterns and/or hierarchical porosities typical for biological materials. Therefore, natural/biological tissues and materials are today becoming significantly important as templates, preforms and/or supports for the development of new types of biomorphous material. Recently, a large variety of bio-inspired inorganic replicas have been manufactured via biomineralization or templating of sacrificial biological templates such as viruses [5], bacteria [6], diatoms [7–11], biopolymers [12], eggshells [13], sea urchins [14], spider silks [15], insects [16], wood [1, 2, 17, 18], and leaves [19–21], yielding absolutely novel combinations of compositions (e.g., TiO_2, Al_2O_3, zeolite, silica.), structures, porosity, and properties.

In general, there exist two main strategies for the design of biomorphous inorganic materials, and these are shown schematically in Figure 14.1. It is possible to differentiate between biotemplating processes using the natural preform for: (i) a direct mineralization/templating process (the "direct replica"), or using it as; (ii) a sacrificial/temporary template (the "sacrificial template-type replica"). Some examples for both processes are listed in Figure 14.1.

14.1.1 The Direct Replica

The so-called "direct replica" is generated via a multi-step process (Fig. 14.1, method A). The replication itself is the mineralization of biological specimens, which can be achieved in two different ways. As shown in Method A-I, it is possible to intercalate the biotemplate with precursors (morpho-synthesis) via a sol–gel, and to subject the material to a thermal treatment, in order to obtain the final porous inorganic replica [22, 23]. Alternatively, as displayed in Method A-II, biological preforms can be subjected to pyrolysis and, in a second step, to reactive infiltration with metal precursors via molten metals or via CVI-CVD processes [17, 24]. This second process will be discussed more in detail in the following paragraphs.

14.1.2 The Sacrificial Template-Type Replica

In the alternative approach shown in Figure 14.1 (Method B), biological/natural structures are utilized as sacrificial (temporary) macro-template or structural directing agents. The process generally consists of a coating procedure, which can be performed via a sol–gel process (pathway B-I) or via a direct/reactive coating, for example an *in-situ* hydrothermal crystallization process (pathway B-II). The latter alternative consists of an *in-situ* growth of a stable and intergrown inorganic film around the biotemplate macrostructure, and does not require any further processing [25]. The former technique generally requires processing of the precursor layer which has been obtained in a previous step via a sol–gel technique [26]. In both cases (pathways B-I and B-II), hybrid organic/inorganic composites are obtained in an intermediate stage, consisting of a biological scaffold that is uniformly coated with an inorganic film. The biotemplate is generally removed by a thermal treatment, where the organic macro-template is thermally degraded, yielding a highly porous and self-supporting inorganic structure that partially replicates the macro and micro structures of the original biological preform [27]. A replica prepared according to this method is therefore an empty, open porous scaffold. In fact, after the removal of the organic template, the volume previously occupied by the biotemplate remains empty, providing the final inorganic replica with further porosity.

14.1.3
Cellular Ceramics

From an engineering viewpoint, the transformation of organic preforms of biological origins into cellular ceramics seems to offer a technically feasible means of adapting to technical devices the outstanding properties of the complex hierarchical-built anatomies of organic preforms. Anisotropic cellular ceramics with hierarchical porosity ranging from the nanometer to the millimeter scale are interesting for applications as filters, catalysts or catalyst supports, membranes and adsorbents. Therefore, biomorphous ceramics manufactured from lignocellulosic raw materials and preforms have become of increasing interest in recent years [28]. Lignocellulosic biocomposites such as timber and natural fibers are intricate materials with great biodiversity, but with a chemical composition that is dominated by monosaccharides (pentoses and hexoses) forming cellulose and hemicelluloses, and p-OH phenylpropanes present in lignin. Cellulose-based templates have been converted into biomorphous ceramics by means of infiltration reaction techniques. Native plant tissue with an evolutionarily controlled vascular pore channel system provides accessibility for vapor or liquid infiltrants, which subsequently react at the cell wall surface to generate carbide- or oxide-based biomorphous ceramics [29]. The hierarchical microstructure of the native template can be retained down to the submicron scale, offering the possibility for increasing toughness by tailoring the local strut microstructure [30].

14.1.3.1 Polysaccharides

Polysaccharides offer an interesting potential as a template material because of their chemical as well as structural variability, and abundance in nature.

Cellulose is one of the most important primary plant products, with an annual worldwide production between 10^{10} and 10^{11} kg. There is general agreement that cellulose synthesis in higher plants occurs at the plasma membrane via membrane-associated enzyme complexes [31], but it can also be synthesized by

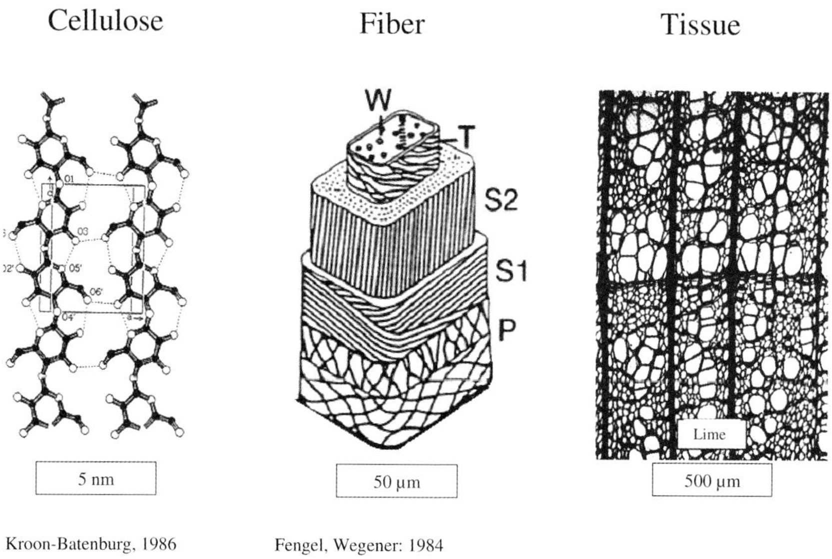

Fig. 14.2 Levels of structural hierarchy of cellulose molecules, cellulose fibers in the cell wall segments, and cellular anatomy of wood tissue.

other organisms including bacteria and fungi, most algae, and some animals. Cellulose is found in the cell walls of natural plants in which it is arranged within a complex assembly of various polysaccharides and proteins. Cellulose consists of unbranched β 1,4 linked glucose residues arranged in linear chains, where every other glucose residue is rotated approximately 180° (Fig. 14.2). As a result, *cellobiose* is the structural repeating unit of the glucan chains in cellulose. The extended glucan chain polymer forms a flat, ribbon-like structure that is further stiffened by van der Waals forces, as well as intra- and intermolecular hydrogen bonds, leading to a regular crystalline arrangement of glucan chains. In nature, cellulose never occurs as a single chain, but rather exists as a partly crystalline array of many parallel, oriented chains (microfibrils) which are the basic units. The degree of polymerization (the number of anhydroglucose units; AGUs), the intermolecular arrangement of individual cellulose chains, and the number of chains (dimensions of the microfibril) is highly variable, depending on the source from which the cellulose is obtained. Several allomorphs, including cellulose I, II, III, and IV have been described which differ in the orientation relationships between vicinal glucose chains.

Figure 14.2 shows, schematically, the various levels of structural hierarchy of polymeric cellulose, which is a major constituent of the plant cell wall and hence of plant tissue. On the molecular level, a large variety of chemical derivatization reactions at the [=CH–OH]-group functionalities have been reported, including etherification [=CH–O–R], esterification [=CH–C(O)–R], acetalization [=C–O–C(R_2)–O–C=], and oxidation [=COOH] [32]. Derivatization provides solubility of cellulose, charge formation, acidic or basic functionalities, and may be used to change the hydrophilic and hydrophobic behavior. The surface structure of cellulose can be modified by treatment with specific enzymes.

Cellulase enzymes have been used to control the elimination of fibrils from the fiber surface in a process known as "biopolishing" [33]. Cellulases form an enzymatic complex which contains three types of enzyme: endoglucanase or endocellulase (β-1,4-D-glucan-4-glucan hydrolase); cellobiohydrolase or exocellulase (1,4-β-D-glucan cellobiohydrolase); and β-glucosidase or cellobiolase (β-D-glucoside glucohydrolase). Through a mechanism of catalytic hydrolysis, all of the components of cellulases break down the β-1,4-glucosidic bonding of the cellulosic polymer, although each one acts specifically [34].

Plant cell walls are involved to provide skeletal support (mechanical stability), but they also play a dominant role in cell growth and morphogenesis, in cell recognition and signaling, in digestibility and in herbivore nutrition. The wall of tracheid cells of higher plants such as wood consists of fibrous cellulose composites, set in a chemically complex matrix of hemicellulose and pectin, often hardened with lignin. Growth in cell volume coincides with deposition of the primary layers of its walls which, after growth in cell dimensions is completed, is followed by the deposition of a secondary wall to the inside of the primary wall. Helicoidal structures of cellulose fibrils are found both in primary and secondary cell walls, and these play key roles in the mechanical behavior of cellular tissue [35]. Cellulose orientation is relevant to commercial applications of fibers in paper pulps, ropes, textiles, and timbers.

Woods are cellular solids, composed of mixed biopolymers (cellulose fibers, lignin and hemicellulose as matrix material), and with a relative density ranging from 0.05 to 1. Hardwoods (e.g., oak) have an average composition of cellulose (40–50 wt%), hemicellulose (20–35 wt%), and lignin (15–35 wt%). The molecular structures and compositions of the major biopolymers of cellulose, hemicellulose and lignin are complex, and may vary for different types of wood. Chemically, the major constituents of wood are carbon (50 wt%), oxygen (44 wt%) and hydrogen (6 wt%), plus trace elements (1 wt%). Extended structure analyses of wood using micro-beam scanning X-ray scattering have revealed a microstructure with highly oriented cellulose fibers at the nanometer level of the cell wall. This is thought to be responsible for the unique mechanical stability (elasticity, strength and toughness) of living trees under harsh mechanical loading conditions [36]. The highly anisotropic and open cellular structure of wood tissue may serve as a hierarchically structured template to generate novel cellular ceramics with micro-, meso-, and macro-structures pseudomorphous to the initial porous tissue. The transportation system of wood tissue, which consists of large-pore channels (vessels) with a diameter of 30 to 45 mm in soft woods and 10 to 400 mm in hard woods, provides rapid access for gaseous or liquid infiltrants which may be used for ceramic conversion. Biological preforms used for conversion into biomorphous ceramics include a variety of different species such as hardwoods (*Dicotyledonous angiosperms*: oak, beech, lime, poplar, maple), softwoods (*Gymnosperms*: pine, fir), and other species including palms (rattan, lianas), or fruits (*Luffa aegyptica*) [27, 37–39].

Instead of naturally grown plant tissue, pre-processed technical products such as wood fibers, fiber boards and paper have also been used as precursor performs for the manufacture of biomorphous ceramics with macroscopic pore channel structures [40], as well as macro-scale sacrificial template for the formation of self-supporting inorganic, for example zeolite-containing replicas with a hierarchical porosity [27]. The pre-processing of lignocellulosics by delignification and the surface treatment of cellulose fibers offers a versatile pool of cellulose fiber macrostructures to be used in the manufacture of lightweight ceramic structures of variable meso- and macro-pore architecture. Non-oxide ceramics based on silicon carbide were prepared either by infiltration of Si-precursors into a biocarbon preform obtained from pyrolyzed paper structures, or by the reaction of Si-powder loaded artificial paper. The highly porous fibrillar strut microstructure offers low weight, high permeability, and good thermal shock resistance, all of which are of particular interest in the application of catalyst carrier systems in chemical engineering.

14.2 Conversion of Lignocellulosics into Ceramic Substrate

Basic approaches for materials manufacturing using cellulose precursor structures, often called the CDC (Cellulose-Derived Composites) processes, generally include three steps: (i) carbonization; (ii) shaping; and (iii) the infiltration/reaction (Fig. 14.3).

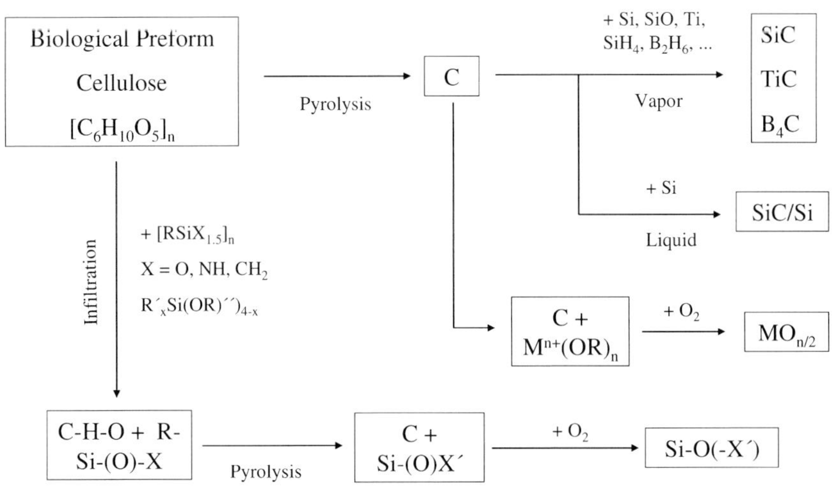

Fig. 14.3 Processing scheme of the conversion of biological preforms into biomorphous ceramics retaining the microstructure of the natural template.

The unique anatomic features of the native plant tissue can be retained during pyrolysis in an inert atmosphere, yielding a porous template composed primarily of carbon. The carbon preform was shown to be easily machineable prior to conversion to a ceramic composite [41]. Thermal degradation of wood biopolymers and the evolution of the atomic/molecular and mesoscopic structure of the carbonaceous material studied by wide-angle X-ray scattering, small-angle X-ray scattering and Raman spectroscopy revealed three distinct temperature regions in the conversion of a natural preform to carbon:

- Evaporation of water and dehydration together with slight depolymerization between room temperature and 250 °C.
- Degradation of all biopolymers and evaporation of low molecular-weight fragments between 250 and 350 °C, leading to major weight loss and dimensional changes.
- Aromatization forming turbostratic carbon layer above 350 °C which grow continuously in size up to >1000 μm.

Pyrolysis occurs in a stepwise manner, with hemicellulose breaking down first at 200–260 °C, cellulose next at 240–350 °C, and lignin at 280–500 °C. Between 200 and 400 °C, almost 80% of the total weight loss occurs, which may vary between 40 wt% (lignin) and about 80 wt% (cellulose) [42]. Between 400 and 800 °C, aromatic reactions occur and the carbon network shrinks (by 15–22% in axial and 22–40% in radial/tangential directions) to accommodate the excessive volume (vacancies) left by the evolving gases. Above 800 °C, thermal-induced decomposition and rearrangement reactions are almost terminated, leaving a carbon template structure. Residual hydrogen is released, defects are healed, and the degree of crystallinity of the carbon units increases with temperature.

The carbon preform may be infiltrated at low temperature with a liquid metal organic precursor ($(M^{n+}(OR)_n)$, or at high temperatures with liquid Si or vapor reactants (Si, SiO, SiH_4, Ti, B_2H_6, ...) to form single or multiphase carbide- or oxide-based reaction products. Instead of Si melt infiltration, which requires temperatures above the melting point of Si at 1410 °C, binary Si–Al as well as ternary alloy melts (Si–Al–Mg) offer significantly lower the infiltration and reaction temperatures [43], which might be favorable for achieving micro-structures with small particle sizes of the SiC-reaction phase. Biomorphous ceramics prepared by reaction with a carbon preform include SiC, SiSiC, SiC–$MoSi_2$, TiC, ZrC, Al_2O_3, ZrO_2, and TiO_2 [29, 37, 38, 44–49]. Alternatively, the non-extracted or extracted biological preform may directly be infiltrated with liquid organic precursor systems including polysiloxanes, -silazanes or -carbosilanes after modification of the native cell wall structure (maleic anhydride) to facilitate penetration of the organic precursor [50]. After curing the polymeric precursor at 100–200 °C, pyrolytic high-temperature annealing above 800 °C finally results in the formation of Si–O–C composite materials. Compared to the infiltration of a biocarbon preform, however, a pronounced anisotropic shrinkage of 15–35% (depending on the direction) was observed.

The kinetics of thermally induced decomposition during pyrolysis are strongly affected by the transport of gaseous decomposition products (H_2O, CO_2, alcohols

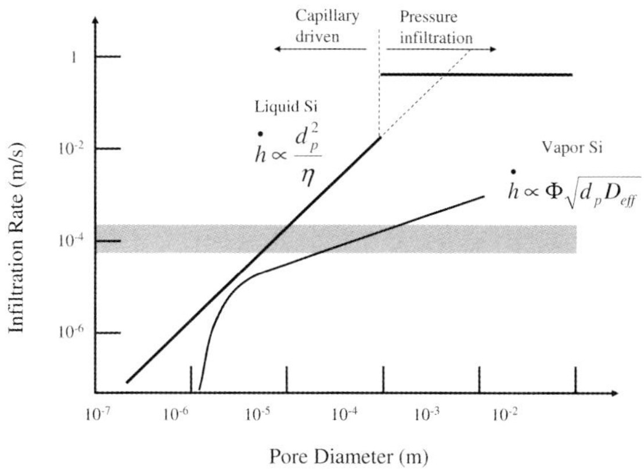

Fig. 14.4 Range of infiltration rates as a function of pore channel diameter d_p calculated for liquid and for vapor Si infiltration. η = melt viscosity; D_{eff} = effective vapor diffusivity.

and a variety of additional species) via the open-pore channel system. Analysis of vapor phase diffusion indicates that a critical pore channel diameter may be derived, below which no effective gas phase transport can occur within reasonable time. For the case of Si and SiO vapor infiltration at 1400–1600 °C this critical pore channel radius was estimated to be in the range of 1 μm [28]. The same lower boundary holds for liquid Si infiltration, whereas an upper boundary limiting capillary driven infiltration lies in the range of approximately 100 μm. Above this threshold, pressure-assisted infiltration may result in accelerated infiltration rates (Fig. 14.4).

Four different stages of reactive Si-melt infiltration in the carbon preform could be distinguished, starting with the heterogeneous nucleation of nano-grained β-SiC on the carbon surface by a vapor phase reaction below the melting point of Si [29]. After spontaneous Si melt infiltration, a stepwise reaction results in the simultaneous formation of a nano-grained (10–80 nm) SiC-layer and a coarse-grained (>10 μm) SiC. Further reaction proceeds slowly by diffusion of the reactants through the formed SiC layer, and the microstructure evolution is dominated by dissolution and re-crystallization processes (Ostwald ripening).

The mechanical properties of cellular ceramic structures derived from biological preforms are controlled by various levels of structural hierarchy [36]:

- Fractional density ρ/ρ_0 (where ρ is the preform density and ρ_0 the strut density) generally scales by $(\rho/\rho_0)^n$, with n ranging from 1 to 3 depending on the property and anisotropic loading direction; as a consequence, periodical variations of density due to seasonal variations of growth conditions will result in corresponding local variations of crack propagation resistance.
- Open- or closed-cell morphology of various shape (e.g., hexagonal, elliptical, hexaeder, tetrakisoctaeder), which

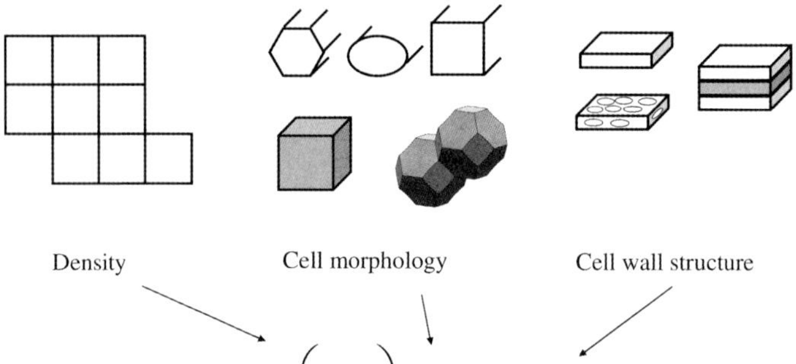

$$\sigma = f\left(\frac{\rho}{\rho_0}\right) f(\Phi)\sigma_0$$

Fig. 14.5 General dependency of a property σ (e.g., strength, toughness, Young's modulus, Poisson ratio) on the various structural hierarchical features of cellular materials.

determines local stress distribution upon loading and critical fracture initiation conditions.

- The cell wall structure may be dense or porous and single layer or multilayer, which is expressed by the intrinsic properties of the strut material and its variations as given by grain size and strut porosity dependence and the Weibull parameters (Fig. 14.5).

Depending on the direction of loading, n may attain characteristic values:

Direction of loading	*Out-of-plane* (axial)	*In-plane* (radial, tangential)
Young's modulus	n = 1	n = 3
Strength	n = 1	n = 2
Toughness	n = 1.5	n = 1.5

FE simulations of stress distribution in radial-loaded (*in-plane*) biomorphous silicon carbide derived from *Pinus sylvestris* showed that substantially higher tensile stress levels are expected to occur in areas of thin strut thickness – for example, low density compared to areas with thicker struts of high density [51]. The stress ratio was found to depend on the pore shape, with cellular pores of square shape being subjected to lower stress levels than pores of elliptical shape under the same external loading. Crack advance is likely to be localized on the highly stressed areas, whereas regions of high density should yield higher fracture-resistant segments in the mechanically loaded structure.

Due to the uni-directed pore morphology, the highly porous biomorphous β-SiC ceramics exhibit an anisotropic mechanical behavior, with the fracture stress in axial direction being 20-fold higher compared to the radial direction [30]. Generally, the *in-plane* stiffness and strength (stress acting perpendicular to cell elongation) are the lowest, because *in-plane* loading makes the cell walls bend. The *out-of-plane* stiffness and strength (stress acting parallel to the cell elongation) are much larger because they require axial extension or compression of the cell walls. Biomorphous ceramics with a cellular micro- and macro-structure pseudomorphous to naturally grown tissue show a complex mechanical behavior, which is governed by the unique arrangement of cells. In some aspects, the fracture behavior in biomorphous ceramics is similar to that of fibrous monolithic ceramics as well as that of laminate composite ceramics showing non-catastrophic stress–strain behavior [30]. A pronounced anisotropy of fracture behavior is a characteristic feature which depends on the loading conditions with respect to the orientation of the cell packing structure. Fracture of the cell wall, as well as the cell interface (intercellular lamella), is supposed to dominate the micromechanical crack propagation. The energy absorption capacity of biomorphous ceramics might be governed by cracking and frictional sliding. Both of these mechanisms are more effective when extensive delamination occurs prior to fracture of the individual cells. Thus, tailoring of the strut microstructure and the interface between the cells by suitable processing techniques seems to play a key role for improving the mechanical properties of low-density biomorphous ceramics.

Three different sources of cellulose templates were applied as preforms to manufacture open cellular SiSiC-ceramic structures (method A). These cellular biomorphous ceramics could be used as monolithic carriers for catalysts layers, and were therefore functionalized, in a separate step, with catalytic active zeolite layers via a coating process (see Section 14.3). The three cellular biotemplates used were:

- Sponges obtained from the matured dried fruits of *Luffa aegyptica* (syn. *Luffa cylindrica*), which consists of a fibrous network composed of cellulose, hemicellulose and small amounts of mannan and galactan. *Luffa aegyptica* is a tropical annual climber vine plant, which belongs to the Cucurbitaceae family.
- Rattan (*Calamus rotang* L.), which is a tropical climbing palm of the subfamily Calamiodeae of the family Arecaceae. Rattan exhibits no branches or seasonal rings. Its pore structure is characterized by a multimodal pore size distribution with vessel diameters up to 500 μm.
- Corrugated cellulose-derived cardboard monoliths, which have been obtained by processing of cellulose fiber-based paper into ceramic products.

In addition to the conventional steps of the liquid silicon infiltration (LSI) process for the biotemplate conversion to SiSiC ceramics, the preparation of cardboard-derived replica involved a pre-processing step consisting of shaping and joining of the paper perform in order to prepare monoliths. The process continued further with the pyrolysis in an inert atmosphere to form the biocarbon replica and the reaction with vapor or liquid precursors at high temperature to convert the biocarbon fiber structure into inorganic ceramic phases. In contrast to the conversion process of natural tissue templates, where vapor and liquid infiltration kinetics are limited by constrained slow diffusion (Knudsen) and viscous frictional drag in pores smaller than approximately 1 μm, the large macroscopic channels in paper-derived preforms (channel porosity > 50%) offer accelerated access for liquid as well as gaseous precursor infiltration and reaction. Thus, large-size components can effectively be converted into ceramic components. Alternatively, to an external precursor source, dip coating of the internal channel surfaces by a low-viscous slurry or incorporation of solid precursor in the paper structure may provide an improved local distribution of the reactants. An example of macrocellular silicon carbide components manufactured from a corrugated paper preform obtained via the biotemplating LSI process mentioned above is illustrated in Figure 14.22 (see Section 14.3).

Compared to capillary infiltration from an external precursor source, the local infiltration offers short transportation distances and hence reduced times of high-temperature treatment. Using a polyalkylsiloxane $(CH_3SiO_{1.5})_n$ with $n = 300–400$ as a binder phase for Si, Si/Al and SiC powders, it was shown that the dip-coated preform may achieve superior mechanical properties after curing at 150 to 250 °C. The polysiloxane may also be used as an adhesive bonding for lamination and joining. During annealing, the polysiloxane decomposed to an amorphous Si–O–C phase above 600 °C, leaving an inorganic residue of more than 70 wt% which at even higher temperatures crystallized to SiO_2, SiC, and C. Incorporation of the reactive filler powders in the paper structure can further reduce transportation paths of the reactants. Pre-ceramic paper with Si-filler fractions up to 80 wt% and of mean particle sizes less than 10 μm could be achieved. The hierarchical porous structure with large-pore channels in the millimeter range and the struts with interconnected pores of effective sizes in the micrometer range provide suitable conditions for designing biomorphous catalyst carrier structures. Due to the fibrillar strut structure mimicking the initial cellulose fiber arrangement, turbulent flow is induced perpendicular to the main flow direction along the macroscopic channels, and hence a high effective surface is available for the catalytic reaction.

14.3
Hierarchical Porous Zeolite-Containing Macrostructures

Zeolites are microporous (pore diameter < 1.3 nm) crystalline aluminosilicates that exhibit molecular-sieving capabilities (an ability to distinguish and limit the access of molecules on the basis of size or structure, useful for sensor and membrane applications) and ion-exchange properties. Their surface contains sites which can be designed as acidic or basic, depending on their structure and the applied modification procedure, providing zeolites with unique and specific catalytic properties [52–54]. Thus, zeolites and zeo-type materials have found application in the fields of ion-exchange, adsorption, separation, and catalysis [52–55]. In technical applications, zeolites are usually mixed to a binder, and then pressed and shaped into pellets of various sizes and forms. In order to provide better accessibility to the micropores and active sites, and to reduce the pressure drop generated by pellet packings, hierarchical (micro-/meso-/macro-) porous zeolite-based structures have been developed [56–63]. As they benefit from each different pore-size regime [19, 21], hierarchical porous structures are indeed of potential major significance in catalysis, separation and adsorption, where diffusion plays a key role in the material performance [53–55].

There exist two types of structured hierarchical porous zeolite-based materials: (i) zeolite-coated composites, and (ii) self-supporting zeolite architectures.

During the past years, *zeolite-coated composites* [56] have been the subject of intense research for the development of vehicle emission control systems [57, 58] structured catalytic packings for novel reactor concepts [59–61] (i.e., monolithic configuration [59, 60], micro-reactor technologies [62]) and adsorption/separation units [62, 63]. In addition, zeolites and zeolitic composites are used increasingly in environmental catalysis [57–60]. Very different substrate and structure types – for example, ceramic honeycombs [58] and foams [48], metal foams [64] and packings [60] and glass [65] – have been employed as the support for the preparation of novel composites, functionalized with layers of various zeolite types (e.g., MFI, BEA, LTA). The aim is to accomplish a hierarchical-porous architecture with a precise functional design in which, for example, the molecular sieving capabilities and/or the catalytic activity of zeolites are matched with the mechanical properties and structural design of the support. However, the range of supports currently available commercially is very limited with respect to the pore network, and the structural and mechanical properties.

Self-supporting zeolite architectures can be prepared via sacrificial templating processes using various structure-directing agents, such as polyurethane (PU) foams, carbon fibers, and polymethyl methacrylate (PMMA) beads [48, 56, 59]. The resulting materials generally consist of a porous scaffold with an enhanced accessibility to the active sites, but generally exhibit quite poor mechanical stability, compared to composites materials coated with zeolite.

Bio-inspired zeolite-based macro-structures (either self-supporting or composite materials) [1–3, 7–9] with novel shapes, complex functional patterns and hierarchical porosity [4], could be of significant importance in the development of adsorbents, membranes, sensors and catalysts with improved performances [1, 4, 19, 21, 54]. In comparison to technically manufactured materials, such as ceramic foams, ceramic and metal honeycombs and packings, the cellular biological structures exhibit hierarchical structures, complex morphologies, and unique porous architectures. Biological cellular materials with the unique molecular sieving and catalytic properties of zeolites, and with their microporosity and high specific surface areas, might lead to a novel class of multifunctional hierarchical (micro-/meso-/macro-) porous materials with a wide range of properties.

Such unique biomorphous multifunctional hierarchical porous structures can be achieved by realizing new processes to replicate biological cellular structures with inorganic replicas containing zeolites. The general scheme shown in Figure 14.1 might also be used to describe the two methods (A and B) for the development of zeolite-based biomorphous materials (composites or self-supporting replicas). The two approaches for the processing of zeolite-containing materials are discussed, based on different examples, in the following sections. The two principal pathways for processing biological templates, in order to obtain zeolite-based biomorphous materials with a hierarchical porosity, are described more specifically in Figure 14.6.

The replication method A is based, initially, on the processing of a cellular ceramic replica, whilst the functionalization of the same with a zeolite layer can be realized, in a second step, by using a coating process (dip-/slurry) [1, 2] or by a direct *in-situ* hydrothermal crystallization [66]. Method B is characterized by a direct coating of the biological templates with zeolite films of tightly intergrown crystals. This leads to an inorganic/organic hybrid composite material, which

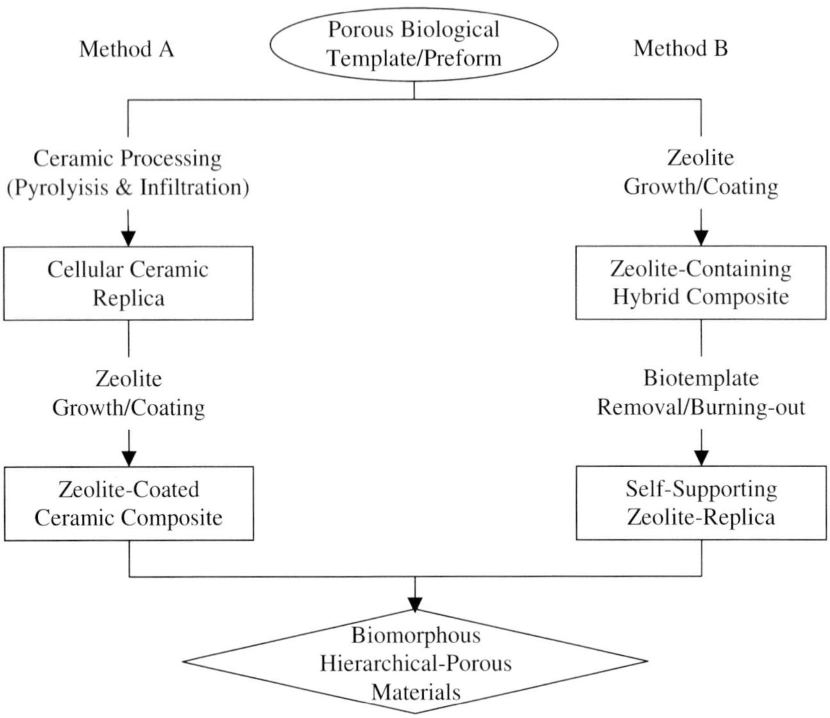

Fig. 14.6 Two possible approaches towards the processing of biomorphous zeolite-based materials with hierarchical porosity.

still contains the biological support. Finally, the biological (organic) template is removed, usually by a thermal treatment (oxidation) or by chemical leaching, leaving a self-supporting zeolite scaffold that resembles the macro and micro-architecture of the original biological template. The biological support acts therefore as a sacrificial template, or macrostructure-directing agent. A further level/volume of porosity is generated during removal of the biotemplate, which might on the one hand reduce the mechanical stability of the replica, but on the other hand provide more accessibility to the single zeolite crystals and their microporous cavities.

14.3.1
Replicating Materials of Biological Origin

According to the general description of the preparation pathways shown in Figure 14.1 (see Section 14.1), biomorphous zeolite replicas [1, 2, 7, 8, 19–21, 27] can be achieved via the growth of zeolite films directly onto the surface of the biological support ("sacrificial template-type replica"). The two key steps of this multi-stage process (method B) are:

- the deposition (adsorption) of pre-synthesized nanocrystals modified by a cationic polyelectrolyte [1, 7], or the direct hydrothermal synthesis [1, 27], or chemical vapor deposition (CVD) treatment [7], leading to the nucleation and growth of zeolite nano-crystals directly (one-step process) on the biological support; and
- secondary crystal growth, during which the zeolite seeds on the biotemplate surface are grown into films of tightly intergrown zeolite crystals, in order to obtain a stable self-supporting zeolitic scaffold [27].

Following this two-stage procedure, a hybrid (organic/inorganic) composite is obtained, which must be treated (calcination at high temperatures) in order to remove the biotemplate as well as the organic template which is used during synthesis to generate the micropore system in the zeolite framework [27].

A large variety of biological specimens with complex morphologies and hierarchical architectures, which can be used as sacrificial templates, are provided in nature. Very recently, Valtchev et al. [19, 21] reported the zeolite imprinting of an *Equisetum arvense* leaf via a biomineral-silica-induced mechanism (the support had an intrinsic biogenic silica source which acted as a promoter for the nucleation of zeolite crystals). Many other examples of self-supporting zeolitic structures have been obtained by using different biological structures as sacrificial templates. Mann et al. prepared monolithic, sponge-like, self-supporting zeolitic structures by using dextran as template [40], while Dong et al. [1] reported the use of wood tissues (notably cedar and bamboo) for the preparation of wood-like zeolite replica. In all cases, it is reported that the obtained zeolitic replica retains both the macro- and also the microstructure of the organic tissues. In a near-perfect manner, the imprinting of all the morphological features of the organic tissues at every length scale could be mimicked in the resulting zeolitic replica.

In a recent investigation, we developed a means of fabricating biomimetic self-supporting MFI-type (Silicalite-1, ZSM-5 and bi-layer) zeolite frameworks with hierarchical porosity, complex architecture and design. Biological template specimens of a *Luffa* sponge have also been used as macro-scale sacrificial structure builder. The practical applicability of such biomorphous, self-supporting zeolitic materials in a technical catalytic process has also been demonstrated [27].

The *Luffa* sponge biotemplate, when used as sacrificial macrotemplate at different magnifications is illustrated in Figure 14.7, and shows the open-porous and

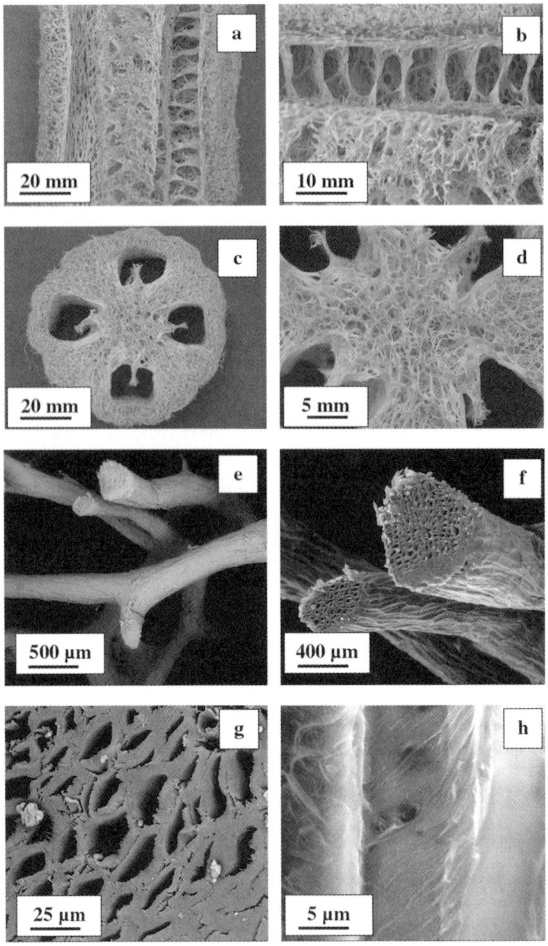

Fig. 14.7 *Luffa* biotemplate at different magnifications. (a,b) Vertical cross-sections; (c,d) (a) and (b), respectively, viewed from above; (e,f) *Luffa* struts at different magnifications; (g,h) microchannels along the struts.

complex structure. In order to accomplish the biomimetic templating of *Luffa* sponges with MFI-type (Silicalite-1, ZSM-5 and bi-layer) zeolite, a two-stage process was developed which consisted of an *in-situ* hydrothermal zeolite seeding, followed by a secondary crystal growth step. Both steps are basically hydrothermal treatments [27]. Briefly, dried samples of *Luffa* sponge were subjected into a concentrated precursor solution (reaction mixture), which was optimized for the *in-situ* nucleation of MFI nanocrystals. During this hydrothermal treatment step, a complete coverage of the biotemplate with nano-sized MFI, Silicalite-1 or ZSM-5 seeds of 200 to 400 nm was achieved. The seeded *Luffa* samples were then treated in a second hydrothermal process with a second dilute precursor solution, which promoted only the growth of the zeolite seeds (secondary crystal growth) into continuous thin intergrown zeolite films on the surface of the biological scaffold. In addition to the biological template (*Luffa* in this case), the MFI-type zeolite crystallization itself required a structure-directing agent on the molecular level (molecular templates) – for example, tetrapropylammonium hydroxide (TPAOH) as the molecular structure-directing agent for the MFI framework formation. The removal of the two different templates was achieved by thermal treatment in air (calcination), which led to a complete residue-free degradation of any organic compound from the hybrid composite, leaving the pure zeolitic self-supporting macro structures.

A self-supporting monolith (Fig. 14.8), up to 5 cm in length and 3 cm in diameter, consisting of a pure Silicalite-1 phase was obtained. It can be noted, that the macrostructure of the *Luffa* sponge is completely preserved after the synthesis procedures and removal of the biotemplate. Comparisons between the scanning electron micrographs of the *Luffa* sponges template (Fig. 14.7) and the calcined free-standing zeolitic replicas (Fig. 14.9) reveal that the inorganic framework mimics perfectly the original *Luffa* sponge micro-/macro-architecture. An external scaffold of well intergrown zeolite crystals supports the entire biomorphic structure. The film thickness of this scaffold is between 5 and 30 μm, depending on the zeolite type, the number of steps, and the synthesis conditions [27].

Furthermore, a continuous bundle of zeolitic microchannels/rods was formed (Fig. 14.9d–f) inside the struts of the zeolitic scaffold. It seems obvious that, during the synthesis, the nutrient solution of the synthesis mixture penetrates deeply into the vascular system of the *Luffa* sponge, most probably driven by capillary force, so that the crystallization takes place also in the inner cellular network (channel diameter 10–20 μm). Such zeolitic microtubes are characterized by crystals tightly packed together and wrapped in a spiral mode (Fig. 14.9g,h). This can be seen as an effect of textural imprinting of the *Luffa* gourd vascular system: the channel walls, in fact, exhibit a coiled texture with the same winding angle (compare Fig. 14.7h and 14.9g,h). Moreover, depending on the synthesis conditions, both hollow and solid architectures of the microtubes could be observed. Therefore, the fabrication of zeolite microrods and hollow microchannels with high aspect ratio (>100), helical texture and both tortuous and straight geometries, could also be achieved by zeolitization of *Luffa* sponges. Thus, the replication of Luffa sponge led to structures where single zeolite crystals form films with a three-

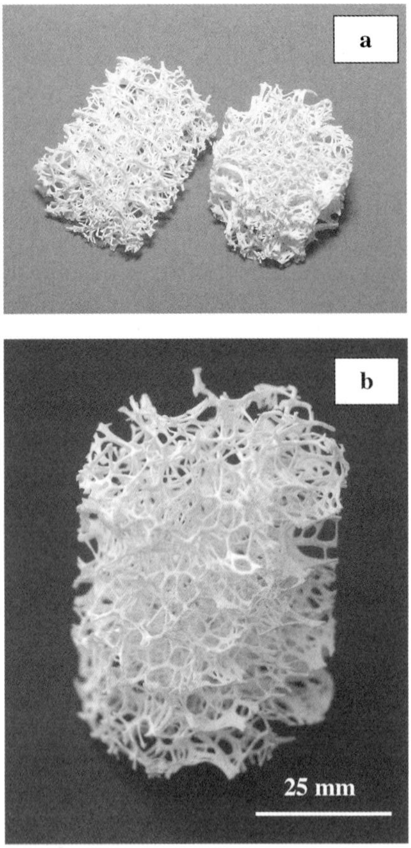

Fig. 14.8 (a,b) Optical images of the self-supporting Silicalite-1 *Luffa* monoliths.

dimensional (3-D) monolithic organization/architecture (Figs. 14.8 and 14.9a,b), where the empty struts of the inorganic scaffold (left by the decomposition of the biotemplate) were filled with bundles of zeolite microtubes/microrods extending continuously along the entire structure (cf. Figs. 14.8 and 14.9).

Owing to its micro-/macro-structure, another interesting sacrificial template for the preparation of hierarchically porous zeolitic replica is cellulose-derived paper. Sano et al. [69] were the first to report the manufacture of pure, self-supporting MFI-type zeolitic structures. This was obtained by a direct one-step hydrothermal crystallization in the presence of filter paper. This one-step synthesis procedure did not allow a fine control of the crystallization process: consequently, zeolite crystals grew randomly on and in between the cellulose fibers. More recently, Dong et al. [1] described the replication of cellulose acetate filter membranes with zeolite. Silicalite-1 replicas were achieved through the adsorption of pre-

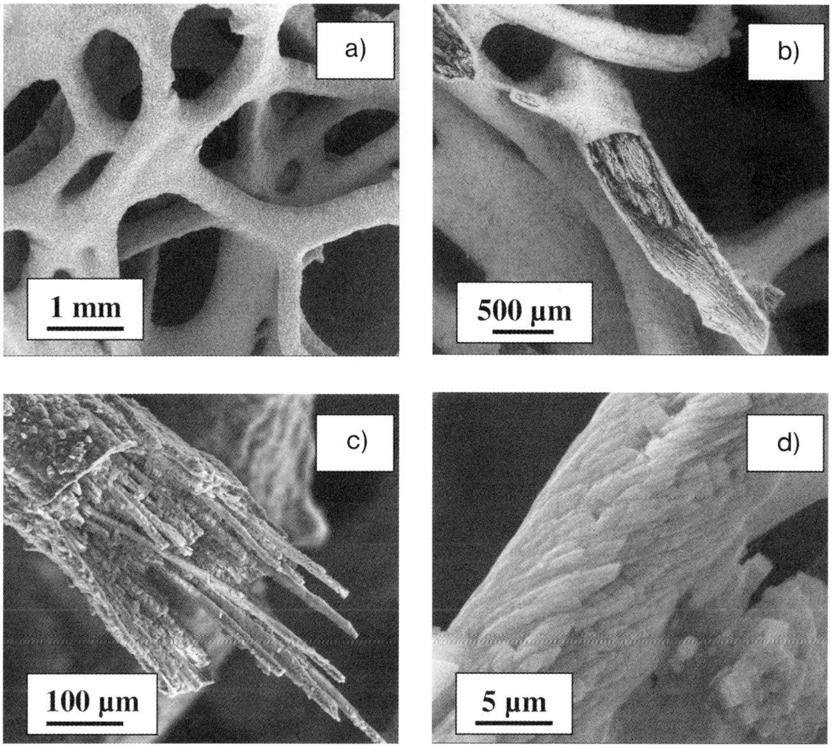

Fig. 14.9 (a–d) Scanning electron microscopy (SEM) images of the self-supporting zeolite replica of *Luffa*.

synthesized zeolite nano-seeds (*ex-situ* seeding) onto the support surface, previously modified by a cationic polyelectrolyte. The crystal growth (to micrometer-size crystals) was attained by a successive hydrothermal or a CVD treatment, which led to replicas of the cellulose-filter structure.

Cellulose paper represents an already processed biological material, and its fine, hierarchically built micro-structures and hydrophilic nature make this material an extremely appealing sacrificial template for the preparation of the above self-supporting zeolite replica. From a technical point of view, the most interesting aspect is that paper can be easily shaped into custom-designed macrostructures, such as slabs, tubes, and corrugated-type monoliths. The fibrous structure of cellulose-derived paper at different magnifications is shown in Figure 14.10.

In such pre-processed biological-based preform, the single fibers act on the molecular level binding the zeolite crystals onto their surface. Figure 14.11 show different cellulose paper macrostructures, including a slab (a), a monolith (b), and a tube (c), and their corresponding self-supporting zeolitic replica (only for Fig. 14.11a and c) [84]. It should be noted from these illustrations that the self-

Fig. 14.10 (a,b) SEM images of cellulose paper at two different magnifications.

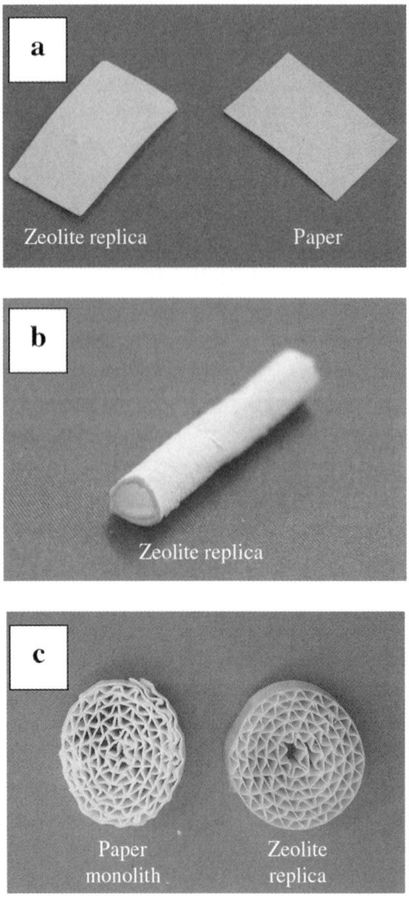

Fig. 14.11 Different shapes and structures of self-supporting zeolite replicas obtained from cellulose paper macrotemplates. (a) Slabs; (b) hollow tubes; (c) cellular monoliths.

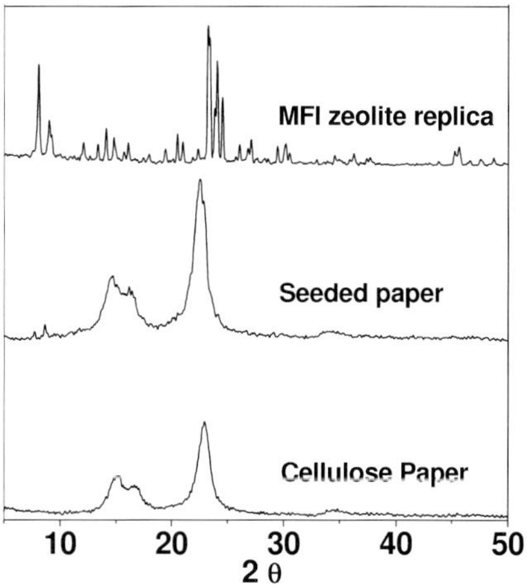

Fig. 14.12 X-ray diffraction (XRD) patterns of the different stages during the replica process of cellulose paper. Cellulose template, seeded zeolite/cellulose hybrid composite, self-supporting Silicalite-1 zeolite replica after calcination.

supporting zeolite replicas retain the macroscopic shapes and geometries of the pre-shaped paper templates.

The X-ray diffraction (XRD) patterns of the samples at the different stages of the templating process (the cellulose template, the seeded zeolite/hybrid composite and the self supporting replica) are shown in Figure 14.12. Following the seeding procedure, the typical reflections of the MFI structure, with two peaks at $2\theta = 7.9$ and 8.9° and a triplet in the range $2\theta = 23$–24°, were observed. After secondary growth and thermal treatment to remove the molecular and macroscopic templates, the XRD pattern showed the typical reflections of a pure MFI-type zeolite.

Scanning electron microscopy (SEM) images of the zeolite-seeded hybrid samples revealed a complete coating of each cellulose fiber by tightly packed films of nanosized Silicalite-1 crystals. SEM micrographs showed that the layer of MFI nanocrystals deposited on the biological substrate was so perfectly intergrown that it was almost impossible to distinguish single zeolite seeds. The SEM analyses of the calcined free-standing zeolitic replicas (Fig. 14.13) revealed the precision how the original cellulose paper micro-/macro-structure had been mimicked. The material showed a complex structural hierarchy which is developed on four different levels:

- molecular scale: template-directed zeolite crystal synthesis;
- microscopic range: single cellulose fiber coating/replication with film of tightly-packed crystals;
- macroscopic range: (i) formation of an interconnected network of tortuous zeolite microchannels yielding self-

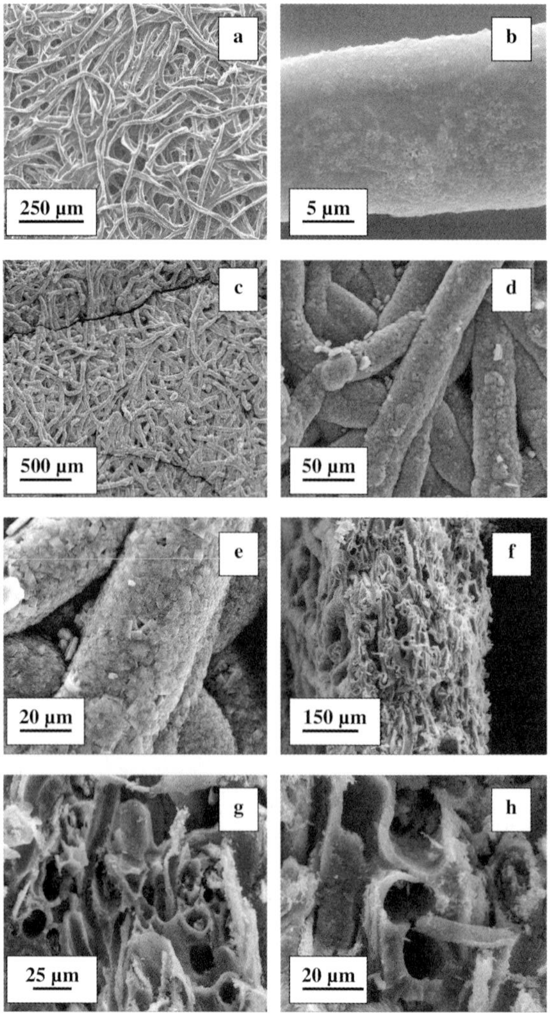

Fig. 14.13 SEM images of the cellulose fibers after seeding (a,b), and of the self-supporting Silicalite-1 zeolite replica of the cellulose paper slab from above (c–e) and cross-section lateral view of the sample (f–h) at different magnifications.

supporting, net-like zeolite macro-architecture; and (ii) development of complex monolithic structures by shaping and replicating the cellulose paper template in the form of cardboard monoliths, tubes, sheets, sandwich composite, curved surfaces, etc.; and

- a film (of thickness ~ 20 μm for the Silicalite-1 and of ~8 μm for the ZSM-5 replica) of tightly intergrown zeolite crystals

> adhered perfectly to each cellulose fiber and maintained its structural integrity after thermal degradation of the biological support.

These hollow zeolitic microchannels with tortuous geometry are well interconnected, yielding a stiff and mechanically stable architecture, after oxidation of the sacrificial from the cellulose substrate. It should be mentioned here that a three-point bending strength mechanical test performed on Silicalite-1 zeolite replica of paper, shaped in the form of planar slabs, showed a crushing stress of 2.5 MPa. This indicates a considerable good mechanical resistance, which is essential for the technical applicability of the material.

To summarize, self-supporting zeolitic replicas could be obtained from biological structures (e.g., *Luffa* or cellulose-derived paper matrixes), via a two-step hydrothermal synthesis, inheriting the complex morphology and the intricate architecture of the biotemplate. As shown previously [27], the biomorphous replica exhibited enough mechanical stability to be used for technical processes (e.g., heterogeneous catalysis) and complex hierarchical (micro-/macro-) porous networks.

14.3.2 Zeolite Functionalization of Biomorphous Cellular Ceramics

The interest in bio-inspired functionalized cellular ceramics for technical applications is rapidly growing, as witnessed by the increasing number of publications in the field. Various biotemplating techniques which allow the conversion of lignocellulosic biocomposites (e.g., timber, wood) into a variety of different inorganic biomorphous ceramics have been employed so far. To date, the preparation of TiO_2 [71], TiC [72], SiO_2 [73], SiOC [50], SiC and SiSiC [28, 45, 67, 74] cellular biomorphous ceramics has been described in the literature. As mentioned in Section 14.2, such bio-inspired inorganic porous materials could be utilized as supports for catalysts and, more specifically, as zeolite coatings. A merging of the zeolite characteristics of ordered uniform microporous frameworks and active centers for catalysis (acidity) with the properties of such biomorphous composites might represent an effective approach for the manufacture of monoliths with molecular-sieving capabilities, hierarchical porosity, and good mechanical stability.

Recently, the present authors' groups have focused on the development and functionalization of wood- and cellulose-derived biomorphous ceramics with zeolite. Figure 14.14 illustrates a scheme describing the preparation of a biomorphous zeolite/ceramic composite starting from a shaped Rattan biotemplate, via pyrolysis and LSI infiltration of the C-preform (as described in Section 14.2). This route can be assigned to method A, the "direct-templating" route, which was previously classified in Figure 14.1 (see Section 14.1). Further functionalization of the biomorphous SiSiC cellular ceramics with zeolite was achieved by a support self-transformation (SST) method, which was developed by the present authors [66, 79, 80] and which, fundamentally, is a direct hydrothermal zeolite crystallization. During this process, the SiSiC support supplies at least one of

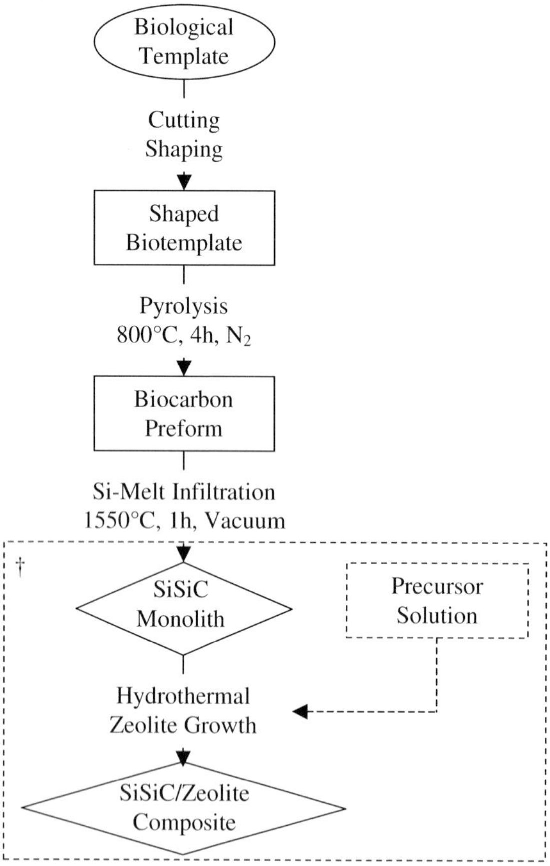

Fig. 14.14 Flow-chart. Three-step biotemplating process for the development of cellular biomorphous SiSiC/zeolite composites. The process consists of a LSI process for the manufacture of SiSiC cellular ceramic supports/monoliths and a subsequent Support Self-Transformation technique for the growth of a zeolite coating on the above-structured ceramic supports.

the framework constituents (Si in the specific case) of the zeolite in the finally formed zeolite layers on the support. In the particular case described, no external (additional) Si-source for the functionalization of the SiSiC support with zeolites was utilized in the reaction mixture.

The inset (†) of the scheme shown in Figure 14.14 is enlarged in Figure 14.15 to provide a more detailed description of the functionalization process of cellular biomorphous SiSiC ceramics with zeolite. Preparation of the cellular ceramic used as carrier for the zeolite coatings was described in Section 14.2, while specific conditions used to prepare the SiSiC supports used for the zeolite coating

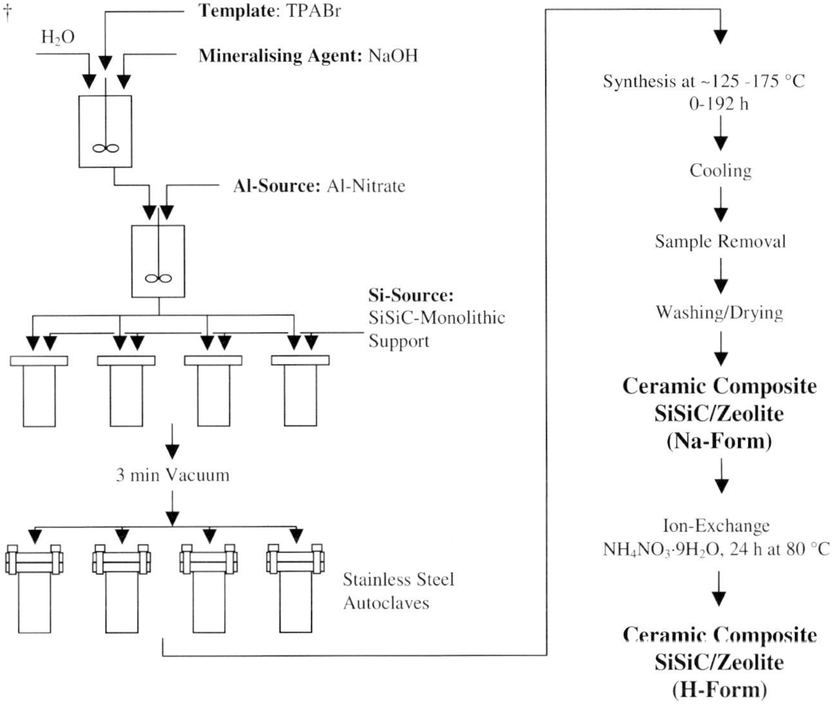

Fig. 14.15 Detailed explanation of the Support Self-Transformation technique for the growth of zeolite onto cellular biomorphous SiSiC monoliths.

process are illustrated in Figure 14.14. The process begins from the biomorphous SiSiC monolith, using it as a substrate for the synthesis of zeolite films directly onto the surface of the cellular support via a hydrothermal synthesis, and results with the catalytically active H-form of the ceramic SiSiC/zeolite composite. In order to achieve the catalytically active form (H-form) of the zeolite coating, a post-synthesis ion-exchange process with ammonium nitrate as the exchange agent must be used. Some of the optimized preparation conditions for the different preparation step are illustrated in the scheme. Figure 14.16 shows the specific monolithic biomorphous structures at the different preparation stages, namely the Rattan biotemplate (a), the carbon-preform (b), and the SiSiC-ceramic after liquid infiltration (c).

The crystallization of Silicalite-1 and ZSM-5 type zeolite (MFI type), which takes place on the walls of channels of the biomorphous SiSiC ceramics, was performed in a one-step hydrothermal synthesis, consisting basically of a partial dissolution–crystallization process. Thus, the SiSiC support was used not only as a carrier for the zeolite layer but also as the unique source of silicon in the batch for the zeolite crystallization process. No external Si-source was used in the pre-

Fig. 14.16 Optical image of the Rattan biotemplate, C-preform after pyrolysis, and SiSiC ceramic after liquid silicon infiltration (LSI).

cursor solution; thus, no aging was required and a clear synthesis mixture was achieved. The hydrothermal crystallization was performed in Teflon-lined autoclaves. The excess of metallic silicon which infiltrated into the support's matrix during the preparation process of the SiSiC ceramic was partially dissolved during hydrothermal crystallization (due to the alkaline environment of the reaction mixture) and re-organized in the presence of template molecules into a tetrahedral Si coordination in the crystalline zeolitic framework [50, 66, 73]. For the Silicalite-1 crystallization, no additional framework constituent was necessary. However, the ZSM-5 crystallization required a second zeolite framework builder (Al) which had to be supplied from the precursor solution during the preparation step of the reaction mixtures. Figure 14.17 shows a series of X-ray diffraction

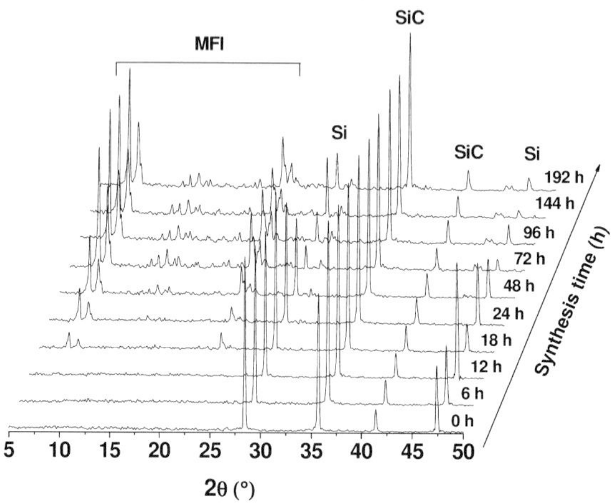

Fig. 14.17 XRD patterns. Kinetics of ZSM-5 crystal growth on the Rattan-derived SiSiC support (composition of the synthesis mixture 1 SiO_2: 0.005 Al_2O_3: 0.04 TPA_2O: 0.15 Na_2O: 75 H_2O) at 175 °C, synthesis temperature 175 °C.

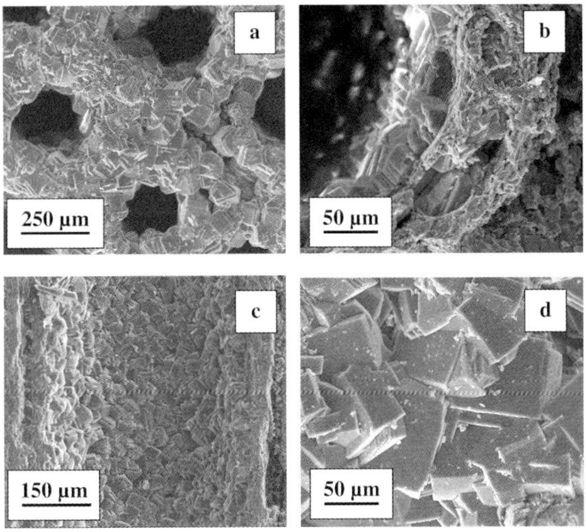

Fig. 14.18 SEM images of the SiSiC/ZSM-5 monolith (a,b) from above and (c,d) vertical cross-section.

patterns representing the crystallization kinetics of a SiSiC/MFI-type-composite starting from the Rattan-derived SiSiC-ceramic support. After only 48 h of synthesis, the XRD patterns were typical of MFI structure, with two peaks at $2\theta = 7.9$ and 8.9°, and a triplet in the range $2\theta = 23$–24° (Fig. 14.17). The XRD peaks identifying the SiSiC support belonged to Si and SiC. The peaks corresponding to the MFI increased with the crystallization time, and the peaks to the Si in the support decreased due its dissolution and consumption for the formation of zeolite, whereas the SiC typical peaks remained near-constant over the entire crystallization period, indicating that the SiC was stable under the applied synthesis conditions.

SEM images of the Rattan-derived SiSiC/ZSM-5 composite sample are shown in Figure 14.18. The ZSM-5-coated sample exhibited a dense and well-intergrown zeolite layer (Fig. 14.18a,d), covering completely the internal surface of the cylindrical channels (Fig. 14.18c). After synthesis, the pores of the composite were still open (Fig. 14.18a), and no pore clogging with zeolites occurred during the described synthesis. The result of the partial dissolution of the Si from the ceramic support and zeolite crystallization process, involving the surface of the ceramic substrate, is therefore the formation of an interpenetrating zeolite/SiC matrix, as can be seen in Figure 14.19.

This observation was also confirmed by the Si, C and O mapping (figure not shown) on a cross-section of SiSiC/MFI (both Silicalite-1 and ZSM-5 coatings) composites obtained by EDX analyses [66, 79]. Figure 14.19 shows the SEM images in backscattering mode of the Rattan-derived cellular SiSiC support (a) and a SiSiC/ZSM-5 composite (b). Zeolite crystals were not merely deposited on the surface but rather grew in the cavities left by the dissolution of Si from the SiSiC

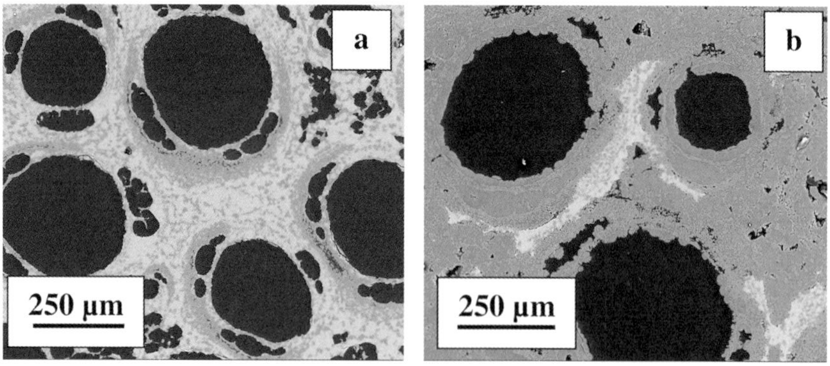

Fig. 14.19 Back-scattered electron-mode SEM images of the (a) Rattan-derived cellular SiSiC support and the (b) SiSiC/ZSM-5 composite after 96 h synthesis at 175 °C.

matrix. Such an interconnection between the zeolite and the SiC and hybrid architecture/composition of the coating might yield two interesting aspects with regard to technical applications: a more effective heat transfer in case of strongly exothermic catalytic reactions, and a faster and more efficient heating of the

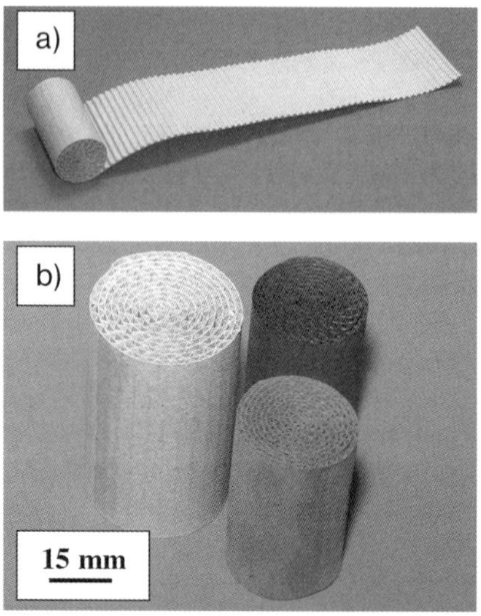

Fig. 14.20 Optical images of: (a) cellulose-derived corrugated cardboard and the forming process to obtain cellulose-derived monoliths; and (b) the three stages of the LSI process for the development of cellular SiSiC ceramic monoliths (cardboard monolith pyrolyzed C-preform and SiSiC ceramic).

zeolite coating for regeneration cycles and adsorption/desorption processes. An EDX mapping (figure not shown) of a zeolite layer in a SiSiC/ZSM-5 composite showed that the Al within the ZSM-5 zeolite coating is mainly concentrated in the surface of the zeolitic layer, while the intermediate layer has a higher Si/Al ratio [79]. This resulted in a gradient of Al distribution over the cross-section of the coated zeolitic layer. Considering the function of the Al in the framework of the zeolite as a "carrier of the active (acidic)" sites, such a gradient might have a direct influence on the catalyst's properties.

A similar process to that applied for the development of Rattan-derived SiSiC/zeolite ceramic composites was used in combination with cellulose-derived corrugated cardboard preforms. This bio-derived template can be easily shaped, obtaining different monolithic structures with different sizes of the channels. From a technical viewpoint, this represents a great advantage compared to wood-derived biomorphous ceramics [80]. Commercially available, one-sided corrugated cardboard sheets consisting of recycled, secondary cellulose fiber papers with channel diameters ranging from 1 to 8 mm, were rolled to produce cylindrical monoliths of the desired size and porosity (Fig. 14.20).

The XRD investigations confirmed the expected phase transformations for the LSI biotemplating process and the zeolite crystallization (Fig. 14.21).

During pyrolysis of the cardboard monoliths, a weight loss of 75% and a linear anisotropic shrinkage of about 20% in the axial and 30% in the radial directions were observed. The micro- and macrostructures, however, were preserved in the carbon preform, despite the pronounced anisotropic shrinkage occurring during pyrolysis (compare Fig. 14.22a,b with Fig 14.22c,d). During the LSI of the carbon preforms, the liquid Si spontaneously infiltrated into the pyrolyzed cardboard

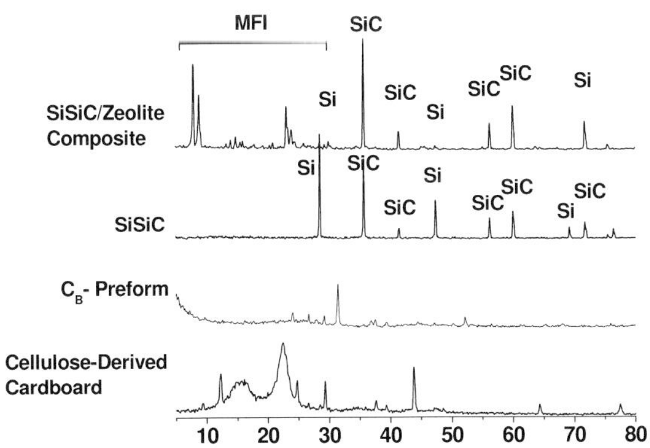

Fig. 14.21 XRD patterns of the three stages of the LSI process for the development of cellular SiSiC ceramic monoliths (cardboard monolith, pyrolyzed C-preform and SiSiC ceramic replica), and of the SiSiC/zeolite ceramic composite.

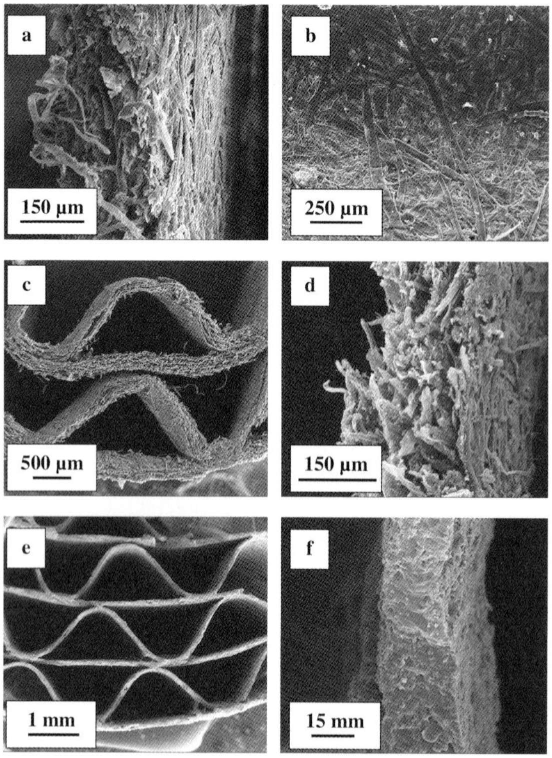

Fig. 14.22 SEM images of the three stages of the LSI process for the development of cellular SiSiC ceramic monoliths. (a,b) Cardboard monolith; (c,d) pyrolyzed C-preform; (e,f) SiSiC ceramic replica at two different magnifications.

char at temperatures above 1420 °C. The Si-melt reacted with carbon to form β-SiC, without changing the macroscopic size and shape of the carbon preform. In the pores and on the surface of the struts of the β-SiC, non-reacted and solidified Si was also present after infiltration, yielding a SiSiC-ceramic composite with ~45 wt% Si and ~55 wt% SiC. Figure 14.22e,f show clearly the infiltration of the C-preform with the Si melt and the formation of a dense SiSiC ceramic strut (compare with the C-preform stage, shown in Fig. 14.22c,d).

SEM images of the cardboard-derived SiSiC/ZSM-5 composite are shown in Figure 14.23a,b. At the start of the crystallization process the surface of the support was coated with zeolite nanocrystals (not shown), and the growth of these crystals with increasing synthesis time led to a uniformly thick zeolite coating. The crystals were randomly oriented, different in size, and highly intergrown. The surface of the pore channel in the SiSiC support was completely coated with well-developed continuous layers of MFI-type zeolite (Fig. 14.23a

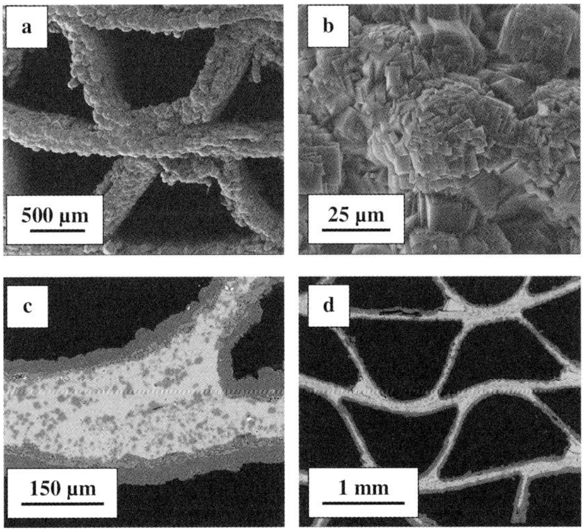

Fig. 14.23 SEM image of the ZSM-5-coated cardboard-derived SiSiC composite in secondary electron mode (a,b) and cross-sectional view in back-scattered mode (c,d).

and b). From polished cross-sectional SEM images in back-scattered electron mode (Fig. 14.23c and d), a thickness of the ZSM-5 layer of about 20 to 50 μm was determined in a sample after a synthesis time of 96 h. The Si present on the surface of the struts of the SiSiC support was dissolved during the hydrothermal zeolite synthesis, while unreacted Si was still present in the core of the struts (Fig. 14.23c). The voids left by the dissolved Si were filled with zeolite during the hydrothermal synthesis, and an interpenetrating SiC/zeolite matrix was again observed, yielding an excellent bonding/contact of the zeolite layer to the SiSiC support. The SiSiC/Silicalite-1 composite was quenched twice in cold water from a temperature of 600 °C, but still no weight loss was observed, which indicated the excellent bonding of the zeolite to the SiSiC support [80]. Zeolite loadings up to 40% of the final composite weight were obtained, which implied a loading of approximately 192 kg_{ZEO} m^{-3} for cardboard-derived ceramic monoliths with 220 cpsi.

In conclusion, a complete route was described which included templating LSI for the preparation of cellular ceramic monoliths and the support of self-transformation for functionalization of the substrate surface with MFI-type zeolite, and the fabrication of structured SiSiC-zeolite composites possessing hierarchical porosity, molecular-sieving capabilities and exhibiting high thermomechanical stability. In future, such composites are expected to find applications in areas such as catalysis (monolithic reactors, catalytic filters and converters, microreactors) and adsorption/separation.

14.4
Conclusion

In conclusion, two approaches have been presented for the preparation of biomorphous cellular ceramics and biomorphous inorganic zeolite-based materials with hierarchical porosity, which exhibit an enormous potential in the fields of catalysis and separation. Some examples of inorganic zeolite-based bio-inspired materials (self-supporting or ceramic composite replicas) have been presented and discussed. Biomorphous, hierarchically porous, self-supporting zeolite replicas of cellulose paper and *Luffa cylindrica* have been successfully produced via a two-step *in-situ* crystallization generally named as sacrificial-template synthesis (see method B-2 in Fig. 14.6). However, by following a different approach (method A in Fig. 14.6), cellular biological supports such as Rattan and corrugated cardboard have been subjected to a three-step biotemplating process in order to produce biomorphous SiSiC/zeolite composites. The process consisted of pyrolysis and LSI of the biological preforms for the development of biomorphous SiSiC cellular ceramic supports and Support Self-Transformation for the growth of zeolite crystals in the inner cavities of the above-mentioned porous substrates.

Further studies are performed at present in our laboratories, in order to study the applicability of such materials in technical catalytic processes and to gain a better understanding of the relationship between properties and structure. More specifically, research is increasingly focused on the influence of different processing parameters on the composition of the zeolite coatings, and on the optimization of the support geometry and properties. Currently, we are studying the performances of biomorphous SiSiC/ZSM-5 composites (Rattan- and cardboard-derived) in environmental catalysis processes (e.g., the selective catalytic reduction of nitrous oxide).

In general, the large number of structures and open-porous networks displayed by biological materials, and the huge numbers of inorganic replicas and zeolite types, offers scientists great flexibility and variety for tailoring these material properties for specific or targeted applications.

Acknowledgments

Financial support from DFG, AiF and FCI is gratefully acknowledged. The authors are also gratefully indebted to T. Selvam, S. Gopalakrishnan, A. Avhale and T. Fey for their help in preparing the chapter.

References

1 A. Dong, Y. Wang, Y. Tang, N. Ren, Y. Zhang, Y. Yue, Z. Gao, *Adv. Mater.* **2002**, *14*, 926–929.
2 R. Daw, *Nature* **2002**, *418*, 491–493.
3 S. Mann, *J. Chem. Soc. Dalton Trans.* **1997**, 3953–3961.
4 S. Mann, *Angew. Chem. Int. Ed.* **2000**, *39*, 3392–3406.

5 S.-W. Lee, S.K. Lee, A.M. Belcher, *Adv. Mater.* **2003**, *15*, 689–692.

6 S.A. Davis, S.L. Burkett, N.H. Mendelson, S. Mann, *Nature* **1997**, *385*, 420–423.

7 M.W. Anderson, S.M. Holmes, N. Hanif, C. Cundy, *Angew. Chem. Int. Ed.* **2000**, *39*, 2707–2710.

8 A. Stein, *Adv. Mater.* **2003**, *15*, 763–775.

9 S. Oliver, A. Kuperman, N. Coombs, A. Lough, G. Ozin, *Nature* **1995**, *378*, 47–50.

10 S. Mann, G.A. Ozin, *Nature* **1996**, *382*, 313–318.

11 S. Mann, *Nature* **1993**, *365*, 499–505.

12 D. Walsch, L. Arcelli, T. Ikoma, J. Tanaka, S. Mann, *Nat. Mater.* **2003**, *2*, 386–390.

13 D. Yang, L. Qi, J. Ma, *Adv. Mater.* **2002**, *14*, 1543–1546.

14 F.C. Meldrum, R. Seshadri, *Chem. Commun.* **2000**, 29–30.

15 L. Huang, H. Wang, C.Y. Hayashi, B. Tian, D. Zhao, Y. Yan, *J. Mater. Chem.* **2003**, *13*, 666–668.

16 G. Cook, P.L. Timms, C. Göltner-Spickermann, *Angew. Chem. Int. Ed.* **2003**, *42*, 557–559.

17 H. Sieber, C. Hoffmann, A. Kaindl, P. Greil, *Adv. Eng. Mater.* **2000**, *2*, 105–109.

18 Y. Shin, J. Liu, J.H. Chang, Z. Nie, G.J. Exarhos, *Adv. Mater.* **2001**, *13*, 728–732.

19 V. Valtchev, M. Smaihi, A.C. Faust, L. Vidal, *Angew. Chem. Int. Ed.* **2003**, *42*, 2782–2785.

20 G. Chin, *Science* **2003**, *301*, 19–21.

21 V. Valtchev, M. Smaihi, A.C. Faust, L. Vidal, *Chem. Mater.* **2004**, *16*, 1350–1355.

22 Y. Shin, J. Liu, J.H. Chang, Z. Nie, G. Exarhos, *Adv. Mater.* **2001**, *13*, 728–732.

23 W. Zhang, D. Zhang, T. Fan, J. Ding, Q. Guo, H. Ogawa, *Microporous Mesoporous Mater.* **2006**, *92*, 227–233.

24 C.R. Rambo, H. Sieber, *Adv. Mater.* **2005**, *17*, 1088–1091.

25 J. Huang, T. Kunitake, *Am. Chem. Soc.* **2003**, *125*, 11834–11835.

26 N. Popovska, D.A. Streitwieser, C. Xu, H. Gerhard, *J. Eur. Ceram. Soc.* **2005**, *25*, 829–836.

27 A. Zampieri, G.T.P. Mabande, T. Selvam, W. Schwieger, A. Rudolph, R. Hermann, H. Sieber, P. Greil, *Mater. Sci. Eng. C* **2006**, *26*, 130–135.

28 P. Greil, *J. Eur. Ceram. Soc.* **2001**, *21*, 105–18.

29 C. Zollfrank, H. Sieber, *J. Am. Ceram. Soc.* **2005**, *88*, 51–58.

30 P. Greil, E. Vogli, T. Fey, A. Betzold, N. Popovska, H. Gerhard, H. Sieber, *J. Eur. Ceram. Soc.* **2002**, *22*, 2697–2707.

31 W.A. Glasser, *Mater. Res. Soc. Bull.* **1994**, *19*, 46.

32 D. Klemm, *Comprehensive Cellulose Chemistry 2. Functionalization of Cellulose.* Wiley-VCH, Weinheim, **1999**.

33 M. Vallet-Regi, *J. Chem. Soc. Dalton. Trans.* **2001**, *2*, 67.

34 A. Hüttermann, C. Mai, A. Kharazipur, *Appl. Microbiol. Biotech.* **2001**, *55*, 387.

35 A.C. Neville, *Biology of fibrous composites.* Cambridge University Press, **1993**.

36 L.J. Gibson, *Metals Mater.* **1992**, *8*, 333–338.

37 E. Vogli, J. Mukerji, C. Hoffmann, R. Kladny, H. Sieber, P. Greil, *J. Am. Ceram. Soc.* **2001**, *8*, 1236–1240.

38 J. Cao, C.R. Rambo, H. Sieber, *Ceram. Int.* **2004**, *30*, 1967–1970.

39 C. Zollfrank, N. Travitzky, H. Sieber, T. Selchert, P. Greil, *Adv. Eng. Mater.* **2005**, *7*, 743–746.

40 D. Walsh, A. Kulak, K. Aoki, T. Ikoma, J. Tanaka, S. Mann, *Angew. Chem.* **2004**, *116*, 6859–6863.

41 C.E. Byrne, D.C. Nagle, *Carbon* **1997**, *35*, 259–266.

42 P. Greil, T. Lifka, A. Kaindl, *J. Eur. Ceram. Soc.* **1998**, *18*, 1961–1973.

43 X. Zheng, R.A. Rapp, *Mater. Sci. Eng.* **1998**, *A255*, 75–80.

44 T. Ota, M. Takahashi, T. Hibi, M. Ozawa, S. Suzuki, Y. Hikichi, H. Suzuki, *J. Am. Ceram. Soc.* **1995**, *78*, 3409–3411.

45 E. Vogli, H. Sieber, P. Greil, *J. Eur. Ceram. Soc.* **2002**, *22*, 2663–2668.

46 H. Sieber, E. Vogli, F. Müller, P. Greil, N. Popovska, H. Gerhard, *Key Eng. Mater.* **2003**, *206–213*, 2013–2016.

47 C.R. Rambo, J. Cao, O. Rusina, H. Sieber, *Carbon* **2005**, *43*, 1174–1183.
48 O. Chakrabarti, L. Weisensel, H. Sieber, *J. Am. Ceram. Soc.* **2005**, *88*, 1792–1798.
49 C.R. Rambo, J. Cao, H. Sieber, *Mater. Chem. Phys.* **2004**, *87*, 345–362.
50 C. Zollfrank, R. Kladny, H. Sieber, P. Greil, *J. Eur. Ceram. Soc.* **2004**, *24*, 479–487.
51 T. Fey, H. Sieber, P. Greil, *J. Eur. Ceram. Soc.* **2005**, *25*, 1015–1024.
52 R. Szostak, *Handbook of Molecular Sieves*. Van Nostrandt Reinhold, New York, **1992**.
53 M.E. Davis, *Nature* **2002**, *417*, 813–821.
54 M.E. Davis, *Science* **2003**, *300*, 438–439.
55 J. Weitkamp, *Solid State Ionics* **2000**, *131*, 175–188.
56 J.C. Jansen, J.H. Koegler, H. van Bekkum, H.P.A. Calis, C.M. van den Bleek, F. Kapteijn, J.A. Moulijn, E.R. Geus, N. van der Puil, *Microporous Mesoporous Mater.* **1998**, *21*, 213–226.
57 R.J. Farrauto, R.M. Heck, *Catal. Today* **1999**, *51*, 351–360.
58 H. Katsuki, S. Furuta, *J. Am. Ceram. Soc.* **2000**, *83*, 1093–1097.
59 A. Cybulski, J.A. Moulijn, in: A. Cybulski, J.A. Moulijn (Eds.), *Structured Catalysts and Reactors*. Marcel Dekker, New York, **1998**, p. 71.
60 G.B.F. Seijger, O.L. Oudshoorn, A. Boekhorst, H. van Bekkum, C.M. van den Bleek, H.P.A. Calis, *Chem. Eng. Sci.* **2001**, *56*, 849–857.
61 E.V. Rebrov, G.B.F. Seijger, H.P.A. Calis, M.H.J.M. de Croon, C.M. van den Bleek, J.C. Schouten, *Appl. Catal. A* **2001**, *206*, 125.
62 Y.S.S. Wan, J.L.H. Chau, A. Gavriilidis, K.L. Yeung, *Microporous Mesoporous Mater.* **2001**, *42*, 157–175.
63 Z. Lai, G. Bonilla, I. Diaz, J.G. Nery, K. Sujaoti, M.A. Amat, E. Kokkoli, O. Terasaki, R.W. Thompson, M. Tsapatsis, D.G. Vlachos, *Science* **2003**, *300*, 456–460.
64 F. Scheffler, R. Herrmann, W. Schwieger, M. Scheffler, *Microporous Mesoporous Mater.* **2004**, *67*, 53–59.
65 F. Scheffler, W. Schwieger, D. Freude, H. Liu, W. Heyer, F. Janowski, *Microporous Mesoporous Mater.* **2002**, *55*, 181–191.
66 A. Zampieri, H. Sieber, T. Selvam, G.T.P. Mabande, W. Schwieger, F. Scheffler, M. Scheffler, P. Greil, *Adv. Mater.* **2005**, *17*, 344–349.
67 H. Sieber, C. Hoffmann, A. Kaindl, P. Greil, *Adv. Funct. Mater.* **2000**, *2*, 105–109.
68 R. Lakes, *Nature* **1993**, *361*, 511–515.
69 T. Sano, Y. Kiyozumi, K. Maeda, M. Toba, S. Niwa, F. Mizukami, in: R. von Balmoos, et al. (Eds.), *Proceedings of the 9th International Zeolite Conference*, Montreal, 1992. Butterworth-Heinemann, **1993**, pp. 239–246.
70 S. Mann, G.A. Ozin, *Nature* **1996**, *382*, 313–318.
71 (a) T. Ota, M. Imaeda, H. Takase, M. Kobayashi, N. Kinoshita, T. Hirashita, H. Miyazaki, Y. Hikichi, *J. Am. Ceram. Soc.* **2000**, *83*, 1521–1523; (b) B. Sun, T. Fan, D. Zhang, *Mater. Lett.* **2004**, *58*, 798–801.
72 B. Sun, T. Fan, D. Zhang, *Mater. Lett.* **2004**, *58*, 798–801.
73 Y. Shin, J. Liu, J.H. Chang, Z. Nie, G.J. Exarhos, *Adv. Mater.* **2001**, *13*, 728–732.
74 C. Zollfrank, H. Sieber, *J. Eur. Ceram. Soc.* **2004**, *24*, 495–506.
75 A. Hofenauer, O. Treusch, F. Tröger, G. Wegener, J. Fromm, M. Gahr, J. Schmidt, W. Krenkel, *Adv. Eng. Mater.* **2003**, *5*, 794–799.
76 M. Singh, B.-M. Yee, *J. Eur. Ceram. Soc.* **2003**, *24*, 209–217.
77 M. Singh, J. Martínez Fernández, A.R. de Arellano-López, *J. Eur. Ceram. Soc.* **2003**, *7*, 247–254.
78 B. Philip, *Nature* **2000**, *409*, 413–416.
79 A. Zampieri, S. Kullmann, T. Selvam, J. Bauer, W. Schwieger, H. Sieber, T. Fey, P. Greil, *Microporous Mesoporous Mater.* **2005**, *90*, 162–174.
80 A. Zampieri, H. Sieber, W. Schwieger, G.T.P. Mabande, T. Selvam, F. Scheffler, P. Greil, *Stud. Surf. Sci. Catal.* **2005**, *158*, 145–152.
81 F. Scheffler, A. Zampieri, W. Schwieger, J. Zeschky, M. Scheffler, P. Greil, *Adv. Appl. Ceram.* **2005**, *104*, 43–48.

15
"Bio-Casting": Biomineralized Skeletons as Templates for Macroporous Structures

Fiona Meldrum

Abstract

A wide range of amorphous, polycrystalline and single crystal macroporous solids with highly regular, sponge-like structures were synthesized using templating routes. Biology provides access to many structures with unique morphologies that would be impossible to synthesize *de novo*, and the investigations described herein profit from an existing macroporous biomineral – sea urchin skeletal plates – to synthesize macroporous solids with bicontinuous structures and highly regular pores of 10–15 μm in size. The application of a range of synthetic techniques varying from sol–gel chemistry to electroless deposition enabled the synthesis of amorphous and polycrystalline solids, the structures of which could be controlled from a double-sided solid, to skeletal structure, and finally to a perfect replica of the sea urchin plate, according to the nature and quantity of the material deposited. Polymer replicas of the sea urchin plates were further employed as environments in which to precipitate single crystals. Whilst biology is expert at producing single crystals with complex morphologies, this is typically achieved by interplay of a number of pathways. Single crystals with identical structures to the original sea urchin plates were successfully produced using these templates, indicating that shape-constraint alone is sufficient to define crystal morphologies. The technique is extremely general, and can be applied to a wide range of crystals.

Key words: calcium carbonate, calcite, templating, macroporous, biomineralization, biomimetic, crystal, strontium sulfate.

15.1
Introduction

Biominerals provide a unique inspiration for materials design. Operating under ambient conditions, Nature has developed synthetic routes and design strategies

Handbook of Biomineralization. Edited by P. Behrens and E. Bäuerlein

ISBN: 978-3-527-31805-6

to produce minerals with structures that are optimized for their particular function [1–4]. Bone, for example, displays remarkable mechanical properties which derive from its inorganic/organic composite character, its complex hierarchical architecture, and the nanoscale arrangement and interactions of the constituents [4–6]. Mollusk shell nacre, despite having a very low organic content ($\approx 1\%$), is superior to most other composite ceramics in stiffness, strength and toughness, and is approximately 3000-fold more resistant to fracture than a single crystal of pure aragonite [7, 8]. These properties can be attributed to the structural organization of nacre, which comprises layers of interlocking aragonite platelets separated by a thin layer of organic material. Magnetotactic bacteria have evolved to profit from the specific properties of a mineral, and precipitate single domain-sized crystals of the magnetic iron oxide magnetite (Fe_3O_4). The crystals are aligned in chains along the long axis of a bacterium, endowing the organism with a permanent magnetic dipole that enables it to navigate in the Earth's magnetic field [9, 10].

Perhaps the characteristic feature of biominerals which most immediately strikes the casual viewer, however, is the astonishing range of morphologies displayed [11]. Many of the biominerals with the most unusual morphologies, such as the silicaceous skeletons of diatoms and radiolaria, are constructed from amorphous minerals [12]. With no preferred morphology, an amorphous material would appear to be the first choice of construction material to build a structure with complex form. Polycrystalline biominerals are again widespread, and exhibit a huge range of morphologies [1, 2], which is again easy to rationalize. Perhaps the most fascinating category of biominerals however, is the single crystals with complex morphologies. A perfect example is provided by the skeletal plates of echinoderms, which exhibit bicontinuous sponge-like structures and curved surfaces (Fig. 15.1). Despite this morphological complexity, each skeletal plate is a single crystal of calcite, the synthetic equivalent of which is a regular rhombohedron [13–15].

One exciting goal of new materials synthesis is to develop routes for the production of ceramics with complex structures, synthetically. Briefly considering biology, a range of strategies is employed to control crystal morphologies, with relatively simple changes in crystal shape being achieved through the interaction of additives with growing crystals. The precipitation of $CaCO_3$ in the presence of macromolecules extracted from within $CaCO_3$ biominerals has demonstrated the interaction of these macromolecules with specific sets of crystal planes, resulting in minor morphological changes [16, 17]. Single crystals with complex forms invariably grow within structured vesicles, which effectively mold the gross morphology of the developing mineral [15, 18]. Yet, it is intriguing that a soft mold can direct the growth of a hard mineral to eliminate stable planar faces and generate curved surfaces. Indeed, some calcite and aragonite biominerals have been shown to form via an amorphous precursor phase [19–22], and it has been suggested that amorphous calcium carbonate (ACC) may provide a route to the morphological control of single crystals [23–25]. Templating approaches are also applied to the formation of polycrystalline and amorphous biominerals with com-

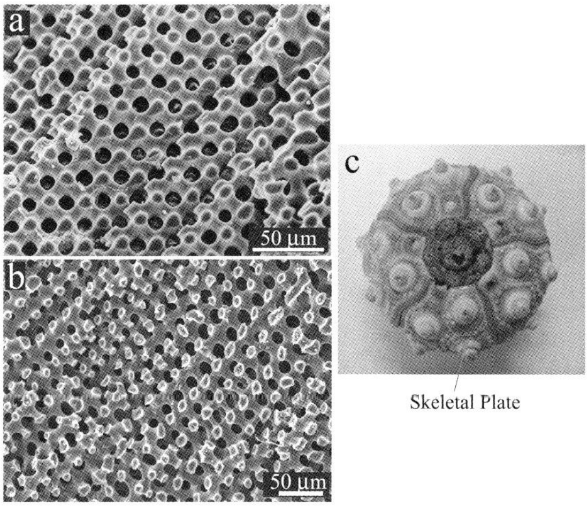

Fig. 15.1 (a,b) Cross-sections through a sea urchin skeletal plate, showing the bicontinuous structure and pores of diameter 10–15 µm. (c) Polymer replica of sea urchin plate.

plex forms, and recent experiments have shown that pattern formation in diatoms results from phase separation and self-assembly processes [26, 27].

In common with biology, synthetic approaches to the formation of inorganic structures with complex morphologies have relied upon templating routes, typically employing organic and (less frequently) inorganic templates. Organic templates including surfactant phases [28–30], polyelectrolyte and polymer capsules [31–34], membrane pores [23, 24], block-polymers [35, 36] and colloidal crystals constructed from polymer spheres [37–39] have been widely exploited as molds. A number of templates taken directly from biology, including pollen grains [40], butterfly wings [41], an organic matrix extracted from a cuttlefish [42], and wood [43], have provided the basis for the formation of more unusual morphologies. Although inorganic templates have received considerably less attention, they clearly offer the potential for significantly greater chemical and thermal stability as compared to their organic counterparts. Porous alumina membranes have been used to prepare nanorods of materials including bismuth and titania [44, 45], and colloidal crystals of silica spheres have been widely used to template porous inorganic solids [46–48]. As a novel approach to preparing ceramics with complex structures, Sandhage has used the direct reaction of silicaceous diatom frustrules [49–51]. A range of porous ceramics, including TiO_2, ZrO_2 and $BaTiO_3$ with morphologies approaching those of the original diatom, were prepared using this method. Sea urchin skeletal plates have also been used as the basis for making photonic solids [52]. A high dielectric contrast three-dimensional (3-D) photonic crystal exhibiting a stop band in the mid-infra-red (IR) range was fabricated by filling the sea urchin plate with poly(dimethyl siloxane) (PDMS), removing the calcium carbonate, heating to shrink the structure by 50%, and finally backfilling with tellurium and removing the residual polymer. Templating therefore provides a powerful route to the formation of inorganic structures with complex morphologies, and is restricted only by the available templates themselves.

This chapter will focus specifically on the application of templating routes to the synthesis of macroporous solids. Macroporous solids are valuable for a range of applications, including catalysis and catalyst supports, chemical filtration and separations, optoelectronics, and cell immobilization. Moreover, there is considerable interest in producing these porous solids with well-defined pore sizes and geometries. While the syntheses of micro- and meso-porous materials are well established, techniques available for synthesizing macroporous solids remain limited. Templating methods have been applied and rely upon the existence of suitable templates. Highly ordered macroporous solids with fully interconnected pores have been prepared by templating colloidal crystals of either polymer or silica spheres [37–39]. Indeed, this has become a well-established route to the preparation of direct and inverse opal structures in simple and ternary oxides, chalcogenides, non-metallic and metallic elements, and organo-silicates. A range of alternative templating approaches have also been investigated, including the templating of hydrogels [53], emulsions [54, 55], composites of block copolymers and surfactants [56], and a viscous polysaccharide matrix [57], although most lead to non-ordered solids. An attractive template-free route to non-ordered macroporous solids has also been developed, based upon solid-state reactions which lead to intimate mixtures of two insoluble phases with distinct chemical reactivities [58–61]. Subsequent dissolution of one of these phases then generates a macroporous monolith of the remaining solid.

Among the studies described in this chapter, we profit from an existing biomineral – namely sea urchin skeletal plates – which are used to prepare amorphous, polycrystalline and single crystal macroporous solids [62–67]. The sea urchin plates offer uniquely ordered, bicontinuous structures with pores of diameter ranging from 10 to 15 μm, which cannot as yet be replicated synthetically *de novo*. As the calcium carbonate and porous fractions of the plates possess identical sizes and shapes, filling of the pores with the target material and subsequent dissolution of the calcium carbonate generates a porous solid with an identical structure to the original template. The generality of this approach has been demonstrated on application of a range of methodologies including electroless deposition and sol–gel chemistry to the synthesis of many diverse systems such as macroporous gold [66, 67], nickel, titania, and silica [65].

The preparation of a polymer replica of the urchin skeletal plate, and the use of this template as an environment in which to precipitate a range of single crystals, has also provided a unique opportunity for investigating the factors involved in controlling single crystal morphologies, and provides insight into the morphological control of crystals in biological systems [62–65, 68]. The growth of a range of minerals, including $CaCO_3$, $SrSO_4$ and $CuSO_4 \cdot 5H_2O$, within the prepared polymer replicas has demonstrated that either large single crystals or polycrystalline particles with complex, porous structures identical to the original polymer membrane can be generated, according to the solution concentrations applied. The surface chemistry of the polymer template also appears to pay a pivotal role in defining the structure of the product particles [68]. These studies provide a striking example of how a rigid mold can dictate the form of a growing crystal, demonstrating that single crystals with intricate morphologies can be produced synthetically, in the absence of complex biological pathways.

15.2 Amorphous and Polycrystalline Macroporous Solids

A wide range of amorphous and polycrystalline macroporous solids were prepared using the sea urchin skeletal plates as structural molds, including gold [66, 67], nickel, silica, and titania [65]. That sea urchin plates have structures where the pore and mineral networks occupy equal volumes and identical geometries allows great flexibility in the product structures that can be synthesized. The thickness of the deposited material can be readily controlled such that a surface coating generates a double-sided product in which the surface separates two non-interconnecting porous networks, while partial filling of the pores can lead to a skeletal structure where the geometry of the porous network is identical to the original urchin plate, but the pores can be significantly larger than 10 to 15 μm in size. The limiting case – in which the pore volume is entirely filled – yields a perfect replica of the urchin plate in the selected material. This represents a 50% space filling, as compared with 26% for face-centered cubic structures of close-packed spheres. The pore diameters and size distributions are highly controlled, with negligible defects present.

Two principal templating procedures were applied to the synthesis of macroporous solids: (i) direct synthesis within the $CaCO_3$ sea urchin skeletal plates themselves; and (ii) templating of a polymer replica of the urchin plate. While $CaCO_3$ is readily dissolved under mild reaction conditions, and can therefore provide an excellent template material for some products, it is for the same reason somewhat restrictive in terms of the reactions that can be carried out within the urchin plates. In contrast, the polymer replica is chemically and mechanically stable and can therefore support chemical reactions at low pH or where calcium chelators are employed, where $CaCO_3$ would be unstable.

15.2.1 Polymer Replicas of Sea Urchin Skeletal Plates

Polymer replicas of the sea urchin skeletal plates were prepared from methyl methacrylate/ethyl acrylate, as this copolymer was sufficiently tough to be cut into thin membranes, and is soluble in chloroform, making it a suitable template material for inorganic solids [62–64]. Infiltration of a clean urchin plate with the monomer solution and subsequent curing led to complete infiltration of the polymer into the urchin plate. Uniform sections of thickness approximately 0.5 mm were then cut through the cured polymer/plate and the $CaCO_3$ was removed from the polymer sections by dissolution in HCl. This procedure yielded sponge-like polymer membranes that were identical in structure to the original sea urchin plates (Fig. 15.1b). These macroporous polymer membranes exhibited interesting structures in their own right, and were valuable as templates for materials produced under reaction conditions where either the $CaCO_3$ urchin plate would have been chemically unstable, where a thin template was preferred, such as when filtration methods were employed and surface blockage of the template could occur, and where the target material was soluble under similar conditions to $CaCO_3$.

15.2.2
Macroporous Gold

The native sea urchin plates were directly employed as templates for the formation of macroporous gold [66, 67]. A simple synthesis was used which exploited the sponge-like structure of the urchin plate. Briefly, one end of a dry urchin plate was immersed in a gold paint comprising organo-stabilized gold particles, and the paint was absorbed into the plate via capillary action, resulting in the formation of a continuous coating of gold over the surface of the plate. The plate was then heated to burn off the protective organic matter in the paint, and the dipping/heating cycle was repeated ten times. The formation of a surface-coverage of gold particles was apparent from a change in the surface structure of the plate from the original smooth, to a highly roughened surface. Finally, the gold-coated plate was annealed at 400 °C, and subsequent gentle dissolution of the $CaCO_3$ yielded the macroporous gold product.

The templated gold solid exhibited a unique macroporous structure, the structure and dimensions of which were precisely defined by the original sea urchin plate (Fig. 15.2a). As the preparation technique deposited a layer of gold onto the surface of the urchin plate, the product material possessed a double-sided struc-

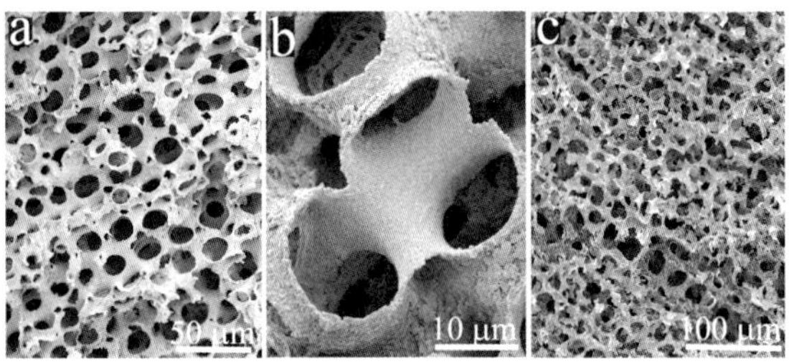

Fig. 15.2 (a) Macroporous gold produced from a sea urchin skeletal plate template. (b) A higher magnification image showing the rough surface originally in contact with the pore and the smooth surface originally in contact with the urchin plate. (c) Macroporous nickel produced via electroless deposition, showing a skeletal structure.

ture, comprising two non-intersecting porous networks. The two sets of networks are readily distinguished, as the gold surface originally in contact with the urchin plate is entirely smooth, while the surface directed into the pore is rough (Fig. 15.2b). An increase in the number of dipping/annealing cycles used did not significantly increase the average thickness of the gold coating due to blockage of the surface pores with gold particles.

15.2.3 Macroporous Nickel

A different experimental method – electroless deposition – was applied to the production of macroporous nickel [65]. In contrast to the two-sided structure achieved for gold, electroless deposition of nickel resulted in partial filling of the pore volume, and the generation of a skeletal macroporous product. Electroless deposition uses non-galvanic reduction of metal cations in solution to coat a surface, and was again applied to the native sea urchin plates [69]. The surface of the urchin plate was initially covered with a thin layer of gold, to enable a high-quality nickel film to be deposited. The plating bath solution was then slowly drawn through the urchin plate under suction, and the plate finally sintered at 500 °C, before dissolving the $CaCO_3$ template. The templated nickel solid produced was again macroporous in structure with pore sizes of ~30 to 40 μm, and was formed as a solid network throughout the pore structure of the template (Fig. 15.2c). This difference in structure as compared with the gold, which maintained surface coverage of the urchin plate throughout the annealing process, may derive from weaker adherence of the nickel than the gold to the calcium carbonate, or from a higher percentage filling of the pore volume with the nickel.

15.2.4 Macroporous Silica

A number of techniques were applied to form macroporous silica replicas of the sea urchin skeletal plates [65]. Termed the "filtration method", the void fraction of an urchin plate was filled with 0.2- to 0.3-μm-diameter pre-prepared silica particles. A dilute sol of the particles was filtered under suction through a skeletal plate which was mounted on a polycarbonate track-etch membrane to trap the silica particles within the plate. The use of a dilute solution of particles was essential to prevent surface blockage. On completion of the filtration process the skeletal plate was annealed at 600 °C, and the calcium carbonate of the skeletal plate removed by dissolution in weak acid. In common with the templated gold, a two-sided macroporous silica structure was produced, deriving from surface coverage of the urchin plate by the silica particles (Fig. 15.3a). Again, the thickness of the silica coating on the plate was not significantly increased on lengthening the filtration time as the surface pores tended to become blocked, stopping the flow of particles through the plate.

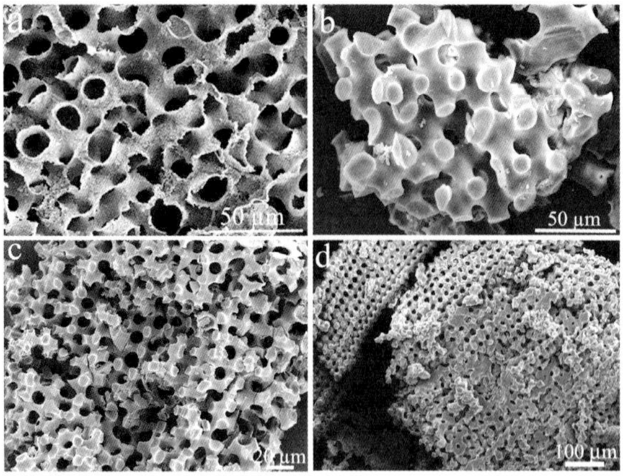

Fig. 15.3 (a) Templated macroporous silica produced by a "filtration method", showing double-sided structure and (b) by a "particle/hydrolysis method", producing a solid replica. (c) Macroporous titania produced by hydrolysis of titanium tetrachloride in the polymer replica and (d) by the "particle/hydrolysis method" in the polymer replica.

Two further methods for producing silica replicas of the urchin skeletal plate structure were also investigated, namely a "hydrolysis method" and a "particle/hydrolysis method". In the first method, either a sea urchin plate or polymer replica was immersed in tetraethoxyothosilicate (TEOS) or silicon tetrachloride and then subjected to controlled hydrolysis. The silica particles precipitated within the urchin plate were isolated by dissolution of the $CaCO_3$ plate in acid, while those precipitated in the polymer membrane were isolated by dissolution of the membrane in chloroform. This technique was in general of limited success, however, and led to poor replication of the template structure for both the polymer membrane and sea urchin plate.

In contrast, the "particle/hydrolysis method" led to excellent templating of both the sea urchin plates and polymer replicas. In this technique, the selected template was initially immersed in a solution of small, pre-prepared silica particles to give partial filling of the pores, and then dried. The hydrolysis method, as described previously, was then applied. The templated particles produced in both the urchin plates and polymer replicas exhibited macroporous, sponge-like structures which were almost identical in morphology to the original template (Fig. 15.3b), indicating that complete filling of the template pore structure had occurred. The average size of the SiO_2 particles produced varied according to the template used, and the 100-µm particles produced in the sea urchin plates were typically significantly larger than the 50 µm or smaller particles produced in the polymer membranes. A powder X-ray diffraction (XRD) analysis of all of the macroporous silicas showed them to be amorphous, as anticipated for room-temperature hydrolysis.

15.2.5
Macroporous Titania

The hydrolysis and particle/hydrolysis methods were also applied to the production of macroporous titania [65]. The basic "hydrolysis method" was investigated with both the polymer membrane and sea urchin plate templates, employing both titanium ethoxide and titanium tetrachloride as reagents, but was found to be unsuited to the $CaCO_3$ template as the low pH values generated during hydrolysis resulted in dissolution of the mineral. This technique was successfully employed with the polymer replicas, however, and yielded large porous particles with both reagents. The structural perfection of the templated solids depended quite heavily on the starting material employed. The use of titanium ethoxide ($Ti(OC_2H_5)_4$) generated porous titania products the gross morphologies of which were clearly based on the template morphology, but which displayed many defects and a rather poor structural match with the original template. In contrast, excellent replication of the template structure was achieved with the titanium tetrachloride, and casts almost identical in morphology to the original sea urchin were produced with sizes often exceeding 200 μm (Fig. 15.3c).

Application of the particle/hydrolysis method to the polymer templates also yielded macroporous titania solids with structures identical to the original sea urchin plates. An analogous experimental method was again applied to that described for synthesis of the silica particles, with the exception that an anatase sol was used to partially infiltrate the polymer template, followed by hydrolysis of either titanium ethoxide or titanium tetrachloride. Very large macroporous monoliths of TiO_2, very close in structure to the original plate and often exhibiting sizes in excess of 1 mm were produced (Fig. 15.3d). An examination of the mechanism by which the particle/hydrolysis method operates showed that the precursor particles initially coated the polymer, partially filling the pores and possibly promoting the subsequent hydrolysis step. The hydrolysis step then led to complete filling of the porous network. The crystalline structures of the synthesized macroporous titania solids were also investigated with powder XRD, to identify the phase of the titania produced. The hydrolysis method, when operating at room temperature, produced entirely amorphous products, while the particle/hydrolysis method showed partial crystallinity due to the presence of the entrapped anatase particles used in the infiltration step. Subsequent calcination of both of these samples at 600 °C resulted in crystallization of the amorphous TiO_2 to give crystalline anatase, as shown by well-defined XRD patterns.

15.3
Macroporous Single Crystals

Whilst it is relatively straightforward to prepare amorphous and polycrystalline solids with complex morphologies via templating methods (as described in Section 15.2), it is intriguing as to whether it is possible to mold the morphologies of single crystals in the same way. Clearly, to grow a single crystal with complex structure demands control over the number of nucleation sites, such that a single crystal rather than a polycrystalline product develops. Furthermore, in order for the product crystal to show high-energy curved surfaces (as opposed to the low-energy planar faces characteristic of crystals grown under equilibrium conditions) suggests that growth must occur under conditions where the crystal impinges on, and is thus limited by, the template. The sponge-like polymer templates described here provide a unique opportunity for investigating control over single crystal morphologies by rigid templates. Morphological control of crystals in biology is clearly a complex process in which the growth environment is uniquely defined, nucleation sites are controlled and a family of soluble organic macromolecules are typically present and interact with the growing crystal [1, 2]. Organisms can also control whether a mineral forms directly within the growth environment from ions or from a precursor material and, importantly, can vary how growth is controlled over time – a factor which would be difficult to effectively mimic synthetically.

The re-growth of calcite crystals within the polymer replica of a sea urchin plate therefore provides an opportunity for investigating the factors involved in controlling single crystal morphologies to produce structures as complex as those created biologically. The sponge-like polymer membrane offers an environment with identical geometry and size to the single crystal calcite urchin plate, and therefore is an ideal template for investigating whether shape constraint alone is sufficient to cast the morphology of single crystals, or whether the interplay of more complex variables such as the presence of organic additives is additionally required. Initial experiments were carried out with calcite, as in the production of sea urchin plates Nature has already demonstrated that it is possible to produce large single crystals of calcite with sponge-like structures. Indeed, it is interesting to note that the majority of large, single crystals with complex morphologies produced in Nature are calcite. The generality of this templating approach to controlling the morphology of single crystals was therefore investigated by growing a range of crystals including strontium sulfate, lead sulfate and copper sulfate within the sponge-like polymer template.

15.3.1 Calcium Carbonate

Calcium carbonate was precipitated within the polymer replicas of sea urchin skeletal plates, produced as described in Section 15.2.1, using a double-diffusion technique [62, 63]. A wetted polymer membrane was placed between two half-U-tube arms, which were then simultaneously filled with solutions of $CaCl_2$ and Na_2CO_3 at identical concentrations. This set-up was maintained for periods varying from 30 min to 4 days, after which time the particles which had formed within the membrane were isolated by dissolving the polymer in chloroform. Evidence of templating of the morphologies of the particles precipitated within the membranes was obtained for all of the solution concentrations used, although

Fig. 15.4 Templated calcite crystals produced after 24 h from: (a) 0.4 M reagents, showing polycrystalline structure; and (b) 0.02 M reagents, showing single crystal structure. (c) Calcium carbonate particle formed as a thin film over the membrane surface from 0.02 M reagents and isolated after 6 h.

the quality of replication of the polymer morphology and the structure of the templated particles was strongly concentration-dependent.

Higher reagent concentrations (>0.1 M) yielded large, porous, polycrystalline calcite networks, the gross structures of which were clearly defined by the polymer template (Fig. 15.4a) [62]. A reduction in the reagent concentrations to values less than ~0.06 M resulted in a marked structural change in the templated particles. In this concentration regime, single crystals of calcite with structures perfectly replicating the morphology of the polymers template were obtained (Fig. 15.4b). These crystals showed both non-crystallographic curved surfaces, which were defined by contact with the polymer mold, as well as planar faces characteristic of crystalline materials, which represented the growth front of the crystal. The single crystal structure of these particles was indicated by the presence of aligned rhombohedral faces present on the perimeter of the particles, and was confirmed using single crystal XRD. Very few particles were produced in the membrane when concentrations were further reduced to <0.01 M. The concentration-dependence of the particle structure is readily rationalized. The suppression of nucleation, as occurs at low solution supersaturations, produces well-separated nuclei which then grow into single crystals, while higher supersaturation values increase the rate of nucleation, resulting in many nuclei and polycrystalline particles.

An investigation of the mechanism of formation of the templated single crystals was carried out by isolating particles precipitated within the membrane with time, and showed that for 0.02 M reagents, templated calcite single crystals and polycrystalline vaterite particles of sizes 20 to 30 μm were produced in approximately equal proportions at early times [63]. Longer incubation times resulted in an increase in the ratio of calcite to vaterite, with the proportions of calcite reaching 60%, 75% and >80% after 2 h, 1 day and 3 days, respectively, and reaching maximal sizes of 150 to 200 μm. The number of particles isolated from the membranes increased up to approximately 8–10 h, and then remained effectively constant after this time. That the proportion of calcite to vaterite particles increased without a change in the total number of particles suggested dissolution of the unstable vaterite phase and reprecipitation as calcite. This is further supported by the observation of particles containing both single crystal calcite and polycrystalline vaterite. Experimental observations also suggested that ACC may have initially coated the polymer membranes at early stages of incubation [63]. In addition to the solid particles, fragile two-sided particles – which formed as a thin film of calcium carbonate over the polymer membrane surface – were also observed (Fig. 15.4c). These particles represented about 50% of all particles at 1–2 h, and examination by XRD at this time showed that they contained polycrystalline calcite. This analysis is also consistent with the observation that ACC was the first phase formed on precipitation of calcium carbonate from bulk solution at room temperature from 0.02 M reagents, and that calcite was the sole transformation product after $\approx$ 45 min.

15.3.2 Strontium Sulfate

Strontium sulfate was also successfully templated to produce large, macroporous single crystals or oligocrystalline (comprising a small number of large particles) particles [64]. These crystals were again produced by a double-diffusion technique, and typically reached sizes of 150 to 200 μm after 1 to 2 days. In common with the $CaCO_3$ system, the formation of templated single crystals of $SrSO_4$ was strongly concentration-dependent, with single crystals being produced at low reagent concentrations, and a transition to polycrystalline structures occurred with increasing concentrations. While solution concentrations between 0.01 M and 0.2 M yielded templated single crystal products (Fig. 15.5a), an increase in the reagent concentration to 0.5 M caused a change to mixed single crystal/polycrystalline structure (Fig. 15.5b) and further increase to 1.0 M resulted in polycrystalline particles (Fig. 15.5c). The crystalline structures of the $SrSO_4$ particles were analyzed using single crystal XRD, and showed that some "single crystal" particles were "oligocrystalline" rather than single crystal in structure. The proportion of the oligocrystalline to single crystal samples investigated was approximately 5:2 for 0.05 M reagents. The isolation of particles produced from 0.05 M reagents over time showed that rapid growth occurred within the first

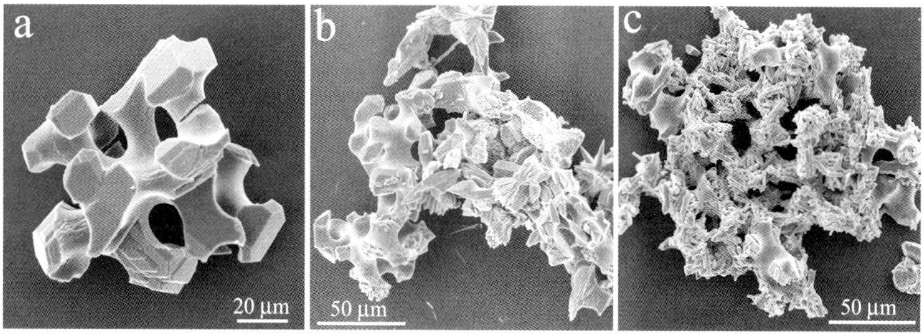

Fig. 15.5 Templated $SrSO_4$ particles isolated from a polymer membrane after 24 h from reagent solutions of concentration: (a) 0.01 M, showing single crystal structure; (b) 0.5 M, showing single crystal and polycrystalline particles; and (c) 1 M, showing polycrystalline structure.

few hours of reaction, but this slowed significantly after 12 h. Templated crystals which were typically under 50 μm in size were isolated from the membrane after 1 h reaction time, while growth to ~80 μm and 100–180 μm occurred after 2 h and 6–12 h, respectively. Sizes of 150 to 200 μm were achieved after 1 to 2 days, with maximum sizes of about 230 μm being reached after 3 days.

15.3.3 Lead Sulfate and Lead Carbonate

Both $PbSO_4$ and $PbCO_3$ could also be templated using the urchin polymer membranes to produce single crystals with sponge-like morphologies and curved surfaces [64]. Templated single crystals were observed for a wide range of concentrations for the lead sulfate, while the conditions required to produce templated single crystals of lead carbonate were more restrictive. Precipitation of $PbSO_4$ from 0.02 M and 0.05 M reagents for 24 h led to templated single crystals with similar sizes and morphologies (Fig. 15.6a), whilst an increase in the reagent concentrations to 0.1 M again resulted in a transition to templated polycrystalline

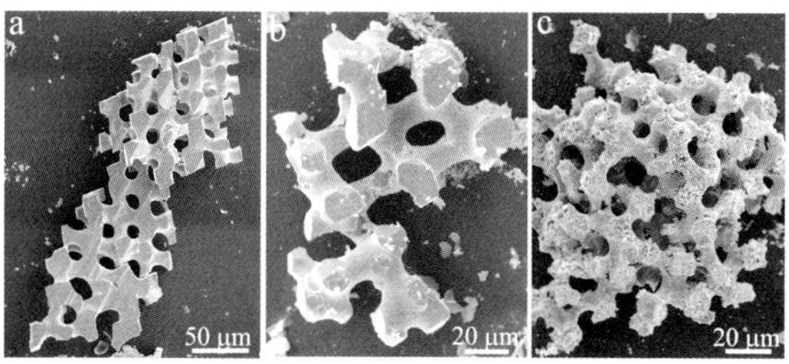

Fig. 15.6 (a) $PbSO_4$ particles isolated from a polymer membrane from 0.05 M reagents after 24 h; (b, c) $PbCO_3$ particles isolated from polymer membrane after 24 h from 0.05 M reagents showing (b) single crystal and (c) polycrystalline structures.

particles. The templating of $PbCO_3$, in contrast, always yielded polycrystalline particles in addition to templated single crystals, and was significantly more limited in the concentration regime under which templated single crystals were produced. Reagent concentrations of 0.05 M yielded both single crystal and polycrystalline particles at a ratio of approximately 2:1, and the polycrystalline particles at >200 μm in diameter were typically significantly larger than the ~100 μm single crystals (Figs. 15.6b and 15.6c). Lower reagent concentrations were anticipated to give a higher proportion of single crystal particles, but few particles were produced at concentrations of 0.02 M and below. Examination of the structures of the "single crystal" $PbSO_4$ and $PbCO_3$ particles precipitated using 0.05 M reagents with single crystal XRD again showed that these particles were oligocrystalline, comprising one or more large crystals. Lower reagent concentrations would be expected to give true single crystals, but the small size of particles produced precluded further investigation by single crystal XRD.

15.3.4 Copper Sulfate and Sodium Chloride

An "evaporation method", in which crystals were precipitated on controlled evaporation of a saturated solution, was applied to precipitate copper sulfate and sodium chloride within the polymer membranes, and templated single crystals of NaCl and $CuSO_4{\cdot}5H_2O$ were successfully produced (Fig. 15.7) [64]. Templating of single crystal morphologies is therefore not restricted either to the double-diffusion technique, or to crystals with low solubilities. In common with the $SrSO_4$, $PbSO_4$ and $PbCO_3$, the particles of NaCl and $CuSO_4{\cdot}5H_2O$ appeared morphologically as single crystals when precipitated under slow growth conditions, but were often shown to be oligocrystalline by single crystal XRD. As the particles of $CuSO_4{\cdot}5H_2O$ and NaCl crystals grew to relatively large sizes (~400 μm) in comparison with the other crystals examined, intergrowth of particles nucleated at separate sites may have occurred in these systems.

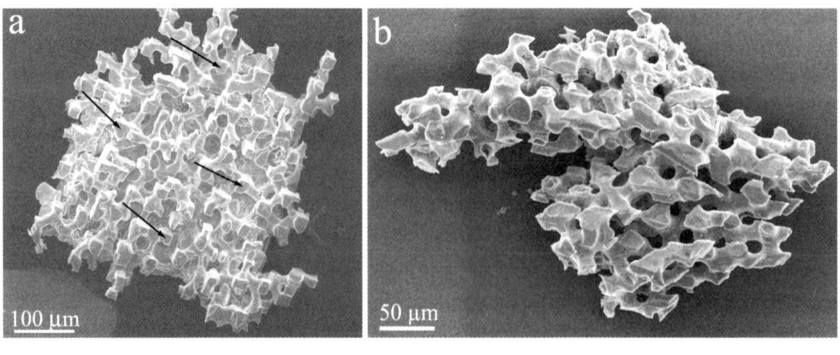

Fig. 15.7 Templated crystals isolated from polymer membrane after 24 h of (a) NaCl and (b) $CuSO_4{\cdot}5H_2O$.

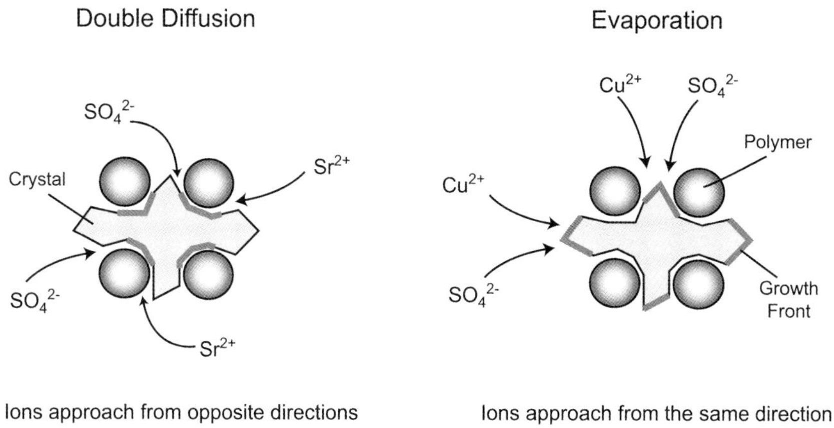

Fig. 15.8 Schematic diagram illustrating the influence of the precipitation route on the surface structure of the templated particles formed.

Although the evaporation method was successful in producing templated macroporous single crystals, the surfaces of the NaCl and $CuSO_4{\cdot}5H_2O$ crystals were typically rougher than those of the single crystals produced using the double-diffusion method, suggesting that the surfaces of the NaCl and $CuSO_4{\cdot}5H_2O$ crystals had not been defined by close contact with the polymer membrane. A key distinction between the double-diffusion and evaporation techniques is in the geometry of the experimental set-up (Fig. 15.8). In the case of double diffusion, crystal growth occurs by diffusion of ions from *opposite directions*, such that growth of the crystal proceeds from the point where the anions and cations meet, providing a driving force for ion diffusion throughout the membrane. The 3-D sponge-like structure of the membrane facilitates access of the ions to the growth front, and pores in the polymer membrane become entirely filled with the crystal.

In contrast, in the evaporation technique, the anions and cations can diffuse to the growth front of the crystal from the *same direction*, such that there is no driving force for ions to diffuse throughout the entire membrane. Crystal growth will therefore continue on the nearest crystal face, rather than diffusing further into the membrane interior and combining with the inner crystal surfaces (Fig. 15.8). The crystal will therefore not continue to grow to impinge upon the polymer membrane, resulting in the rougher crystal faces observed.

15.3.5 Polycrystalline Systems

A range of other materials, including $MgCO_3$, $ZnCO_3$, $BaCO_3$ and $SrCO_3$, were also precipitated within the polymer templates, but all yielded polycrystalline structures under the experimental conditions investigated [64]. The structures of the $MgCO_3$ and $ZnCO_3$ products were very similar, taking the form of two-sided,

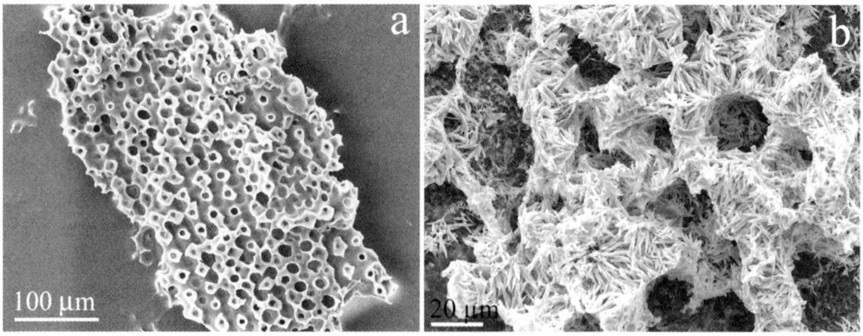

Fig. 15.9 Templated particles of: (a) $MgCO_3$ and (b) $BaCO_3$ particles, produced from 0.05 M reagents after 24 h reaction time.

bicontinuous particles which had formed as a thin sheet of small, close-packed particles over the polymer membrane (Fig. 15.9a). The products obtained from the $SrCO_3$ and $BaCO_3$ were also very similar in structure to each other, showing a macroporous, skeletal structure and comprising a network of ~20 μm needle-like crystals (Fig. 15.9b). These systems illustrate a fundamental and intuitive requirement for producing templated single crystals – that the equilibrium size of the crystal under the selected growth conditions exceeds the length scale of the template. All of these minerals precipitate with crystal sizes significantly smaller than the template pores, resulting in a polycrystalline product.

15.3.6
Controlling Crystal Nucleation: Influence of the Polymer Surface Chemistry

In order to produce single crystals of complex morphology, control must be exerted not only during growth but also at the point of nucleation. Single crystals must derive from a single nucleation site, and growth can then occur by extension through the porous network. The crystal nuclei must also be spatially well-separated such that the intergrowth of neighboring particles does not occur to produce a polycrystalline product structure. As heterogeneous nucleation on a suitable substrate occurs at a lower driving force than homogeneous nucleation, it was expected that the polymer surface chemistry would itself play a role in directing crystal nucleation. The influence of the surface chemistry of the polymer in determining the structure of the templated particles was investigated through surface modification of the polymer, either through treatment with an oxygen plasma, or through coating with gold and subsequent modification with functionalized self-assembled monolayers (SAMs) [68]. The untreated membranes offer few ionizable groups and were therefore expected to exert little influence on crystal nucleation. In contrast, the introduction of ionizable groups would be anticipated to generate surfaces which would actively promote nucleation. The generality of the approach was investigated by studying the growth of $CaCO_3$ (calcite),

Fig. 15.10 Calcite particles grown within surface-treated polymer membranes on: (a) an untreated membrane; and (b) an Au/mercaptohexadecanoic acid-treated membrane. (c) A $SrSO_4$ single crystal precipitated within a Au/mercaptopropanesulfonic acid-treated polymer membrane. (d) $PbSO_4$ single crystal precipitated within a Au/hexadecanethiol-treated membrane.

$SrSO_4$ and $PbSO_4$ crystals within the surface-treated membranes, all of which form templated single crystal within the untreated polymer membranes [62–64].

Functionalization of the gold-coated polymer membrane with a range of SAMs had a marked influence on the structure of the particles grown within the polymer membrane. The Au/hexadecanethiol SAM offered the most hydrophobic environment, and supported the growth of templated single crystals of calcite, as did the alcohol-terminated SAMs (Fig. 15.10a). The charged SAMs (Au/mercaptohexadecanoic acid and Au/mercaptopropanesulfonic acid), in contrast, directed the formation of polycrystalline calcite particles (Fig. 15.10b). These particles showed sponge-like morphologies, and principally comprised intergrown calcite rhombohedra of size ~ 10 µm which were randomly oriented over the volume of the particle, indicating multiple nucleation sites. The oxygen plasma-treated polymer membranes also supported the formation of polycrystalline particles comprising intergrown, rhombohedral calcite particles. These data therefore suggest that the formation of large, templated single crystals within the membrane relied upon the presence of limited nucleation sites. This observation was further examined by investigating $SrSO_4$ and $PbSO_4$ growth within the surface-modified polymer membranes. Interestingly, the precipitation of $SrSO_4$ from 0.05 M reagents and $PbSO_4$ from 0.02 M reagents yielded single crystals, independent of the membrane surface chemistry (Fig. 15.10c).

That the precipitation of $PbSO_4$ within the polymer was independent of the membrane surface chemistry can be attributed to the pH of the reagent solution. The $PbSO_4$ precipitated at pH 5.7, a significantly lower pH value than the pH 8 of the $CaCO_3$ and $SrSO_4$. The degree of ionization of the SAM will be lower at this pH value, resulting in weaker association of the Pb^{2+} cations with the membrane surface, and reducing the potential nucleation density on the membrane. That a surface effect was obtained for nucleation of the calcite crystals in the membrane, but not for the $SrSO_4$, is intriguing. Experiments were carried out at similar pH values (8.3 for calcite and 7.9 for the $SrSO_4$) under which conditions the degree of ionization of the surface and interactions with the cations in solution would have been similar. Consideration of the initial supersaturations of the $CaCO_3$ and $SrSO_4$ solutions and the interfacial energies of calcite and $SrSO_4$ show that crystallization of both of these minerals occurred under very similar conditions. Indeed, under the very high supersaturation conditions applied in these reactions, homogeneous nucleation may be expected to predominate.

A significant difference does exist between $CaCO_3$ and $SrSO_4$, however, and this occurs as the rich polymorphism of $CaCO_3$. ACC is anticipated to be the first phase precipitated in the calcite system (see Section 15.3.1) and is likely to associate with the membrane surface, either via heterogeneous nucleation, or through adsorption of particles formed by homogeneous nucleation in solution. The ACC precursor phase will generate a high density of Ca^{2+} and $CO_3{}^{2-}$ ions at the membrane/solution interface, and subsequent recrystallization may be directed by the membrane. That little influence of the membrane over $SrSO_4$ precipitation occurs can therefore be attributed to the high supersaturation values. The strong surface-dependence observed for the calcite system, in contrast, may be due to the formation of an ACC precursor phase which is intimately associated with the membrane and mediates the recrystallization process, resulting in multiple nucleation sites and a polycrystalline product when ionizable surface groups are present.

15.4 Summary

Templating provides a route to complex morphologies which cannot currently be accessed by any other synthetic method, and is limited only by the availability of suitable templates. The experiments described here employ a structured biomineral – namely, sea urchin skeletal plates – as the basis for forming macroporous solids with amorphous, polycrystalline and single crystal structures with uniquely regular, bicontinuous morphologies. The technique is extremely general, being used with synthetic techniques including electroless deposition and sol–gel chemistry, and its versatility can be further extended through the production of a chemically stable polymer replica of the original urchin plate. The intrinsic bicontinuous structure of the template also affords control over the ultimate structure of the templated solid. While complete filling of the pores with the selected material results in perfect casts of the urchin plate structure, surface coverage of the plate provides a double-sided surface, and partial filling of the porous network can lead to a skeletal structure with pores larger than in the original template.

The growth of single crystals within the polymer replica of the sea urchin skeletal plates provides a unique opportunity for investigating the factors involved in controlling the morphologies of single crystals. The present results demonstrate that templating provides a general approach to producing single crystals with complex morphologies and curved surfaces, such as those displayed by sea urchin skeletal elements. The technique can be applied to a wide range of crystals, with the provision that the "natural" size of the crystal under the growth conditions selected must exceed the length scale of the template. The growth conditions must also be selected to give limited nucleation sites, such that an individual nucleus can grow throughout the template without impinging on neighboring particles. The formation of single crystals was therefore favored at low solution superstaurations, and within a template environment which did not promote heterogeneous nucleation. These results show that such complex single crystal morphologies are not restricted to biology, but can be produced synthetically by external definition of the form only.

References

1 H.A. Lowenstam, S. Weiner, *On Biomineralization*. Oxford University Press, **1989**.

2 S. Mann, *Biomineralization: Principles and Concepts in Bioinorganic Materials Chemistry*. Oxford University Press, **2001**.

3 J. Aizenberg, J.C. Weaver, M.S. Thanawala, V.C. Sundar, D.E. Morse, P. Fratzl, *Science* **2005**, *309*, 275.

4 G.E. Fantner, T. Hassenkam, J.H. Kindt, J.C. Weaver, H. Birkedal, L. Pechenik, J.A. Cutroni, G.A.G. Cidade, G.D. Stucky, D.E. Morse, P.K. Hansma, *Nat. Mater.* **2005**, *4*, 612.

5 J.-Y. Rho, L. Kuhn-Spearing, P. Zioupos, *Med. Eng. Phys.* **1998**, *20*, 92.

6 J.D. Currey, *J. Exp. Biol.* **1999**, *202*, 3285.

7 A.P. Jackson, J.F.V. Vincent, R.M. Turner, *J. Mater. Sci.* **1990**, *25*, 3173.

8 A.P. Jackson, J.F.V. Vincent, R.M. Turner, *Proc. R. Soc. Lond. B* **1988**, *234*, 415.

9 D.A. Bazylinski, R.B. Frankel, *Nature Rev. Microbiol.* **2004**, *2*, 217.

10 D.A. Bazylinski, R.B. Frankel, in: P.M. Dove, J.J. De-Yoreo, S. Weiner (Eds.), *Reviews in Mineralogy and Geochemistry, Vol. 54*. Mineral Society of America, Washington, **2003**, p. 217.

11 D.A. Thompson, *On Growth and Form*. Cambridge University Press, Cambridge, **2004**.

12 E. Bäuerlein, *Angew. Chem. Int. Ed.* **2003**, *42*, 614.

13 G. Donnay, D.L. Pawson, *Science* **1969**, *166*, 1147.

14 D.M. Raup, *J. Geol.* **1959**, *67*, 661.

15 P. Dubois, C.P. Chen, in: M. Jangoux, J. Lawrence (Eds.), *Echinoderm Studies, Vol. 3*. AA Balkema, Rotterdam, **1989**, p. 109.

16 S. Albeck, L. Addadi, S. Weiner, *Connective Tissue Res.* **1996**, *35*, 365.

17 S. Albeck, J. Aizenberg, L. Addadi, S. Weiner, *J. Am. Chem. Soc.* **1993**, *115*, 11691.

18 J.R. Young, K. Henriksen, in: S. Weiner, P.M. Dove, J.J. De-Yoreo (Eds.), *Reviews in Mineralogy and Geochemistry, Vol. 54*. Mineral Society of America, Washington, **2003**, p. 189.

19 E. Beniash, L. Addadi, S. Weiner, *J. Struct. Biol.* **1999**, *125*, 50.

20 I.M. Weiss, N. Tuross, L. Addadi, S. Weiner, *J. Exp. Zool. B* **2002**, *293*, 478.

21 Y. Politi, T. Arad, E. Klein, S. Weiner, L. Addadi, *Science* **2004**, *306*, 1161.

22 L. Addadi, S. Raz, S. Weiner, *Adv. Mater.* **2003**, *15*, 959.

23 E. Loste, F.C. Meldrum, *Chem. Commun.* **2001**, *10*, 901.
24 E. Loste, R.J. Park, J. Warren, F.C. Meldrum, *Adv. Funct. Mater.* **2004**, *14*, 1211.
25 J. Aizenberg, D.A. Muller, J.L. Grazul, D.R. Hamann, *Science* **2003**, *299*, 1205.
26 N. Poulsen, M. Sumper, N. Kroger, *Proc. Natl. Acad. Sci. USA* **2003**, *100*, 12075.
27 M. Sumper, *Science* **2002**, *295*, 2430.
28 Y.W. Tan, E.M.P. Steinmiller, K.S. Choi, *Langmuir* **2005**, *21*, 9618.
29 Q.M. Ji, T. Shimizu, *Chem. Commun.* **2005**, *35*, 4411.
30 K.J.C. van-Bommel, A. Friggeri, S. Shinkai, *Angew. Chem. Int. Ed.* **2003**, *42*, 980.
31 K.K. Perkin, J.L. Turner, K.L. Wooley, S. Mann, *Nano Lett.* **2005**, *5*, 1457.
32 M. Yang, J. Ma, Z.W. Niu, X. Dong, H.F. Xu, Z.K. Meng, Z.G. Jin, Y.F. Lu, Z.B. Hu, Z.Z. Yang, *Adv. Funct. Mater.* **2005**, *15*, 1523.
33 J.S. Jan, S.J. Lee, C.S. Carr, D.F. Shantz, *Chem. Mater.* **2005**, *17*, 4310.
34 D.G. Shchukin, G.B. Sukhorukov, H. Mohwald, *Angew. Chem. Int. Ed.* **2003**, *42*, 4472.
35 R.C. Hayward, B.F. Chmelka, E.J. Kramer, *Adv. Mater.* **2005**, *17*, 2591.
36 D.H. Kim, Z.C. Sun, T.P. Russell, W. Knoll, J.S. Gutmann, *Adv. Funct. Mater.* **2005**, *15*, 1160.
37 A. Stein, R.C. Schroden, *Curr. Opin. Solid State Mater. Sci.* **2001**, *6*, 553.
38 O.D. Velev, A.M. Lenhof, *Curr. Opin. Coll. Int. Sci.* **2000**, *5*, 56.
39 M.A. Carreon, V.V. Guliants, *Eur. J. Inorg. Chem.* **2005**, 27.
40 S.R. Hall, H. Bolger, S. Mann, *Chem. Commun.* **2003**, *22*, 2784.
41 J. Silver, R. Withnall, T.G. Ireland, G.R. Fern, *J. Mod. Optics* **2005**, *52*, 999.
42 W. Ogasawara, W. Shenton, S.A. Davis, S. Mann, *Chem. Mater.* **2000**, *12*, 2835.
43 Z.T. Liu, T.X. Fan, W. Zhang, D. Zhang, *Microporous Mesoporous Mater.* **2005**, *85*, 82.
44 L. Li, Y. Zhang, G.H. Li, L.D. Zhang, *Chem. Phys. Letts.* **2003**, *378*, 244.
45 I.S. Park, S.R. Jang, J.S. Hong, R. Vittal, K.J. Kim, *Chem. Mater.* **2003**, *15*, 4633.
46 F.B. Su, X.S. Zhao, Y. Wang, J.H. Zeng, Z.C. Zhou, J.Y. Lee, *J. Phys. Chem. B* **2005**, *109*, 20200.
47 S. Kuai, S. Badilescu, G. Bader, R. Bruning, X.F. Hu, V.V. Truong, *Adv. Mater.* **2003**, *15*, 73.
48 F. Yan, W.A. Goedel, *Angew. Chem. Int. Ed.* **2005**, *44*, 2084.
49 Y. Cai, S.M. Allan, K.H. Sandhage, *J. Am. Ceram. Soc.* **2005**, *88*, 2005.
50 M.R. Weatherspoon, S.M. Allan, E. Hunt, Y. Cai, K.H. Sandhage, *Chem. Commun.* **2005**, *5*, 651.
51 S. Shian, Y. Cai, M.R. Weatherspoon, S.M. Allan, K.H. Sandhage, *J. Am. Ceram. Soc.* **2006**, *89*, 694.
52 Y.H. Ha, R.A. Vaia, W.F. Lynn, J.P. Costantino, J. Shin, A.B. Smith, P.T. Matsudaira, E.L. Thomas, *Adv. Mater.* **2004**, *16*, 1091.
53 R.H. Jin, J.J. Yuan, *J. Mater. Chem.* **2005**, *15*, 4513.
54 R. Ravikrishna, R. Green, K. Valsaraj, *J. Sol Gel. Sci. Technol.* **2005**, *34*, 111.
55 A. Imhof, D.J. Pine, *Adv. Mater.* **1999**, *11*, 311.
56 H.R. Chen, J.L. Gu, J.L. Shi, Z.C. Liu, J.H. Gao, M.L. Ruan, D.S. Yan, *Adv. Mater.* **2005**, *17*, 2010.
57 R. Gonzalez-McQuire, D. Green, D. Walsh, S.R. Hall, J.Y. Chane-Ching, R.O.C. Oreffo, S. Mann, *Biomaterials* **2005**, *26*, 6652.
58 E.S. Toberer, A. Joshi, R. Seshadri, *Chem. Mater.* **2005**, *17*, 2142.
59 E.S. Toberer, R. Seshadri, *Adv. Mater.* **2005**, *17*, 2244.
60 M. Rajamathi, S. Thimmaiah, P.E.D. Morgan, R. Seshadri, *J. Mater. Chem.* **2001**, *11*, 2489.
61 E.S. Toberer, T.D. Schladt, R. Seshadri, *J. Am. Chem. Soc.* **2006**, *128*, 1462.
62 R.J. Park, F.C. Meldrum, *Adv. Mater.* **2002**, *14*, 1167.
63 R.J. Park, F.C. Meldrum, *J. Mater. Chem.* **2004**, *14*, 2291.

64 W. Yue, A.N. Kulak, F.C. Meldrum, *J. Mater. Chem.* **2006**, *16*, 408.

65 W. Yue, R.J. Park, A.N. Kulak, F.C. Meldrum, *J. Cryst. Growth* **2006**, *294*, 69–77.

66 R. Seshadri, F.C. Meldrum, *Adv. Mater.* **2000**, *12*, 1149.

67 F.C. Meldrum, R. Seshadri, *Chem. Commun.* **2000**, 29.

68 B. Wucher, W. Yue, A.N. Kulak, F.C. Meldrum, *Chem. Mater.* **2006** (in press).

69 G.G. Gawrilov, *Chemical (Electroless) Nickel-Plating*. Portcullis Press Ltd, Redhill, Surrey, **1979**.

Part IV
Protein Cages as Size-Constrained Reaction Vessels

Handbook of Biomineralization. Edited by P. Behrens and E. Bäuerlein

ISBN: 978-3-527-31805-6

16
Constrained Metal Oxide Mineralization: Lessons from Ferritin Applied to other Protein Cage Architectures

Mark A. Allen, M. Matthew Prissel, Mark J. Young, and Trevor Douglas

Abstract

The process of biomineralization is characterized by control over mineral morphology, phase, orientation, and size. Spherical protein cages, similar to ferritin, that present an electrostatically distinct interior and exterior surface serve as model systems for biomineralization and biomimetic templated materials synthesis and encapsulation. Ferritin, ferritin-like proteins, and spherical viruses can serve as nano-containers that direct mineralization, which isolates stable particles inside a protein cage. The mineralization of metal oxide materials is dominated by the electrostatic characteristics of the interior of the protein cage, leading to the development of a model for biomimetic synthesis. The electrostatic model has been described using the Gouy–Chapman theory of charged interfaces to determine the electrostatic surface potential of the interior of the protein cage and to determine the effect that these nucleation sites have on incoming ions and forming the mineral core. This electrostatic model can be probed by genetic modification of the protein cages. The plant virus, Cowpea chlorotic mottle virus, has a positively charged interior surface for condensation and packaging of viral nucleic acid. Using site-directed mutagenesis, the positive charges can be altered to negatively charged glutamic acid residues and thus form a ferritin-like protein capable of mineralizing a range of transition-metal oxides.

Key words: biomineralization, biomimetic synthesis, ferritin, virus, cowpea chlorotic mottle virus (CCMV), protein cage, Dps, heat shock protein.

16.1
Introduction

Biominerals are usually composites of hard (inorganic) and soft (organic) materials. The interaction between these provides the basis for controlled morphology,

Handbook of Biomineralization. Edited by P. Behrens and E. Bäuerlein

ISBN: 978-3-527-31805-6

polymorph selection, and spatial localization in biological systems. Understanding the basis for the interactions at the interface between hard and soft materials has also been significant in the design and implementation of synthetic biomimetic material systems.

The iron storage protein ferritin is a unique biomineralization system, and is the inspiration for the studies described in this chapter. It is superficially a simple system with only a single protein component, which directs biomineralization of iron oxide at the protein–solution interface. The protein however, forms a closed shell architecture which incorporates all the control elements necessary for biomineralization. These include an enzymatic catalyst for molecular transformation of precursor ions, a mineral nucleation site, and an architecture that defines and constrains the overall morphology of the biomineral. In addition, the colloidal nature of the protein cage renders the final biomineral soluble and mobile, yet biochemically inert. Many of these properties and control elements are highly desired in the fabrication of synthetic materials and we (and others) have incorporated some of our understanding of ferritin biomineralization towards biomimetic synthesis.

Supramolecular assemblies of protein subunits into a cage-like architecture are not unique to ferritins, and from a synthetic biomimetic standpoint they represent novel environments by which materials can be synthesized in a size-constraining mode of encapsulation. There are a number of protein cage architectures that, like ferritin, are all assembled from a distinct number of subunits to form a precisely defined molecular container in the 5- to 100-nm size regime. Other examples of these cage-like architectures are chaperonins [1–3], DNA-binding proteins [4–9], and a very large class of protein cages, the viruses [10, 11]. Typically, protein cages are almost spherical in nature, and represent a range of relatively simple symmetries including tetrahedral, octahedral, and icosahedral. The library of functional protein cage architectures that serve as platforms for such purposes as biomimetic material synthesis, magnetic resonance imaging (MRI) contrast agents, gene therapy, drug encapsulation, cell-specific targeting and catalysis is under development. Several protein cages that have been used for encapsulated metal oxide nano-material synthesis [12–17] are illustrated in Figure 16.1. All of the protein cages represented in Figure 16.1, as well as many others, have also been probed by both chemical and genetic modification for adding non-native functionality and exploiting the great versatility of protein cage architectures.

Conceptually, there are three different interfaces presented by all protein cage architectures. These are the interior and exterior surfaces as well as the interface between subunits (Fig. 16.2). Here, we present three particular protein cage platforms ferritin, Dps, and CCMV – that serve as size-constraining reaction vessels for nano-material synthesis specifically using the *interior* surface of the protein cage. From the understanding of directed biomineralization in ferritin we have developed a model for surface-induced metal oxide formation, and have used this as a guiding principle for the synthesis of metal oxide nano-particles in other, natural or engineered, protein cage architectures. In this way, we can demon-

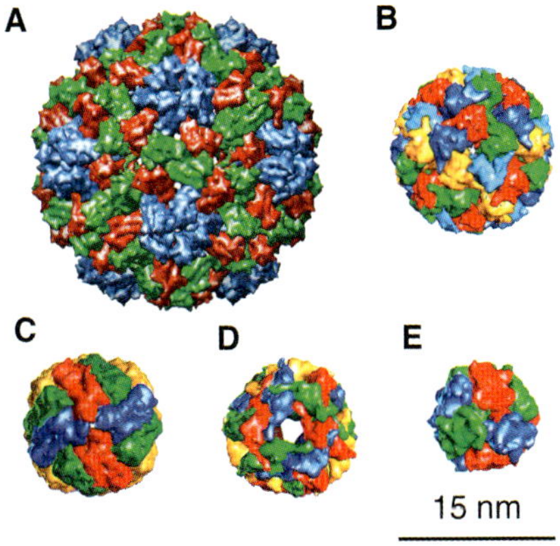

Fig. 16.1 Library of protein cage architectures, all of which have been used for the spatially controlled mineralization of metal oxide nanoparticles. A = cowpea chlorotic mottle virus (CCMV) [64]; B = lumazine synthase [70]; C = human ferritin (Fn) [71]; D = small heat shock protein from *Methanococcus jannaschii* (sHSP) [3]; E = DNA-binding protein for starved cells (Dps) [5].

strate control over composition, polymorph selection, and overall morphology using synthetic reactions. The principles outlined here are not limited to the three protein cage systems described, but rather serve as a model for protein-encapsulated biomimetic synthesis [14].

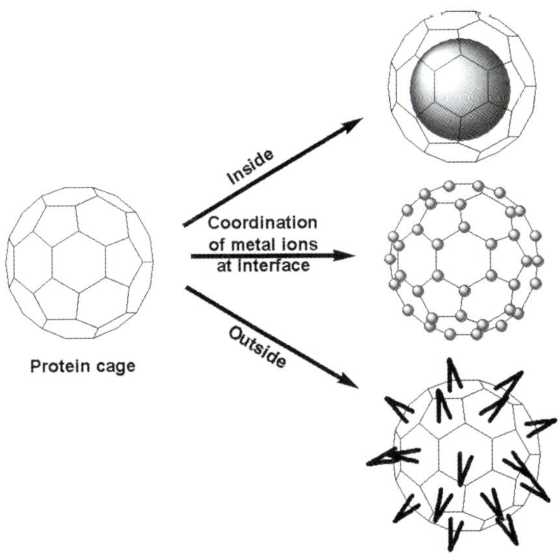

Fig. 16.2 Schematic representation of the three crucial protein interfaces, the inside, the outside, and the interface between subunits.

16.2
Biomineralization of Iron Oxide in Mammalian Ferritin

Ferritin is a spherical protein cage architecture that is almost ubiquitous in biology, where it functions to direct the biomineralization of iron as a mechanism for maintaining iron homeostasis [18, 19]. While the primary amino acid sequences of ferritins show little homology, the structural homology (at the 2°, 3°, and 4° levels) is very highly conserved. All ferritins are composed of 24 structurally identical subunits that assemble into a very robust protein cage with octahedral (432) symmetry (Fig. 16.3A and B). The external diameter of these assembled protein cages is 12 nm, and the architecture defines an internal cavity that is 6 to 8 nm in diameter. The structural motif of the ferritin subunit consists of a four-helix bundle with a fifth C-terminal helix (helix E) oriented at 60° to the four-helix bundle axis (Fig. 16.3C). In the 4° structure of the assembled protein cage the fifth helix forms the fourfold axis through assembly of an intersubunit four-helix bundle [20–22].

Mammalian ferritin is comprised of two classes of subunits that are structurally near-identical, although they differ in their 1° sequence. These two forms of subunits – the H-chain (heavy) and the L-chain (light) – self assemble to form hetero-24-mers with different ratios of each subunit, depending upon the organ from which the ferritin is isolated. The designations of H and L were made based on their differences in subunit electrophoretic mobility with molecular masses of 21 and 19 kDa, respectively [18]. H-chain ferritin has a conserved enzymatic activity known as the ferroxidase site, and is known to catalyze the oxidation of Fe^{2+}, with molecular O_2, more rapidly than the L-chain. The L-chain subunit has a greater negative charge which, in the assembled Fn, is presented on the interior surface as clusters of acidic residues (Glu and Asp) that comprise the mineral nucleation site. H-chain ferritin also has a nucleation site that is in close proximity to the ferroxidase site, with one glutamate residue shared between the two sites [20].

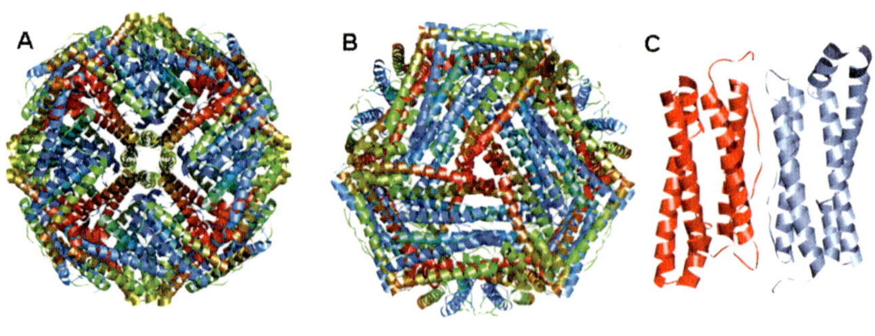

Fig. 16.3 Ribbon diagrams of human ferritin (pdb file:1fha) (A) The assembled 24-subunit protein cage, looking down the fourfold axis. (B) The assembled 24-subunit protein looking down the threefold axis. (C) A dimer of protein subunits arranged anti-parallel, making up the twofold axis of the assembled 24-mer.

Whilst iron is a necessary element for life, it has a paradoxical relationship in biology due to its reactivity in forming reactive oxygen species. When iron is stored as a nanoparticle of iron oxide (ferrihydrite) inside the protein cage ferritin (Fn), it is completely sequestered and rendered inert [23]. The encapsulation and sequestration of the iron oxide nanoparticle in biological systems highlights its potential for use as a synthetic platform for materials synthesis. The cage-like property of Fn provides an ideal size-constrained reaction environment for nanomaterial synthesis where the protein shell acts both to direct mineralization and as a passivating layer preventing unwanted particle–particle interactions.

16.3 Mineralization

In vivo, Fn is responsible for sequestering and storing toxic iron as an innocuous mineral of iron oxide (ferrihydrite) through an overall protein-mediated reaction represented in Eq. (1) [23]:

$$4Fe^{2+} + O_2 + 6H_2O \rightarrow 4FeOOH + 8H^+ \quad (1)$$

The actual biological process of iron oxidation and encapsulation is considerably more complex than Eq. (1) indicates, and some of the intimate steps remain unresolved. In the presence of Fn, potentially toxic iron is sequestered and stored inside the cavity of the protein cage. The mineralized particles are electron-dense, and are the approximate dimensions of the interior of the protein cage (5–7 nm diameter) (Fig. 16.4A). When iron is allowed to undergo oxidative hydrolysis *in vitro* in the absence of Fn, an uncontrolled homogeneous nucleation results in the mineralization and precipitation of iron oxide. There are approximately 15 common polymorphs of iron oxide or iron oxy-hydroxide [24], but under the

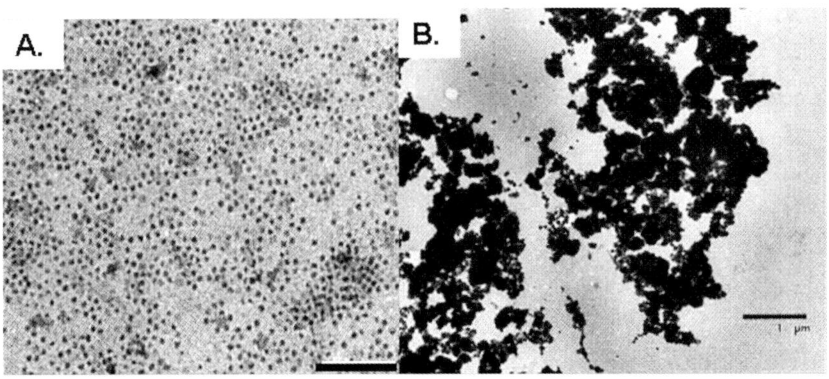

Fig. 16.4 (A) Unstained transmission electron microscopy (TEM) image of the iron oxide mineral cores present in native cow spleen ferritin. (B) Unstained TEM image of bulk, unconstrained iron oxide precipitation resulting from the oxidative hydrolysis of Fe(II).

narrow range of conditions compatible with biology and in the absence of macromolecular directing agent, the mineral phases of lepidocrocite [γ-FeO(OH)) or goethite (α-FeO(OH)] will be formed [18] (Fig. 16.4B). In the presence of a directing agent, almost all common iron oxide polymorphs have been observed, with the exception of hematite [24]. It is interesting to note that in the presence of Fn, only a particular phase of iron oxide (ferrihydrite) is formed. Ferrihydrite is less crystalline than lepidocrocite or goethite, and is characterized by electron or X-ray diffraction studies to commonly have either two or six diffraction lines. This kinetically trapped phase of iron oxide is not usually a particularly stable phase, but is stable when prepared inside Fn; this indicates the ability of biomolecules to direct and stabilize a particular polymorph of mineral.

The mechanism by which iron is incorporated into Fn *in vitro* can be described by four major events: iron entry; iron oxidation; iron oxide nucleation; and iron oxide particle growth. Iron entry into the cage-like architecture occurs via the channel (threefold symmetry) formed at the interface between subunits [25, 26]. Fe(II) oxidation is enzymatically catalyzed by reaction at the ferroxidase center, resulting in the formation of Fe(III). The nucleation of an iron oxide material from this insoluble ion is facilitated at the interior protein interface, and the particle grows from this nucleus but is limited by the size constraints of the cage.

Conserved acidic residues along the threefold channel in eukaryotes have been shown to bind metals. Electrostatic calculations on the recombinant human H-chain Fn reveal electrostatic gradients at the threefold axes that act as a guiding force directing cations through the channel toward the interior of the protein cage [25] (Fig. 16.5). Specifically, the electrostatic guidance suggests a pathway

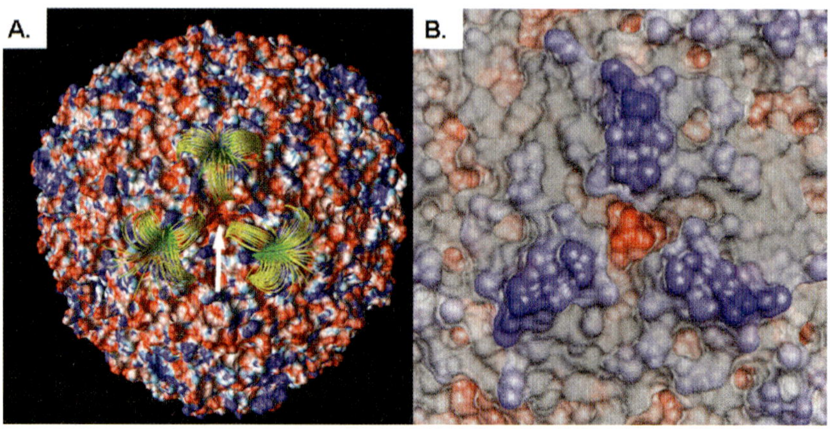

Fig. 16.5 (A) Computed gradients in the electrostatic potential at the threefold axis of human ferritin, showing the cation guidance down the channel [25]. (B) A view of the threefold channel; basic residues surrounding the channel are shown in blue, and the acidic residues lining the channel in red.

between the channel and ferroxidase center. It has also been suggested that this channel may be dynamic, thus modulating the dimensions of the opening to the cage interior. This dynamic breathing of the Fn protein cage has been confirmed by the permeation of 7- to 9-Å-sized electron paramagnetic resonance (EPR) spin labels into the interior of the protein cage directed predominantly by charge effects of the threefold axis [27].

The biomineralization of iron oxide in Fn occurs through several well-defined steps, although the exact pathway can vary depending on the ratio of iron to Fn protein cage. Many of these steps have been characterized, and it is apparent that at low iron loading ratios (<48 Fe/cage) the mineralization is dominated by the enzymatic oxidation of Fe(II), while at higher iron loadings the process is dominated by the mineral itself [28, 29]. The multi-step mineralization process is mediated at all levels by the protein cage, which facilitates electrostatic cation guidance, an enzymatic ferroxidase activity, as well as guiding a spatially selective nucleation and size constrained mineral particle growth.

16.4 Iron oxidation

The enzymatic activity of H-chain mammalian Fn directs the oxidation of Fe^{2+} to Fe^{3+}, at a site known as the ferroxidase site. The ferroxidase site is located within the four-helix bundle of the H-chain subunit, and involves amino acids E27, Y34, E61, E62, H65, and Q141 (Fig. 16.6). An electrostatic guiding force directs metal cations that enter the protein cage towards this site. Initially, a single iron binds to this site followed by a second iron, both of which have been observed spectroscopically [30]. The two Fe(II) ions are linked at the ferroxidase site via a bridging

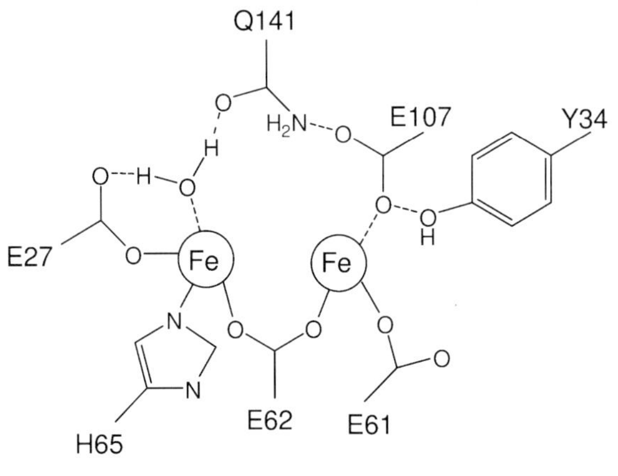

Fig. 16.6 The ferroxidase site of H-chain Fn, with two Fe^{2+} bound at by amino acids E27, Y34, E62, H65, E107, and Q141.

carboxylate group from a glutamic acid (E62) and establishes the metal binding motif as part of a protein superfamily known as the diiron carboxylate superfamily [31]. The coordination of two ferrous ions at the ferroxidase site precedes the binding of molecular oxygen, which undergoes a two-electron reduction to form a μ-peroxo-diferric complex. This intermediate decays to form an antiferromagnetically coupled μ-oxo-bridged diferric species (and subsequently either a hydroxy or dihydroxy-bridged diferric complex) [27, 28] with the release of H_2O_2. The hydroxy-bridged species is released from the ferroxidase site and bound at the nucleation site on the interior of the protein cage, and eventually forms an iron oxide cluster and the mineral core [28]. Peroxide produced in the initial reaction can be subsequently used as an oxidant in mineral formation [29, 32].

16.5
Iron Oxide Nucleation and Mineral Growth

Ferric ions released from the ferroxidase site are thought to pass to a cluster of negatively charged amino acids (the nucleation site) in close proximity. One of the glutamic acid residues (E61) that comprises the ferroxidase site is believed to facilitate this movement. This was proposed based on the proximity and disorder of this residue in the X-ray structure. This cluster of amino acids that are the nucleation sites are found in both H- and L-chain subunits [18, 33], and are located on the interior of the protein cage, causing a net negatively charged interior surface (Fig. 16.7). Ferric ions (as mono or dihydroxy species) are thought to bind to this cluster of negatively charged residues, which facilitates the oligomerization

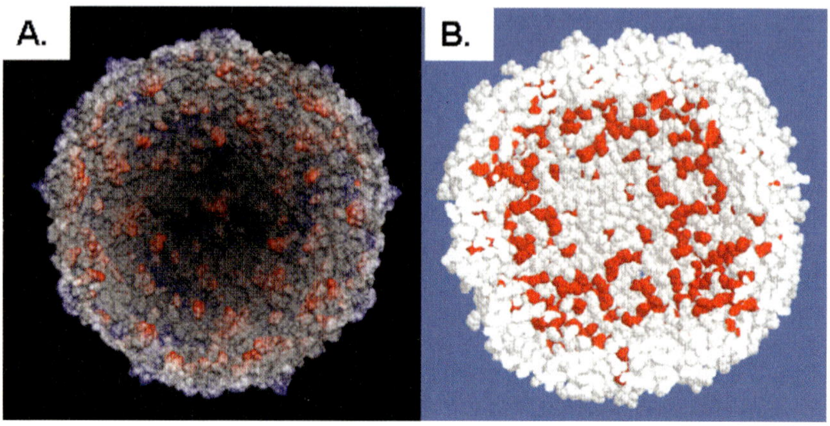

Fig. 16.7 (A) The interior surface showing the distribution of charged residues in mammalian ferritin (red is negative charge; blue is positive charge). (B) Space-filling model showing the cluster of glutamic acid amino acids responsible for the nucleation site (shown in red).

of Fe(III) units, eventually forming a critical nucleus from which the ferrihydrite particle can grow. Early intermediates have been observed using NMR relaxometry consistent with antiferromagnetic precursor species as the cluster matures into the mineral core [34]. The developing mineral particle has been suggested to act as catalyst for the additional oxidation of Fe(II) with O_2, which directly contributes to the accumulation and growth of the mineral particle, independent of the enzymatic ferroxidase reaction. During *in-vitro* mineral formation in Fn this pathway was found to be operative at high Fe:protein ratios, which may not be physiologically relevant but certainly provides a unique control mechanism for synthetic mineralization reactions and in the development of a biomimetic model for the protein-mediated reactions [28].

16.6 Summary of Ferritin Mineralization Reaction

The iron oxide mineralization processes in mammalian Fn can be summarized by a series of reactions [Eqs. (2)–(6)] which result in the formation of a spatially constrained iron oxide nanoparticle core [35]:

$$\mathrm{Pr} + 2\mathrm{Fe}^{2+} \rightarrow \mathrm{Pr} - [\mathrm{Fe}_2]^{4+} \tag{2}$$

$$\mathrm{Pr} - [\mathrm{Fe}_2]^{4+} + \mathrm{O}_2 \rightarrow \mathrm{Pr} - [\mathrm{Fe}_2^{4+}\mathrm{O}_2]^{4+} \tag{3}$$

$$\mathrm{Pr} - [\mathrm{Fe}_2^{4+}\mathrm{O}_2]^{4+} \rightarrow \mathrm{Pr} - [\mathrm{Fe}_2^{6+}\mathrm{O}_2^{2-}]^{4+} \tag{4}$$

$$\mathrm{Pr} - [\mathrm{Fe}_2^{6+}\mathrm{O}_2^{2-}]^{4+} + \mathrm{H_2O} \rightarrow \mathrm{Pr} - [\mathrm{Fe_2O}]^{4+} + \mathrm{H_2O_2} \tag{5}$$

$$\mathrm{Pr} - [\mathrm{Fe_2O}]^{4+} + \mathrm{H_2O} \rightarrow \mathrm{Pr} + 2\mathrm{FeOOH}_{(\mathrm{core})} + 2\mathrm{H}^+ \tag{6}$$

In this scheme, Pr is the protein and the charge is balanced by the presence of either hydroxide ions or bound glutamate residues of the protein. Equation (2) demonstrates an initial binding of two Fe^{2+} to the protein cage (at the ferroxidase center); Eq. (3) is the binding of molecular O_2 to form a dioxygen complex; Eq. (4) is oxidation of the diferrous complex to form the diferric complex; Eq. (5) is formation of the peroxo intermediate followed by degradation to the μ-oxo-bridged complex; and Eq. (6) is the release of the iron complex from the ferroxidase site to form the FeOOH core at the nucleation site [35]. This mechanism leads to the overall reaction presented in the balanced Eq. (1).

The ferroxidase reaction suggests that Fn biomineralization is specific for iron *in vivo*. However, *in-vitro* experiments with homopolymers of L-chain Fn, having no ferroxidase activity, have been shown to encapsulate iron oxide nanoparticles at nearly equivalent efficiencies [33]. H-ferritin mutants, where the nucleation sites have been deleted but the ferroxidase sites remain intact, are also able to spatially direct mineralization within the confines of the protein cage. In mutant L-ferritins, lacking both the ferroxidase and the nucleation sites, there was a loss of spatial control in directing mineral formation, and bulk precipitation of iron oxides occurs [33, 36]. What then are the roles of the ferroxidase and nucleation sites that make them individually dispensable but together an absolute requirement? The ferroxidase site converts Fe(II) to Fe(III). The Fe(II) is orders of magnitude more soluble, at physiologically relevant pH, than the Fe(III). Thus, the ferroxidase center converts an under-saturated condition to a supersaturated condition *inside the protein cage* – this is sufficient to achieve the spatially directed mineralization observed. The high negative charge density on the interior surface of the assembled Fn protein cage that constitutes the nucleation sites serve to aggregate ions at the protein interface. This perhaps increases the local concentration at the interface and facilitates oxidative mineralization. Each of these components is sufficient to direct mineralization, but when they are both absent all control is removed.

16.7
Model for Synthetic Nucleation-Driven Mineralization

Observing successful nucleation site-driven mineralization in Fn suggests that the oxidation and mineralization reaction may not exhibit a high degree of specificity for Fe, and that mineralization could be expected to occur with a range of transition metal ions. The data indicate that Fn mineralization can be driven through purely electrostatic effects at the interior protein interface. A model for this activity can be approximated using the Gouy–Chapman theory of charged surfaces. Briefly, Gouy–Chapman theory states that the charge on a surface influences the ion distribution of electrolyte proximally to the surface through Coulombic interactions (Fig. 16.8) [37, 38]. The surface charge potential decays exponentially as a function of distance from surface, according to Eq. (7):

$$\psi_x = \psi_o e^{-\kappa x} \tag{7}$$

where ψ_x is the potential at a distance x, ψ_0 is the surface potential, and κ is a Debye parameter measured in reciprocal distance. The distance (x) from the surface where ψ_x is $(1/e)\psi_0$ defines the thickness of the diffuse double layer. The width of the double layer changes as a function of ionic strength of the media and the charge of the interacting ion. This distance can be determined using Eq. (8), which determines the electric potential profile of the diffuse double layer.

$$\kappa = \left(\frac{e^2 \sum n_i^0 z_i^2}{\varepsilon_0 \varepsilon_\gamma k_B T} \right)^{1/2} \tag{8}$$

In Eq. (8), e is the fundamental charge (C), n_i is the ionic strength, z_i is the charge of the interacting ion, ε_0 is the dielectric constant of the solution, ε_γ is the permittivity in a vacuum, k_B is Boltzmann's constant, and T is absolute temperature [39]. Assuming an ionic strength of approximately 0.1 M, the thickness of

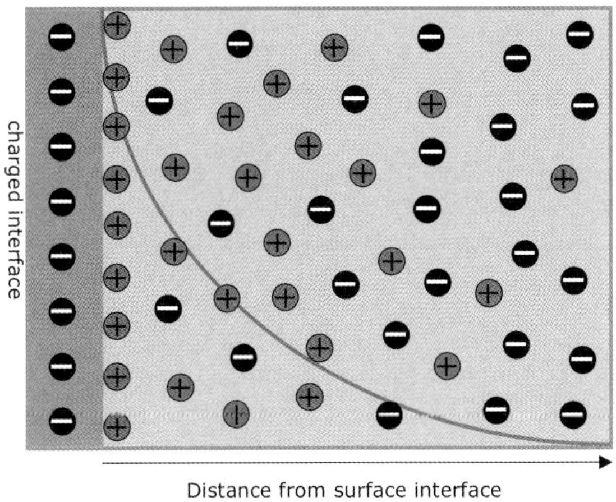

Fig. 16.8 A schematic of a charged surface showing the double layer, according to the Gouy–Chapman model, and the distribution of counterions relative to the surface. The exponential decrease in counterion concentration is shown by the curve.

the diffuse double layer for divalent cations (Fe^{2+}, Co^{2+}, Mn^{2+}, etc.) is 0.758 nm, and for a trivalent cation is 0.339 nm. The exponential decay of the surface potential with distance is illustrated in Figure 16.8. The concentration of counterions follows the surface potential and exhibits maximal concentration at the interface, but drops off exponentially and approaches bulk concentration a few nanometers from the surface. For divalent metal species the potential ψ_x drops to 10% of the surface potential ψ_0 within 1.74 nm of the surface of the cage, and to 0.78 nm for trivalent metal species. According to this model, the very negatively charged interior surface of Fn will accumulate counterions, (Fe(II)), at concentrations significantly higher than bulk concentration, in close proximity to this surface. This could potentially have two effects, both of which facilitate the oxidative mineralization process that results in the formation of the iron oxide core in Fn. The first is that binding of Fe(II) to negatively charged carboxyl species greatly lowers the reduction potential, making oxidation a favorable process. As the nucleation sites in Fn are comprised of clusters of glutamic acid residues, accumulation of Fe(II) at this interface is expected to increase the oxidation of Fe(II) to form Fe(III). Second, the highly charged surface which accommodates the accumulation of Fe(II/III) ions in close proximity acts as a substrate for stabilizing highly charged clusters; the precursors of the nucleation site of L-chain Fn consists of residues E57, E60, E61, E64 and E67, all of which reside on or close to the interior surface of the protein [36]. This model for surface charge-directed oxidative mineralization suggests that there is very little specificity for Fe. In biological systems, the Fe specificity most probably arises from the specific transport of Fe to ferritin, as there is almost certainly no free Fe in the cell. The lack of specificity for Fe has, however, been used very successfully for directing the formation of other metal oxide particles in Fn and other protein cage architectures represented in Figure 16.1.

Using the surface-directed electrostatic model as a guiding principle, Fn has been used as a template for the synthesis of non-native minerals, including cobalt oxides ($CoOOH$, Co_3O_4) [40–42] and manganese oxides (Mn_2O_3, Mn_3O_4) [43, 44], amongst other materials [45–50]. The fact that these mineralization reactions are not specific to iron suggests that the electrostatic character of the interior surface of the protein cage plays an important role in mineralization.

Protein cage-directed mineralization appears to require three essential features: (i) pores in the protein shell that allow molecular access to the interior the protein cage from the bulk solution; (ii) chemically distinct interior and exterior surfaces; and (iii) protein cage stability under the conditions required for the synthesis. The protein cages represented in Figure 16.1 satisfy these three basic criteria. This model has provided a rational approach for synthetic biomimetic mineralization that relies on the electrostatic character of the interior of a protein cage for directing material synthesis.

16.8 Mineralization in Dps: A 12-Subunit Protein Cage

In order to test the electrostatic model for protein-directed mineralization, synthetic reactions were performed using a protein cage, Dps, with similar characteristics as Fn but without the clear biological mineralization function of ferritin. There is a clear relationship between Fn and the Dps class of proteins. Dps (DNA binding protein from nutrient starved cells) was originally isolated from *Escherichia coli* [51], although since its discovery structural and functional homologues have been isolated in many other bacteria [52–54] as well as archaea [7, 8, 55]. While there is some structural similarity between Fn and Dps proteins, the Dps only assemble into 12-subunit cages with tetrahechal (23) symmetry (Fig. 16.9) [53]. The

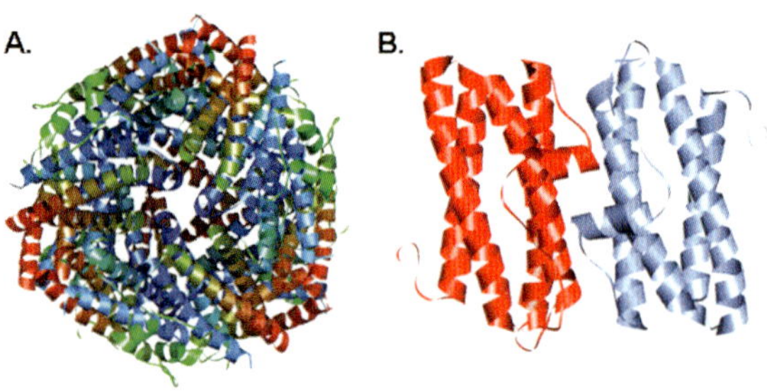

Fig. 16.9 (A) Ribbon diagram of the assembled 12-subunit Dps protein cage from *Listeria innocua* (pdb file:1qgh). (B) The dimer interface between two subunits, which contain the diiron binding motif of bacterial Dps proteins.

subunit structure has a four-helix bundle core, while a fifth helix sits in a loop connecting the B and C helices perpendicular to the four-helix bundle (Fig. 16.9B). Figure 16.9B shows a subunit dimer of a Dps from *Listeria innocua* (LDps), which also contains a ferroxidase-like binding site that is dissimilar to Fn in that it is found at the dimer interface between subunits rather than within the four-helix bundle [4, 56, 57].

The Dps architecture has two types of threefold symmetry pores. One of these channels (see Fig. 16.9A) is lined with hydrophilic amino acids that can donate cations from bulk solution allowing molecular access to the interior surface. The size of this pore is approximately similar to the threefold pore of Fn, at about 0.7–0.9 nm [4, 56]. It is proposed that once metal cations are inside the protein cage, interaction with the very negatively charged interior surface can occur, which facilitates a similar oxidative mineralization reaction described for Fn. The electrostatic surface of the interior of LDps is similar to the interior surface of Fn (see Fig. 16.7), with clusters of glutamic acid residues that can be involved in mineral core nucleation. These structural characteristics of the Dps cage have been used to direct mineralization of Co_3O_4 and Co(O)OH, as well as the cubic phase of γ-Fe_2O_3 (maghemite) [12, 58]. These synthesis reactions were all performed using a similar approach. Briefly outlined, to the empty Dps protein cage two solutions – one containing the M(II) ion and the other with oxidant (H_2O_2) – were added at a defined rate. The bulk reaction involving metal ion oxidation and subsequent precipitation was slow relative to the protein-catalyzed reaction, and in all cases no bulk precipitation was observed. The resultant nanoparticles are extremely small (3.75 ± 0.88 nm) and contain only a few hundred metal ions (Fig. 16.10) [12]. Thus, using the Dps cage as a template has allowed us a controlled synthetic access to nanomaterials that approach molecular dimensions. Recently, a Dps structure was reported in which the protein was crystallized while containing varying amounts of iron. Small clusters of Fe ions were identifiable, and several iron ions were seen bound to specific, negatively charged amino acids in the protein shell. This is the first such example of a precursor to

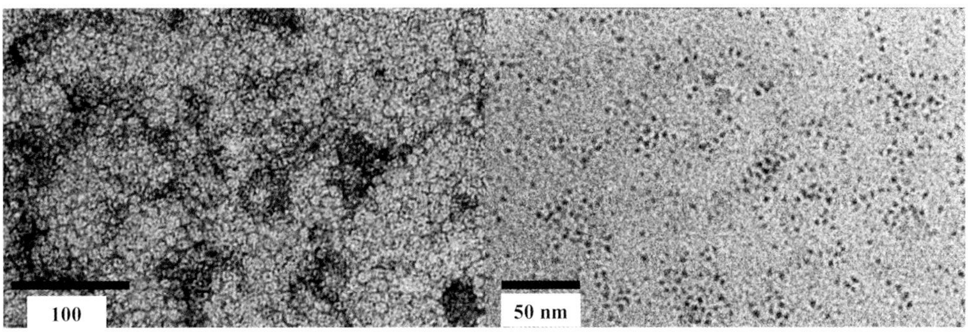

Fig. 16.10 Negatively stained and unstained TEM images of mineralized LDps.

the mineral core that has been structurally identified [59]. The potential for using the Dps proteins to both understand their role in biomineralization and biomimetic materials synthesis has been enhanced by recent discovery and structural characterization of two Dps proteins from hyperthermophilic microorganisms, both of which exhibit elevated thermal stability – an ideal property for synthetic applications [7, 8].

16.9
Icosahedral Protein Cages: Viruses

Viruses and virus-like particles (VLPs) are exquisite examples of supramolecular assembly that represent the organization of subunits into a precisely defined, stable protein cage. All viruses function by using protein or protein/lipid capsids in order to transport their nucleic acid to a host cell. They are metastable structures poised between the stability required for packaging and transport and the necessary instability associated with cargo release and infection. Many viruses package their genomic nucleic acid through non-covalent electrostatic interactions with the protein cage interface, which indicates the unique distinction between the interior and exterior surfaces.

Viruses are inherently highly symmetrical structures, and are assembled from multiple copies of a limited number of subunits. The subunits can be organized into several different architectures, including filaments, rods, spindles, helices, and spherical capsids with icosahedral symmetry [60]. The last two virus types – helical and icosahedral – represent the largest proportion of viruses, examples of which can be found in hosts from all domains of life. The highly symmetrical assembly of subunits is favorable because it allows for a maximum number of subunit contacts while still assembling into a capsid that is capable of encapsulating a genome large enough to encode its capsid protein [61].

In 1956, Watson and Crick proposed that the structure of spherical viruses is based on the aggregation of identical protein subunits around nucleic acid [11]. They based the symmetry of these viruses on the number of rotational axes, as well as multiples of 12 subunits to form the assembled virus. Since 1956 it has been confirmed that the icosahedral symmetry accounts for the largest number of spherical viruses. An icosahedron is defined as a structure with 20 triangular faces and 12 vertices, resulting in 60 identical points akin to a Buckminster fullerene (C_{60}). In Buckminster fullerenes, each carbon atom is identical, although the environment in which it sits is related by 532 symmetry. A spherical virus that assembles into a true icosahedron could be built from 60 copies of an identical coat protein subunit, although this severely limits the size of the genome that can be packaged inside the assembled virion [61], and very few functional viruses actually form from only 60 identical subunits. Icosahedral viruses are naturally found with architectures that range in size from 18 to 750 nm, comprising between 60 and several thousand individual coat protein subunits [62, 63].

16.10
Cowpea Chlorotic Mottle Virus: A Model Protein Cage

CCMV is a plant virus that assembles from 180 identical subunits into an icosahedral protein cage (Fig. 16.11). In the icosahedral architecture, each subunit, though chemically identical, is present in three slightly different chemical environments designated A, B, and C, which correspond to the colors in Figure 16.1 as blue, green, and red, respectively. CCMV is a member of the Bromoviridae, which are distinguished from other classes of plant viruses by encapsulating a tripartite genome that consists of three unique, single-stranded, positive-sense RNA molecules that are encapsulated independently of each other in separate viral capsids [64]. The total size of the genome is about 8.2 kb, comprising RNA 1 (3.2 kb), RNA 2 (2.9 kb), and RNA 3 (2.1 kb); included in the RNA 3 portion is a small fourth strand of RNA [65].

The near-atomic resolution X-ray structure of CCMV has been solved, but the N-terminal regions of the protein subunit are disordered in the structure and not (yet) resolved. It has been shown that the N-terminus of the subunit is very positively charged and is responsible for condensing and encapsulating the viral nucleic acid [64]. In mutants where the first 25 amino acids (containing nine positively charged lysine and arginine amino acids) are deleted, the capsid maintains assembly characteristics but is not capable of encapsulating viral nucleic acid. However, if only the first seven amino acids (not containing positive charge) are deleted, the virus still forms infectious particles indistinguishable from the native virus [64]. This indicates that the very positively charged portion of the N-terminus drives the encapsulation of viral nucleic acid within the protein cage, and is therefore likely located on the interior of the assembled cage. CCMV was the first icosahedral virus that could be reassembled *in vitro* to form empty protein cages with no encapsulated nucleic acid [66]. Furthermore, the overall architecture of CCMV is dynamic and undergoes a pH and metal ion-dependent swelling at the quasi-threefold axis, which allows molecular access to the interior surface of the protein cage. This process is known as "gating" or "swelling",

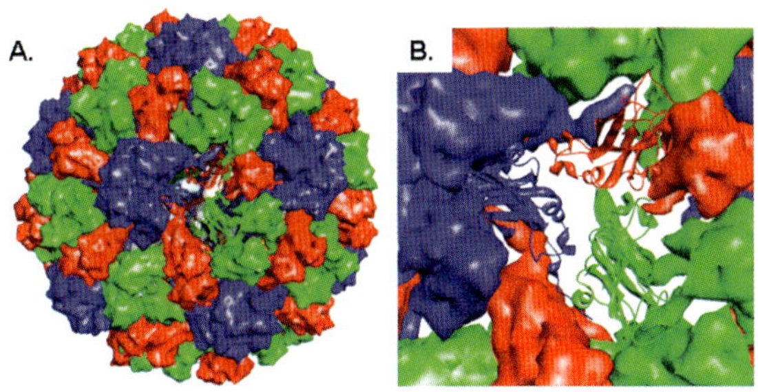

Fig. 16.11 A model of cowpea chlorotic mottle virus (CCMV) based on the X-ray structure (pdb file:1cwp) with three subunits, which constitute the asymmetric unit, shown as ribbons.

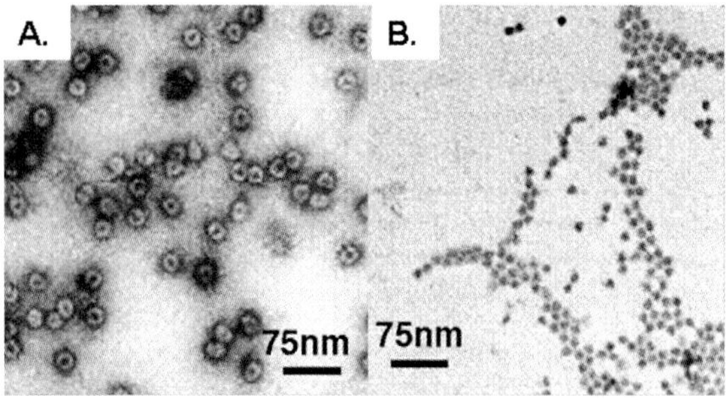

Fig. 16.12 Unstained TEM image of electron-dense polyoxotungstate cores formed at the positively charged interior protein interface of wild-type CCMV [15].

where there is an increase in the overall dimensions of the exterior of the protein cage by about 10% [67, 68].

CCMV therefore represents an ideal system for evaluating the validity of our electrostatic surface-directed mineralization model. The cage is stable, the interior interface is highly charged, and there are pores that allow molecular access between the interior and exterior environments. Although the surface charge on the inside of the assembled virus is the opposite of Fn, empty CCMV represents a protein cage with an electrostatically distinct positively charged interior surface. Taking advantage of the properties of the native CCMV cage, purified empty viral cages were mineralized by allowing the accumulation of polyoxometallates at the interior surface of the protein cage. Incubation of metal-oxo precursors (WO_4^{2-}) under conditions of high pH (pH 7.5, swollen conditions), where the interior surface is expected to be positively charged, would result in aggregation of the negatively charged metal-oxo anions at the interface according to the double-layer model. When the pH is lowered, a number of changes to the system occur. An acid-catalyzed polymerization of the metal-oxo ions results in the formation of large polyoxometallate anions, which readily crystallize at the interface to form nanoparticles, entrapped within the CCMV protein cage [15]. This occurs in part due to the higher charge, and therefore greater Coulombic interaction with the protein interface. Also, as the pH drops the CCMV gating transition is induced and the large pores in the cage are significantly closed down. This is analogous to a "ship-in-a-bottle" which, once formed, cannot exit because the pore size and the entrapped object are incommensurate [15] (Fig. 16.12).

16.11
Redesigning CCMV to Make a Fn Mimic

The CCMV subunit can be readily modified by PCR-mediated site-directed mutagenesis. Mutant CCMV subunits can be expressed in a heterologous expression

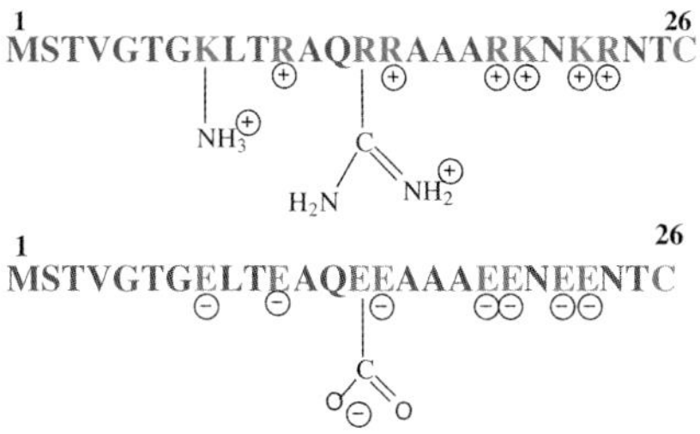

Fig. 16.13 The sequence of the disordered N-terminal region of wild-type CCMV and the subE mutant in which all basic residues (R,K) have been genetically replaced with glutamic acid residues (E).

system where they self-assemble to form stable icosahedral protein cages [69]. In this way, particles can be formed independent of their original biological activity (viral infection). To further test the electrostatic surface model, we designed mutations to CCMV to make the interior of the cage more like that of Fn. To this end, each of the nine positively charged amino acids along the N-terminus were changed to glutamic acid residues (Fig. 16.13). This mutant – termed subE – assembles into an apparently empty protein cage that is morphologically indistinguishable from wild-type CCMV, as determined by transmission electron microscopy (TEM) (Fig. 16.14) and size-exclusion chromatography (SEC). The incubation of SubE, with its altered interior surface charge, with Fe(II) ions in the presence of air results in the spatially selective mineralization of Fe_2O_3, without bulk precipitation [14]. In contrast, the unaltered CCMV cage exhibits no tendency to directed Fe_2O_3 mineralization and only the bulk reaction is observed. Thus, the modified CCMV cage, subE, has been shown to act as a Fn mimic for mineralization of iron oxide nanoparticles (Fig. 16.15).

The inspiration for CCMV-templated mineralization has been the spatially selective and polymorph specific biomineralization in Fn, resulting in the forma-

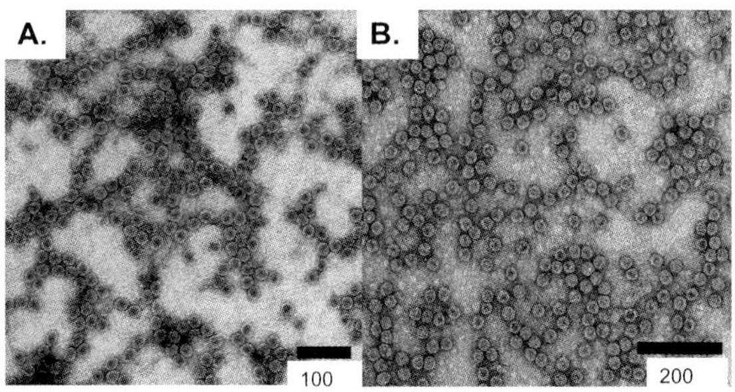

Fig. 16.14 Negatively stained TEM image of wild-type CCMV (A) and subE (B). Morphologically, these two protein cages are identical, but the two cages differ in overall charge by 3240 units.

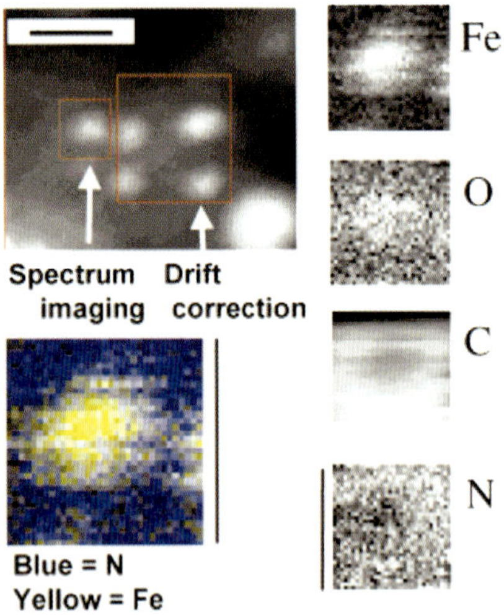

Fig. 16.15 TEM image of iron oxide nanoparticles synthesized inside the subE protein cage, which acts as a ferritin mimic. The data are provided from high-angle annular dark-field image (top left) and element specific spectral imaging of a single particle showing signal from C, N, O, and Fe with an overlay of the N (specific to the protein cage) and Fe (specific to the mineral) shown in blue and yellow, respectively.

tion of ferrihydrite cores. Bulk mineralization, when performed under identical conditions in the absence of Fn, results in lepidocrocite formation. Whilst the electrostatic surface-directed re-design of CCMV is very effective at mimicking the spatially selective mineralization, it does not reproduce the polymorph selection, because under these same conditions with subE as a template lepidocrocite is formed [14].

16.12 Conclusions

The mammalian Fn protein cage directs a unique biomineralization reaction that results in the spatially constrained formation of a specific iron oxide polymorph. Whilst the overall processes involved in the biochemical mineralization reactions of Fn are not fully elucidated, certain principles can be identified as being dominant factors driving the reaction. We have focused on the dominance of Coulombic effects in directing the mineralization reactions at the interior interface of Fn and other protein cage architectures, and have adopted a biomimetic approach to elaborate the extent to which mineralization reactivity can be predicted and controlled by applying a Gouy–Chapman model to the role of surface charge in these ionic mineral formation reactions.

Protein cage architectures represent a versatile class of nanoscale reaction vessels that can be modified, both chemically and genetically, in order to introduce functionality that allows control over the synthesis of a range of materials. The properties of nanomaterials are dependent upon size, and protein cages bridge a range of sizes allowing for tailoring of the properties of encapsulated material by selecting the protein cage reaction vessel that will yield materials of a desired size. Material synthesis inside protein cages has been shown to rely on three critical aspects: (i) pores in the protein shell that allow molecular access to the interior of the cage; (ii) a chemically distinct interior environment that favors ionic aggregation; and (iii) stability of the protein cage for the conditions necessary for preparing a particular material.

Using the virus capsid of CCMV is a clear example of how the electrostatic model for mineral formation can be utilized in a protein cage architecture. The native function of packaging RNA can be subverted and the CCMV cage redesigned and modified to perform specific material synthesis well beyond the scope of its natural activity. Through both genetic and chemical modification this cage shows how a desired functionality can be introduced into a protein cage by satisfying all three requirements necessary for use as a reaction vessel for nanomaterial synthesis.

The properties that drive the biomimetic mineralization reactions described in this chapter are not unique to the three classes of cage described here, and can – in principle – be introduced into other architectures. Rather, it is the chemical and genetic plasticity of these systems towards modification that makes them so versatile towards being used as both model systems and templates for purely synthetic reactivity.

References

1 P.J.B. Koeck, H.K. Kagawa, M.J. Ellis, H. Hebert, J.D. Trent, *Biochim. Biophys. Acta* **1998**, *1429*, 40–44.

2 J.D. Trent, *FEMS Microbiol. Rev.* **1996**, *18*, 249–258.

3 K.K. Kim, R. Kim, S.H. Kim, *FASEB J.* **1998**, *12*, A1329.

4 M. Bozzi, G. Mignogna, S. Stefanini, D. Barra, C. Longhi, P. Valenti, et al., *J. Biol. Chem.* **1997**, *272*, 3259–3265.

5 R.A. Grant, D.J. Filman, S.E. Finkel, R. Kolter, J.M. Hogle, *Nat. Struct. Biol.* **1998**, *5*, 294–303.

6 A. Grove, S.P. Wilkinson, *J. Mol. Biol.* **2005**, *347*, 495–508.

7 R. Ramsay, B. Wiedenheft, M. Alle, G.H. Gauss, C.M. Lawrence, M. Young, et al., *J. Inorg. Biochem.* **2006**, *100*, 1061–1068.

8 B. Wiedenheft, J. Mosolf, D. Willits, M. Yeager, K.A. Dryden, M. Young, et al., *Proc. Natl. Acad. Sci. USA* **2005**, *102*, 10551–10556.

9 G.H. Zhao, P. Ceci, A. Ilari, L. Giangiacomo, T.M. Laue, E. Chiancone, et al., *J. Biol. Chem.* **2002**, *277*, 27689–27696.

10 J.B. Bancroft, C.E. Bracker, G.W. Wagner, *Virology* **1969**, *38*, 324–325.

11 F.H.C. Crick, J.D. Watson, *Nature* **1956**, *177*, 473–476.

12 M. Allen, D. Willits, J. Mosolf, M. Young, T. Douglas, *Adv. Mater.* **2002**, *14*, 1562–1565.

13 T. Douglas, M. Allen, M. Young, Self-assembling Protein Cage Systems and Applications in Nanotechnology,

in: S.R. Fahnestock, A. Steinbuchel (Eds.), *Polyamides and Complex Proteinaceous Materials I.* Wiley-VCH, Weinheim, **2002**, p. 517.

14 T. Douglas, E. Strable, D. Willits, A. Aitouchen, M. Libera, M. Young, *Adv. Mater.* **2002**, *14*, 415.

15 T. Douglas, M. Young, *Nature* **1998**, *393*, 152–155.

16 M.L. Flenniken, D.A. Willits, S. Brumfield, M.J. Young, T. Douglas, *Nano Lett.* **2003**, *3*, 1573–1576.

17 W. Shenton, S. Mann, H. Colfen, A. Bacher, M. Fischer, *Angew. Chem., Int. Ed.* **2001**, *40*, 442–445.

18 P.M. Harrison, P. Arosio, *Biochim. Biophys. Acta* **1996**, *1275*, 161–203.

19 P.M. Matias, J. Tatur, M.A. Carrondo, W.R. Hagen, *Acta Crystallogr. F. Struct. Biol. Cryst. Commun.* **2005**, *61*, 503–506.

20 P.D. Hempstead, S.J. Yewdall, A.R. Fernie, D.M. Lawson, P.J. Artymiuk, D.W. Rice, et al., *J. Mol. Biol.* **1997**, *268*, 424–448.

21 D.M. Kurtz, Jr., *J. Biol. Inorg. Chem.* **1997**, *2*, 159–167.

22 D.M. Lawson, P.J. Artymiuk, S.J. Yewdall, J.M.A. Smith, J.C. Livingstone, A. Treffry, et al., *Nature* **1991**, *349*, 541–544.

23 D.E. Mayer, J.S. Rohrer, D.A. Schoeller, D.C. Harris, *Biochemistry* **1983**, *22*, 876–880.

24 R.M. Cornell, U. Schwertmann, *The Iron Oxides.* 2nd edn. Wiley-VCH, Weinheim, **2003**.

25 T. Douglas, D.R. Ripoll, *Protein Sci.* **1998**, *7*, 1083–1091.

26 E.C. Theil, H. Takagi, G.W. Small, L. He, A.R. Tipton, D. Danger, *Inorg. Chim. Acta* **2000**, *297*, 242–251.

27 X. Yang, P. Arosio, N.D. Chasteen, *Biophys. J.* **2000**, *78*, 2049–2059.

28 E.R. Bauminger, A. Treffry, M.A. Quail, Z. Zhao, I. Nowik, P.M. Harrison, *Biochemistry* **1999**, *38*, 7791–7802.

29 S. Lindsay, D. Brosnahan, T.J. Lowery, K. Crawford, G.D. Watt, *Biochim. Biophys. Acta* **2003**, *1621*, 57–66.

30 F. Bou-Abdallah, G. Biasiotto, P. Arosio, N.D. Chasteen, *Biochemistry* **2004**, *43*, 4332–4337.

31 G. Zhao, F. Bou-Abdallah, P. Arosio, S. Levi, C. Janus-Chandler, N.D. Chasteen, *Biochemistry* **2003**, *42*, 3142–3150.

32 G. Zhao, P. Arosio, N.D. Chasteen, *Biochemistry* **2006**, *45*, 3429–3436.

33 S.H. Juan, S.D. Aust, *Arch. Biochem. Biophys.* **1998**, *350*, 259–265.

34 V. Herynek, J.W.M. Bulte, T. Douglas, R.A. Brooks, *J. Biol. Inorg. Chem.* **2000**, *5*, 51–56.

35 N.D. Chasteen, P.M. Harrison, *J. Struct. Biol.* **1999**, *126(3)*, 182–194.

36 P. Santambrogio, S. Levi, A. Cozzi, B. Corsi, P. Arosio, *J. Biol. Chem.* **1996**, *314*, 139–144.

37 P. Atkins, *Physical Chemistry.* 6th edn. W.H. Freeman & Co/Oxford English Press, New York, **1998**.

38 T.B. Kinraide, U. Yermiyahu, G. Rytwo, *Plant Physiol.* **1998**, *118*, 505–512.

39 A.J. Bard, L.R. Faulkner, *Electrochemical Methods Fundamentals and Applications.* John Wiley & Sons, New York, **1980**.

40 T. Douglas, V.T. Stark, *Inorg. Chem.* **2000**, *39*, 1828–1830.

41 J.W. Kim, S.H. Choi, P.T. Lillehei, S.H. Chu, G.C. King, G.D. Watt, *Chem. Commun.* **2005**, *32*, 4101–4103.

42 R. Tsukamoto, K. Iwahor, M. Muraoka, I. Yamashita, *Bull. Chem. Soc. Jpn.* **2005**, *78*, 2075–2081.

43 S. Mann, F.C. Meldrum, *Adv. Mater.* **1991**, *3*, 316–318.

44 F.C. Meldrum, T. Douglas, S. Levi, P. Arosio, S. Mann, *J. Inorg. Biochem.* **1995**, *58*, 59–68.

45 T. Douglas, D.P.E. Dickson, S. Betteridge, J. Charnock, C.D. Garner, S. Mann, *Science* **1995**, *269*, 54–57.

46 D. Ensign, M. Young, T. Douglas, *Inorg. Chem.* **2004**, *43*, 3441–3446.

47 R.M. Kramer, C. Li, D.C. Carter, M.O. Stone, R.R. Naik, *J. Am. Chem. Soc.* **2004**, *126*, 13282–13286.

48 F.C. Meldrum, B.R. Heywood, S. Mann, *Science* **1992**, *257*, 522–523.

49 K. Sano, H. Sasaki, K. Shiba, *J. Am. Chem. Soc.* **2006**, *128*, 1717–1722.

50 K.K.W. Wong, S. Mann, *Adv. Mater.* **1996**, *8*, 928–932.

51 M. Almiron, A.J. Link, D. Furlong, R. Kolter, *Genes Dev.* **1992**, *6*, 2646–2654.
52 P. Ceci, A. Ilari, E. Falvo, E. Chiancone, *J. Biol. Chem.* **2003**, *278*, 20319–20326.
53 A. Ilari, C. Savino, S. Stefanini, E. Chiancone, D. Tsernoglou, *Acta Crystallogr. Sect. D. Biol. Crystallogr.* **1999**, *55*, 552–553.
54 M. Marjorette, O. Pena, G.S. Bullerjahn, *J. Biol. Chem.* **1995**, *270*, 22478–22482.
55 S. Reindel, C.L. Schmidt, S. Anemuller, B.F. Matzanke, *Biochem. Soc. Trans.* **2002**, *30*, 713–715.
56 A. Ilari, S. Stefanini, E. Chiancone, D. Tsernoglou, *Nat. Struct. Biol.* **2000**, *7*, 38–43.
57 S. Stefanini, S. Cavallo, B. Montagnini, E. Chiancone, *J. Biol. Chem.* **1999**, *338*, 71–75.
58 M. Allen, D. Willits, M. Young, T. Douglas, *Inorg. Chem.* **2003**, *42*, 6300–6305.
59 K. Zeth, S. Offerman, L.O. Essen, D. Oesterhelt, *Proc. Natl. Acad. Sci. USA* **2004**, *101*, 13780–13785.
60 S.C. Harrison, J.J. Skehel, D.C. Wiley, Virus structure, in: B.N. Fields, D.M. Knipe, P.K. Howley (Eds.), *Fundamental Virology*. Lippincott-Raven, New York, **1996**, p. 59–99.
61 T.S. Baker, J.E. Johnson, Principles of virus structure, in: W. Chiu, R.M. Burnett, R.L. Garcia (Eds.), *Structural Biology of Viruses*. Oxford University Press, New York, **1997**, pp. 38–79.
62 S.B. Larson, J. Day, A. Greenwood, A. McPherson, *J. Mol. Biol.* **1998**, *277*, 37–59.
63 C. Xiao, P.R. Chipman, A.J. Battisti, V.D. Bowman, P. Renesto, D. Raoult, et al., *J. Mol. Biol.* **2005**, *353*, 493–496.
64 J.A. Speir, S. Munshi, T.S. Baker, J.E. Johnson, *Virology* **1993**, *193*, 234–241.
65 R. Allison, M. Janda, P. Ahlquist, *J. Virol.* **1988**, *62*, 3581–3588.
66 J.B. Bancroft, E. Hiebert, *Virology* **1967**, *32*, 354–356.
67 J.A. Speir, S. Munshi, G. Wang, T.S. Baker, J.E. Johnson, *Structure* **1995**, *3*, 63–78.
68 F. Tama, C.L. Brooks, *J. Mol. Biol.* **2002**, *318*, 733–747.
69 S. Brumfield, D. Willits, L. Tang, J.E. Johnson, T. Douglas, M. Young, *J. Gen. Virol.* **2004**, *85*, 1049–1053.
70 K. Schott, R. Ladenstein, A. Konig, A. Bacher, *J. Biol. Chem.* **1990**, *265*, 12686–12689.
71 R.R. Crichton, *Angew. Chem. Int. Ed. Engl.* **1973**, *12*, 57–65.

17
The Tobacco Mosaic Virus as Template

Alexander M. Bittner

Abstract

The tobacco mosaic virus (TMV) is a protein–RNA tube that can either be propagated in plants or assembled from coat proteins and RNA strands. While TMV is the standard example for natural self-assembly (of proteins), and its biology and biochemistry are very well known, the concept of using TMV as scaffold for inorganic structures is relatively new. Reasons for using TMV are manifold. First, proteins offer enormously variable and programmable surfaces, but only aggregates of proteins (such as virus particles) reach lengths of nanometers and even tens of nanometers, whereby useful nano-objects can be assembled. The TMV structure is also one of the best-characterized protein aggregates. The tube shape allows materials to be bound either to the outer wall, forming tubes filled with TMV, or to the inner channel, creating strings of clusters and wires, with such syntheses proceeding in suspension under ambient conditions. The initial driving force to create new and more complex nanoscale materials is technology-based, with typical aims being to reduce the size of computing, memory storage, and sensor devices, to store more functional elements in smaller volumes. A secondary benefit is that nano-objects exhibit physical properties which differ from those of the bulk material.

Key words: tobacco mosaic virus, nanoparticles, nanowires, electroless deposition, self-assembly.

17.1
Introduction

This chapter provides a general introduction to the subject of biomolecules as templates, and expands the subject from tobacco mosaic virus (TMV) to other one-dimensional (1-D) biomacromolecules (fibers and tubes) that are used as templates. Relevant aspects of TMV surface chemistry with regard to the exterior

Handbook of Biomineralization. Edited by P. Behrens and E. Bäuerlein

ISBN: 978-3-527-31805-6

coat and interior channel wall are also discussed. Inorganic structures that are adsorbed onto, or grown on, the exterior surface provide tubes of 18 nm inner diameter, are currently the subject of major investigations. The ensuing potential applications are increasingly the focal point of studies also with DNA- or bacteriophage-based systems. The internal channel in TMV, when used as a template, is only 4 nm in diameter, and permits the creation of true nanoscale structures such as aligned clusters or wires. When exploring the use of other biotemplates, the ultimate goal is the self-assembly of a two- or three-dimensional network [1], in which each molecule is multifunctional and cooperates with at least its neighbors, via a large number of 1-D interconnections.

17.2
Biomolecules as Templates for Nanostructures

Biomolecular fibers and tubes (Fig. 17.1) combine two unique properties: (i) they usually self-assemble from small subunits (proteins, nucleic acids, or sugars); and (ii) they allow for chemical modifications on the nanoscale. When this results in modifications capable of binding inorganic material it can be referred to as "biomolecular mineralization", as opposed to biomineralization and biomimetic processes. Thus, the aim is to exploit the special physical properties of 1-D nanoobjects (for a review, see Ref. [2]), and here we concentrate more on single objects than on large-scale ordered assemblies that may be advantageously synthesized in inorganic pores or oriented after their synthesis [3]. A notable feature of biomolecule/inorganic composites is that, in most cases, artificial nanostructures exhibit a size distribution, whereby not all objects among a sample are of the

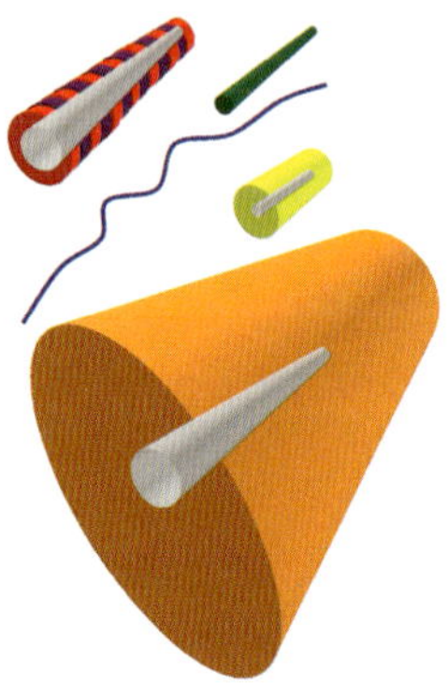

Fig. 17.1 Some fiber- or tube-shaped biomolecular templates, drawn approximately to scale. Yellow: 18 nm-wide, 300 nm-long tubular tobacco mosaic virus (TMV; without nucleic acid), 4 nm interior channel diameter. Blue/red: Microtubulus; the colors signify schematically that microtubuli form from protein dimers. Orange: Peptide tube, e.g., built from Phe-Phe (FF). The interior channels are depicted in gray. Green: M13 bacteriophage (without nucleic acid). Blue: DNA.

same size. In contrast, biomolecules (in a purified sample) are of identical size, (complex) shape, and chemical group content. The point in question is whether the use of a template leads to the creation of structures that are better defined than those synthesized without a template. This is especially important for 1-D objects, where inhomogeneities in diameter are typical of inorganic structures.

Tubes, wires and chains of clusters all show effects of reduced dimensionality, and all physical properties are potentially exploitable, optical (e.g., anisotropic and energy-shifted luminescence), magnetic (e.g., loss of ferromagnetic coupling or anisotropy), electronic (e.g., change of conductivity) and mechanical (change of shape). It should be noted that these properties are rarely unique in their own right – they can be found in macroscale structures and in non-biological nanostructures – but they are nearly always unique when in *combination*. Properties such as the conduction of electrical current or magnetism are usually associated with inorganic structures, which should in principle lead to disadvantages in using biomolecules for nanoscale procedures. This means that biomolecules are often of limited interest *per se*, and serve merely as *templates* for other (inorganic) nanostructures. In this way a bio-inorganic "nanocomposite" is formed, with the biomolecules simply providing their shape as a mold or mask for the structures, and being removed (and even possibly recycled) when the process is complete. Whilst it is possible to produce both micro- and nanoscale devices by using conventional top-down techniques (especially electron-beam lithography), the production of structures with diameters < 20 nm is very difficult, and consequently growth at or in a template offers an elegant alternative. A possible example for a complete self-assembling electronic device on the nanoscale is illustrated in Figure 17.2. This route – which incorporates self-assembly from small subunits, followed by binding, nucleation and growth of the desired material at the desired location, and integration into a functional device – should be planned to tolerate errors.

The reasons for using TMV in this respect are manifold. Although the best-known linear biotemplate is DNA, others are becoming increasingly important, with TMV as perhaps the most important alternative. In the case of DNA, one

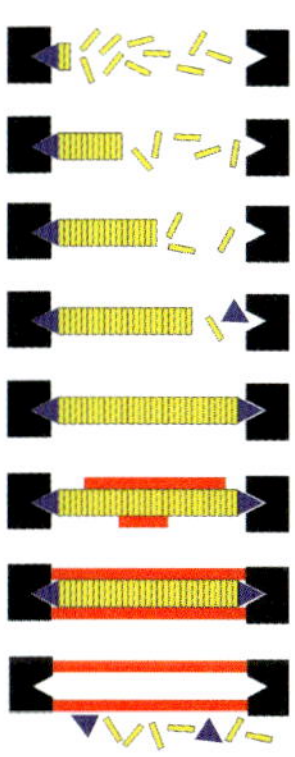

Fig. 17.2 Construction of an electronic nanodevice by self-assembly. The yellow rectangles symbolize self-assembling proteins (or other molecules). By chemical or antigen-antibody linkages (blue), these grow between microcontacts (black). The functional electronic material (red) is selectively deposited on the protein wire. The protein can be recycled.

important point is its simple handling and minimal biohazard, and this is equally valid for TMV as it can be produced simply and in large quantities (in plants). However, reconstitution from artificial RNA and protein produced by ectopic expression may be favorable as the length of the "artificial virus" can be finely tuned, and no infection is required. The yield of virions (virus particles) from infected plants can be heavily influenced by otherwise very welcome (bio)chemical changes of the coat protein. Indeed, the reconstitution of RNA and coat proteins to a functional virus was shown as early as 1955! [4].

We now focus on the templated inorganic nanostructures, and mainly on why these structures have created such interest that ever-increasing numbers of laboratories are currently conducting investigations in these areas. The driving force to create new and more complex nanostructures has two main components. The first component is technology-based, and relates to the scaling-down in areas such as computing, memory storage, and sensor devices, the aim being to store ever-greater numbers of functional elements within ever-smaller volumes. This situation is well demonstrated by the ever-decreasing size of transistors and interconnections in computer chips. The second component is that nano-objects generally exhibit physical properties which differ from those of the bulk material, including the quantization of electrical conductance, the (dis)appearance of ferromagnetic coupling, or changes in light absorption and emission. In this respect, inorganic structures (e.g., metals, metal compounds and ionic crystals) and organic/inorganic composites are often more attractive than organic molecules [5]; good examples here include semiconductor clusters, with their unique and finely tunable light emission. Although nanoscale objects have been produced for many decades (and in some cases for centuries!), one relatively new aspect is the controlled fabrication and integration into devices, with nanometer precision. Unfortunately, until now the fabrication has been much better controlled than the precise integration, and in this respect linear nano-objects (fibers, tubes, chains of clusters) will undoubtedly play a major role as spatial directors and interconnections [2].

The construction of nano-objects is based on the standard inorganic chemistry of metal ions, such as precipitation and redox reactions (Table 17.1). The most interesting aspect here is the interface to the biomolecule [5]: a future nanotechnology (by strict definition) could be built very elegantly on a combination of biology, chemistry, and nanoscale physics [6], the key being to combine biochemical synthetic methods with inorganic methods (Fig. 17.3). The advantages of employing biomolecules as scaffolds are twofold: first, there is a better definition of the nanostructures, compared to template-free syntheses; and second, principally superior methods such as synthesis from small precursors or even by atom manipulation are far from practical. For example, the growth of metal wires of <5 nm thickness (see Section 17.5) is extremely difficult to achieve when using lithographical methods on surfaces. The highly complex chemistry of biomolecules allows the building of nanodevices with various (even multiple) functions and physical properties, which can be integrated into micro- or macroscale devices.

Table 17.1 Examples of chemical deposition reactions of functional materials (in bold type) on TMV.

1. $Au(III) \rightarrow Au(III)_{ads}$ 2. $Au(III)_{ads} + 1.5\ BH_3 + 1.5\ H_2O \rightarrow \mathbf{Au} + 1.5\ BH_2(OH) + 3\ H^+$ (adsorption followed by chemical reduction)
$Ag^+ + e^- + h\nu \rightarrow \mathbf{Ag}$ (photoreduction)
1. $Pd(II) \rightarrow Pd(II)_{ads}$ 2. $Ni(II) + BH_3 + H_2O \rightarrow \mathbf{Ni} + BH_2(OH) + 2\ H^+$ (electroless deposition, catalyzed by Pd and Ni)
$Si(OC_2H_5)_4 + 2\ H_2O \rightarrow \mathbf{SiO_2} + 4\ C_2H_5OH$ (hydrolysis, sol-gel procedure)
$Ti(CH_3)_4 + 2\ H_2O \rightarrow \mathbf{TiO_2} + 4\ CH_4$ (repeated; atomic layer deposition)
$Zn(II) + 2\ OH^- \rightarrow \mathbf{ZnO} + H_2O$ (catalyzed by Pd in presence of NO_3^-)

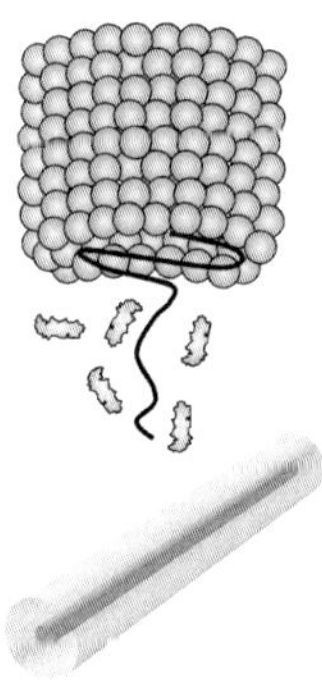

Fig. 17.3 Self-assembly of 6400-bp ss-RNA and 2100 coat proteins to the helical TMV. The RNA is deeply buried inside the proteins (diameter of the RNA helix: 8 nm, diameter of TMV's interior channel: 4 nm). Shorter or longer artificial RNA will result in shorter or longer particles.

In time, this should trigger an exciting competition between sequential "top-down" structuring and parallel "bottom-up" self-assembly: The conventional scaling down of devices (as is well known from microelectronics) can only work sequentially – that is, slowly, on the sub-10-nm scale – whereas molecular recognition and self-assembly permit the building of complex molecules and nanostructures very quickly from smaller units, which assemble in a well-defined manner. It is for these reasons that 1-D (linear) biomolecules as templates are currently a "hot topic" of research [7]. Thus, nanoscale science and nanotechnology will undoubtedly help to tackle some of today's most pressing issues such as placing nano-objects with nanometer precision [1], programmed electrical contacting, and the transition from assembly on surfaces to assembly in three dimensions.

17.3 The Surface Chemistry of TMV

Proteins serve as the prime example of highly complex and extremely variable nanostructures, despite being based on simple principles. In addition, their construction is programmable in DNA, which makes protein synthesis by expression in bacteria or yeast far simpler than organic synthesis. However, it is still impossible – for chemistry and biochemistry alike – to predict all properties of a nanosized molecule, despite highly developed modeling procedures. The simple structure of virions should allow their chemical behavior to be understood and exploited. Virions consist of a nucleic acid strand and a protein cage or tube (the capsid) that is composed of a number of identical proteins, often assembled in helically wound (tubular) fashion, as in TMV (see Fig. 17.3). Viruses are usually specialized for certain hosts: TMV does not infect mammals, and consequently bacteriophages and plant viruses can be handled in the laboratory using only basic safety measures. It should be noted that an infection with plant viruses requires the presence of either living organisms or lesions of the plant tissues (or direct uptake by the root system), while mechanical infection pathways usually rely on the chemical and mechanical stability of the virions, a situation which is highly advantageous in nanoscale science.

Although TMV has been known to exist since 1898, its chemical composition was first determined more recently, during the 1930s [8]. The initial investigations into viruses on the nanoscale were conducted using transmission electron microscopy (TEM) [9–11], but only when TEM was combined with the viral structure (which was determined during the 1930s by using X-ray diffraction [8]) was the full potential of this highly complex pattern of functional groups realized. Subsequently, by using organic synthesis and/or modern genetic engineering,

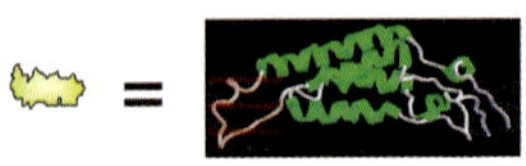

H_2N-SYSITTPS-QFVFLS-SAWA
DPIELINLCTNALGN
QFQ
TQQARTVVQRQFSEVW
KPSPQVTVRFPDSDFKVYRY
NAVLDPLVTALLGA
FDTRNRIIEVENQANPTTAETLDA
TRRVDDATVAIRSAINNLIVELIRG
TGSY-NRSSFES-SSGLVWTSGPAT-COOH

Fig. 17.4 Structure (amino acids 1 to 154) and amino acid sequence (all 158 amino acids) of a single coat protein (top) of the *vulgare* strain. Note that the inner loop that clads the channel (red) is negatively charged, while the adjacent parts of the four-barrel helices (green) are positively charged to accommodate the RNA. The outer surface (blue) accessibility is difficult to estimate as the groove (see Fig. 17.3) exposes additional moieties; however, the surface is clearly hydrophilic and does not exhibit a high charge density.

full use has been made of such groups by controlled modification of single amino acid residues. A model of a single protein is shown in Figure 17.4. In the *vulgare* strain, no His is present, and the only Cys is buried deep in the hydrophobic four-barrel structure; hence, the best ligands for transition metals are not available. In total, 2130 protein molecules are arranged helically, with 16.3 units comprising one turn. The helical RNA strand (8 nm diameter) is buried deep inside the protein structure, and is not exposed to the central channel of 4 nm diameter, nor to the outer surface [12]. The protein loop that forms the wall of the central channel shows less conformational stability [13]. This loop and the exterior surface may be responsible for the relatively large mobility of the C atoms inside the proteins that can be demonstrated using nuclear magnetic resonance (NMR) [14]. The particle length is 300 nm, with 18 nm exterior diameter [15, 16], but linear end-to-end alignment frequently occurs, yielding multiples of 300 nm in length. The pure protein can also self-assemble into long, 18-nm-wide tubular structures which are, of course, non-infective. TMV is able to tolerate ethanol, aqueous dimethyl sulfoxide and temperatures up to 90 °C; it is not affected by pH values from 3.5 up to about 9 for at least several hours, and it retains its infectivity when in dried form, for example in cigarettes. Site-directed mutations such as Glu → Gln at position 50 ("E50Q"), E95Q and D109N can enhance the stability by reducing the negative charge [17]. The hydrophilic exterior TMV surface provides Ser and Thr moieties, as well as the C- and the N-termini of the coat proteins, and it is slightly negatively charged when the pH in a suspension is higher than the pI value of 3.5. Consequently, binding to hydrophobic, uncharged graphite is slow and weak [18]. Cd(II) can induce side-by side assembly also in two dimensions, presumably by forming interviral salt bridges [19].

In order to preserve the chemical structure, binding molecules or metal cations (or complexes) from aqueous solutions is preferable, while in principle gas-phase reactions pose no problem [20, 21]. Strategies or recipes for site-specific binding of metal ions or complexes exist only in rare cases [22]; they are better developed for the special case of isomorphic heavy atom substitutions in protein crystals. More important are *mutations*, which can be used to create a certain number of reactive groups at preselected sites of a protein or of a protein assembly [23]. In terms of applications, transition-metal ions might be the most attractive (see Section 17.4); consequently, strongly binding ligands such as Cys are preferred for mutations [24, 25]. His, with its ability to bind many ions strongly (Ni(II) is used mostly), should be a viable alternative. Schlick et al. have shown that some organic reactions are also site-specific and very efficient [26]; the simplicity of these reactions will clearly open new opportunities for TMV research.

The M13 bacteriophage, as shown using phage display techniques, can easily be handled and modified in the laboratory, and has been used in nanoscale science for some years. The combinatorial synthesis of random amino acid sequences paved the way to the highly selective immobilization and growth of various salts, and even metallic alloys [27]. Another approach, again based on combinatorial screening, utilizes antibodies; an example of this is immunostaining with gold (Au) clusters bound to antibodies, selectively attached to the TMV surface or to one of its end faces [28]. Nonetheless, the rational design of protein surfaces of equal complexity to TMV remains a major challenge.

17.4
Nanostructures on the Exterior TMV Surface

Nanocomposites of biomolecule and inorganic materials demonstrate the advantage of biomolecules, even in application-oriented systems. Few nanoscale templates are available in organic and inorganic chemistry, and their shape, size and chemistry are in most cases neither well defined nor easily varied. The template function of biomolecules cannot always yield perfectly defined products, however, and unfortunately TMV proved to be no exception to this rule. Yet, in view of the technological and even scientific applications, a perfectly defined composite mass might be less important than a well-defined shape and size; in other words, it often suffices that only *one* physical property is well controlled (e.g., length *or* diameter *or* magnetism). The plethora of chemical groups on TMV provides enormous advantage over inorganic and even organic polymer structures, and although spatially selective binding or the synthesis of inorganic materials is usually not achieved, antibodies can be engineered to bind with extreme specificity – for example, exclusively to one end face of TMV [28]. The problem of antibody production can be circumvented elegantly by employing citrate-covered Au clusters (Fig. 17.5) [29]. In contrast, cationic Au clusters seem to assemble preferentially on the exterior surface [30]. The binding of metal clusters to TMV was first demonstrated during the late 1930s [7, 9–11], and inspired many pioneering investigations which involved the electron microscopic examination and staining of biological samples.

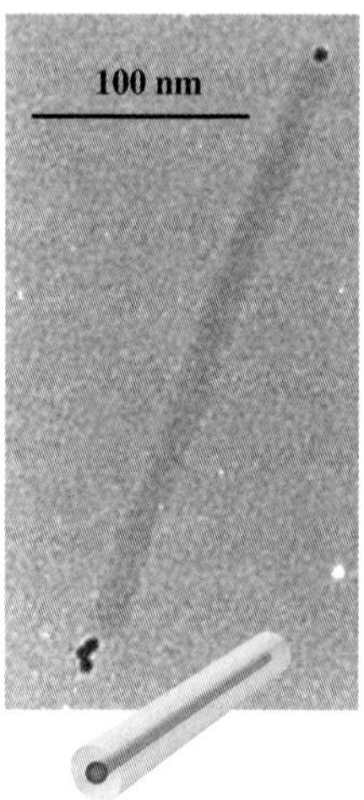

Fig. 17.5 Transmission electron microscopy (TEM) image of Au clusters selectively bound to the ends of TMV. The surface chemistry of TMV requires merely citrate-covered Au, a standard reagent. Interactions between Au and exposed RNA stabilize the structure. TMV *vulgare* with the mutation G155S (which does not influence the end faces) was used here and in all following TEM images.

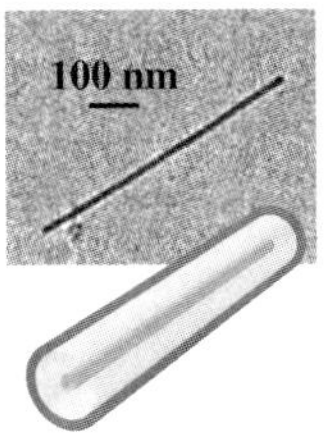

Fig. 17.6 TEM image, example of a typical nucleation scenario on TMV (two virions). After immersion in Pd(II) solution and washing, the Pd-TMV suspension was reduced with the reductant dimethylamine borane. The surface is covered with small Pd clusters.

Various mutation and chemical modification strategies concentrate on residues on the exterior surface of TMV [23–26]. In passing, it should be noted that a variety of metal ions [Ca(II), Mg(II), Al(III), Zn(II)] is present, but these are bound firmly to the proteins [31]. Depending on the material to be produced, either counterions can be offered to induce the precipitation of a salt, or reductants to transfer the ion to the zero-valent state.

Let us first investigate the better-known synthesis of metals on TMV. Various stages of coating with Co, Ni, Ag, Pd, and Au clusters can be achieved without special ligands (see Fig. 17.6). Clearly, the presence of carboxylate and OH groups is beneficial, but in general the methods listed in Table 17.1 have proved to be useful in both the presence [24, 25, 32] and absence [33–37] of surface modifications. It should be noted that, during the reduction of ions to metal clusters, the ions or atoms must coalesce (which is thermodynamically much favored), and so some mobility on the surface is necessary. When Cys is present, the bonds not only to the metal ions but also to many metal clusters (especially for noble metals) will be much enhanced, and this was indeed observed for the product [24, 25]. Surprisingly, similar coverages can be obtained without surface modification; in the case of Pd(II), the presence of $HPO_4{}^{2-}$ likely aids the binding. One major drawback of this reduction method is that the clusters remain isolated: continued metallization to thick and continuous structures cannot be easily achieved, although tubular structures would be more attractive, for example in terms of electrical conductivity or magnetic moment. Metal evaporation results in the coating of only a small part of the virion, albeit with a very well controllable thin layer [20].

A good alternative approach here is the *electroless deposition* of metals, which is comparable to the "enhancement" of latent photographic patterns. Here, a solution with the metal ions, complexants and a reductant is contacted with TMV that has been pretreated with noble metal complexes such as Pt(II), Pd(II) or Au(I). After exchanging the solution to the electroless deposition bath, the deposition begins, catalyzed by the noble metal clusters. The crucial point here is that the deposited metal catalyzes its own reductive deposition, and consequently the metal clusters will *continue* to grow. The structure shown in Figure 17.7 was obtained in this way. Electroless deposition is very well suited to the creation of continuous wires of many micrometers length on DNA [1] and on end-to-end aggregated virions [35], although shape control is often unsatisfactory.

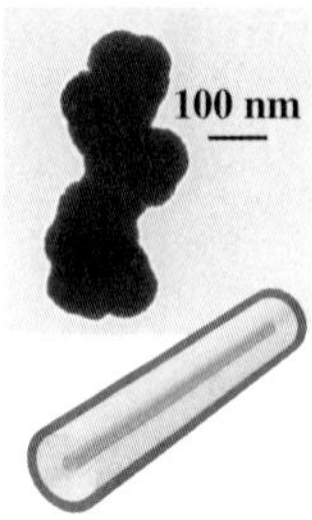

Fig. 17.7 Electroless deposition of Ni on a single virion (TEM). TMV was first immersed in Pd(II), then contacted with a bath containing Ni(II) and dimethylamine borane. Ni forms first on Pd (see Fig. 17.6), thereafter on the growing Ni layer, until the virion is completely covered.

Although the production of salts and oxides were amongst the first reactions to be tested on TMV [38], these processes are gaining new momentum. The production methods used include hydrolysis or similar substitution reactions (Table 17.1) that produce insoluble compounds. In contrast to metallizations, the nucleation can be better controlled, and smooth TiO_2 layers of only a few nanometers thickness can be produced by the hydrolysis of alkoxy derivatives of Ti(IV) [39]. Controlled cycles of adsorption and exposure to water vapor comprise the method of atomic layer deposition, which is based on the same reaction, although alkyl derivatives can also be used [21]. SiO_2 coatings, based on the hydrolysis of alkoxy derivates of Si(IV), are further typical examples [38, 40]. Other compounds such as iron oxides, PbS and CdS proved more challenging, but continuous layers can be produced with these materials [38, 41]. A catalytic process for the deposition of ZnO layers can also be employed for TMV (Fig. 17.8). As with electroless metal deposition, the virion is first activated with Pd(II), the deposition reaction apparently being based on the reduction of nitrate at Pd, which results in a local increase in pH and subsequent precipitation of ZnO [29].

What is the basis for a successful mineralization? In order to direct the synthesis to the TMV surface, binding of the metal cations is required. It appears that – in analogy to the Cd(II) salt bridges [19] – many types of metal ions can bind quite well. Given the dense coverage with OH groups, this behavior is well understood. It should be noted that, on a larger scale, the interstitial space between densely arranged virions can be used as a template for SiO_2 [42], but this is more comparable with biomineralizations. The control of growth, although superior to that achieved with metals, is far from perfect, and it is not quite clear why SiO_2 layers are especially thin and smooth. Certainly, atomic layer deposition

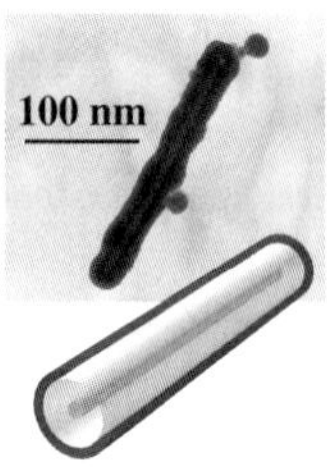

Fig. 17.8 TEM images: mineralization of TMV with ZnO. First, the virus suspension was activated with Pd(II), and then contacted with $Zn(NO_3)_2$ and dimethylamine borane, producing a thick layer of ZnO. Two spheres have nucleated outside the ZnO coating.

offers a much higher degree of process control (essentially monolayer control), whereas catalytic schemes benefit from the simple approach (synthesis at ambient conditions).

With regards to applications of the produced composites, it is important to control the orientation in order to facilitate contacting (see Fig. 17.2) or to orient the composite. It should be noted that, with current technology, such assembly tasks require in almost all cases a supporting flat substrate and careful control of the surface's chemistry. One very important example here is the improvement of information storage in magnetic structures. Nano-objects allow for an extremely high density, as exemplified by a ferritin derivative [43]. It is intriguing that typical sizes of virions are close to minimum sizes for ferromagnetic crystals (ranging from a few nanometers up to about 20 nm). In this respect TMV, with its intrinsically anisotropic structure, is even more interesting. Magnetic characterizations of Fe_xO_y [38] or Co structures on TMV [35] are not yet available, but tubes with walls of more than 10 nm thickness can be expected to show normal ferromagnetism. The anisotropy of such tubes is most likely very attractive, as the magnetization direction is probably aligned with the tube axis, and the switching field may be quite large. Another example is provided by SiO_2: calculations have shown that the optical properties (phonon frequencies) of SiO_2 tubes depend on the presence of a virion inside the tube [44]. The measurement of such effects requires either large amounts of extraordinarily pure composites, or nano-optical techniques in order to effect a local analysis. Islam et al. recorded a local near-field optical signal as early as 1997 [45], while Brehm et al. later detected several infra-red bands with <10 nm spatial resolution [46]. A third application example concerns electrical conductivity which, as yet, has been demonstrated only for Pt-covered TMV with a Cys surface modification [24]. Clearly, the presence of Pt induces some conductivity, but the nature of the transport phenomenon awaits further clarification. Uncovered virions are non-conductive [24, 29], as would be expected. In order to make full use of the self-assembly, a TMV-like structure should be used, as suggested by Figure 17.2. One-dimensional nanostructures bind to microscopic contacts, self-assemble into a long wire, and are then chemically modified with extremely high specificity by inorganic and organic substances, comparable to the DNA-templated nanotube transistor [1].

The possibility of orienting the nanocomposites is crucial, as only then can defined electrical connections be built [1], and the photons interact with structures in a directional manner. Figure 17.9 shows that this can be simply achieved by the application of directional forces during adsorption onto a surface [29, 47]. An alternative here might be dip-pen lithography, which can create ordered patterns of only one or few virions [48]. One basic question for future applications is whether an as-yet not constructed *complete network* of nanostructures is required (see [1]). The use of self-assembly represents probably the most elegant and also efficient means to build such complex structures, and at least for the end-to-end assembly this is viable for TMV (see Fig. 17.3). However, self-assembly implies also *self-disassembly*, and so a certain fraction of nanostructures will disintegrate. Remedies for this could be post-treatments such as covalent linkages

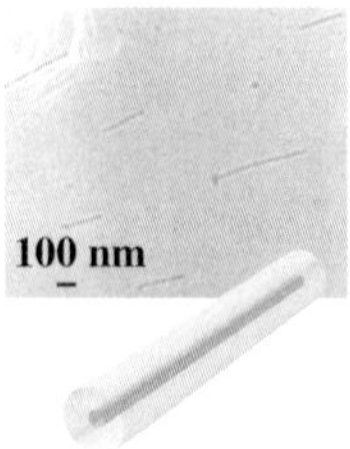

Fig. 17.9 Simple orientation of TMV on a surface (here on a carbon/polymer-coated electron microscopy grid imaged by TEM); the sample was deposited from a droplet of suspension that was removed by suction in the direction of the axes. Some virions are end-to-end aligned (600 nm length).

after self-assembly, perhaps with glutaraldehyde (a standard fixation method). Mutations may also be helpful in this respect; for example, the Glu50 → Gln50 (E50Q) mutation of the TMV coat protein increases inter-protein interactions and results in a higher stability of the virion [17]. Alternatively, benefit might even be gained from disassembly by recycling valuable biomolecules (see Fig. 17.2).

17.5
Clusters and Wires inside the 4-nm-Wide Channel of TMV

With regards to uniformity and shape control, confinement inside the templates represents the most successful route towards the creation of sub-10 nm composites. This scale is very difficult to attain with standard lithography methods, let alone control of the composite's shape. Usually, pores in aluminum oxide, tracks etched in polymers, or the inner surfaces of zeolites and carbon tubes are used [3]. For TMV, and also for other systems (microtubuli, peptide tubes), a detailed understanding of the chemistry of the channel (pore) walls is lacking, although several impressive experiments have been demonstrated in this respect.

First, the synthesis of various metal clusters can be directed inside the interior channel [33–35]. To this end, the presence of salts must be excluded by careful dialysis, and the pH must also be adjusted to neutral or slightly basic. The deprotonation of COOH groups should now be especially effective in the inner loop (see Fig. 17.4), as the outer parts have only one COOH group per protein unit, and since all other parts cannot be accessed (the TMV walls are impenetrable, as shown by uranyl staining of the channel). The carboxylate groups can now enter the ligand sphere of typical precursor complexes such as $PdCl_4{}^{2-}$, $PtCl_4{}^{2-}$, or $AuCl_4{}^{-}$. However, binding to amine or amide residues should also be possible. Here, the most powerful analysis method might be mass spectroscopy of the complete virion, as demonstrated by Fuerstenau et al. [49].

In a second step, the photochemical reduction or electroless deposition of a second metal (Ni) yielded several 3- to 4-nm clusters aligned inside the channel. The use of stronger reductants (dimethylamine borane instead of light or hypophosphite) increases the electroless deposition kinetics so much that probably only one or two nucleation centers [adsorbed Pd(II) or Pt(II)] serve as starting points. The autocatalytic growth produces metal wires of only ~3 nm diameter (Fig. 17.10), and this has been demonstrated for Ni, Co and Cu [35, 47, 50, 51]. In the

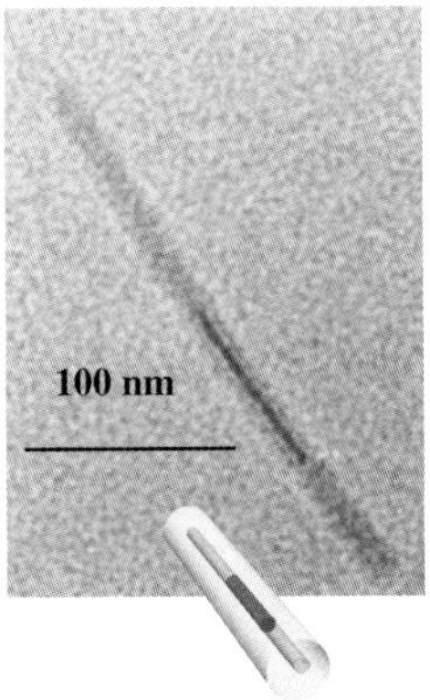

Fig. 17.10 Ni wire of ~3 nm diameter in the interior channel of a TMV. The virus suspension was treated as for Figure 17.7, but first carefully dialyzed to remove phosphate (TEM image).

case of end-to-end aligned virions, the channels connect, and up to 600-nm-long wires can be observed [50]; hence, a rather large flow of the reactants is required. It should be noted that the number of metal ions in the solution confined to the 300-nm-long channel corresponds only to about 75 – which explains why intact tubes never contain more than two wires. This example proves the ability of electroless deposition to access extremely small cavities, which had already been hinted at by electroless deposition of down to about 20-nm-wide Ag wires inside peptide tubes [52]. Generally, electroless deposition is advantageous for high-aspect ratio structures (hence for tubes) as they can be filled or decorated without clogging up or closing an orifice. As shown more clearly in Figures 17.5 to 17.10, the analysis of such small structures is usually based on TEM. A combination of TEM imaging with an analysis of the energy loss yields local chemical information (absorption edges of elements) on the 1-nm scale. In this way, the chemical identity of the wires can be conclusively proven [51].

17.6 Perspectives

Leaving aside the by-now demonstrated suitability of TMV and other 1-D structures (DNA and the M13 bacteriophage) as templates, one can attempt to determine which other biomolecules might qualify as templates. The most important requirements for the nanotechnological application of biomolecules when building a nanodevice are that they should be mechanically stable, chemically resistant against solvents or gases, and ideally also stable at elevated temperatures. Indeed, biomolecules such as DNA, some viruses, membrane proteins of archaea, prions and prion-related peptides, and also carbohydrates, show thermal and chemical stabilities that may be acceptable for building stable devices. Moreover, any potential nanodevice might be prepared under biocompatible conditions; today's Cu interconnects in computer chips are fabricated electrochemically at ambient pressure and temperature. As-yet uncommon biomolecules in nanoscale science include carbohydrates (e.g., saccharides), viroids (pure RNA loops which are highly infective for certain plants), cells (which may be too complex and not sufficiently stable), and the larger parts of cells (e.g., organelles). These all offer chemical and structural functionalities that are, as yet, largely unexploited for nanoscale science, and this applies also to the widely used single molecules and self-assembled molecular structures. In other cases, biomolecules are already investigated in detail for possible applications – for example, actin and myosin should be promising elements for artificial linear motors, while hemolysin can form tunable molecular pores or channels. The future may also favor the use of pure organic molecules or biomolecules: a *single* molecule can – in analogy to composites – exhibit electrical conductivity, an interesting electronic structure, and a defined mechanical behavior. The integration of single molecules in devices may once again require 1-D composites, for example as nanoscale wires.

References

1 E. Braun, K. Keren, *Adv. Physics* **2004**, *53*, 441.
2 Z. Tang, N.A. Kotov, *Adv. Mater.* **2005**, *17*, 951.
3 Y. Xia, P. Yang, Y. Sun, Y. Wu, B. Mayers, B. Gates, Y. Yin, F. Kim, H. Yan, *Adv. Mater.* **2003**, *15*, 353.
4 H. Fraenkel-Conrat, R.C. Williams, *Proc. Natl. Acad. Sci. USA* **1955**, *41*, 690.
5 A.M. Bittner, *Surf. Sci. Rep.* **2006**, *61*, 383.
6 R.A.L. Jones, *Soft Machines*. Oxford University Press, New York, **2004**.
7 A.M. Bittner, *Naturwissenschaften* **2005**, *92*, 51.
8 B.D. Harrison, T.M.A. Wilson, *Phil. Trans. R. Soc. Lond. B* **1999**, *354*, 521.
9 H. Ruska, *Archiv. ges. Virusforsch.* **1943**, *2*, 480.
10 G.A. Kausche, E. Pfankuch, H. Ruska, *Naturwissenschaften* **1939**, *27*, 292.
11 G.A. Kausche, *Biol. Zbl.* **1940**, *60*, 179.
12 M. Zaitlin, *AAB Descriptions of Plant Viruses* **2000**, *370*, 8.
13 H. Wang, G. Stubbs, *Acta Crystallogr.* **1993**, *A49*, 504.
14 J.L. DeWit, N.C.M. Alma-Zeestraten, M.A. Hemminga, T.J. Schaafsma, *Biochemistry* **1979**, *18*, 3973.
15 A. Klug, *Phil. Trans. R. Soc. Lond. B* **1999**, *354*, 531.
16 G. Stubbs, *Phil. Trans. R. Soc. Lond. B* **1999**, *354*, 551.
17 B. Lu, G. Stubbs, J.N. Culver, *Virology* **1996**, *225*, 11.
18 M. Knez, M.P. Sumser, A.M. Bittner, C. Wege, H. Jeske, D.M.P. Hoffmann, K. Kuhnke, K. Kern, *Langmuir* **2004**, *20*, 441.
19 A. Nedoluzhko, T. Douglas, *J. Inorg. Biochem.* **2001**, *84*, 233.
20 H. Kahler, B.J. Lloyd, Jr., *J. Appl. Phys.* **1950**, *21*, 699.
21 M. Knez, A. Kadri, C. Wege, U. Gösele, H. Jeske, K. Nielsch, *Nano Lett.* **2006**, *6*, 1172.
22 S. Mann (Ed.), in: *Biomimetic Materials Chemistry*. VCH, New York, **1996**.
23 H. Yi, S. Nisar, S.-Y. Lee, M.A. Powers, W.E. Bentley, G.F. Payne, R. Ghodssi, G.W. Rubloff, M.T. Harris, J.N. Culver, *Nano Lett.* **2005**, *5*, 1931.
24 S.-Y. Lee, J. Choi, E. Royston, D.B. Janes, J.N. Culver, M.T. Harris, *J. Nanosci. Nanotechnol.* **2006**, *6*, 974.
25 S.-Y. Lee, E. Royston, J.N. Culver, M.T. Harris, *Nanotechnology* **2005**, *16*, 8435.
26 T.L. Schlick, Z. Ding, E.W. Kovacs, M.B. Francis, *J. Am. Chem. Soc.* **2005**, *127*, 3718.

27 C.E. Flynn, S.-W. Lee, B.R. Peelle, A.M. Belcher, *Acta Materialia* **2003**, *51*, 5867.

28 M.H.V. v. Regenmortel, *Phil. Trans. R. Soc. Lond. B* **1999**, *354*, 559.

29 S. Balci, PhD thesis. EPF, Lausanne, **2006**.

30 T. Yonezawa, S.-Y. Onoue, N. Kimizuka, *Chem. Lett.* **2005**, *34*, 1498.

31 W.E.C. Wacker, M.P. Gordon, J.W. Huff, *Biochemistry* **1963**, *2*, 716.

32 M. Demir, M.H.B. Stowell, *Nanotechnology* **2002**, *13*, 541.

33 E. Dujardin, C. Peet, G. Stubbs, J.N. Culver, S. Mann, *Nano Lett.* **2003**, *3*, 413.

34 M. Knez, M. Sumser, A.M. Bittner, C. Wege, H. Jeske, S. Kooi, M. Burghard, K. Kern, *J. Electroanal. Chem.* **2002**, *522*, 70.

35 M. Knez, M. Sumser, A.M. Bittner, C. Wege, H. Jeske, T.P. Martin, K. Kern, *Adv. Funct. Mater.* **2004**, *14*, 116.

36 W.L. Liu, K. Alim, A.A. Balandin, *Appl. Phys. Lett.* **2005**, *86*, 253108.

37 S.-Y. Lee, J.N. Culver, M.T. Harris, *J. Colloid Interf. Sci.* **2006**, *297*, 554.

38 W. Shenton, T. Douglas, M. Young, G. Stubbs, S. Mann, *Adv. Mater.* **1999**, *11*, 253.

39 S. Fujikawa, T. Kunitake, *Langmuir* **2003**, *19*, 6545.

40 E. Royston, S.-Y. Lee, J.N. Culver, M.T. Harris, *J. Colloid Interf. Sci.* **2006**, *298*, 706.

41 T. Douglas, M. Young, *Adv. Mater.* **1999**, *11*, 679.

42 C.E. Fowler, W. Shenton, G. Stubbs, S. Mann, *Adv. Mater.* **2001**, *13*, 1266.

43 E. Mayes, A. Bewick, D. Gleeson, J. Hoinville, R. Jones, O. Kasyutich, A. Nartowski, B. Warne, J. Wiggins, K.K.W. Wong, *IEEE Trans. Magn.* **2003**, *39*, 624.

44 V.A. Fonoberov, A.A. Balandin, *Nano Lett.* **2005**, *5*, 1920.

45 M.N. Islam, X.K. Zhao, A.A. Said, S.S. Mickel, C.F. Vail, *Appl. Phys. Lett.* **1997**, *71*, 2886.

46 M. Brehm, T. Taubner, R. Hillenbrand, F. Keilmann, *Nano Lett.* **2006**, *6*, 1307.

47 A.M. Bittner, X.C. Wu, S. Balci, M. Knez, A. Kadri, K. Kern, *Eur. J. Inorg. Chem.* **2005**, 3717.

48 R.A. Vega, D. Maspoch, K. Salaita, C.A. Mirkin, *Angew. Chem. Int. Ed.* **2005**, *44*, 6013.

49 S.D. Fuerstenau, W.H. Benner, J.J. Thomas, C. Brugidou, B. Bothner, G. Siuzdak, *Angew. Chem. Int. Ed.* **2001**, *40*, 542.

50 M. Knez, A.M. Bittner, F. Boes, C. Wege, H. Jeske, E. Maiß, K. Kern, *Nano Lett.* **2003**, *3*, 1079.

51 S. Balci, A.M. Bittner, K. Hahn, C. Scheu, M. Knez, A. Kadri, C. Wege, H. Jeske, K. Kern, *Electrochim. Acta* **2006**, *51*, 6251.

52 M. Reches, E. Gazit, *Science* **2003**, *300*, 625.

Part V
Encapsulation

Handbook of Biomineralization. Edited by P. Behrens and E. Bäuerlein

ISBN: 978-3-527-31805-6

18
Biomimetic Biopolymer/Silica Capsules for Biomedical Applications

Michel Boissière, Joachim Allouche, and Thibaud Coradin

Abstract

Several living organisms build up silica shells as protections against their environment. Following a biomimetic approach, biopolymer/silica hybrid macro-, micro- and nanocapsules were designed and their potential application for encapsulation and drug release systems evaluated. In the case of alginate, silica deposition involves a first coating of poly-L-lysine and leads to dense stable membranes. In contrast, gelatin capsules interact directly with silicates to form core/shell particles with particulate surfaces. In both cases, the presence of the silica coating enhances the mechanical and/or thermal stability of the materials. Moreover, cellular uptake experiments indicate that these hybrid particles can be internalized without inducing cell death, a first indication of their biocompatibility. The possible extension of this approach to a wide range of synthetic and biological macromolecules can be envisioned, suggesting that such a biomimetic approach is a promising route for the design of bio-functional hybrid nanomaterials.

Key words: silica, biopolymer, hybrid materials, biomimetism, core/shell particles, encapsulation, drug release systems, alginate, gelatin.

18.1
Introduction

In Nature, biogenic silica coatings mainly play a protective role for the cells that they entrap [1]. Thus, silica shells can protect living organisms against biotic (predator (diatoms) [2] or parasite (plants) [3]) and abiotic (acidic environment (bacteria) [4], UV radiation (archea) [5]) stresses.

This ability of silica to stabilize organic or biological systems has been widely used for the design of hybrid materials [6]. However, even if these materials are obtained via soft chemistry routes, very few of them are formed in biomimetic conditions, as the reaction conditions often involve organic solvents and organo-

Handbook of Biomineralization. Edited by P. Behrens and E. Bäuerlein

ISBN: 978-3-527-31805-6

metallic precursors (silicon alkoxides). Even in the case of bioencapsulation [7], where the preservation of the biological activity of entrapped systems requires synthetic conditions very close to natural environments, examples of materials inspired from biogenic silica are scarce. For instance, diatoms- and sponge-related peptides were used for enzyme encapsulation in silica gels [8, 9]. In the case of cell encapsulation, it was found that silicate and colloidal silica precursors, corresponding to the form of silica that living organisms have learned to cope with, were more suitable for bacteria survival than silicon alkoxides [10, 11].

Indeed, when designing biomimetic materials, one often tends to focus on a single aspect of the biomineralization process, such as templating molecules [12, 13], confined media or self-assembly processes [14, 15]. In this context, biopolymer/silica hybrid materials may be considered as suitable models to examine the influence of silicification on biological systems. More specifically, as biopolymers are widely used for the design of biomaterials, several groups have evaluated the possible interest of combining bio-macromolecules with silica for biomedical applications [16].

In this chapter, these approaches are illustrated through the examples of two different systems – that is, alginate- and gelatin-based hybrid capsules. Both materials could be developed as macro-, micro- and nanocapsules, and were evaluated for their potential biotechonological and biomedical applications. Moreover, they were elaborated via biomimetic routes in the sense that they involve silica formation from natural inorganic precursors (i.e., soluble silicates) in the presence of biosilicification-related template molecules. A comparison between these two systems reveals the determining influence of biopolymer–silicate interactions on material properties and provides fruitful insights for the future developments of biomimetic silica-based hybrid biomaterials.

18.2
Biomimetic Alginate/Silica Hybrid Capsules

18.2.1
Alginate Capsules in Biotechnology and Medicine

Alginic acid is a polysaccharide mainly extracted from brown algae. It is a linear co-polymer of β-D-mannuronic and α-L-guluronic acid (two carboxylated sugars) that is usually recovered in its deprotonated form, alginate, associated with sodium ions. The addition of divalent cations (e.g., Ca^{2+}) induces cross-linking of the polymer via coordination of the metal ions by carboxylate groups of the guluronic segments [17]. Alginate gels are commonly prepared as beads, in the 0.5- to 5-mm-diameter range, by using a dripping procedure [18]. In short, alginate solutions are added dropwise to a calcium chloride (or other divalent cation salt) solution. When the drop reaches the solution, rapid gel formation occurs at its surface, freezing the alginate in its initial shape – that is, a sphere. These beads are widely used for enzyme and cell encapsulation in biotechnological applications, especially for bioreactor design [19].

Alginate can also be associated with polyelectrolytes, such as polyamines, that are deposited on the capsule surface following a layer-by-layer approach [20]. This leads to the formation of a membrane the diffusion properties of which can be tailored by the nature of the added polyelectrolyte. Moreover, due to the weak alginate–cation bond strength, bead cores can be liquefied by the addition of strong chelates, such as sodium citrate, to form hollow particles [21]. In this context, poly-L-lysine (PLL)/alginate microcapsules have been largely studied for cell encapsulation applications in medicine [22]. More specifically, these capsules have been used for the design of so-called "artificial organs", a promising alternative to cellular grafts where immobilized active cells are protected from the immune response system by the membrane barrier [23]. At this time, the most successful devices are PLL/alginate-encapsulated islets of Langerhans; these are used as artificial pancreas in diabetes therapy and currently are in clinical phase evaluation [24].

PLL/alginate particles have also been studied as drug carriers, more specifically for oligonucleotide delivery [25]. In this case, diluted solutions of alginate and calcium ions are mixed to form a pre-gel, after which PLL is then added, allowing the recovery of micro- and nanoparticles [26].

Different inorganic phases, such as calcium phosphate [27] or bioglass particles [28], have been associated with alginate for biomedical applications. In the case of silica, several composite gels have been described [16], but attention was mostly paid to hybrid capsules.

18.2.2 Alginate/Silica Hybrid Capsules

Two main routes have been developed to design alginate/silica hybrid capsules [29]. The first route relies on the addition of silica precursors (colloidal silica, silicon alkoxide) to the alginate solution before capsule formation [30, 31]. Alternatively, calcium-cross-linked alginate beads can be formed in a first step and then placed in contact with silica precursors [32]. These hybrid capsules were evaluated for enzyme and cell encapsulation, and demonstrated enhanced stability when compared to their purely bio-organic equivalents [33, 34].

Among these different approaches, three have led to silica-coated capsules. The major challenge is to anchor the silica deposit on the capsule surface as both materials bear a negative charge for $pH > 4–5$. In this context, Sakai et al. proposed a procedure involving 3-aminopropyltrimethoxysilane $[(CH_3O)_3Si(CH_2)_3NH_2]$ [35]. The interest of this approach lies in the twofold functionality of this precursor: while $Si(OCH_3)_3$ groups are available for hydrolysis/condensation, the cationic amino groups can interact with alginate. Thus, it is possible to directly deposit a silica layer on an alginate capsule surface. Moreover, all amino groups are not involved in the binding of silica with the bead surface, so that the deposition of an additional alginate layer is possible to improve capsule biocompatibility. The viability of islets of Langerhans encapsulated in these hybrid capsules was successfully maintained after implantation, with insulin secretion activity detected over 100 days (Fig. 18.1a) [36].

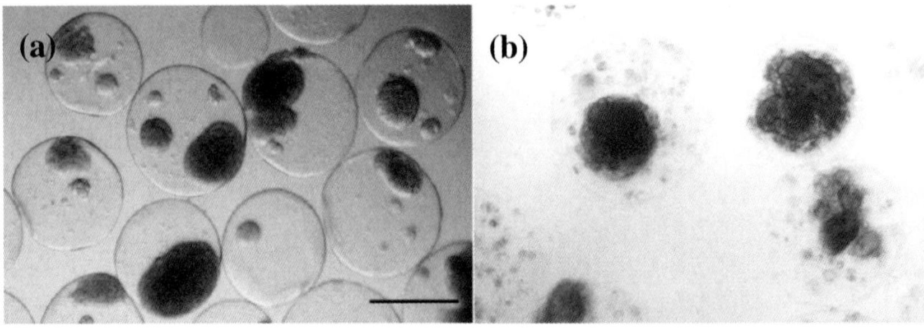

Fig. 18.1 Islets of Langerhans encapsulated in silica-coated alginate beads as reported by (a) the Sakai group (original magnification ×60; reprinted from Ref. [36], with permission from Elsevier), and (b) the Carturan group (original magnification ×100; reproduced from Ref. [39], with permission from the Royal Society of Chemistry).

An alternative approach was reported by Carturan et al., the procedure being based on the Biosil process. This involved a pre-immobilization of cells on a biopolymer scaffold placed under a gas flow consisting of an inert gas vector saturated with vapors of volatile silicon alkoxides [37]. Silicon alkoxide hydrolysis occurs at the contact with the humid scaffold surface, but the released alcohol molecules are carried away by the gas flow and do not reach the encapsulated cells. A thin layer of silica is deposited, the properties of which can be tailored by the reaction conditions. Interestingly, the silica deposition process occurs before the gelation of the alginate bead so that it strongly influences the size of the hybrid capsules [38]. This procedure was successfully applied to alginate capsule-entrapped Langerhans islets (Fig. 18.1b) [39].

Even if these two procedures lead to the formation of silica outer shells, they cannot be considered as biomimetic. In contrast, Coradin et al. reported an alternative procedure based on PLL/alginate capsules that was directly inspired by biomimetic considerations and could be further scaled down to hybrid nanocapsules [40].

18.2.3
Biomimetic Approaches

This procedure is based on the fact that PLL chains can activate silica condensation from sodium silicate solutions [41, 42]. It was therefore proposed that PLL-coated alginate bead surfaces could be suitable templates for silica formation. Such a coating was achieved by immersing PLL/alginate capsules in sodium silicate solutions at neutral pH [40]. Diluted silica solutions were used to limit condensation on the capsule surface and to avoid gelation in the whole reaction volume. As a result, a thin membrane, of 1–2 μm thickness, was formed (Fig. 18.2). Interestingly, this coating appears smoother than for silica obtained from PLL-sodium silicate mixtures in solution, suggesting an effect of PLL conformation/packing on the capsule surface. After liquefaction of the bead

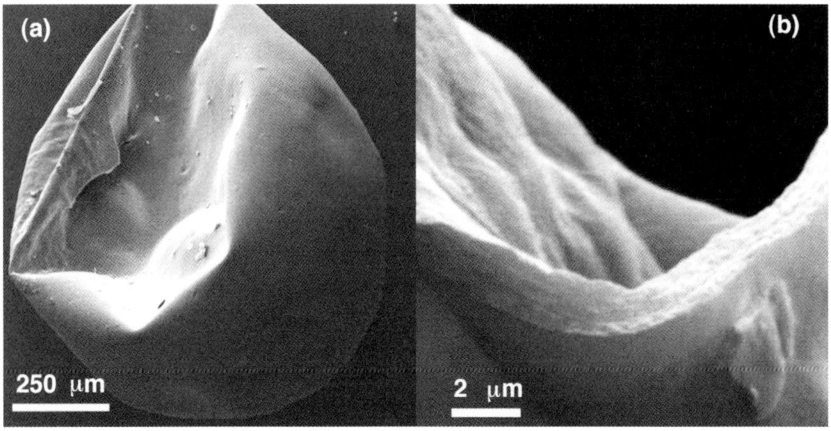

Fig. 18.2 Scanning electron microscopy (SEM) images of alginate/silica macrocapsules after core liquefaction. (a) Overview; (b) magnification showing the smooth surface and the silicified membrane structure.

core, the silica coating was shown to enhance the PLL-alginate membrane mechanical stability but did not significantly modify its permeability. Additionally, similar experiments were performed with colloidal silica, that were also shown to interact with PLL. In this case, however, the resulting coating was very unstable, due to the absence of any strong chemical bonds between the deposited colloids. In term of biomimetics, this suggests that silica nanostructures formation should result from a simultaneous nanoparticle growth/assembly process rather from the packing of pre-formed colloids. These materials were evaluated for the β-galactosidase enzyme encapsulation, and showed a good preservation of its catalytic activity [40]. The cyanobacterium *Nostoc calcicola* was also successfully encapsulated in such capsules, and used for the removal of heavy metals from polluted water by biosorption (S. Ramachandran, personal communication).

If macrocapsules are to be suitable for bioreactor design, then the particles for drug delivery via intravenous injection require dimensions below 100 nm [43], and the possibility of scaling-down the process of silica/PLL/alginate capsule formation was studied in this context. In a first step, PLL/alginate microparticles were obtained using the above-mentioned pre-gel route [26]. Similar to the macrocapsule coating approach, these particles were placed in contact with diluted sodium silicate solutions at near-neutral pH. In this case, hybrid capsules in the range of 1 to 2 μm diameter were recovered, consisting of an alginate core coated with a 50- to 100-nm-thick silica shell (Fig. 18.3) [44].

These particles are still too large for the targeted applications however, and an alternative procedure was developed using a spray-drying approach [44]. When an alginate, calcium and PLL mixture was used, nano- to microspheres (100–1000 nm) could be recovered. However, these exhibited a limited stability in solution and could not be used for further coating. Thus, silicates were introduced in the mixture before spray-drying. In that case, hybrid nanoparticles of 200–300 nm

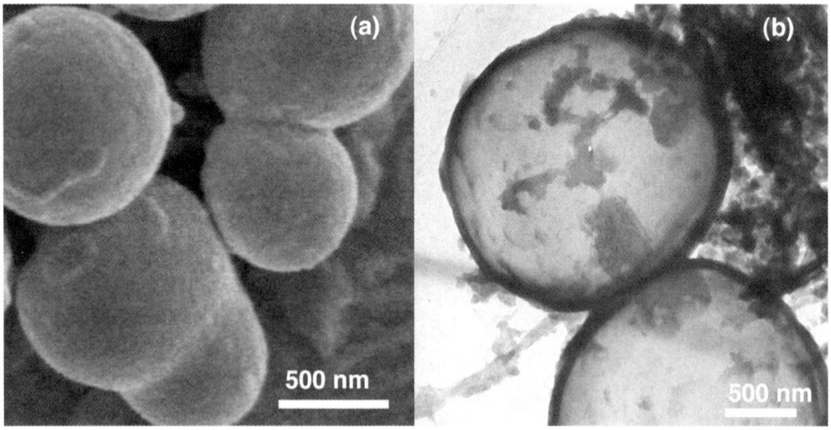

Fig. 18.3 Scanning electron microscopy (a) and transmission electron microscopy (b) images of alginate/silica microparticles.

diameter were obtained (Fig. 18.4a); these did not show any core/shell structure, but rather consisted of an interpenetrating network (IPN) of polymer and silica. Preliminary studies of cellular responses to these nanoparticles were also performed [44]. The hybrid capsules were found to be taken up by 3T3 fibroblasts, without causing cell death, and appeared to be sequestered into specific compartments where they were degraded. Interestingly, during the first 48 h after incor-

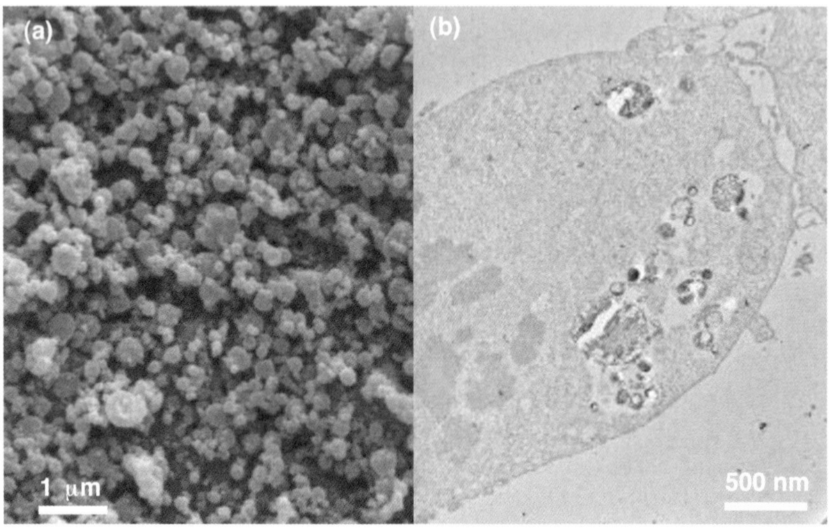

Fig. 18.4 Scanning electron microscopy and transmission electron microscopy images of alginate/silica nanoparticles. (a) As synthesized; (b) after internalization by fibroblasts. (Adapted from Ref. [44].)

poration, only the bio-organic component of the hybrid network had disappeared, leaving highly porous silica nanoparticles (Fig. 18.4b). This finding opens up the possibility of designing two-step delivery systems, with a rapid release of the alginate-encapsulated drug followed by a slow release of the silica-encapsulated drug.

18.2.4 Concluding Remarks

Alginate/silica biomimetic associations within capsules of various sizes are possible via the addition of PLL. Poly-L-lysine has a twofold function: (i) it acts as a bridge between the polymer and the mineral phase; and (ii) it activates silica formation, allowing condensation to occur only at the capsule surface. However, this is only possible if the bio-organic template is stable enough, as demonstrated for nanoparticles. In this case, hybrid capsules are achievable but farther from biomimetic conditions.

In many cases, the presence of silica leads to an enhancement of the stability of the materials, without being detrimental to associated biological functions. However, the biomimetic approaches described here are not always suitable for a given application. For example, in the case of artificial organs, the silicon alkoxide routes developed by Sakai [35] and Carturan [39] appear more promising than the sodium silicate approach, due to a larger flexibility of the process in terms of concentration or chemical functions, as well as to a better stability of the silica network. In contrast, silicates may be more suitable for drug delivery applications where the presence of residual alcohol resulting from incomplete alkoxide hydrolysis would be a strongly limiting factor.

On the basis of these results, it is possible to imagine many different other biopolymer/silica hybrid capsules. When these polymers are negatively charged, the presence of PLL is expected to play a similar role as for alginate, and the first steps in that direction were recently taken with carrageenan [45]. However, if the macromolecules bear a positive charge, then direct electrostatic interactions are possible with silicates. Previously reported results with chitosan [46], lysozyme [47] and polyamines [48, 49] have provided much more data for gelatin-based capsules developed in parallel to alginate/silica materials, and a basis for fruitful comparison.

18.3 Biomimetic Gelatin/Silica Hybrid Capsules

18.3.1 Gelatin Capsules for Biomedical Applications

Gelatin is obtained by the partial denaturation of collagen [50]. Collagen is a generic term for proteins that form triple-helix structures composed of three polypeptide chains, each of which exhibits a glycine (Gly) residue in every third position of the amino acid sequence [51]. Other amino acids are mainly proline and hydroxyproline residues, together with lysine and hydroxylysine. As a result of denaturation, the main constituents of gelatin are large and complex polypeptide molecules of the same amino acid composition as the parent collagen, covering a broad molecular weight distribution range. Type A gelatin produced by acid processing has a typical isoelectric point between pH 7 and 9. Type B gelatin produced by alkaline or lime processing exhibits more reproducible amphoteric characteristics, reaching an isoelectric point of 4.5–5.

Due to their common properties, gelatin is sometimes used as an alternative to collagen in tissue engineering [52]. For bone repair materials, it is often associated with calcium phosphate [53]. Cell-adhesion properties and bioactivity of some gelatin/silica membranes, obtained using silicon alkoxide precursors, were also evaluated [16].

However, the most useful property of gelatin solutions is their ability to form thermotropic gels, via the partial restoration of collagen-like triple helices. For gelatin of animal origin, the gel temperature is close to body conditions, and consequently gelatin-based materials have long been used in pharmaceutical science [54], more specifically as gelatin macrocapsules for oral drug delivery [55]. These capsules are usually obtained via molding processes from melted protein solutions. Micro- and nanocapsules have also been developed as drug delivery carriers, mainly via water-in-oil emulsion routes [56].

18.3.2 Gelatin–Silica Interactions

Just as the development of alginate/silica hybrid capsules was triggered by the discovery of the ability of PLL to activate silica formation, the design of gelatin-based particles originates from biomimetic studies of gelatin–silica interactions. In fact, such studies were first described at the end of the 19th century when it was observed that silicic acid $Si(OH)_4$ combined with gelatin to form insoluble precipitates [57]. Since then, this system has been investigated in much more detail [16], and shown that two different types of interaction could arise, depending on the pH conditions. In acidic media (pH 1–4), below gelatin's isoelectric point, silicates are only weakly charged either positively or negatively, so that hydrogen bond formation is possible. This results in the formation of a gelatin/silica precipitate. In the pH range of 5 to 8, where type A gelatin is still positively charged, silicates bears a negative charge and attractive electrostatic interactions may occur, also leading to precipitation.

More recently, silicate–gelatin interactions have been revisited in the context of a biomimetic approach of biosilicification. Coradin et al. showed that, by controlling reagent concentrations and pH conditions, it was possible to form silica nanoparticles with a narrow size range in the presence of type A gelatin [58]. A model was proposed, involving electrostatic interactions between the biopolymer and the silicates, together with the control of particle growth and packing via the self-association of the gelatin chains upon gelation.

This hypothesis was recently confirmed when gelatin thin films were used as templates for silica particle formation [59]. When gelatin films were placed in contact with silicate solutions at pH 5, two processes were found to occur simultaneously. The fast deposition of a dense monolayer of 10- to 100-nm silica particles was observed at the film surface, followed by the delayed formation of larger particles aggregates (200–2000 nm) that appeared stuck together by polymer fibers. Because gelatin gels tends to dissolve at the contact of water, protein brushes can be formed that extend several micrometers away from the surface. It was therefore proposed that the smallest particles were formed at the contact with the dense gelatin network coating, whereas the largest particles resulted from the interactions of silicates with the loose network of protein brushes. This is in agreement with the hypothesis that the size of the silica particles decreases as the density of templating chains increases.

Overall, these data suggest that gelatin surfaces are suitable for the deposition of silica. Thus, when compared to alginate, a direct reaction between gelatin capsules and silicates could be envisioned. On this basis, the design of gelatin/silica core/shell macro-, micro- and nanoparticles was studied.

18.3.3
Gelatin/Silica Hybrid Capsules

The first attempts to design gelatin/silica capsules were performed on macroparticles [60]. In similarity with alginate, these particles were obtained using a dripping procedure, but as gelatin gelation is thermotropic (instead of ionotropic as for alginate), bead formation was induced by dropping the hot (37 °C) protein solution into a cold (3 °C) water bath. Capsules in the millimeter size range could be recovered and placed into contact with diluted sodium silicate solutions at neutral pH, whereupon a white deposit was formed instantly. Interestingly, hybrid capsules showed no significant variation in diameter, whereas gelatin beads placed in contact with demineralized water were approximately 1 mm larger. This suggested that the silica coating stabilizes the gelatin network via its confinement, avoiding its swelling by additional water uptake. Observations of the mineralized capsules revealed the formation of a silica coating which consisted of sub-micrometric particles entangled with polymer fibers (Fig. 18.5). This coating was very similar to the large aggregates deposited on gelatin thin films, suggesting that its formation proceeds following a similar mechanism – that is, via electrostatic interactions of silicates with protein brushes resulting from gelatin surface dissolution. It was also observed that an increase in silica concentration did not influence particle size or packing but did control the deposited layer thickness over a 5- to 20-μm range. As the density and spatial extension of the gelatin brushes did not vary with silicate concentration, it was suggested that the silica shell thickness was only limited by the available amount of silica precursors with which they can interact.

In order to evaluate the potential of these capsules as drug delivery systems, the influence of the silica coating on molecule and macromolecule internal diffusion

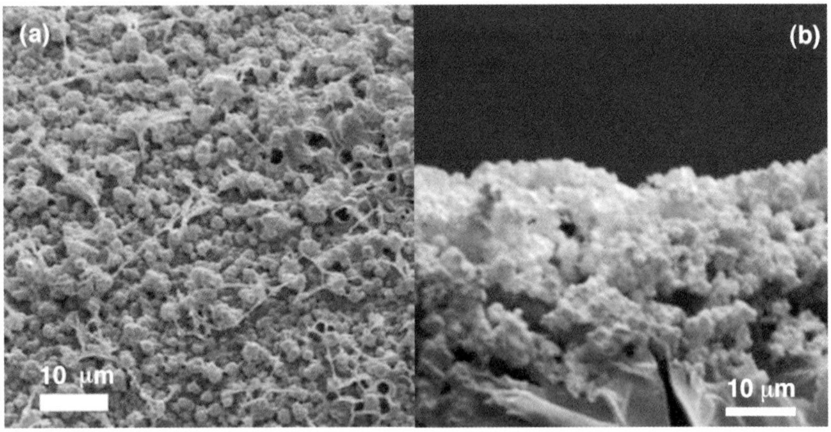

Fig. 18.5 Scanning electron microscopy images of gelatin/silica macroparticles surface (a) and membrane (b).

was studied [60]. At 20 °C, a significant decrease in mass transfer kinetics was observed for silicified capsules. However, this effect could be partly attributed to adsorption of the diffusing molecules on the silica particle surface. Similar experiments performed at 35 °C showed that both types of capsule dissolved rapidly, although this process was delayed over a few minutes in the presence of silica.

This approach was extended to micro- and nanoparticles [61]. The gelatin capsules were obtained via a novel water-in-oil emulsion route. In fact, these capsules are generally obtained at room temperature where the protein network must be stabilized by chemical cross-linking. As an alternative, this route involves the formation of the emulsion at 40 °C, followed by its rapid freezing at low temperature. This process also avoids the gel desolvation that usually occurs when acetone is added to the emulsion to precipitate gelatin particles. Additionally, the size of the resulting particles can be adjusted by stirring conditions (i.e., magnetic stirring for microcapsules and sonication for nanocapsules).

Once again, the coating of these particles was achieved by immersing them in neutral sodium silicate solution at room temperature. In the case of microparticles, a coating of a few micrometers in thickness is observed, consisting of aggregated silica nanoparticles with sizes in the 50- to 200-nm range (Fig. 18.6a). For nanoparticles, silica shells with rough, uneven surfaces are formed on the gelatin surface, the thickness of which increases in the 20- to 50-nm range with increasing silicate concentrations (Fig. 18.6b). Thus, it appears that the silica coating always consists of a layer of aggregated nanoparticles. However, for a given silicate concentration, the nanoparticle size and the silica layer thickness decrease with gelatin capsule dimensions. One explanation for this lies in the fact that smaller surfaces should lead to rapid confluence of growing silica particles, thus limiting their size. As far as shell thickness is concerned, it can be suggested that

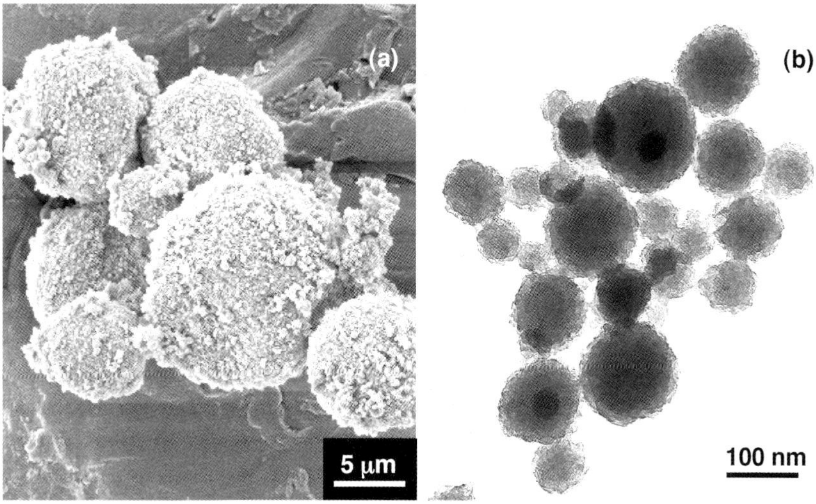

Fig. 18.6 Scanning electron microscopy and transmission electron microscopy images of gelatin/silica microparticles (a) and nanoparticles (b). (Adapted from Ref. [61].)

the surface brushes spatial extension decreases with gelatin particle size, reducing their ability to attract silicate species. In fact, while gelatin fibers are clearly visible within the silica particle network formed on the macrocapsule surface, they are barely discernable on the microsphere outer shell, and absent from nanoparticle coatings.

Additionally, in the case of nanoparticles, if the silicate concentration is too low, well-defined hybrid capsules cannot be recovered, probably due to the poor stability of the silica shell that cannot prevent gelatin dissolution when nanoparticles are dispersed in water at room temperature. Attempts to overcome this problem revealed two important points. First, the silica coating formation was undertaken at low temperature (5 °C) to limit gelatin dissolution. If this approach successfully increased nanoparticle stability, it also significantly slowed the silica condensation process. Thus, a compromise must still be found between the two phenomena. The stabilization of gelatin particles was also achieved via chemical cross-linking using glutaraldehyde, but in this case no silica deposition could be observed, even after several hours of reaction. In fact, glutaraldehyde is a dialdehyde that bridges protein chains together via imine bond formation with amino acid amine groups. However, these amine functions, present as positively charged ammonium at pH 7, are also responsible for the electrostatic interactions between gelatin and silicates. Therefore, once involved in the cross-linking bond formation, they are no longer available for silica condensation activation. The possibility for fibroblast cells to internalize gelatin/silica nanoparticles was also studied [60]. Similar to alginate-based nanocomposites, the biopolymer component is rapidly degraded intracellularly, leaving hollow silica shells (Fig. 18.7).

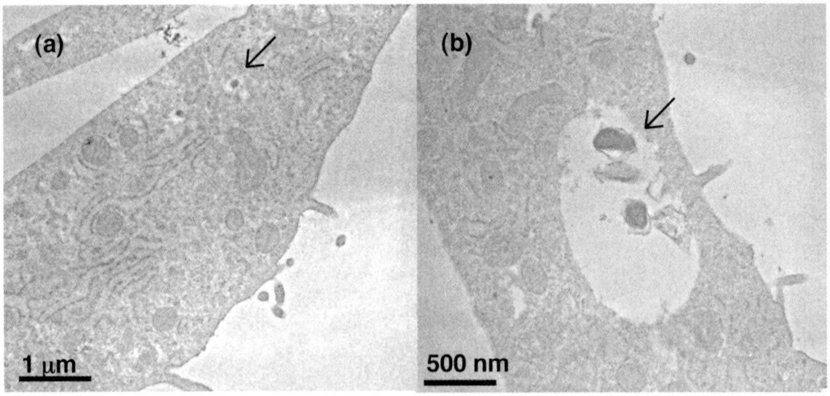

Fig. 18.7 Transmission electron microscopy images of gelatin/silica nanoparticles (a) internalized and (b) degraded by fibroblast cells. (Adapted from Ref. [61].)

No influence of gelatin/silica nanocapsules on fibroblast viability was observed for concentrations up to 30 mg mL^{-1}. Hence, these results confirm that the presence of silica does not hinder nanoparticle uptake by living cells, and that internalization does not induce rapid cell death. These two findings are very important for the future development of these hybrid capsules in drug delivery applications.

18.4
Alginate Versus Gelatin

Alginate and gelatin differ in several major aspects that impact upon capsule preparation, properties, and applications. A summary of differences relevant to this discussion are listed in Table 18.1. First, alginate is a polysaccharide of vegetal origin, whereas gelatin is a protein usually obtained from bovine or porcine sources. The latter point has become a major concern during the past few years due to the possibility of bovine spongiform encephalopathy (BSE) transmission. The alternative use of fish gelatin is still to be explored, but its main limitation lies in its lower gelation temperature. A second important difference between the two polymers is their gelation mechanism: alginate gels are formed ionotropically and involve chemical cross-linking, whereas gelatin networks are thermally driven physical gels. In many instances, the case of gelatin is more favorable as no additives are needed to induce gel formation. Moreover, this temperature-dependent stability can be used if heat-sensitive materials are needed. In contrast, alginate gels require inorganic salts addition, though the resultant networks are much more stable. As a consequence, gelatin is preferred for drug-delivery applications where polymer degradation is required, whereas alginate is selected for biocatalysts and bioreactor design for which the gel integrity is necessary on a

Table 18.1 Comparison of alginate and gelatin intrinsic and hybrid capsules properties.

	Alginate	Gelatin
Intrinsic properties		
Nature	Polysaccharide	Protein
Origin	Vegetal	Animal
Gel type	Chemical	Physical
Gel formation	Ionotropic	Thermotropic
Charge (pH 7)	Negative	Positive
Capsule properties		
Silica formation	Indirect (2 steps)	Direct (1 step)
Shell structure	Dense	Particulate
Core/shell formation	Macro- and microparticles	All scales
Potential application	Encapsulation	Drug release

long term. Finally, the two polymers differ in charge in the pH range from 5 to 8, with alginate being negatively charged and (type A) gelatin positively charged. In the context of biomaterials, it is known that neutral or weakly negative surfaces are more suitable, as positive surfaces appear to favor opsonization (protein adsorption triggering phagocytosis) [62].

Indeed, in the case of silica-coated particles, the material surface will always be negatively charged under physiological conditions. However, the biopolymer charge has a strong effect on its ability to favor silica deposition, as only positive surfaces can directly activate silicate condensation, as shown for gelatin. The possibility of coating negatively charged particles with poly-cations, such as PLL on alginate, allows the formation of suitable surface, but this implies another step in the preparation procedure. Overall, the synthesis of alginate/silica particles is much more complex than that of their gelatin equivalents. As a result, at present nanomaterials with core/shell structures are only formed from the protein, whereas IPN particles can be obtained from the polysaccharide. Moreover, these particles were obtained via a spray-drying method which is difficult to control and involves a heating step. This contrasts with the gelatin/silica nanocapsules preparation, which is conducted under mild conditions and via an adaptable emulsion route.

This difference has many implications for the future applications of these materials. For example, it is often desirable to graft the drug nanocarrier's surface either with hydrophilic polymers to improve their stealth, or with specific surface-recognition molecules to allow organ targeting [63]. As such grafting procedures are easily available from organically modified silicon alkoxides, the gelatin/silica core/shell structures appear much more suitable than their alginate counterpart. Another drawback of the alginate hybrid nanoparticles lies in the spray-drying procedure, which may lead to the partial degradation of any material to be encapsulated. As an example, the possibility of incorporating magnetite Fe_3O_4 nanoparticles within biopolymer/silica nanocapsules was recently studied for the design of HYbrid MAgnetic Carriers (HYMAC) that may find applications in magnetically controlled drug-release systems or hyperthermia [64]. When the magnetic nanoparticles were introduced in the alginate/PLL/silicate mixture, lepidocrocite (FeOOH) and fayalite (Fe_2SiO_4) were found in the resulting capsules, indicating a partial release of Fe^{2+} ions that was attributable to the spray-drying conditions. In contrast, magnetite nanoparticles could be entrapped within gelatin/silica without any loss of their integrity.

At this stage, the balance clearly appears in favor of gelatin-based capsules in the context of drug-delivery nanocarrier development. In contrast, for microencapsulation approaches, which require the long-term stability of the host material, alginate/silica particles appear more suitable. This is due not only to the fact that alginate gels are intrinsically more stable than gelatin networks but also to the better stability of dense PLL-templated silica shells when compared to gelatin-induced particular coatings. Although previously the biomimetic approach was not considered the most suitable for macrocapsule design for artificial organ applications, preliminary studies demonstrating successful cyanobacteria encapsulation and bioreactor design suggest promise for future developments in biotechnological devices.

Based on these two examples, it can be suggested that this biomimetic approach could be extended to any charged polymer (including synthetic ones), provided that they can be elaborated as capsules. The possibility of applying this methodology to neutral macromolecules can also be envisioned, though this would most likely imply a preliminary covalent grafting of suitable functions on the templating particle surface. On this basis, the route described here can be considered as a very general method to create silica-coated hybrid capsules.

18.5 Perspectives

To date, biogenic silica shells such as diatom frustules have mainly attracted attention due to their highly sophisticated morphology, which has led to the development of nanostructured silica, the exhibition of multi-scale hierarchically organized pore networks, and potential applications in catalysis, optics, or membrane science [14, 15, 65]. The examples presented here indicate, however, that more simple biomimetic silica-based materials might also be relevant to the design of biotechnological and biomedical devices. Because these materials are prepared following biomimetic routes, their elaboration should be compatible with the preservation of the biological system's integrity. Moreover, these conditions meet many requirements of the so-called "green chemistry" routes (biorenewable raw materials, room temperature conditions, no toxic reagents or by-products), in agreement with today's concerns for sustainable development.

Much is still to be achieved before economically viable devices can be obtained, however, with one major limitation being the absence of extensive data on the long-term *in-vivo* fate of silica-based materials. In the case of macrocapsules and hybrid artificial organs, the long-term stability data suggest that whilst these materials may be well tolerated in animals, as yet no human evaluation data exist [36, 39]. However, some initial findings on silica-based nanoparticles for drug and gene delivery are available, and these have created an air of optimism [66, 67]. Because a biomimetic approach can be applied to a wide range of polymers, and is compatible with many different particle preparation processes, it will undoubtedly play a major role in the future development of hybrid encapsulation hosts and drug-delivery systems.

References

1 T.L. Simpson, B.E. Volcani (Eds.), *Silicon and Siliceous Structures in Biological Systems*. Springer-Verlag, New York, **1981**.

2 C.E. Hamm, R. Merkel, O. Springer, P. Jurkojc, C. Maier, K. Prechtel, V. Smetacek, *Nature* **2003**, *421*, 841–843.

3 T.L.W. Carver, M.P. Robbins, B.J. Thomas, K. Troth, N. Raistick, R.J. Zeven, *Phys. Mol. Plant Pathol.* **1998**, *52*, 245–257.

4 R. Asada, K. Tazaki, *The Canadian Mineralogist* **2001**, *39*, 1–16.

5 V.R. Phoenix, K.O. Konhauser, D.G. Adams, S.H. Bottrell, *Geology* **2001**, *29*, 823–826.

6 P. Gomez-Romero, C. Sanchez (Eds.), *Functional Hybrid Materials*. Wiley-VCH, Weinheim, **2004**.

7 D. Avnir, T. Coradin, O. Lev, J. Livage, *J. Mater. Chem.* **2006**, *16*, 1013–1030.

8 H.R. Luckarift, J.C. Spain, R.R. Naik, M.O. Stone, *Nat. Biotechnol.* **2004**, *22*, 211–213.

9 K.M. Roth, Y. Zhou, W. Yang, D.E. Morse, *J. Am. Chem. Soc.* **2005**, *127*, 325–330.

10 A. Coiffier, T. Coradin, C. Roux, O.M.M. Bouvet, J. Livage, *J. Mater. Chem.* **2001**, *11*, 2039–2044.

11 N. Nassif, O. Bouvet, M.-N. Rager, C. Roux, T. Coradin, J. Livage, *Nat. Mater.* **2002**, *1*, 42–44.

12 S.V. Patwardhan, S.J. Clarson, C.C. Perry, *Chem. Commun.* **2005**, 1113–1121.

13 P.J. Lopez, C. Gautier, J. Livage, T. Coradin, *Curr. Nanosci.* **2005**, *1*, 73–83.

14 M. Sumper, E. Brunner, *Adv. Funct. Mater.* **2006**, *16*, 17–26.

15 Q. Sun, E.G. Vrieling, R.A. van Santen, N.A.J.M. Sommerdjik, *Curr. Opin. Solid State Mater. Sci.* **2004**, *8*, 111–120.

16 T. Coradin, J. Allouche, M. Boissière, J. Livage, *Curr. Nanosci.* **2006**, *2*, 219–230.

17 A. Smidsrod, A. Haug, *Acta Chem. Scand.* **1965**, *19*, 329–351.

18 D. Poncelet, C. Dulieu, M. Jacquot, in: R. Wijffels (Ed.), *Immobilized Cells.* Springer Laboratory Manual, Heidelberg, **2000**, pp. 15–30.

19 J.N. Barbotin, J.E. Nava Saucedo, in: S. Dimitriu (Ed.), *Poly-saccharides: structural diversity and functional versatility.* Marcel Dekker, New York, **1996**, pp. 749–774.

20 Y.J. Wang, *Mater. Sci. Eng. C*, **2000**, *13*, 59–63.

21 F. Lim, A.M. Sun, *Science* **1980**, *210*, 908–910.

22 H. Uludag, P. De Vos, P.A. Tresco, *Adv. Drug Deliv. Rev.* **2000**, *42*, 29–64.

23 G. Orive, R.M. Hernandez, A. Rodriguez-Gascon, R. Calafiore, T.M.S. Chang, P. De Vos, G. Hortelano, D. Hunkeler, I. Lacik, J.L. Pedraz, *Trends Biotechnol.* **2004**, *22*, 87–92.

24 P. De Vos, P. Marchetti, *Trends Mol. Med.* **2002**, *8*, 363–366.

25 G. Lambert, E. Fattal, P. Couvreur, *Adv. Drug Deliv. Rev.* **2001**, *47*, 99–112.

26 M. Rajaonarivony, C. Vauthier, G. Couarraze, F. Puisieux, P. Couvreur, *J. Pharm. Sci.* **1993**, *82*, 912–917.

27 C.C. Ribeiro, C.C. Barrias, M.A. Barbosa, *Biomaterials* **2004**, *25*, 4363–4373.

28 H. Keshawa, A. Forbes, R.M. Daya, *Biomaterials* **2005**, *26*, 4171–4179.

29 T. Coradin, N. Nassif, J. Livage, *Appl. Microbiol. Biotechnol.* **2003**, *61*, 429–434.

30 Y. Fukushima, K. Okamura, K. Imai, H. Motai, *Biotechnol. Bioeng.* **1988**, *32*, 584–594.

31 Z. Konsoula, M. Liakopoulou-Kyriadikes, *Process Biochem.* **2006**, *41*, 343–349.

32 O. Heichal-Segal, S. Rappoport, S. Braun, *Bio/technology* **1995**, *13*, 798–800.

33 E. Bressler, O. Pines, I. Goldberg, S. Braun, *Biotechnol. Prog.* **2002**, *18*, 445–450.

34 S. Xu, Z. Jiang, Y. Lu, H. Wu, W.-K. Yuan, *Ind. Eng. Chem. Res.* **2006**, *45*, 511–517.

35 S. Sakai, T. Ono, H. Ijima, K. Kawakami, *Biomaterials* **2001**, *22*, 2827–2834.

36 S. Sakai, T. Ono, H. Ijima, K. Kawakami, *Biomaterials* **2002**, *23*, 4177–4183.

37 G. Carturan, R. Dal Monte, M. Muraca, *Mater. Res. Soc. Symp. Proc.* **2000**, *628*, CC10.1.1–14.

38 S. Boninsegna, R. Dal Toso, R. Dal Monte, G. Carturan, *J. Sol-Gel Sci. Technol.* **2003**, *26*, 1154–1157.

39 G. Carturan, R. Dal Toso, S. Boninsegna, R. Dal Monte, *J. Mater. Chem.* **2004**, *14*, 2087–2098.

40 T. Coradin, E. Mercey, L. Lisnard, J. Livage, *Chem. Commun.* **2001**, 2496–2497.

41 T. Mizutani, H. Nagase, N. Fujiwara, H. Ogosh, *Bull. Chem. Soc. Jpn.* **1998**, *71*, 2017–2019.

42 T. Coradin, O. Durupthy, J. Livage, *Langmuir* **2002**, *18*, 2331–2336.

43 K. Kostarelos, *Adv. Colloid Interf. Sci.* **2003**, *106*, 147–168.

44 M. Boissière, P.J. Meadows, R. Brayner, C. Hélary, J. Livage, T. Coradin, *J. Mater. Chem.* **2006**, *16*, 1178–1182.

45 M. Boissière, A. Tourette, J.M. Devoisselle, F. Di Renzo, F. Quignard, *J. Colloid Interf. Sci.* **2006**, *294*, 109–116.

46 K. Molvinger, F. Quignard, D. Brunel, M. Boissière, J.M. Devoisselle, *Chem. Mater.* **2004**, *16*, 3367–3372.

47 T. Shiomi, T. Tsunoda, A. Kawai, H. Chiku, F. Mizukami, K. Sakaguchi, *Chem. Commun*, **2005**, 5325–5327.

48 J.N. Cha, H. Birkedal, L.E. Euliss, M.H. Bartl, M.S. Wong, T.J. Deming, G.D. Stucky, *J. Am. Chem. Soc.* **2003**, *125*, 8285–8269.

49 K.J.C. van Bommel, J.H. Jung, S. Shinkai, *Adv. Mater.* **2001**, *13*, 1472–1476.

50 A. Veis, *The Macromolecular Chemistry of Gelatin*. Academic Press, New York, **1964**.

51 K. Gelse, E. Pöschl, T. Aigner, *Adv. Drug Deliv. Rev.* **2003**, *55*, 1531–1546.

52 X. Li, L. Jin, G. Balian, C.T. Laurencin, D.G. Anderson, *Biomaterials* **2006**, *27*, 2426–2433.

53 H.-W. Kim, H.-E. Kim, V. Salih, *Biomaterials* **2005**, *26*, 5221–5230.

54 D. Olsen, C. Yang, M. Bodo, R. Chang, S. Leigh, J. Baez, D. Carmichael, M. Perälä, E.-R. Hämäläinen, M. Jarvinen, J. Polarek, *Adv. Drug Deliv. Rev.* **2003**, *55*, 1547–1567.

55 S.V. Sastry, J.R. Nyshadham, J.A. Fix, *Pharm. Sci. Technol. Today* **2000**, *3*, 138–145.

56 A.K. Gupta, M. Gupta, S.J. Yarwood, A.S.G. Curtis, *J. Control. Release* **2004**, *95*, 197–207.

57 T. Graham, *J. Chem. Soc.* **1862**, *15*, 216–270.

58 T. Coradin, S. Bah, J. Livage, *Colloids Surf. B* **2004**, *35*, 53–58.

59 T. Coradin, A. Marchal, N. Abdoul-Aribi, J. Livage, *Colloids Surf. B* **2005**, *44*, 191–196.

60 T. Coradin, J. Livage, *Mat. Sci. Eng. C.* **2005**, *25*, 201–205.

61 J. Allouche, M. Boissière, C. Hélary, J. Livage, T. Coradin, *J. Mater. Chem.* **2006**, *16*, 3120–3125.

62 A. Vonarbourg, C. Passirani, P. Saulnier, J.-P. Benoit, *Biomaterials* **2006**, *27*, 4356–4373.

63 D.E. Owens III, N.A. Peppas, *Int. J. Pharm.* **2006**, *307*, 93–102.

64 M. Boissière, J. Allouche, R. Brayner, C. Chanéac, J. Livage, T. Coradin, *J. Nanosci. Nanotechnol.* in press.

65 G.J. de A.A. Soler-Illia, C. Sanchez, B. Lebeau, J. Patarin, *Chem. Rev.* **2002**, *102*, 4093–4138.

66 C. Barbé, J. Bartlett, L. Kong, K. Finnie, H.Q. Lin, M. Larkin, S. Calleja, A. Bush, G. Calleja, *Adv. Mater.* **2004**, *16*, 1959–1966.

67 I. Roy, T. Y. Ohulchanskyy, D.J. Bharali, H.E. Pudavar, R.A. Mistretta, N. Kaur, P.N. Prasad, *Proc. Natl. Acad. Sci. USA* **2005**, *102*, 279–284.

Part VI
Imaging of Internal Nanostructures of Biominerals

Handbook of Biomineralization. Edited by P. Behrens and E. Bäuerlein

ISBN: 978-3-527-31805-6

19
Energy-Variable X-Ray Diffraction with High Depth Resolution Used for Mollusk Shell Analysis

Emil Zolotoyabko

Abstract

Recently, we have developed a novel X-ray diffraction (XRD) technique which allows us to obtain structural and microstructural characteristics in polycrystalline materials with high depth resolution. The method is based on the analysis of the XRD peaks collected at various X-ray energies, the latter being controllably changed by sufficiently small steps. By increasing energy and, correspondingly, the X-ray penetration length, the progressively deeper material's layers are probed. A theoretical analysis of the problem showed that the depth resolution is determined primarily by the energy steps applied and, thus, the sub-micrometer range for the depth steps is easily achievable. Application of this technique to some aragonitic mollusk shells resulted in the depth resolved microstructural characteristics, such as preferred orientation, lamella size, averaged microstrain fluctuations and residual strains, which previously were not accessible.

Key words: biomineralization, mollusk shells, microstructure, X-ray diffraction, synchrotron radiation, grain size, residual strains, microstrain fluctuations, preferred orientation.

19.1
Introduction

Characterization of the microstructure of materials with high spatial resolution is one of the key issues of nanoscience and nanotechnology. At present, the most powerful tools for microstructural characterization are supplied by electron microscopy methods, but these are only surface-sensitive or function within ultrathin layers. Moreover, they become destructive if three-dimensional information is required relating to the sample interior. Another well-established tool in this field is X-ray diffraction (XRD), which non-destructively provides rich microstructural information (preferred orientation of crystallite blocks, crystallite or

Handbook of Biomineralization. Edited by P. Behrens and E. Bäuerlein

ISBN: 978-3-527-31805-6

grain size, averaged magnitude of microstrain fluctuations, and residual strains/stresses) through the analysis of diffraction intensities and shapes of diffraction profiles [1]. One longstanding problem of the XRD methods is that they provide much worse spatial resolution than does electron microscopy. During the past few years, many remarkable successes have been reported in the two-dimensional (2-D) mapping of materials by X-ray microscopy, which has currently achieved a lateral resolution of few tens of nanometers (see e.g. [2–4] and references therein). Despite substantial progress in using X-ray microscopy for the three-dimensional (3-D) mapping of crystalline grains [5], high depth resolution in XRD measurements remains problematic, as X-rays interact much more weakly with materials than do electrons, and their penetration lengths are much larger. As a partial solution for polycrystalline materials, we have developed a novel XRD technique, referred to as energy-variable diffraction (EVD) [6–14].

The basic idea behind this method is rather simple – that is, to change the energy of synchrotron radiation in small steps by means of a double-crystal monochromator and, thus, to precisely control the X-ray penetration into the sample. With increasing energy, the X-rays penetrate into progressively deeper layers of the sample and provide depth-resolved information on its crystalline structure and microstructure. Less trivial is the fact that by this method it is possible to achieve information with sub-micron and even higher depth resolution, towards a nanometer range.

In this chapter we first describe the basis of the method – that is, the origin of its high depth resolution. Theoretical considerations are illustrated by selected examples taken with artificial multilayers, after which applications of the EVD technique to microstructural characterization of the mollusk shells are described.

19.2 The Theory of EVD

Comprehensive mathematical analyses of EVD in an infinitely thick sample and in samples of finite thickness (including thin films) have been performed in our papers [12] and [14]. It is shown, that the quasi-parallel synchrotron X-rays diffracted at various depths in the sample enter the detecting system at slightly different angles with respect to its axis. As a result, these X-rays are registered in the detector with different probabilities. Considering these probabilities together with the in-depth exponential attenuation of the X-ray beams having different energies, allowed us analytically to obtain the shapes of diffraction profiles and to conclude that the peak intensity recorded in the detector originates at a certain sample depth, $Z_c(E)$, which depends on X-ray energy, E. This finding opens the way to microstructural analysis with high depth resolution because how it is possible to precisely change the characteristic depth, $Z_c(E)$, on the basis of its analytic dependence on the X-ray energy, E. In practical terms, the diffraction peaks must be accurately measured at different X-ray energies that are varied in sufficiently small

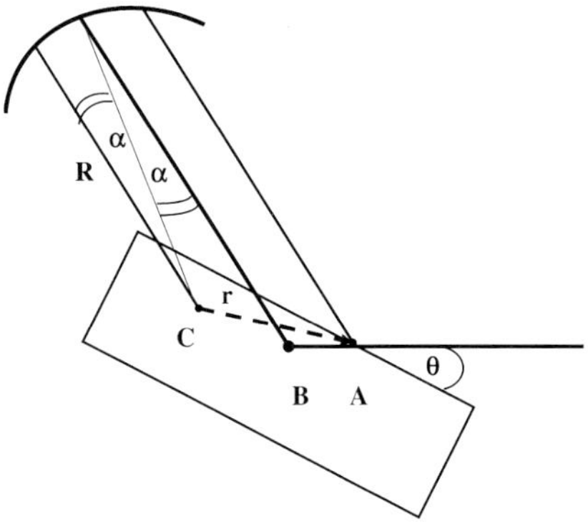

Fig. 19.1 Schematic illustration of the chromatic aberration effect in energy-variable diffraction (EVD). For details, see the text.

steps, after which the extracted microstructural characteristics are related to the characteristic depth, $Z_c(E)$.

We can briefly illustrate these ideas with calculations which focus on depth-resolved residual strains extracted from the diffraction peak positions. We begin with the basic concept of the developed theory – the chromatic aberration effect – which is illustrated graphically in Figure 19.1. It must be emphasized that in the EVD method, a specific point (B) within the sample, where the X-ray scattering occurs, does not generally coincide with the immovable center of rotation (C) of the detection system (see Fig. 19.1). For this reason, the detector's reading will differ from the true angle of diffraction, 2Θ, by some extra quantity, α, which is energy-dependent. This is referred to as the "chromatic aberration effect" [12]. Thus [12]:

$$\alpha(Z) = \beta - \frac{2Z\cos\Theta}{R} \tag{1}$$

where R is the distance between the sample and the receiving slit of the detector, Z is the depth at point B counted along the normal to the surface of the sample, and β is the Z-independent part of the angular deviation, which depends on the Bragg angle and on the vector, r, connecting the point of the X-ray entrance, A, on the crystal surface and point C (see Fig. 19.1). Note that β-value is energy-dependent; its exact expression is given in Ref. [12]. It follows from Eq. (1) that the diffracted X-rays, which originated at different depths, Z, indeed enter the detector at different angles, α, with respect to the axis of the detector's collimating system. They will therefore register with different probabilities, P. In standard Gaussian approximation:

$$P = \frac{1}{\sigma\sqrt{2\pi}} e^{-[\Delta 2\theta - \alpha(Z)]^2/2\sigma^2} \tag{2}$$

where $\Delta 2\Theta$ is the deviation of the detector's angle from the Bragg position, 2Θ, and σ is the dispersion of Gaussian distribution.

Another effect that should be taken into account in our analysis is the exponential attenuation of the X-ray beams in depth,

$$W = \exp(-Z/\Lambda) \tag{3}$$

with the characteristic length

$$\Lambda = \frac{\sin\Theta}{2\mu} \tag{4}$$

being determined in the Bragg scattering geometry by the angle of diffraction, Θ, and the linear absorption coefficient, μ, both of which are energy-dependent.

By combining Eqns. (2) and (3), we obtain the "density" of the diffraction intensity originated at depth Z and registered by the detector:

$$I(Z) = SPW \tag{5}$$

where S is the material's scattering power per unit length. In order to find the angular distribution of the diffraction intensity – that is, the shape of the diffraction profile $I(\Delta 2\Theta)$ – Eq. (5) must be integrated over the sample thickness, T. If S is only weakly dependent on Z, the diffraction intensity, $I(\Delta 2\Theta)$, can be expressed as:

$$I(\Delta 2\Theta) = S\int_0^T (WP)\,dZ = \frac{S}{\sigma\sqrt{2\pi}}\int_0^T dZe^{-Z/\Lambda}e^{-[\Delta 2\Theta-\alpha(Z)]^2/2\sigma^2} \tag{6}$$

Finally, by substituting Eq. (1) into Eq. (6) and integrating over Z (for details, see Refs. [12, 14]), yields:

$$I(x) = \frac{(\sigma R)S}{2\sqrt{2\pi}\cos\Theta}[F(x) - e^{-T/\Lambda}F(x+\varepsilon)], \tag{7}$$

where

$$F(x) = \frac{e^{-x^2/2\sigma^2}}{x + \dfrac{\sigma^2 R}{2\Lambda\cos\Theta}}; \quad x = \Delta(2\Theta) - \beta; \quad \varepsilon = \frac{2T\cos\Theta}{R} \tag{8}$$

Maximum intensity is achieved for angular deviation $(\Delta 2\Theta)_m$, which satisfies the condition:

$$\frac{\partial I(x)}{\partial(\Delta 2\Theta)} = \frac{\partial I(x)}{\partial x} = 0 \tag{9}$$

Differentiating in Eq. (9), yields

$$x = \Delta(2\Theta) - \beta = \frac{2\cos\Theta}{R}\left[-\Lambda + \frac{T\exp(-T/\Lambda)}{1-\exp(-T/\Lambda)}\right] \quad (10)$$

finally providing the peak position at

$$\Delta(2\Theta)_m = \beta - \frac{2\cos\Theta}{R}\left[\Lambda - \frac{T\exp(-T/\Lambda)}{1-\exp(-T/\Lambda)}\right] \quad (11)$$

By comparing Eqns. (1) and (11), it can be concluded that the peak intensity of the registered diffraction profile comes from the characteristic depth:

$$Z_c = \Lambda - T\frac{\exp(-T/\Lambda)}{1-\exp(-T/\Lambda)} \quad (12)$$

For an infinitely thick sample ($T \to \infty$), $Z_c = \Lambda$; that is, it is equal to the X-ray penetration depth at a given energy [see Eq. (4)]. At $T \ll \Lambda$, $Z_c \approx T/2$ and, as expected, tends to zero, if $T \to 0$.

It must be stressed that, in the general case, the characteristic depth, Z_c, is energy-dependent [through Λ, see Eq. (12)], which allows us to conduct the *d*-spacing or strain measurements with high depth resolution, if the energy steps applied are sufficiently small. The experimental procedure is based on the accurate measurements of diffraction profiles at different X-ray energies, determination of the angular peak positions, applying the "chromatic aberration" correction [Eq. (1) with $Z = Z_c$], extracting the *d*-spacing values by using the Bragg law, and relating the obtained changes in *d*-spacings (converted to the strain values) to the energy-dependent depth, Z_c.

19.3
Experimental Results for Artificial Multilayers

In order to explore the capabilities of the EVD technique – and especially the depth-resolution aspect – the method was applied to depth-resolved strain measurements in two dissimilar, but well-defined, artificial multilayer structures. The first structure was a thick sample (ca. 2 mm thickness), which comprised 45 bilayers of Al_2O_3/ZrO_2 with nominal thicknesses of 30 and 20 μm for individual alumina and zirconia sub-layers, respectively. The sample was produced by electrophoretic deposition followed by high-temperature sintering. In fact, some amount of alumina particles had been added to the zirconia layers and vice versa at the deposition stage in order to reduce the thermal stresses caused by difference in the thermal expansion coefficients of the materials. The second sample was in a form of thin film (ca. 9 μm thickness), and consisted of Co/Cu alternating layers with nominal thicknesses of 200 and 100 nm, respectively, electrode-

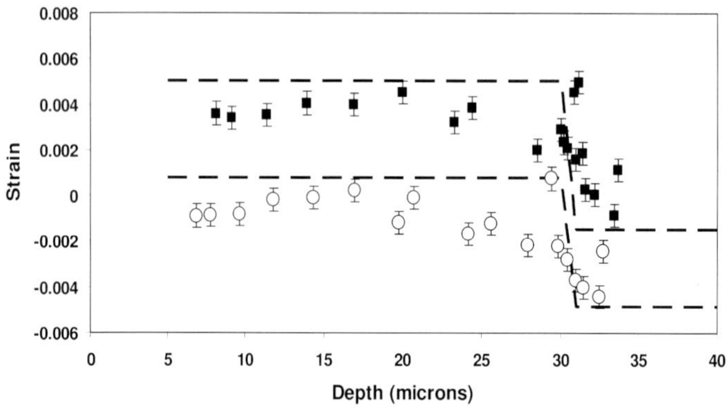

Fig. 19.2 Residual strains in the alumina (filled squares) and zirconia (open circles) particles, as functions of the X-ray penetration depth. The dashed lines represent the simulated strain values in this material system.

posited onto a Si substrate. Diffraction measurements with both samples were carried out at the 5 BMD beam line of the Advanced Photon Source (APS) of Argonne National Laboratory. The experimental details are described in Refs. [12] and [14].

In the case of the alumina/zirconia multilayer, the depth dependence of the normal strain component was measured by tracking the (012)Al_2O_3 and (101)ZrO_2 diffraction profiles in the energy range of 7 to 40 keV. Due to the presence of Zr atoms (K-edge at $E = 17.997$ keV), the X-ray penetration depth, $\Lambda(E)$, in the energy range used is confined to between 6 and 34 μm. It must be remembered that for thick sample $Z_c = \Lambda$. As the first alumina/zirconia interface is buried at depth of 30 μm (as confirmed by scanning electron microscopy, SEM), the primary goal of these measurements was to detect the strain peculiarity when crossing the interface.

Residual strains measured in both types of particle, alumina and zirconia, as a function of the X-ray penetration depth (see Fig. 19.2), provide clear evidence of strain jumps at 30 μm. The dashed lines in Figure 19.2 represent the results of model calculations of depth-dependent strain distributions caused by two sources: (i) biaxial stress arising at the alumina/zirconia interface due to the differences in thermal expansion coefficients; and (ii) isotropic hydrostatic pressure around an isolated particle of the secondary phase within a homogeneous matrix. The simulated strain values were in reasonable agreement with the experimental data [12].

For the Co/Cu multilayer, the (111) diffraction profiles were taken in the energy range of 10 to 40 keV – that is, above the *K*-edges of Co and Cu, which are 7.7089 and 8.9789 keV, respectively. In this energy range and for a given film thickness, the characteristic depth, Z_c, is confined between 0.9 and 4 μm.

The relative difference, $\Delta d/d$, between the (111) Co and Cu *d*-spacings is plotted in Figure 19.3, as a function of Z_c. The solid line in Figure 19.3 represents the

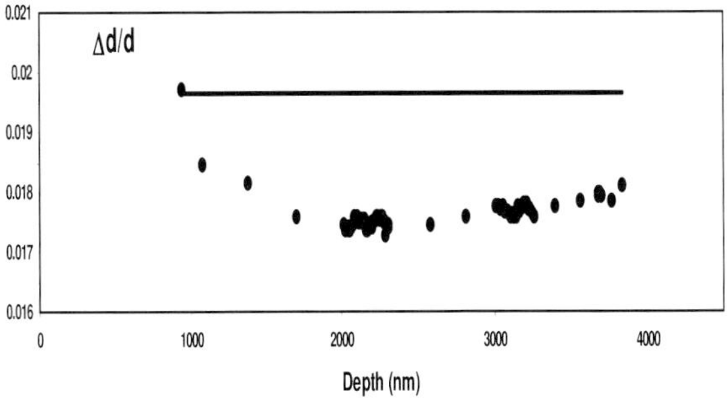

Fig. 19.3 Depth-dependent variation of the relative difference, Δd/d, between the Co/Cu (111) *d*-spacings, measured in the entire range of the probed depths. Note that deeper layers are located closer to the Si substrate. The solid line represents the bulk Δd/d-value.

magnitude of Δd/d calculated for the bulk Co and Cu *d*-spacings. The deviations of experimental points from the calculated line reflect the residual strain distribution across the multilayer. The experimental points are lower than the bulk value as the interfacial strain acts towards equalizing the lattice parameters of the Cu and Co sub-layers and hence reducing their difference. The residual strains are inhomogeneously distributed over a multilayer. Strain values are higher in the deeper layers – that is, those closer to the substrate – and they disappear near the outer surface of the sample (near a depth of 1000 nm; see Fig. 19.3), where the difference between Co and Cu lattice parameters is very close to the bulk value. The observed inhomogeneity reflects the difficulties in producing high-quality Co/Cu multilayers by electrodeposition [14].

A detailed study of the selected depth intervals with high depth resolution (i.e., applying small energy steps) revealed the oscillating character of the strain values (see Fig. 19.4), which reflects the spatial periodicity of the layered structure [14].

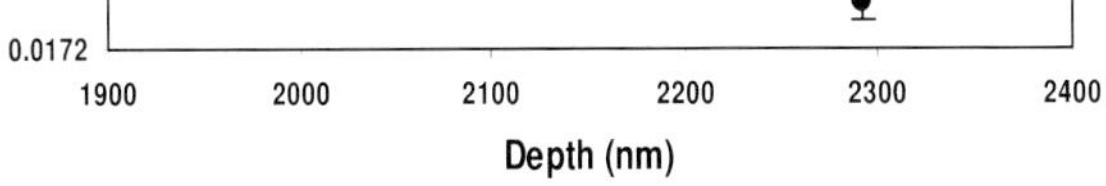

Fig. 19.4 Periodic variation of the relative difference, Δd/d, between the Co/Cu (111) *d*-spacings, measured in the vicinity of a depth, $Z = 2200$ nm. The solid line represents the fit to the *sin*-function with period, $t_s = 135$ nm.

The data shown in Figure 19.4, measured in the energy range between 17 and 20 keV (corresponding to characteristic depths of ca. 2200 nm) with energy steps of 50 eV, show a periodicity of 135 nm. Since far from interfaces (i.e., at the centers of individual sub-layers) a minimum strain is expected, the strain periodicity, t_S, should be half the multilayer period; that is, $t_S = \frac{1}{2} \cdot 300 = 150$ nm in our case. In fact, the measured periodicity is rather close to expectation.

The two examples given here provide convincing evidence of the capabilities of EVD for the study of polycrystalline structures with high depth resolution. The results of EVD applications to inhomogeneous polycrystalline structures produced by Nature, namely mollusk shells, are detailed in the following sections.

19.4
Studies with Mollusk Shells: Strain Analysis

During the past few decades, the biomineralization process has been intensively studied in order to clarify the interrelation between the structure of biogenic crystals and their functioning [15, 16]. In particular, much attention has been paid to mollusk shells, due mainly to their outstanding mechanical characteristics, which are achieved within hierarchically ordered and lightweight structures [10]. In fact, mollusk shells are composite materials that comprise a ceramic matrix and 0.1–5 wt% organic phase, which is located mainly within the intercrystalline boundaries. In addition, some organic macromolecules (intracrystalline) are presumably confined within the crystalline lattice [17]. These molecules are potential sources of local deformation fields which can strongly affect the mechanical properties of biogenic crystals. Bearing this in mind, we applied the EVD technique to probe local residual strains in selected aragonitic shells. All measurements were performed at the 5 BMD beamline of APS at X-ray energies between 7 and 45 keV which, for thick samples, corresponds to the X-ray characteristic depths, $5\ \mu m < Z_c < 110\ \mu m$.

Two different types of aragonitic shell – *Acanthocardia tuberculata* (bivalve) and *Strombus decorus persicus* (gastropod) – were used as samples in these studies [13]. *Acanthocardia tuberculata* has a thick nacre layer, which expands hundreds of microns from the shiny inner surface adjacent to the mollusk mantle [7]. According to SEM and XRD results, the nacre layer is composed of the [001]-oriented lamellae, some 200 to 600 nm thick [8]. *Strombus decorus persicus* has a very thin (a few microns) prismatic layer at the inner surface of the shell, followed by the crossed-lamellar layer, which occupies the rest of the shell thickness [10].

Depth-dependent strain values, $\varepsilon(Z_c)$, in the nacre layer of four shell pieces, which were cut from different *A. tuberculata* seashells, are shown in Figure 19.5. Zero depth corresponds to the inner shell surface, adjacent to the mollusk mantle. The measured *d*-spacings were extracted from the angular positions of the intense (002) diffraction peaks and compared with the tabulated aragonite values. In other words, the data in Figure 19.5 show the depth-dependence of the strain component along the *c*-axis of the orthorhombic unit cell of aragonite lattice.

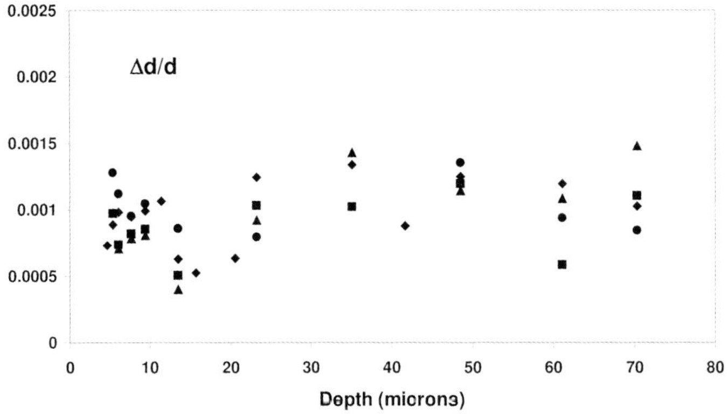

Fig. 19.5 Residual strains measured as a function of the X-ray penetration depth in four specimens of the *Acanthocardia tuberculata* bivalve shells. Data taken from each specific specimen are indicated by circles, triangles, diamonds and squares, respectively.

Despite some scattering of the experimental data, it can be said that the strain values, $\Delta c/c$, have a positive sign, indicating lattice swelling of about 0.001. More precisely, the value averaged over four samples is $(\Delta c/c)_{av} = 0.0009$. Similar results were obtained for *S. decorus persicus* shells, wherein the strain component along the *c*-axis is positive and scattered around the mean value, $(\Delta c/c)_{av} = 0.0008$.

Since residual strains in the investigated shells did not reveal clear depth-dependence, it was assumed that they are of the same order in different structural units. If so, the only question is to completely characterize the strain tensor. For this purpose we used high-resolution X-ray powder diffraction at fixed energy.

High-resolution X-ray powder diffraction measurements were carried out at dedicated beam lines: the ID-31 beam line of ESRF and the 32-ID beam line of APS, both equipped with a crystal-monochromator and crystal-analyzer optical elements.

The use of the advanced analyzing optics resulted in diffraction spectra of superior quality and allowed us to extract the lattice parameters with an accuracy of 10 ppm.

By using high-resolution powder diffraction it was found that the lattice parameters, *a*, *b*, and *c*, in aragonitic shells, *A. tuberculata* and *S. decorus persicus*, differed slightly from those measured accurately by the same technique in geological aragonite [18]. The relative deviations were $\Delta a/a = 0.0008$ and 0.0003, $\Delta b/b = -0.0004$ and -0.0006, $\Delta c/c = 0.0012$ and 0.0010, respectively. Lattice distortions along the *c*-axis agreed well with those obtained in *A. tuberculata* and *S. decorus persicus* by EVD [13]. Further studies revealed similar lattice distortions in many other aragonitic shells from different habitats and classes. Comprehensive measurements of lattice relaxation under mild annealing allowed us to conclude that lattice distortions are caused by organic macromolecules entering aragonite lattice during biomineralization [19, 20]. A deeper understanding of this phenomenon will aid in the development of new approaches for growing biogenic composites and controlling their properties on the molecular level.

19.5
Studies with Mollusk Shells: Preferred Orientation

Similar to strain analysis, the depth-resolved preferred orientation could be retrieved from the diffraction peak intensities measured at different energies. Here, we show experimental results collected for two types of mollusk shells: *A. tuberculata* [7] and *S. decorus persicus* [10]. Measurements were performed at the 5 BMD beam line of APS at energies ranging between 7 and 30 keV, which for thick samples corresponds to $5 < Z_c < 70$ μm. As a measure of the [001]-preferred orientation, in both cases we took the ratio of the (002) and (012) diffraction intensities, which for a random powder equals $\eta_p = 0.042$.

19.5.1
A. tuberculata

The intensity ratio, $\eta = I(002)/I(012)$, taken from one of the *A. tuberculata* shells, is plotted in Figure 19.6, as a function of characteristic depth, Z_c. It should be noted that, even at maximum X-ray penetration depth of 70 μm the nacre layer was still probed. The dependence, $\eta(Z_c)$, has characteristic shape, typical for all investigated specimens of *A. tuberculata*, namely with a gentle ascent at small thicknesses up to a maximum at depth of the order of 10 μm, followed by gradual in-depth decreases of the η-values. The degree of preferred orientation in the na-

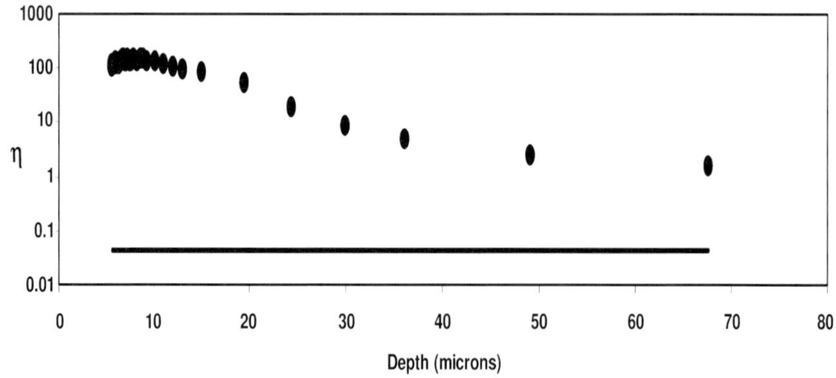

Fig. 19.6 Intensity ratio, $\eta = I(002)/I(012)$, measured in one of the *Acanthocardia tuberculata* shells, as a function of the X-ray penetration depth. The straight solid line represents the η-value for random powder, $\eta_p = 0.042$.

cre layer may be very high, resulting in maximal η/η_p-ratios of about 10^3. Some decrease in the degree of preferred orientation near the shiny surface of the shell is related to the existence of a thin, less-ordered prismatic layer at that location. Maximum texture is achieved in the "fresh" (closest to the mollusk mantle) nacre layer, just adjacent to the prismatic layer. The most interesting point was a gradual degradation of the preferred orientation in the deeper nacre layer. As will be shown below, this degradation is fully correlated with a gradual thinning of the [001]-oriented lamellae. As deeper nacre layers are older (i.e., they are produced by the mollusk during earlier stages of its development), these findings shed additional light on mollusk functioning during its life cycle (see Section 19.6).

19.5.2
S. decorus persicus

The microstructure of *S. decorus persicus* is completely different, in that it has no nacre layer and is built of crossed-lamellar layer with a very thin prismatic layer adjacent to the mollusk mantle [10]. Correspondingly, the in-depth distribution of the degree of preferred orientation (see Fig. 19.7), measured again via the diffraction intensity ratio, $\eta = I(002)/I(012)$, appears very different from that observed for *A. tuberculata* (see Fig. 19.6). In fact, within first 5 μm near the shell surface (i.e., in the prismatic layer itself) the preferred orientation is of the [001]-type. In deeper sub-layers (up to ca. 10 μm depth) the preferred orientation is switched to the [012]-type, which is also confirmed by direct SEM observations. Below 10 μm, the crossed-lamellar layer is located and the intensity ratio approaches that in a random powder. Also in this case, the behavior of the degree of preferred orienta-

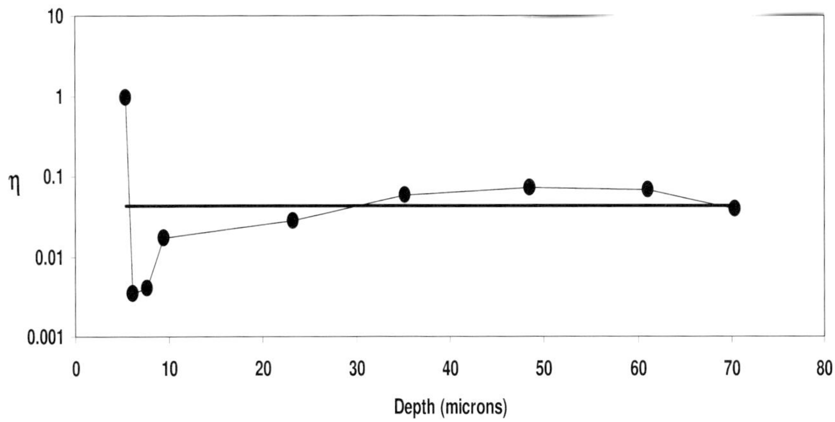

Fig. 19.7 Intensity ratio, $\eta = I(002)/I(012)$, for one of the *Strombus decorus persicus* shells, as a function of the X-ray penetration depth. The straight solid line represents the η-value for random powder, $\eta_p = 0.042$.

tion correlates well with the thickness of the [001]-oriented crystal blocks, as discussed in Section 19.6.

19.6
Studies with Mollusk Shells: Diffraction Profile Analysis

Important microstructural information on grain size and microstrain fluctuations is routinely extracted from the XRD data by analyzing the shapes of the diffraction profiles [1]. In many cases, it is possible to separate the Gaussian-type contribution to the peak broadening due to microstrain fluctuations from the Lorentzian-type contribution due to the finite crystallite size (grain size). We applied this procedure to the diffraction profiles taken at different energies from the mollusk shells of *A. tuberculata* [8, 9] and *S. persicus decorus* [10], with measurements being performed at the 5 BMD beam line of APS for X-ray energies in the range 7 to 30 keV. In order to extract the magnitudes of microstrain fluctuations and crystallite size with depth resolution, the diffraction profiles measured at different energies were fitted to a Voigt function, which is a convolution of Gaussian and Lorentzian functions.

By using the Voigt function, we were able separately to extract the widths of the Gaussian, W_G, and Lorentzian, W_L, contributions to the broadening of the diffraction profiles. On this basis, the average microstrain fluctuations (dispersion, σ_m, of the $\Delta d/d$ distribution) and the crystallite size, L, are expressed as follows [8]:

$$\sigma_m = \frac{\sqrt{W_G^2 - W_I^2}}{4\sqrt{2\ln 2}\tan\Theta} \tag{13}$$

where W_I is the instrument contribution to the broadening of the diffraction profiles, and

$$L = \frac{2d\tan\Theta}{W_L} \tag{14}$$

A typical example of the depth-resolved lamella thickness in the nacre layer of *A. tuberculata* is shown in Figure 19.8. The data were obtained from the same shell piece used to measure preferred orientation (see Fig. 19.6). In fact, by comparing Figures 19.6 and 19.8, it can be concluded that the behavior of the lamella thickness fully correlates with that for preferred orientation: a gentle ascent in the near-surface layer up to maximum value (of ca. 600 nm) at approximately 10 μm beneath the shiny surface of the shell, followed by a gradual decrease (down to 400 nm) in the deeper shell. Certainly, large, well-developed lamellae imply a higher degree of preferred orientation. It is known that the oriented growth of lamella in the nacre layer is controlled by organic macromolecules, which are provided by the mollusk according its genetic program. It is reasonable to assume that a mature animal supplies organic macromolecules more intensively in order to grow larger building blocks (lamellae) for its home. If this were the case, we

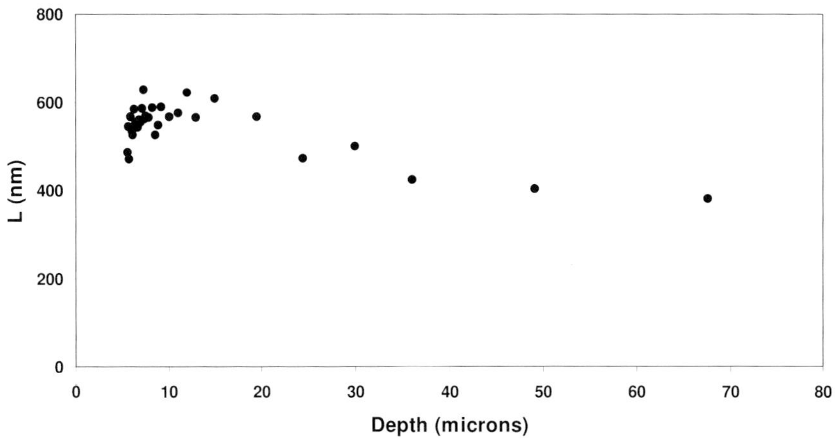

Fig. 19.8 Typical depth-dependence of the thickness of the [001]-oriented lamellae in the nacre layer of *Acanthocardia tuberculata* shells.

would expect to find thinner lamellae in the deeper nacre layers which are built at earlier stages of shell formation. An experimentally observed in-depth decrease in lamella thickness, and hence in the degree of preferred orientation, supported these suggestions.

A decrease in lamella thickness is accompanied by a gradual increase in microstrain fluctuations as a function of depth (see Fig. 19.9). Previously [8], this behavior was explained on a quantitative basis by relating the decreasing lamella thickness to the growing number of inter-lamella boundaries, which are the sources of inhomogeneous strain fields.

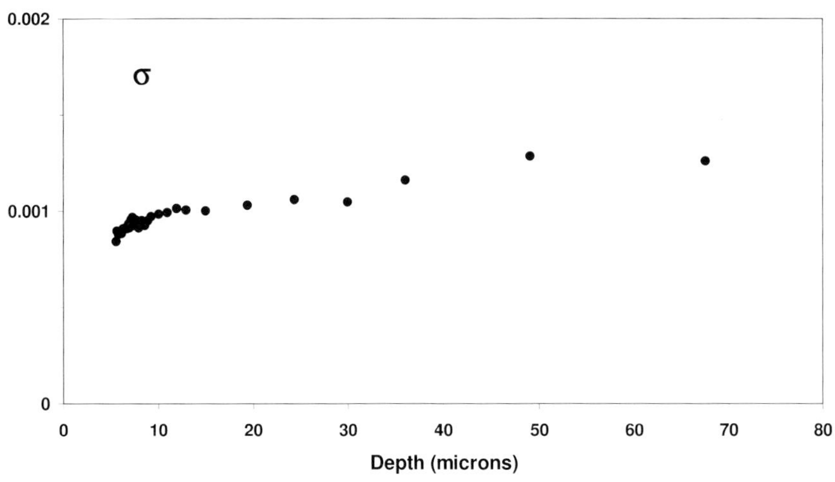

Fig. 19.9 Typical depth-dependence of the averaged microstrain fluctuations in the nacre layer of *Acanthocardia tuberculata* shells.

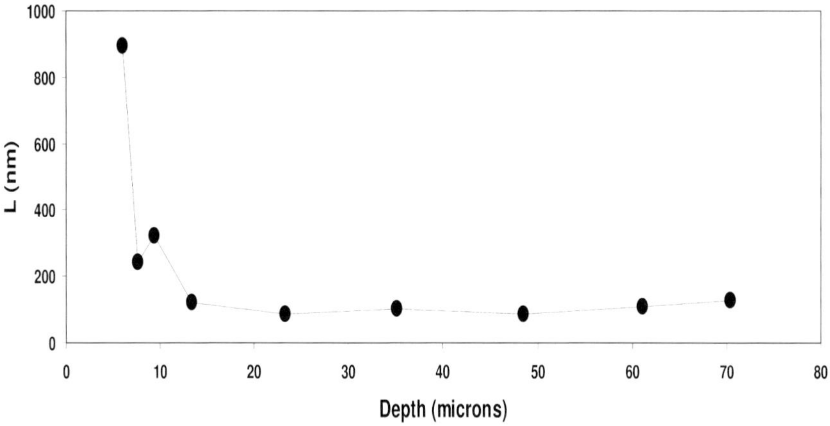

Fig. 19.10 Typical depth-dependence of the thickness of the [001]-oriented lamellae in *Strombus decorus persicus* shells.

The gastropod shell, *S. decorus persicus*, has no nacre, and data on the depth-resolved crystallite size (see Fig. 19.10) allow us to follow the transformation of the prismatic microstructure into the crossed-lamellar layer. This transformation takes place on an impressively short scale: within first 10 μm beneath the surface the crystallite size is changed from about 1 μm down to 100 nm – that is, by one order of magnitude. This was confirmed by direct SEM observations [10]. Large crystalline blocks of the prismatic layer have the [001]-type preferred orientation, as is clearly seen in Figure 19.7. Consequently, the microstrain fluctuations are much smaller in the prismatic layer compared to those in the crossed-lamellar layer (see Fig. 19.11).

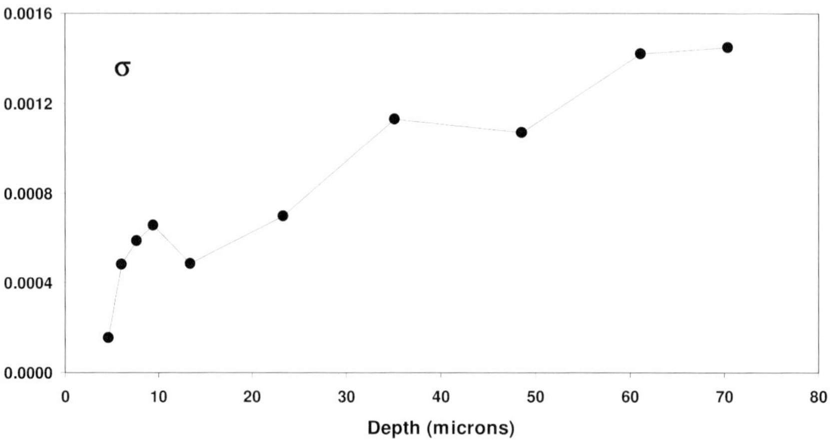

Fig. 19.11 Typical depth-dependence of the averaged microstrain fluctuations in *Strombus decorus persicus* shells.

19.7 Conclusion

The results obtained with artificially grown and natural polycrystalline multilayer structures allow us to conclude that EVD represents a powerful method for the study of microstructural characteristics, with high depth resolution. Experiments with model systems have demonstrated an ability to investigate the structural features down to a 100-nm range. Likewise, it is possible to analyze the layers buried up to 100 μm beneath the sample surface.

The application of EVD to mollusk shells produced a number of interesting – and even unexpected – results. First, the discovery of lattice distortions in biogenic aragonite (as compared to geological aragonite crystals) highlighted an important role for intracrystalline proteins in biomineralization. The EVD technique allows us to measure residual strains/stresses in mollusk shells with depth resolution, as shown for *A. tuberculata* and *S. decorus persicus* mollusk shells, and also by the Daresbury group for the shell of the bivalve, *Ensis siliqua* [21]. We were also able to follow the in-depth behavior of preferred orientation, either very close to (within a few microns) or up to 100 μm beneath the surface. Changes in the type of preferred orientation, as for example at the interface between prismatic and crossed-lamellar layers in the shell of the gastropod *S. decorus persicus*, are clearly detectable. The use of EVD highlighted the remarkable in-depth inhomogeneity of the nacre layer in the shell of the bivalve, *A. tuberculata*, this being revealed through the EVD measurements of the preferred orientation, lamella thickness, and microstrain fluctuations, as extracted from the XRD data. Moreover, modifications of all three parameters, as functions of depth, were consistent one with another, and demonstrated the changes that occur in mollusk activity during the animal's life cycle.

Finally, EVD may be used to characterize many other natural or artificial polycrystalline structures which play a key role in various fields of modern materials science and engineering.

Acknowledgments

The contributions of J.P. Quintana (APS) and B. Pokroy (Technion) to the development of the energy-variable diffraction and data collection are gratefully acknowledged. The author also thanks the Israel Science Foundation, founded by the Israel Academy of Sciences and Humanities, for financial support. Most of the measurements described in this chapter were performed at the DuPont-Northwestern-Dow Collaborative Access Team (DND-CAT) Synchrotron Research Center located at Sector 5 of the Advanced Photon Source of Argonne National Laboratory. DND-CAT is supported by the E.I. DuPont de Nemours & Co., the Dow Chemical Company, the U.S. National Science Foundation through grant DMR-9304725, and the State of Illinois through the Department of Commerce and the Board of Higher Education's Grant IBHE HECA NWU 96. Use of the

Advanced Photon Source was supported by the U.S. Department of Energy, Basic Energy Sciences, Office of Energy Research, under contract No. W-31-109-Eng-38.

References

1 B.E. Warren, *X-Ray Diffraction*. Dover Publications, New York, **1990**.
2 G.E. Ice, B.C. Larsen, *Adv. Eng. Mater.* **2000**, *2*, 643.
3 H.F. Poulsen, D.J. Jensen, G.B.M. Vaughan, *MRS Bulletin* **2004**, *29*, 166.
4 I. Snigireva, A. Snigirev, *J. Evironment. Monitor.* **2006**, *8*, 33.
5 H.F. Poulsen, *Three-Dimensional X-Ray Diffraction Microscopy: Mapping Polycrystals and Their Dynamics.* Springer-Verlag, Berlin, **2004**.
6 E. Zolotoyabko, J.P. Quintana, *MRS Symposia Proc.* **2001**, *678*, EE 3.7.1.
7 E. Zolotoyabko, J.P. Quintana, *Rev. Sci. Instr.* **2002**, *73*, 1663.
8 E. Zolotoyabko, J.P. Quintana, *J. Appl. Crystallogr.* **2002**, *35*, 594.
9 E. Zolotoyabko, J.P. Quintana, *Nucl. Instr. Methods Phys. Res. B* **2003**, *200*, 382.
10 B. Pokroy, E. Zolotoyabko, *J. Mater. Chem.* **2003**, *13*, 682.
11 E. Zolotoyabko, B. Pokroy, J.P. Quintana. *MRS Symp. Proc.* **2004**, *840*, Q7.7.
12 E. Zolotoyabko, B. Pokroy, J.P. Quintana, *J. Synchrotron Radiat.* **2004**, *11*, 309.
13 B. Pokroy, J.P. Quintana, E. Zolotoyabko, *MRS Symp. Proc.* **2005**, *874*, L.7.1.
14 E. Zolotoyabko, B. Pokroy, T. Cohen-Hyams, J.P. Quintana. *Nucl. Instr. Methods Phys. Res. B* **2006**, *246*, 244.
15 H.A. Lowenstam, S. Weiner, *On Biomineralization*. Oxford University Press, New York, **1989**.
16 S. Mann, *Biomineralization: Principles and Concepts in Bioinorganic Materials Chemistry.* Oxford University Press, Oxford, **2001**.
17 A. Berman, L. Addadi, S. Weiner, *Nature* **1988**, *331*, 546.
18 E.N. Caspi, B. Pokroy, P.L. Lee, J.P. Quintana, E. Zolotoyabko, *Acta Crystallogr. B* **2005**, *61*, 129.
19 B. Pokroy, J.P. Quintana, E.N. Caspi, A. Berner, E. Zolotoyabko, *Nat. Mater.* **2004**, *3*, 900.
20 B. Pokroy, A.N. Fitch, P.L. Lee, J.P. Quintana, E.N. Caspi, E. Zolotoyabko, *J. Struct. Biol.* **2006**, *153*, 145.
21 S.J. Eichhorn, D.J. Scurr, P.M. Mummery, M. Golshan, S.P. Thompson, R.J. Cernik, *J. Mater. Chem.* **2005**, *15*, 947.

20
X-Ray Phase Microradiography and X-Ray Absorption Micro-Computed Tomography, Compared in Studies of Biominerals

Stuart R. Stock

Abstract

X-ray absorption micro-computed tomography (microCT) and X-ray phase imaging each provide non-destructive views of the interiors of specimens. Here, the two imaging modalities are applied to study biomineralization in sea urchins, which are heavily mineralized invertebrates from the phylum Echinodermata. Following an introduction of the physical bases of the two imaging modalities, laboratory and synchrotron microCT techniques are contrasted, and four methods of X-ray phase imaging briefly described. Examples are presented from the teeth of *Lytechnius variegatus* and from the spines of *Diadema setosum*. The value of complementing X-ray imaging with destructive analysis techniques is also discussed.

Key words: X-ray imaging, microCT (microComputed Tomography), X-ray phase, synchrotron X-radiation, bone, sea urchin ossicles, calcite.

20.1
Introduction

During the past decade, commercial laboratory micro-computed tomography (microCT) systems have become commercially available, and microCT facilities at the new generation of X-ray synchrotron X-radiation sources have begun to operate in a production – as opposed to a commissioning – mode. This has led to a plethora of reports in which microCT has been used in the analysis of specimen microstructures, both non-destructively and in three-dimensional (3-D) mode. Likewise, the use of X-ray phase imaging increases each year.

In this chapter the techniques of X-ray (absorption) microCT imaging and of X-ray phase imaging (which can and has been used for microCT) are briefly discussed. The examples concentrate on the mineralized tissue of the sea urchin, an invertebrate from the phylum Echinodermata; these millimeter-sized skeletal

Handbook of Biomineralization. Edited by P. Behrens and E. Bäuerlein

ISBN: 978-3-527-31805-6

elements are termed "ossicles", and possess very complex hierarchical structures of single crystalline calcite. Here, laboratory and synchrotron microCT are compared to illustrate how they can provide complementary information. The contrast in X-ray phase radiography is formed by a different mechanism than in absorption-based methods, with the result that phase imaging is much more sensitive to differences in soft tissue than are absorption modalities.

20.2
Absorption MicroCT

Soon after their discovery, X-rays found practical use in the form of radiography, and the shadowgraphs produced during the early part of the 20th century had a profound impact on clinical medicine. Radiographs recorded from different viewing directions and simple triangulation, for example, allowed one to infer the 3-D position of objects with sharp edges, but this approach to 3-D X-ray imaging was of only limited utility. Although Radon developed the mathematics underlying CT early during the 20th century [1], and Cormack showed reconstruction from projections to be a practical proposition during the early 1960s [2], the development of digital computers was a prerequisite for the medical CT systems pioneered by Hounsfield [3]. Within a decade of Hounsfield's innovation, microCT reconstructions had been reported, and, within a few years after that synchrotron microCT results began to emerge (for a review through the 1990s, see Ref. [4]). Subsequently, commercial microCT systems began to appear during the mid-to-late 1990s, such that today up to 500 systems are available worldwide.

Most – but not all – of the X-ray microCT process relies on contrast from differences in the X-ray absorption of different constituents within the specimen. X-rays are absorbed according to the well-known equation: $I/I_0 = \exp(-\mu x)$, where I_0 is the intensity of the incident X-ray beam of wavelength λ, and I is the beam's intensity after traversing x thicknesses of (homogeneous) material with linear attenuation coefficient μ. More fundamentally, X-ray absorption depends on the mass attenuation coefficient μ/ρ (units cm^2 g^{-1}): $I/I_0 = \exp[-(\mu/\rho)\rho x]$, where ρ is density. The mass attenuation coefficient – and hence transmitted intensity – depends heavily on the atomic number Z and the X-ray wavelength λ: $\mu/\rho \sim \lambda^3 Z^3$.

Figure 20.1 illustrates, in cartoon fashion, how specimens can be reconstructed with microCT: multiple views of the sample are recombined into a quantitative map of the interior of the specimen. Consider a cylindrical object with μ_{cyl} and in which a higher absorption rectangle ($\mu_{rect} > \mu_{cyl}$) is embedded. Along viewing direction 1, the rectangle makes a spatially narrow but deep "valley" in the absorption profile P_θ. Along direction 2, the profile has a spatially wide, but shallow "valley". From these two observations, one infers the presence of the high-absorption rectangle within the circular cross-section. Real reconstruction is more complicated than this, but most microCT users can treat this process as a "black box" producing the stacks of cross-sections (termed slices) that he or she will analyze. More details on reconstructions appear elsewhere [5].

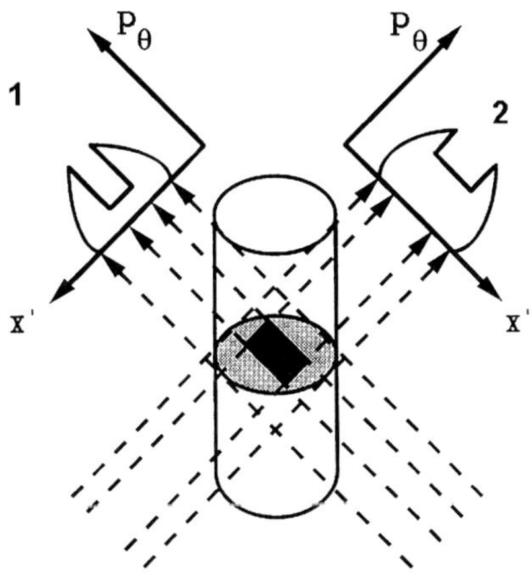

Fig. 20.1 Cartoon view of computed tomography reconstruction showing two viewing directions and the resulting absorption profiles P_θ. For details, see the text.

It is important to remember that voxel (volume element) size in virtually all microCT instruments depends on the diameter of the object being imaged, and the spatial sampling with which each view through the object is recorded (an adequate number of views also must be recorded [5]). If radiographs of a 1 mm-diameter object are recorded with 10^3 detector elements, one can expect to reconstruct the object with 1 μm voxels (ignoring pneumbral blurring, etc.). MicroCT in the author's laboratory is not dissimilar to what can be performed on other commercial systems, and is limited to specimens ~ 30 mm in diameter (2×10^3 detector pixels in the plane of reconstruction allowing 15 μm voxels), while synchrotron microCT is normally performed on specimens ~ 5 mm or smaller diameter. (at the author's institution; see below)

20.3 Phase Radiography

X-rays are very slightly refracted when passing through solids (the indices of refraction differ from one by a few parts per million), but sufficiently that the X-ray wavefronts distort when passing through regions of different electron density (for an introduction to the subject, see Ref. [6]). With a suitable X-ray source – that is, one with adequate spatial coherence – it is possible to detect changes in contrast resulting from X-rays traversing volumes with different electron densities. Most frequently, phase imaging is performed at a synchrotron radiation source such as the Advanced Photon Source (APS); imaging can also be performed with X-ray tubc sources (e.g. [7]).

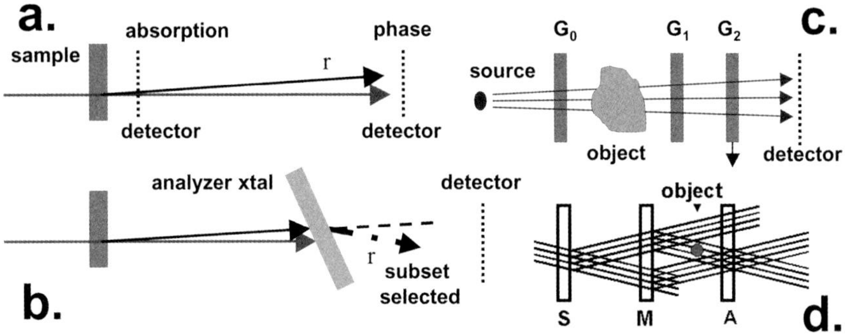

Fig. 20.2 Methods of X-ray phase imaging. (a) Propagation method. Images with the detector near the sample are dominated by absorption contrast, but placing the detector far from the specimens allows refracted X-rays "r" to interfere with transmitted X-rays and to produce edge contrast. (b) Diffraction-enhanced imaging (DEI). (c) Grating enhanced imaging. (d) Bonse–Hart interferometer for imaging. "S", "M" and "A" represent the crystal beam splitters, mirror and analyzer, respectively.

Figure 20.2 illustrates, schematically, four methods where phase effects are used to produce contrast in X-ray images.

- In the *propagation method* (Fig. 20.2a), the detector is placed much farther away from the sample than is normal for X-ray imaging (~1 m versus ~1 cm); refracted X-rays "r" diverge and interfere with other X-rays at the detector plane, producing detectable fringes in the image at external and internal boundaries between materials with different electron densities. Here, contrast is provided by differences in the second derivative of the X-ray phase.
- In *diffraction-enhanced imaging* (DEI; Fig. 20.2b), an analyzer crystal is placed in the X-ray beam after the sample; images recorded with different settings of the analyzer isolate changes in the phase angle, and this method produces image contrast based on changes in the first derivative of the X-ray phase. Essentially, the analyzer selects only a small angular fraction of the refracted radiation.
- The *grating enhanced imaging* method (Fig. 20.2c) is analogous to DEI, except that contrast from changes in the first derivative of phase is provided by translation of one analyzer grating relative to a second instead of by rotation of the analyzer crystal and its periodic array (the crystal lattice).
- *Interferometry* for phase imaging is illustrated in Figure 20.2d. In the Bonse–Hart geometry, a beam splitter "S" produces a reference beam and an imaging beam, the mirror "M" redirects the beams together, the object is placed in one of the beams exiting the mirror and the analyzer "A"

recombines the reference and object-modified beams. Interferometers allow changes in the X-ray phase to be measured directly, not merely its derivatives.

20.4 Sea Urchin Ossicles

Sea urchins are heavily mineralized marine animals. As mentioned above, virtually all of the calcium carbonate comprising the ossicles is single crystal calcite. Ossicles have complex geometries which reflect their different specialized functions. The design motif of sea urchin calcite (except in teeth) is a highly porous structure (on the order of 50% being occupied by soft tissue and fluid) that is termed stereom (Fig. 20.3), and stereom in turn has many different fabrics (dimensions and arrangements of struts) [8]. Only the coarsest stereom can be resolved by laboratory microCT, although synchrotron microCT and new small field-of-view laboratory nanoCT systems allow the individual elements of fine stereom to be imaged.

Most of a sea urchin's volume is enclosed in a protective globe, termed the *test* (Fig. 20.3). The test is shielded from predators (and from wave action) by a large array of spines which project from the test's surface. Depending on the phylogenic order and family, the spines can be either thick or thin, massive or slight, long or short, and hollow or solid. During the process of evolution, amazing combinations of these characters have developed.

Located within the test is the sea urchins' oral apparatus, which is known as Aristotle's lantern. This consists of five identical jaws or pyramids, and is designed for rasping food from hard substrates. Each pyramid is composed of a

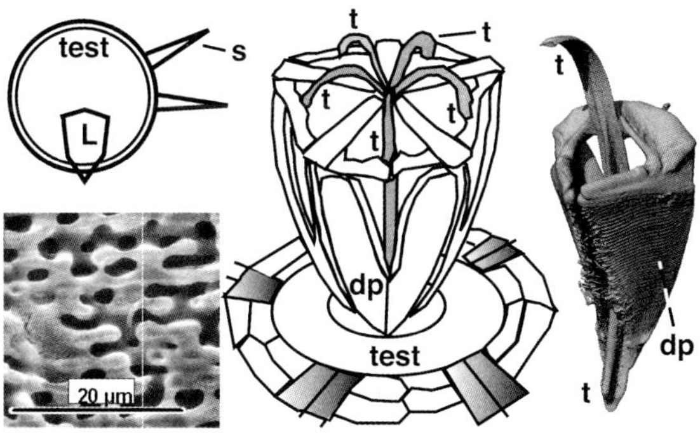

Fig. 20.3 Sea urchin mineralized tissue. Upper left: Schematic of test, spine "s" and jaw structure or Aristotle's lantern "L". Lower left: Scanning electron microscopy (SEM) image of *L. variegatus* stereom. Center: Schematic of lantern showing the five pyramids (two demipyramids "dp" plus one tooth "t"). Right: MicroCT-derived rendering of a *L. variegatus* pyramid (modified from Ref. [12]). In this schematic, the adoral tooth direction is down and the aboral up; the tooth grows from top to bottom.

pair of mirrored demipyramids that are otherwise identical, and a single long but slender tooth attaches to each pair of demipyramids (Fig. 21.3). Tooth cross-sections have evolved to resist bending during grazing [9], and the example discussed below (*Lytechinus variegatus*) is from a camarodont sea urchin – that is, those with teeth shaped like a "T"-girder. The bar across the top of the "T" is called the *flange*, and the leg is termed the *keel*. The flange provides the cutting surface, and the keel prevents the tooth from buckling.

Each tooth grows continuously and contains the entire history of development from unmineralized plumula (aboral end of the tooth) to the adoral dense cutting edge. The keel begins to form midway along the length of the tooth. The microstructure of camarodont (and other types of) sea urchin teeth is very different from stereom elsewhere in the skeleton, and serves not only to reinforce the T-girder structure but also to provide self-sharpening as the tooth wears. The outer volumes of the flange are comprised of plates (primary and secondary) which run more-or-less parallel to the top and bottom flange surfaces, respectively, and parallel to the tooth's adoral–aboral axis; these plates are aligned to resist compression (see below). The center of the flange contains very thin needles which provide a hard, wear-resistant cutting surface. The needles thicken aborally into prisms that fill the tensile-loaded keel: axially aligned fiber reinforcement is the normal composites engineering design for components to be loaded in tension (for further details, see Refs. [9–11]). In *L. variegatus* and other camarodont sea urchins, carinar process plates extend from the flange's secondary plates and bind the prisms in the keel; carinar process plate orientation is thought to provide resistance to secondary bending axes [13].

20.5 Methods

20.5.1 Specimens

The teeth of the sea urchin *L. variegatus* were dissected from the pyramid; the tooth length was approximately 20 mm is these mature specimens. The teeth examined here were divided into the soft aboral end and the hard adoral end. The hard ends were dried and analyzed below; the structure of the soft end is discussed elsewhere [13–15]. A dried, primary spine of sea urchin *Diadema setosum* was also examined; spines of this sea urchin species, from a different phylogenic order than *L. variegatus*, have well-defined structural hierarchies, making them of great interest.

20.5.2 Absorption MicroCT

One *L. variegatus* tooth was imaged with a Scanco MicroCT-20 system operated at 50 kVp (effective X-ray photon energy 26 keV [15]). Data collection parameters were 1024 samples per projection, 500 projections per slice, and 0.3 s per projection, the reconstructed slice thickness was 25 μm, and the in-plane voxel size was 11 μm. The data presented here focus on the portion of the tooth where the keel was just beginning to form.

A matching volume of a second *L. variegatus* tooth was imaged with synchrotron microCT at station 2-BM at APS (described elsewhere [16]). The imaging conditions were: 14 keV photons, views every 0.25° over 180°, data acquisition with a $(1K)^2$ detector and $(1K)^2$ reconstruction with 1.66 μm isotropic voxels. In addition, a fragment of the keel of a third (mature) *L. variegatus* tooth was imaged at station 2-BM (1.33 μm voxels; other details as above).

20.5.3 Phase Radiography

Phase imaging was performed at station 1-ID at APS using a single-crystal phosphor coupled optically to a CCD detector, and the DEI imaging was performed with 30 keV photons. A 333-Si analyzer crystal was used, and single-crystal phosphor-optical lens-CCD system produced images with ~1.5 μm pixels.

20.6 Examples

20.6.1 Absorption MicroCT

Figure 20.4 compares the laboratory and synchrotron microCT slices through the midshaft region of two different *L. variegatus* teeth. The keel and flange are labeled "K" and "F", respectively. The individual structural elements cannot be resolved in the slice from the laboratory microCT (Fig. 20.4a), but there is enough difference in X-ray attenuation to reveal the different structural regions (low absorption central region "LA" running laterally across the flange, the moderately

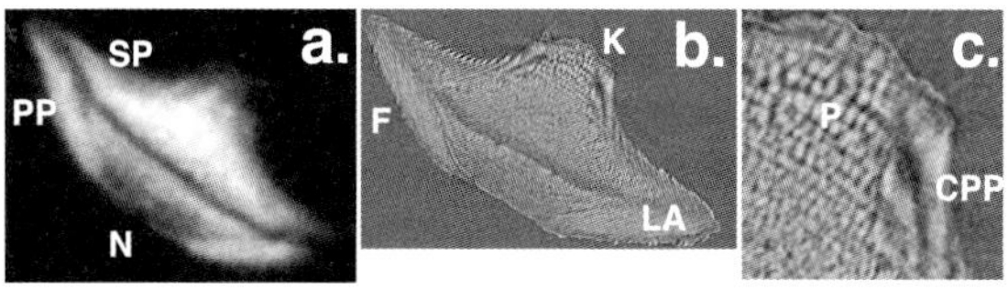

Fig. 20.4 MicroCT slices from the midshaft portions of two *L. variegatus* teeth. (a) Laboratory (from the dataset in Ref. [15]) and (b) synchrotron slices (from the dataset in Ref. [13]) reproduced at the same magnification. (c) Enlargement of keel in (b). The vertical field of view is 0.66 mm in (b) and 0.25 mm in (c). The lighter the voxel, the greater the linear attenuation coefficient. For definitions of symbols, see the text.

attenuating zone "N" at the center of the abaxial surface of the flange and opposite the keel, the primary plate "PP" area, the secondary plate "SP" zone, the area containing the carinar process plates "CPP" and the central region of the keel containing the prisms "P"). In the synchrotron data (Fig. 20.4b,c), the voxel size and resolving power allow the individual plates and prisms to be differentiated.

It is not always necessary to resolve microstructure in order to infer useful information about spatially varying structure within a specimen. Synchrotron microCT (5 μm voxels) is barely able to resolve the stereom elements in a demipyramid of *Asthenosoma varium* [17]; laboratory microCT (9 μm voxels) cannot resolve stereom elements in the demipyramid of *L. variegatus* [12] (data not shown). Note that the voxel sizes in both synchrotron and laboratory microCT were the smallest practical consistent with keeping the intact demipyramid in the field of view for all projection directions, which is a requirement of CT reconstruction. Gradients in mineral density within the *L. variegatus* demipyramid can, however, be measured by converting local values of into volume fractions of calcite [12]: values less than that of dense calcite reflect voxels partially occupied by calcite and partially by empty space.

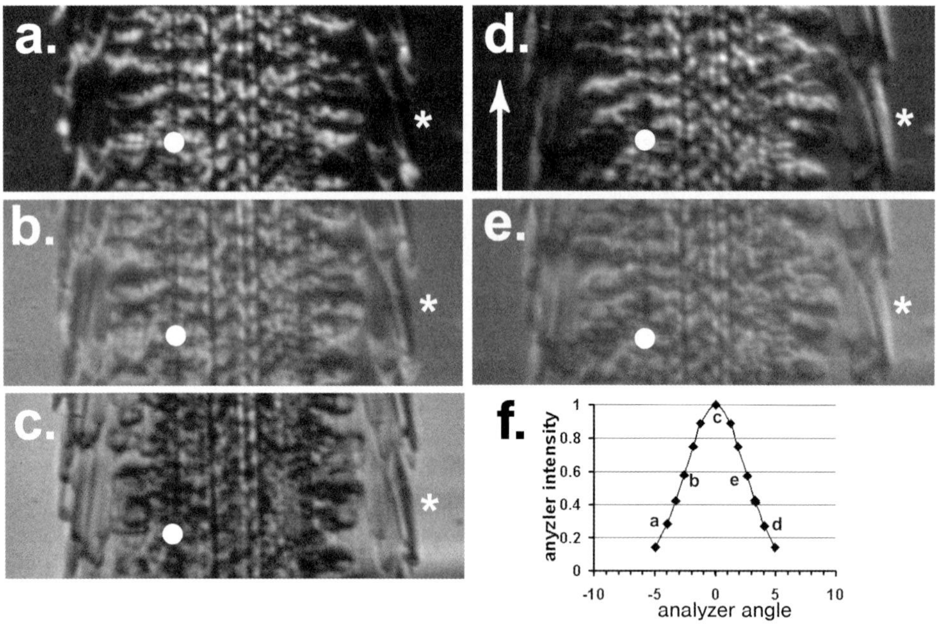

Fig. 20.5 (a–e) Diffraction-enhanced images from a spine of *D. setosum*. The positions on the rocking curve at which the images were recorded are labeled in (f). The horizontal field of view is 1.35 mm, and the symbols label positions discussed in the text. The lighter the pixel, the greater the X-ray intensity [18].

20.6.2
Phase Radiography

Diffraction-enhanced images of a *D. setosum* spine (Fig. 20.5a–e) are used to illustrate changes of contrast produced by orienting the analyzer crystal at different points on its rocking curve (Fig. 20.5f). Images recorded at the peak of the rocking curve (Fig. 20.5c) show contrast dominated by absorption. Images recorded on opposite flanks of the rocking curve (e.g., Fig 20.5a,d) show complementary contrast; one example is to the left and slightly above the white disk, and the second is the narrow feature above "*" on the right edge of the spine. The complementary DEI images can be recombined in a variety of ways to bring out different aspects of the specimen [19]. Note that the images in Figure 20.5 only reflect phase gradients along the vertical direction (arrow); 90° rotation of the spine relative to the analyzer's diffraction vector would be required to reveal gradients along the orthogonal direction. It should be noted that the contrast in these DEI radiographs is produced by the X-ray beam passing through the radial wedges, bridges between wedges, and the perforated central cylinder of the spine (Fig. 20.6a shows three orthogonal cuts through the synchrotron microCT-reconstructed volume of the *D. setosum* spine; the arrow shows one X-ray viewing direction, and adjacent rays clearly pass through different amounts of calcite, producing the phase contrast seen in Fig. 20.5).

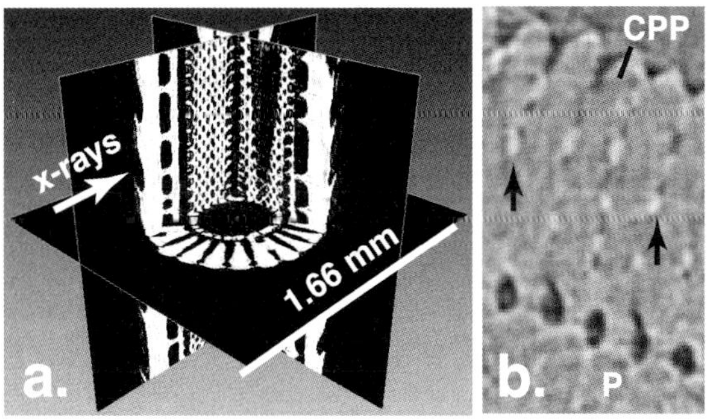

Fig. 20.6 (a) Three orthogonal numerical sections through the reconstructed volume of the *D. setosum* spine with calcite shown in white. (b) Portion of a synchrotron microCT slice of a keel fragment from a *L. variegatus* tooth. The contrast is the same as in Figure 20.4b, and the labels are identified in the text. The horizontal field of view is 275 μm.

20.7
Discussion and Future Directions

X-ray phase imaging has received the most attention in the context of the superior contrast (relative to absorption-based modalities) produced for different soft tissue types, for example tumors in normal tissue [20] or articular cartilage, ligaments and tendons in bone joints [21]. When studying soft tissue microstructure, however, it is essential to maintain dimensional stability, and drying or fixation can produce serious distortions. When imaging soft tissues such as the plumula of the sea urchin tooth, it is easy to keep the sample in fluid, but there is little difference in absorption between soft tissue and water or water-based solutions such as formalin. As a result, microCT absent phase contrast is of little value for soft tissue samples in fluid. Phase contrast, which has a different physical origin than absorption contrast, allows discrimination to be made between fluid and soft tissue. Phase-enhanced radiographs (propagation method) of the soft tissue plumula in *L. variegatus* reveal structures that would not be seen in absorption-based X-ray images [14]; these results suggest that to conduct phase microCT on the plumula would be very valuable for understanding the 3-D structures of mineral-forming cells that subsequently form the tooth's primary plates. Very little is known with regard to what controls the shapes of the cell arrays and their mineral-forming syncytia, and previous studies are quite limited both in number and scope [22]. This no doubt is due to the laboriousness and difficulty of preparing dimensionally stable thin sections for optical and transmission electron microscopy.

Phase effects (hot edges) influence almost all synchrotron microCT reconstructions produced at APS 2-BM (and similar facilities elsewhere). Figure 20.6b shows part of a slice of a fragment of a *L. variegatus* tooth keel [23] and, as will be seen below, phase effects are an important feature in this reconstruction. The central portion of the keel (occupied by unresolved prisms) is labeled "P", and four stacked carinar process plates (one labeled "CPP") are cut perpendicularly. The wavy character of the plates may have evolved because it minimizes lateral sliding of the plates (due to feeding on oblique surfaces, i.e., bending along secondary axes; see [13]) and thereby decreases the likelihood of catastrophic tooth failure. An interesting pattern of contrast is seen in the channels between the carinar process plates (white features labeled by arrows in Fig. 20.6b). Before interpreting this contrast, the stages of mineralization must be described as well as the calcite compositions encountered.

Sea urchin teeth grow continuously, and the constituent high-Mg crystal elements move adorally as the incisal edge frets during feeding. Up to the mid-point along the tooth shaft, the high-Mg calcite mineral elements are spatially separated but parallel crystallographically; these are the primary structure of plates, needles and prisms. Thereafter (adorally), a very high Mg calcite phase grows and cements adjacent primary elements into a true, albeit chemically modulated single crystal. In *L. variegatus*, the very high Mg phase – that is, the secondary mineralized structure – consists of 30–35 mol% Mg versus 10–15 mol% for the high-Mg calcite [15].

The unusual contrast in the channels between carinar process plates in Figure 20.6b reveals the secondary, very high-Mg phase, but this phase may not fill the inter-plate volume completely in this specimen. The majority of the fragment in Figure 20.6b has values of linear attenuation coefficient in the range expected for the energy and calcite composition (high-Mg). The white areas (arrows) have linear attenuation coefficients which are three- to fourfold higher than elsewhere, and this cannot be due to a change in Mg composition: Mg atoms are less absorbing than Ca atoms, and very high Mg calcite (between carinar process plates) has, therefore, a lower value of linear attenuation coefficient compared to the high-Mg calcite of the plates. Prior laboratory microCT of this fragment revealed no anomalous contrast, and careful handling precluded the possibility of contamination producing the extreme values of linear attenuation coefficient. The origin of the unusual contrast is, in fact, the hot edges at the internal interfaces [13]; because the absorption CT reconstruction algorithm used (Gridrec [24]) does not explicitly allow for this effect, physically unrealistic values of linear attenuation coefficient can result.

Interest in sea urchin teeth extends beyond those studying echinoderms. For example, the proteins in sea urchin teeth react strongly with antibodies raised to mammalian teeth [25], despite the difference in mineral formed (calcite in sea urchins versus carbonated apatite in mammals). In contrasting mammalian and sea urchin mineralization proteins, their spatial distributions and products provide a very valuable window into our understanding of vertebrate bones and teeth. The value of combining non-destructive X-ray methods (not only imaging but also microbeam diffraction mapping [15, 23]) with other chemical mapping techniques should not be underestimated. Immunohistochemistry [25, 26], transmission electron microscopy [27] (of macromolecular filled inclusions in tooth plates) and secondary ion mass spectroscopy (SIMS) mapping of protein fragment types and locations [28] are examples for *L. variegatus* teeth, and future combined modes studies offer considerable promise for understanding the relationship between proteins controlling calcite mineral formation and the remarkable geometries in ossicles.

Acknowledgments

The author thanks F. De Carlo (APS) for his collaboration on synchrotron microCT, W.K. Lee and K. Fezzaa (APS) for their collaboration on X-ray phase imaging, A. Veis (Northwestern University) for his collaboration on *L. variegatus* tooth structure, and T.A. Ebert (Oregon State University) for providing the sea urchin spine. Use of the APS was supported by the US Department of Energy, Office of Science, Office of Basic Energy Science, under contract No. W-109E-ENG-38.

References

1 J. Radon, *Ber. Sächsischen. Akad. Wiss.* **1986**, 69, 262–277 in German; translated into English in *IEEE Trans. Med. Imaging* **1986**, *MI-5*, 170–176.

2 A.M. Cormack, *J. Appl. Phys.* **1963**, *34*, 2722–2727.

3 G.N. Hounsfield, *Br. J. Radiol.* **1973**, *46*, 1016–1022.

4 S.R. Stock, *Int. Mater. Rev.* **1999**, *44*, 141–164.
5 A.C. Kak, M. Slaney, *Principles of Computerized Tomographic Imaging*. SIAM, Philadelphia, **2001**.
6 R. Fitzgerald, *Phys. Today* **2000**, *53*, 23–26.
7 F. Pfeiffer, T. Weitkamp, O. Bunk, C. David, *Nature Phys.* **2006**, *2*, 258–261.
8 A.B. Smith, *Special papers in palaeontology, No. 25*. The Palaeontology Association, London, **1980**.
9 K. Märkel, P. Gorny, K. Abraham, *Fort. Zool.* **1976**, *24*, 103–114.
10 K. Märkel, *Z. Morph. Tiere.* **1969**, *66*, 1–50.
11 R.Z. Wang, L. Addadi, S. Weiner, *Phil. Trans. Roy. Soc. (Lond.) B* **1997**, *352*, 469–480.
12 S.R. Stock, S. Nagaraja, J. Barss, T. Dahl, A. Veis, *J. Struct. Biol.* **2003**, *141*, 9–21.
13 S.R. Stock, K.I. Ignatiev, T. Dahl, A. Veis, F. De Carlo, *J. Struct. Biol.* **2003**, *144*, 282–300.
14 S.R. Stock, T. Dahl, J. Barss, A. Veis, K. Fezzaa, W.K. Lee, *Adv. X-Ray Analysis* **2002**, *45*, 133–138.
15 S.R. Stock, J. Barss, T. Dahl, A. Veis, J.D. Almer, *J. Struct. Biol.* **2002**, *139*, 1–12.
16 Y. Wang, F. De Carlo, D. Mancini, I. McNulty, B. Tieman, J. Bresnahan, I. Foster, J. Insley, P. Lane, G. von Laszewski, C. Kesselman, M.-H. Su, M. Thiebaux, *Rev. Sci. Instrum.* **2001**, *72*, 2062–2068.
17 S.R. Stock, K. Ignatiev, F. De Carlo, in: T. Heinzeller, J.H. Nebelsick (Eds.), *Echinoderms: München*. AA Balkema, Leiden, **2004**, pp. 353–358.
18 S.R. Stock, K. Fezzaa, W.H. Lee, unpublished data recorded at station 1-ID of APS.
19 D. Paganin, T.E. Gureyev, S.C. Mayo, A.W. Stevenson, Y.I. Nesterets, S.W. Wilkins, *J. Microsc.* **2004**, *214*, 315–327.
20 A. Momose, T. Takeda, Y. Itai, A. Yoneyama, K. Hirano, *J. Synchrotron. Rad.* **1998**, *5*, 309–314.
21 J. Li, Z. Zhong, R. Lidtke, K.E. Kuettner, C. Peterfy, E. Aliyeva, C. Muehleman, *J. Anat.* **2003**, *202*, 463–470.
22 K. Märkel, U. Röser, U. Mackenstedt, M. Klostermann, *Zoomorph* **1986**, *106*, 232–243.
23 S.R. Stock, J. Barss, T. Dahl, A. Veis, J.D. Almer, F. De Carlo, *Calcif. Tiss. Int.* **2003**, *72*, 555–566.
24 B.A. Dowd, G.H. Campbell, R.B. Marr, V. Nagarkar, S. Tipnis, L. Axe, D.P. Siddons, *SPIE Proc.* **1999**, *3772*, 224–236.
25 D.J. Veis, T.M. Alberger, J. Clohisy, M. Rahima, B. Sabsay, A. Veis, *J. Exp. Zool.* **1986**, *240*, 35–46.
26 A. Veis, J. Barss, M. Rahima, S. Stock, *Microsc. Res. Tech.* **2002**, *59*, 342–351.
27 J.S. Robach, S.R. Stock, A. Veis, *J. Struct. Biol.* **2005**, *151*, 18–29.
28 J.S. Robach, S.R. Stock, A. Veis, *J. Struct. Biol.* **2006**, *155*, 87–95.

Index

Handbook of Biomineralization. Edited by P. Behrens and E. Bäuerlein

ISBN: 978-3-527-31805-6

Further Reading

Bäuerlein, E. (Ed.)

Biomineralization

Progress in Biology, Molecular Biology, and Application

Second, Completely Revised and Extended Edition
2004
Hardcover
ISBN 978-3-527-31065-4

Konigsberger, E.

Biomineralization – Medical Aspects of Solubility

2006
Hardcover
ISBN 978-0-470-09209-5

Kumar, C. S. S. R. (Ed.)

Nanotechnologies for the Life Sciences

10 volumes
2007
Hardcover
ISBN 978-3-527-31301-3

Willner, I., Katz, E. (Eds.)

Bioelectronics

From Theory to Applications

2005
Hardcover
ISBN 978-3-527-30690-9

Niemeyer, C. M., Mirkin, C. A. (Eds.)

Nanobiotechnology

Concepts, Applications and Perspectives

2004
Hardcover
ISBN 978-3-527-30658-9